Green Engineering

Green Engineering
Innovation, Entrepreneurship and Design

Riadh Habash

CRC Press
Taylor & Francis Group
Boca Raton London New York

CRC Press is an imprint of the
Taylor & Francis Group, an **informa** business

CRC Press
Taylor & Francis Group
6000 Broken Sound Parkway NW, Suite 300
Boca Raton, FL 33487-2742

Printed on acid-free paper

International Standard Book Number-13: 978-1-138-30522-9 (Hardback)
International Standard Book Number-13: 978-1-138-03588-1 (Paperback)

Library of Congress Cataloging-in-Publication Data

Names: Habash, Riadh W. Y., author.
Title: Green engineering : innovation, entrepreneurship and design / Riadh Habash.
Description: Boca Raton : Taylor & Francis, CRC Press, 2017. | Includes bibliographical references and index.
Identifiers: LCCN 2017022189 | ISBN 9781138035881 (pbk. : alk. paper) | ISBN 9781138305229 (hardback)
Subjects: LCSH: Technological innovations. | Sustainable engineering.
Classification: LCC T173.8 .H287 2017 | DDC 620--dc23
LC record available at https://lccn.loc.gov/2017022189

Visit the Taylor & Francis Web site at
http://www.taylorandfrancis.com

and the CRC Press Web site at
http://www.crcpress.com

Printed and bound in the United States of America by Sheridan

Dedication

Life can be understood backward; but it must be lived forward.
Gibran Kahlil Gibran

To my parents; and to my family

Contents

Part II The Creativity Landscape

Preface

Though the terms "green" and "sustainable" are often used interchangeably, there are few differences between them. This book is more about sustainability and its three dimensions "social, environmental, and economics." Engineering institutions around the world have begun to integrate sustainability into the curriculum. Topics like innovation, entrepreneurship, and design are highly correlated with sustainability goals, although, such concepts are viewed as management- and/or market driven. The benefits of this combination will be unique; readers will be cross-trained, drawing upon the resources of a world class education.

The content of this book has been developed with emphasis on clarity, with stress on the basic concepts as well as emerging ideas. To enhance the presentation, every chapter of the book includes a case activity with research questions; two types of end of chapter questions based on knowledge acquisition and knowledge possession as well as a set of collaborative knowledge creation activities.

This educational textbook is suitable for leadership, innovation, and entrepreneurship courses with an emphasis on sustainability or for a course devoted entirely to sustainable engineering. On the other hand, the book suites for a course on engineering product design including sustainability in engineering design where sustainability thinking and practice are now evident and increasingly visible.

Organization

The book is divided into an introductory chapter and three parts, each of three chapters. The first chapter establishes the concept of "greening engineering and embracing sustainability," which is about connecting, integrating, and collaborating across functions, disciplines, and people. It defines engineering and explores its transdisciplinary domains. Through history, this chapter shows how engineering is wheeling within the seventh industrial revolution and discusses lessons learned about engineering education. It discusses today's challenges of engineering education and presents approaches for greening engineering by considering various sustainability dimensions. Finally, this chapter develops a framework on contemporary pedagogies for learning and teaching in a technology-rich world with emphasis on the notion of "learner-centered and entrepreneurial environment pedagogy" that promotes open-ended problem-solving and learning-by-doing to enhance innovation and design mind-sets. The chapter is equally important for educators as well as for students for being the stakeholders.

Part I (Chapters 2 through 4) deals with the landscape of sustainability, technology, engineering ethics, and public policy. Chapter 2 traces the roots of sustainability and discusses the progress timeline of sustainable development. This chapter outlines the eight millennium development goals, 17 UN sustainable development goals, and the 12 principles of engineering for sustainable development. Chapter 3 tracks the history of technology and explores technological milestones from the beginning of the nineteenth century. It focuses on challenges related to leveraging applications of technology for sustainable development and explores the impact of appropriate technologies with regard to development problems. Technology transfer through university entrepreneurship and start-up spins is also discussed. Chapter 4 introduces the notions of engineering ethics and public policy with emphasis on ethical theories, engineering profession ethics, ethic codes, engineering public policy, and how they relate to each other as well as to sustainability and education.

Part II (Chapters 5 through 7) deals with creativity and innovation, leadership, and entrepreneurship landscape. This part presents innovation and entrepreneurship as a practice and discipline. It does talk of the psychology and the character traits of entrepreneurs as well as of their actions and behavior. It uses cases, but primarily to exemplify a point. The content is discussed under three main headings. Chapter 5 provides an extensive historical perspective of creativity and innovation and summarizes innovation milestones throughout history. It presents knowledge about creativity, creative thinking, and creativity components in individuals. This chapter also discusses types of innovation, forms of innovation, benefits and risks of innovation, and innovation diffusion to the market. It presents the 16 habits of mind and 6 engineering habits of mind that aid in stimulating creativity and innovation in individuals. Chapter 6 provides a historical perspective to the concept of leadership, discusses several leadership theories and qualities of leadership for meeting sustainability development goals, and explores the role of leadership in engineering education. Chapter 7 introduces the concept of entrepreneurship, its historical roots and development, and discusses the emergence of entrepreneurship and its relation to innovation. Most books on entrepreneurship are business related; however, throughout this chapter the emphasis is on how we think entrepreneurially and the role of entrepreneurship mind-set. This chapter briefly covers almost every aspect of entrepreneurship that is needed by engineering students.

Part III (Chapters 8 through 10) covers the entire domain of engineering design. This part of the book is designed for use in engineering design courses, and as a reference for professionals learning about design methodology and process with a focus on sustainable design as the driving force behind sustainable products. Engineers in the future will need to design for sustainable development, for energy efficiency, mass efficiency, and low environmental emissions. Chapter 8 investigates the "what" of engineering design and its relation to innovation and entrepreneurship. It discusses engineering design language and methodology. This chapter also discusses various design theories. Chapter 9 explores the "how" of engineering design with emphasis on the design process. It provides an experience in successfully designing and developing products. Yet the design elements introduced in this chapter encapsulate the richness of the full design experience of much larger and more complex engineering systems. Chapter 10 introduces the concept of sustainable design and explains the application of sustainable methods to the engineering design process. It also covers important design tools including triple bottom line, cradle to cradle approaches, and life cycle sustainability in the product remanufacturing process.

Because this book distills years of observation, study, and practice, it was possible to use actual case examples and illustrations from years of teaching. Throughout this book many wide-ranging, illustrative cases, questions, and activities have been incorporated that provide useful information for practical applications.

Objectives

"Greening" in this book goes beyond the traditional meaning of becoming more active in protecting the environment. It encompasses the need to enrich engineering education and practice with detailed knowledge from a broad range of disciplines, including natural and social sciences, art, and business. This potent mix of skills can be utilized to develop new approaches to some of our most challenging global problems. In fact, the influence of new forces on the engineering profession is already visible. Today, many leading educational and research institutions have introduced these priorities within their programs, focusing on key areas that include efficiency, green building and transportation, hydrogen as fuel, emissions reduction, sequestration, renewable energy, and life cycle analysis.

This book is distinguished by extensive descriptions of concepts in sustainability, its principles, and its relevance to environment, economy, and society. Reflecting the transdisciplinary approach from several different intellectual streams, this book is a venture into the seemingly disparate studies as means to access modern engineering education. The book will permit a broad range of readers with a reasonable background to learn about various fields of knowledge, including the following:

- Historical thinking by learning the wonders of the past and fostering the ability to make judgments about the present and future
- Contemporary pedagogies for learning and teaching in our technology-rich world
- Cutting-edge contributions from various fields and disciplines into engineering
- Scientific knowledge for strategic planning toward a sustainable society
- Engineering ethics and public policy
- Innovation, creativity, and leadership skills
- Rapidly emerging area of engineering entrepreneurship
- Engineering design methodology and process as well as sustainability in engineering design.

Key Pedagogical Features

To help students develop their understanding, several features have been built into the textbook.

- Each chapter opens with chapter objectives to inform the student of the subject and scope of the topics to be covered.
- The book in general introduces sustainability concepts and principles in the context of every subject under consideration.
- All chapters open with historical perspective that allows readers to think historically.
- All chapters incorporate aspects of engineering education, including the appropriate approaches and pedagogies.
- Each part of the book is developed in a way that it can stand alone as a text or reference to a course or in combination with other parts or chapters according to the needs.
- Each chapter concludes with a cutting-edge case activity that mirrors the types of situations students and engineers face in the field.
- The book provides open-ended tasks for students and instructors at the end of each chapter. These tasks rely on three main approaches of pedagogical skill development, namely knowledge acquisition, knowledge possession, and knowledge creation. In general, the given tasks have no unique answers or solutions and can be adapted to every engineering discipline easily.

Audience

The book is an outgrowth of more than 30 years of teaching, research, and consulting in various areas of engineering. It is designed as an educational textbook for students enrolled in universities and colleges, who are studying engineering and probably other related fields. The book can be used as a learning resource for varied undergraduate courses. Because of its comprehensive coverage and large number of detailed subjects, this book provides a first-level introduction to topics like sustainability and sustainable development; innovation, leadership, and entrepreneurship; and engineering design. This book can be read by all engineers regardless of their specializations. It can also be read by all students of engineering as they would be future designers of products and systems.

To the Instructor

As indicated in Chapter 1, this book intends to promote learner-centered environment pedagogy, from among several other pedagogies, that call for engaging students in the process of learning and knowledge creation. To accomplish this goal, all questions and tasks provided at the end of each chapter are transdisciplinary, open to debate, and mostly open-ended. Accordingly, the adopted pedagogy in this book calls instructors to develop their course online libraries that may be shared with others and with public. Such an approach gives the instructor the opportunity to treat the process of teaching as a design task and makes students more innovative in investigating problems, developing solutions, and creating new knowledge. Instructors may benefit from the following technologies, among many more, in developing their classroom online libraries or blogs for multimedia presentations:

YouTube	Teachers can create YouTube channels to help student leverage videos they create to educate, engage, and inspire other students.
SlideShare	A web-based slide hosting service: users can upload files privately or publicly in various formats such as documents, PDF, slides, videos, and webinars.
VoiceThread	A free cloud application that allows professors to upload lessons and documents to discuss with students via microphone, webcam, text, phone, or audio.
Vidyo	A software-based video conferencing tool that can run on existing hardware.
Skype	An often-free way to chat with students via phone or video call
Blackboard	A web-based tool used by many instructors for online and hybrid courses: it provides discussion boards, calendars, quizzes, and tracks student progress.
Web	Instructors can develop their own online learning resources. You may see two such examples developed by the author: www.g9toengineering.com and www.greenengineers.ca.

Acknowledgment

I am glad I started this journey; I have had a great time doing so. I would like to express my gratitude to the various people who have assisted in the preparation of this book. In particular, I would like to thank my students and colleagues at the University of Ottawa who have provided the reason and environment for writing the book. Also I would like to give special thanks to my family for their help and support during the writing of this book. In particular, I thank my wife Najat and children Gandhi, Mara, Marina, and Mikeli for their support in terms of encouragement and content review.

Author

Riadh Habash is an engineering educator and researcher at the University of Ottawa in Ontario, Canada, author of several books and research papers, and recipient of the National Wighton Fellowship and five university interdisciplinary initiatives awards. He taught for more than 35 years at various universities worldwide. He initiated learning activities that provide learning environment for inquiring minds through collaboration with industry to develop student design projects and entrepreneurial activities.

Glossary of Terms and Abbreviations

Term or Abbreviation	Definition
Chapter 1	
ABET	Accreditation Board for Engineering and Technology
CEP	Closed-ended problems
EHOM	Engineering habits of mind
HOM	Habits of mind
IoT	Internet of things
IT	Information technology
LCE	Learner-centered education
MIT	Massachusetts Institute of Technology
OEP	Open-ended problems
PBL	Project-based learning
SD	Sustainability development
ST	Systems thinking
STEAM	Science, technology, engineering, art, math
STEM	Science, technology, engineering, math
TCE	Teacher-centered education
Chapter 2	
AT	Appropriate technology
CCS	Carbon capture and sequestration
CE	Circular economy
CED	Cumulative energy demand
CO	Carbon monoxide
EPP	Ethical purchasing policy
EROI	Energy return of investment
GDP	Gross domestic product
GHG	Greenhouse gas

HC	Hydrocarbon
IISD	International Institute for Sustainable Development
IUCN	International Union for Conservation of Nature
LEAP	Long-range energy alternatives planning
LEED	Leadership in energy and environmental design
MDG	Millennium Development Goals
NEA	Net energy analysis
NOx	Nitrogen oxides
OECD	Organisation for Economic Co-operation and Development
PNAS	Proceedings of the National Academy of Sciences
PPP	Public–private partnership
SA	Sustainability assessment
SD	Sustainable development
SOx	Sulfur oxides
SWP	State water project
TDS	Total dissolved solids
TM	Toxic metals
UN	United Nations
VOC	Volatile organic compounds
WCED	World Commission on Environment and Development
WEAP	Water evaluation and planning

Chapter 3

AT	Appropriate technology
CFL	Compact fluorescent light
CH_4	Methane
CO2	Carbon dioxide
EMCS	Energy management control systems
GPS	Global positioning system
GPS	Global positioning satellite
GT	Green technology
IP	Intellectual property
LED	Light emitting diode
N_2O	Nitrous oxide
R&D	Research and development
SCOT	Social construction of technology
SME	Small and medium enterprise

SOC	State of charge
STI	Science, technology, and innovation
TEB	Transit-elevated bus
TP	Technology planning
TT	Technology transfer
TTO	Technology transfer office
UC	Ultra-capacitor
WWW	World Wide Web

Chapter 4

AIChE	American Institute of Chemical Engineers
AIEE	American Institute of Electrical Engineers
ASCE	American Society of Civil Engineers
CEES	Center for Engineering Ethics and Society
DG	Distributed generation
ECPD	Engineers' Council for Professional Development
EPP	Ethical purchasing policy
ICT	Information and communication technology
NAE	National Academy of Engineering
NSERC	National Science and Engineering Research Corporation
NSF	National Science Foundation
NSPE	National Society of Professional Engineers
PP	Public Policy
SEEPP	Sustainable Environmental and Ethical Procurement Policy
SG	Smart grid
STS	Sociotechnological system

Chapter 5

ac	Alternating current
BEV	Battery electric vehicle
CFL	Compact fluorescent lamp
dc	Direct current
DRL	Dichroic reflector lamp
EHoM	Engineering habits of mind
EVI	Electric Vehicles Initiative
FCEV	Fuel cell electric vehicles
GI	Green innovation
GM	General motors
HEV	Hybrid electric vehicles
HoM	Habits of mind

IEA	International Energy Agency
IPR	Intellectual property right
IQ	Intelligence quotient
NIS	National Innovation System
PAR	Parabolic aluminized reflector
SDC	Self-driving car
TV	Television
WWW	World Wide Web

Chapter 6

AASHE	Association for the Advancement of Sustainability in Higher Education
ACCE	Algonquin Centre for Construction Excellence
APA	American Psychological Association
EQ	Emotional intelligence
Gt	Gigatonnes
IDP	Integrated design process
LEED	Leadership in energy and environmental design
MIT	Massachusetts Institute of Technology
POB	Positive organizational behavior
ROI	Return on investment
TBL	Triple bottom line
USGBC	US Green Building Council
WICS	Wisdom, intelligence, and creativity, synthesized

Chapter 7

EM	Entrepreneurial marketing
KEEN	Kern Entrepreneurship Education Network
LSVCC	Canadian Labor-Sponsored Venture Capital Corporation
OECD	Canadian Organization for Economic Co-operation and Development

Chapter 8

AR	Augmented reality
CAD	Computer-aided design
CAE	Computer-aided engineering
CAM	Computer-aided manufacturing
CD	Concurrent design
CE	Concurrent engineering

CEAB	Canadian Engineering Accreditation Board
C-K	Concept knowledge
DD	Deterministic design
DfA	Design for assembly
DfD	Design for disassembly
DfL	Design for logistics
DfM	Design for manufacturing
DfMa	Design for maintainability
DfN	Design for network
DfO	Design for obsolescence
DfQ	Design for quality
DfR	Design for reliability
DfS	Design for supportability
DfSC	Design for supply chain
DfV	Design for variety
DfX	Design for X
DT	Design thinking
EDC	Engineering design communication
EDM	Engineering design method
GD	Green design
HFE	Human factors engineering
IPD	Integrated product development
MBD	Model-based design
MD	Modular design
PM	Permanent magnet
PREP	Peer-review evaluation process
RD	Robust design
SD	Systematic design
SDM	Scientific design method
SE	System engineering
SPD	Sustainable product development
VR	Virtual reality

Chapter 9

CDIO	Conceive-Design-Implement-Operate
CM	Configuration management
DD	Detailed design
DFC	Design for cost
DfQ	Design for quality
EDMS	Electronic data management systems
FDM	Finite difference method
FEM	Finite element method
LAN	Local area network

NDP	New product development
PARC	Physical Activity Research Centre
PCD	Product concept design
PD	Product development
PDM	Product data management
PLC	Product life cycle
PLCM	Product life cycle management
RP	Rapid prototyping
RP&M	Rapid prototyping and manufacturing
UCD	User-centered design

Chapter 10

AM	Additive manufacturing
BOM	Bill of materials
CCM	Closed-cycle manufacturing
CDP	Conventional design process
CF	Carbon footprint
CLMC	Closed-loop material cycle
CP	Cleaner production
DfE	Design for environment
DfRem	Design for remanufacturing
DfS	Design for sustainability
EOL	End of life
FU	Functional unit
HDD	Computer hard disk drives
LCA	Life cycle assessment
LCC	Life cycle cost
LCC	Life cycle costing
LCD	Life cycle design
LCE	Life cycle engineering
LCI	Life cycle inventory
LCIA	Life cycle impact assessment
LCSA	Life cycle sustainability assessment
LCT	Life cycle thinking
OEM	Original equipment manufacturer
PEL	Product end of life
PLC	Product life cycle
PSS	Product service system
SED	Sustainable engineering design
SM	Sustainable manufacturing
UNCED	UN Conference on Environment and Development
VDI	Association of German Engineers

WBCSD	World Business Council for Sustainable Development
WF	Water footprint
WFA	Water footprint assessment
WFN	Water Footprint Network

chapter one

Greening engineering and embracing sustainability

I am quite into the idea of engineering being beautiful.

Sean Booth

1.1 Objectives

- Introduce the concept of "greening engineering" which is about connecting, integrating, and collaborating across functions, disciplines, and people.
- Understand the key features that define the broad fields of engineering and engineering profession.
- Highlight the classroom and experiential aspects that make engineering green such as innovation, entrepreneurship, design, leadership, and professionalism.
- Explore the concept of "transdisciplinary" (beyond disciplines) as being the direction the engineering curriculum should take.
- Through history, show how engineering is wheeling within the seventh Industrial Revolution.
- Deepen an understanding of key historical and scientific concepts and lessons learned, and use these in constructing explanations for future engineering education.
- Present major historical eras in engineering education.
- Briefly introduce the main disciplines of engineering.
- Have a strong understanding of what an engineer is and what skills and knowledge are required to be a professional engineer.
- Discuss the challenges that the engineering profession and education face today.
- Discuss the approaches of greening engineering and embracing sustainability through deep consideration of various sustainability dimensions.
- Understand the roles of engineers in an entrepreneurial context.
- Explore the fact of broadening engineering education through enriching the breadth and depth of knowledge and skills.
- Highlight the urgent need of society and industry for entrepreneurial and enterprising engineers.

- Discuss the transition to learner-centered environment pedagogy, where many of the teaching strategies that have been advocated for at least a century by the likes of Dewey, Piaget, Montessori, and Vygotsky are beginning to emerge and be embraced.
- Discuss the effective learning and teaching styles to pave the path toward innovation, entrepreneurship, and design.
- Develop a framework that recommends the aspects in entrepreneurship and design education to be emphasized during each year of engineering study process.
- Discuss pedagogic strategies that may be used as a guideline for designing and implementing future engineering curricula.
- Emphasize on the concept of "learning by doing" through facilitating experiential learning, design entrepreneurial spaces, university–industry collaboration, student competition opportunities, and rewarding faculty innovation and entrepreneurship.
- Provide three end of chapter pedagogical knowledge strategies, namely knowledge acquisition, knowledge possession, and knowledge creation to help understand the above topics and generate new and alternative ideas and solutions.

1.2 Greening engineering

> You cannot understand or solve complex problems without the knowledge and tools of multiple disciplines.
>
> **Kamp (2014)**

1.2.1 Engineering defined

> Science is about knowing, engineering is about doing.
>
> **Henry Petroski**

In this book, "greening" is about connecting, integrating, and collaborating across functions, disciplines, and people. Engineering is that green endeavor that combines information and blends in core capabilities, including systems, products, and services, together in a mutually reinforcing and synergistic way with science, technology, education, art, business, and markets. Engineering is one of the oldest professions, along with medicine and law. However, it differs from the other professions in that doctors and lawyers generally provide their services to specific individuals or corporations. In contrast, engineers tend to design and build things as well as provide services. Their responsibility is more often to society than to specific people. The engineering profession, as with others, is an occupation based upon specialized education and training, as providers of professional advice, skills, and services.

The words "ingenuity" and "engineering" in English and *ingéniosité* and *ingénierie* in French are linked to the same Latin root word and the verb "to engineer" means "to be ingenious." The term "engineering" was used in the 1300s for a person who operated a military engine or machine such as a catapult or, later, cannon. The word "engine," in turn, is derived from the Latin word *ingenium* for ingenuity or cleverness and invention. The terms "art" and "technical" are important because engineering arranges elements in a way that may, or may not, appeal to human senses or emotions, and is also linked with the Greek word *technikos* relating to art, craft, skill, and practical knowledge and language regarding a mechanical or scientific subject. Prior to the development of engineering different fields, engineering and technical were originally closely connected (UNESCO 2010).

Engineering also connects to the natural sciences, and to the social and human sciences. Science, derived from the Latin word *scientia*, meaning "knowledge," relates broadly in a systematic approach to the observation of phenomena and the development of hypothesis, experimentation, and theory regarding these phenomena. In this broad sense, science includes engineering as a highly skilled technique or practice. In a narrower, contemporary sense, science is distinguished into the basic and applied sciences, following the linear model of innovation that research in the basic sciences leads through applied research and development in engineering to technological application, innovation, and diffusion. Engineering in the modern sense relates to art, even though it may not commonly be regarded as artistic; artistic qualities can be appreciated in the creativity and elegance of many engineered objects and structures. Almost every area of human interest, activity, and endeavor has a branch of engineering associated with it (UNESCO 2010).

Practically, engineering is problem identification, formulation, and solving. It is concerned with the transformation of knowledge to value, by establishing the knowledge in some physical and functional forms. At its domain, engineering engages principles of science and mathematics, and domains of design, art, and business to practical ends. Engineering as a broad discipline, practice, and profession has a major role in the creation and implementation of materials, components, machines, structures, systems and processes, and organizations for well-defined purposes. It encompasses a range of specialized subdisciplines that focus on developing a specific kind of product and service, or using a certain type of technology. Engineering has been a key driver of human development and a force in the improvement of economic well-being, health, and quality of life.

1.2.2 Sustainability in engineering

The term "sustainability" is the capacity to endure (continue to exist). Basically, sustainability entails consideration of multiple objectives including reductions in consumption, minimization of negative externalities, and reduction of burdens on future generations that may seem at odds with an immediate-term economic profit motive. It is often misunderstood as a goal to which we should together desire. In fact, it is not an end state that we can reach; rather, it is a path of an evolving system. In addition, system thinking (ST) offers a potential means to overcome the involving barriers.

Engineering, through its role in the creation and implementation of technology, has been a key force in the improvement of economic well-being, health, and quality of life. Three hundred years ago, the average life span was 37 years, the primary effort of the majority of humans was focused on provisioning their tables, and the threat of sudden demise due to disease was a lurking reality (Kagan et al. 2001). Today, human life expectancy is approaching 80 years in many parts of the world as fundamental advances in medicine and technology have greatly suppressed the occurrence of and mortality rates for previously fatal diseases, and the efforts of humankind are focused largely on enhanced quality of life (CIA 2001).

Engineers have generally been trained to work toward the maximization of economic profit. While effective in achieving this objective, the tools of the field can also be employed in the service of other goals that have been less frequently considered in the past. These other objectives such as minimizing resource consumption or maximizing the health of a population can be combined with the profit objective to achieve sustainable systems and processes that still accomplish their economic goals. The best way to achieve this transformation in mind-set, and to prepare industrial engineers to implement sustainability in their careers, is through their education (Nazzal et al. 2015).

Engineering foundations support the development of new knowledge and the creation of safe, reliable, and innovative technologies that advance society and the human condition. Solutions of societal problems require that these technologies be applied in innovative ways with consideration of cultural differences, historical perspectives, and legal and economic constraints, among other issues (NAE 2004). Chapter 2 is devoted to the topic of sustainability and sustainable development (SD), whereas Chapter 10 discusses the topic of sustainability in engineering design. In addition, the entire book is oriented toward this important and global notion.

1.2.3 Creativity, innovation, and entrepreneurship

> Education is a process involving two sets of participants who supposedly play different roles: teachers

> who impart knowledge to students, and students who absorb knowledge from teachers. In fact, as every open-minded teacher discovers, education is also about students imparting knowledge to their teachers, by challenging the teachers' assumptions and by asking questions the teachers hadn't previously thought of.
>
> **Diamond (2005)**

Engineering is a profoundly creative process. Creativity (invention, innovation, and thinking outside the box) is an indispensable quality for engineering, and given the growing scope of the challenges ahead and the complexity and diversity of the technologies of the twenty-first century, creativity will grow in importance. The creativity requisite for engineering will change only in the sense that the problems to be solved may require synthesis of a broader range of interdisciplinary knowledge and a greater focus on systemic constructs and outcomes (NAE 2004).

Creativity and innovation can be stimulated by taking students far out of their comfort zone, by bombarding them with things they have never encountered before, and by challenging them in design projects. Innovation requires whole-brain thinking: right-brain thinking for creativity, imagination, and holistic ST, and left-brain thinking for logical reasoning, analytical thinking, and planning (Kamp 2014).

Creativity training in engineering programs linked with applications in integrative projects could entirely transform how engineers and the public view the engineering profession. Learning the basics of how the creative process functions and how it can be developed may advance and conclude engineers' development. The topic of creativity and innovation is intensively discussed in Chapter 5.

In traditional engineering education, theoretical understanding is required prior to practical experimentation. As a consequence, engineering competences is needed before one can be creative and innovative (MacLeod 2009). However, entrepreneurship in the modern turbulent world often means simultaneously creating and learning new knowledge; therefore, traditional analytic and systematic approaches may not be adequate. Thus, it is interesting to discuss how to promote entrepreneurial mind-sets and behaviors within engineering education (Mäkimurto-Koivumaa and Belt 2015). The wide topic of engineering entrepreneurship is discussed in Chapter 7.

1.2.4 Leadership, professionalism, and ethics

In preparation for innovative and entrepreneurial opportunities, engineers must understand the principles of leadership and be able to practice them in growing proportions as their careers advance. They must also be willing to acknowledge the significance and importance of public service and its place in society, stretching their traditional comfort zone and accepting the challenge of bridging public policy and technology well beyond the roles accepted in the past (NAE 2004). The topic of leadership is discussed in Chapter 6.

In addition to strong leadership, there is a need to acquire a working framework upon which high ethical standards and strong sense of professionalism can be developed. In many ways, the roles that engineers take on have always extended beyond the scope of technology. And as technology becomes increasingly imbedded into every facet of life, the convergence between engineering and public policy will also increase. This new level of relationship requires that engineering develops a stronger sense of how technology and public policy relate. For example, engineers will need to understand the policy by-products of new technologies, and public servants will need to recognize the engineering impact of policy decisions.

To capture student interest and respect, training in ethical responsibilities should be more interwoven with subjects that are already taught and should no longer be on the margins of the curriculum. To develop a good sense of ethical accountability and social responsibility, students need to come in closer contact with senior engineering professionals with whom they can identify and try to emulate (Kamp 2014). The engineering profession recognizes that engineers need to work in teams, communicate with multiple audiences, and immerse themselves in public policy (PP) debates and will need to do so more effectively in the future. Topics of ethics and public policy (PP) are discussed in Chapter 4.

1.2.5 System and design perspectives

Engineering needs system-wide engineering skills more than ever due to current challenges from complex manufacturing designs of networked devices that require a systems perspective. High-technology industries that develop complex systems and machines increasingly look for engineers who have the capabilities to develop the outline for an integral design and keep the overview and take care of the consistency of the system design. These system architects not only need solid fundamental knowledge but also must understand "the big picture" and have a sense of the multidisciplinary problem domain and a good awareness of the business and human context (Kamp 2014).

A most refined depiction is that engineering is about design under control. The engineer designs devices, components, subsystems, and systems, and, to create a successful design, in the sense that it leads directly or indirectly to an improvement in quality of life, must work within the constraints provided by technical, economic, business, political, social, and ethical issues (NAE 2004). Today's products and services need rethinking of traditional design principles and greater fusion with engineering.

Design engineering starts from understanding the customers' needs and their relationship with products. It then works across various functions to define product and its components (mechanical, electrical, software, connectivity, cloud, etc.) and interfaces, services, processes, and activities together into delivering the desired outcomes. The topics of engineering design, engineering product design, and sustainability in engineering design are discussed in Part III of this book.

1.3 Transdisciplinary engineering

> The ideal engineer is a composite... He is not a scientist, he is not a mathematician, he is not a sociologist or a writer, but he may use the knowledge and techniques of any or all of these disciplines in solving engineering problems.
>
> **Nathan Washington Dougherty**

1.3.1 Transdisciplinary model

The concept of "transdisciplinary" (beyond disciplines) was resurrected by Ertas et al. (2003) who argued that it was the direction the engineering curriculum should take. Transdisciplinary education literally transcends the narrow focus of one or more disciplines and is not constrained to adopt pre-existing models for problem definition or solution. By their nature, transdisciplinary approaches synthesize and integrate concepts whose origins are found in different (Ashford 2004). Following Kozmetsky (1997), transdisciplinarity was defined as the integrated use of tools, techniques, and methods from various disciplines. Such thinking forces one to reflect across, beyond, and through the academic disciplines to encompass all types of knowledge about an idea or subjects.

The common thread of all disciplines is design and process science (Tanik and Ertas 1997), which provides the patterns, insight, and logic necessary to apply knowledge and skills to any problem. As shown in Figure 1.1, technologies that are the products of engineering interact at all levels of society. The model is also indicative of the interactions that engineers must make and the languages they must learn to speak (Heywood and Mina 2015).

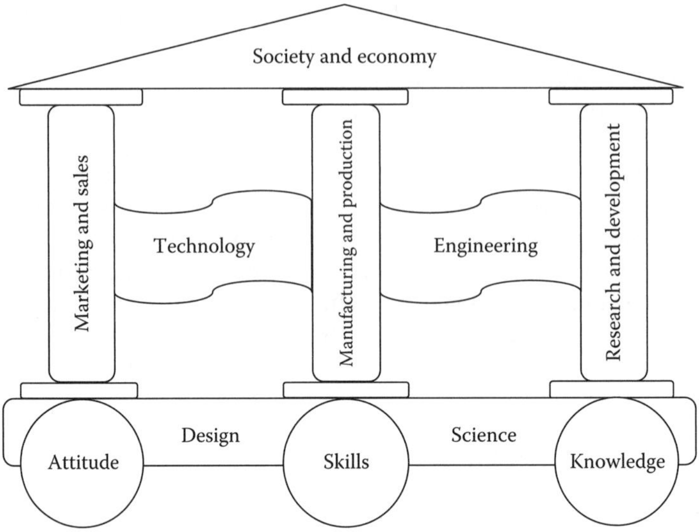

Figure 1.1 Convergence of fields to facilitating transdisciplinary engineering.

While this model endures with scientists on grounds of simplicity, many observers regard the model as descriptively inaccurate and normatively undesirable partly because many innovations were neither based on nor the result of basic scientific research. The social and human sciences emulate the natural sciences in the use of empirical methods. Technological change and innovation is one of the major drivers of economic, social, and human change, so engineering and technology and the social sciences are more closely connected (UNESCO 2010). The design profile may be conveniently supported in terms of three components: knowledge, the known facts, and understood concepts; the used skills in managing and applying knowledge, such as computation, experimentation, analysis, synthesis/design, evaluation, communication, leadership, and teamwork; and the attitudes that dictate the goals toward which skills and knowledge will be directed, including personal values, concerns, preferences, and biases. Knowledge is the database of a professional engineer; skills are the tools used to manipulate the knowledge in order to meet a goal dictated or strongly influenced by the attitudes (Rugarcia et al. 2000).

1.3.2 Transdisciplinary nature of engineering

Engineering appears to be at a turning point. It is evolving from an occupation that provides clients with competent technical advice to a profession that serves the community in a socially responsible manner (Beder 1999). Therefore, transdisciplinary engineering education and research is the logical environment for future development and naturally is an extension of interdisciplinary and multidisciplinary programs. In the multidisciplinary model, communication paths are established between disciplines, but these communication paths may not link all the disciplines directly and they may not be entirely adequate. The transdisciplinary model has only fuzzy interfaces, no barriers. The core design and process activities encompass all of the topical areas and some of the topical areas overlap (Ertas et al. 2000). Communication and interaction paths are easily accomplished. The transdisciplinary nature of engineering is shown in Figure 1.2.

The key message of convergence of various fields, however, is that merging ideas, approaches, and technologies from widely diverse fields of knowledge at a high level of integration is one crucial strategy for solving complex problems and addressing multifaceted intellectual questions underlying emerging disciplines and new technologies (NAE 2004). In fact, engineering is a multifaceted activity that requires engineers to have knowledge and skills that extend well beyond those traditionally associated with the engineering curricular. The reflection of the engineering curricular within programs of engineering and technological literacy necessarily requires a transdisciplinary approach to their teaching (Heywood and Mina 2015).

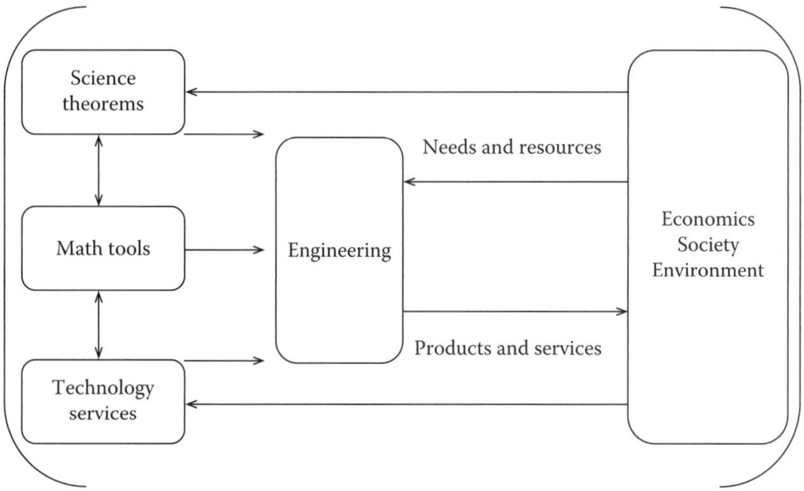

Figure 1.2 The transdisciplinary nature of engineering.

1.4 Historical perspective: Wheeling within the seventh Industrial Revolution

Study the past if you would define the future.

Confucius

1.4.1 Thinking historically

History, like all subjects, represents a systematic way of thinking. A key insight necessary for deep learning of history is that it should be understood as an organized, integrated way of thinking (Elder et al. 2012). History lets us see how decisions made in the past affected societies and civilizations.

Thinking historically introduces students and engineers to the wonders of the past and fosters the ability to make judgments about the present. To think historically is not straightforward, but need to be able to establish sum of concepts according to the "Historical Thinking" project (www.historicalthinking.ca). These include historical significance (events that resulted in great change over long periods of time for large numbers of people), use primary source evidence (not source of information only), identify continuity and change (history is not a list of events, but it is a complex mix of continuity and change), analyze cause and consequence (how and why), take historical perspectives (understanding the social, cultural, intellectual, and emotional settings that shaped people's lives and actions in the past), and understand the ethical dimension of historical interpretations (learn something from the past that helps to face the ethical issues of today).

While approaching history as historical thinking, the historical dimension of other subjects will be understood as well. For example, history of engineering in the context of the way people live and interact with nature as well as with one another is very much the history of humanity itself. Throughout history, developments of knowledge represented by engineers using scientific laws, mathematical equations, and inspired artistry has allowed us to both realize engineering principles and provide a platform for expanding advances in engineering. Inventions such as the wheel and lever exploited basic mechanical principles to develop useful tools. Human beings are partly defined as tool designers and users, and it is this innovation and the design and use of tools that accounts for so much of the direction and pace of change of history. Most of civilization's history in economic and social relations is also the history of engineering applications and innovation (UNESCO 2010).

Importantly, students and engineers need to be made aware of the history of engineering and technology development. For example, they should be aware of how their specialized field evolved over the centuries and of any dangers involved in future radical change or disruptive technologies (Christensen 1997).

1.4.2 Early history

Engineering has been around as a concept for a long time, because the history of human advancement has been one of technological development. In ancient times, many of the wonders of the world can be attributed to the skill and inventiveness of what were, essentially, civil engineers. After the introduction of farming, powerful civilizations in a few geographical situations such as the Persians, Romans, and Mongols exploited and developed long-distance trade routes to expand their regional influence. The Roman aqueducts, Greek temples, the Great Wall of China, the Egyptian pyramids in the Nile Valley, and Sumer's wheels and carts present us with enduring models of the elegant art and science of problem solving, that is, the power of engineering.

First civilization is that of Mesopotamia, Babylon, and Assyrians between Tigres and Euphrates rivers. Sumerians are credited to connecting cities and dwellings on river banks in an effort to build empire called Sumer. They are known to have scripts, and had invented wheels and carts. It is likely that the Sumer first used the wheel in making pottery in 3500 BC and then used it for their chariots in around 3200 BC. They also used irrigation to get water to their crops; they used bronze metals to make strong tools and weapons. The Walls of Babylon were once considered one of the seventh Ancient Wonders of the world. The Assyrians developed glasswork as well as glazes for pottery. Different canals have been found that brought water to the Assyrian Empire capital Nineveh.

Egyptians used surveying to predict Nile River flood waters. Romans learned from Egyptians and Greeks the importance of surveying. The Pyramids were constructed in Egypt during 2800–2400 BC and may be considered as the first large structure construction ever. Labor combined with imagination produced those marvelous structures.

The Greeks, the inventors, made significant contributions in the 1000 years that spanned the BC-AD divide. The Greek cities reached great levels of success that resulted in an exceptional cultural boom, expressed in architecture, drama, science, and philosophy, and the first attempt at democracy. Their inspiration was felt in the whole Mediterranean basin. A scientific approach to the physical sciences concerning civil and mechanical engineering was implemented by Archimedes in the 3rd century BC, by utilizing the Archimedes Principle concerning buoyancy and the Archimedes screw for raising water. Some of Archimedes' inventions required sophisticated knowledge of differential gearing or epicyclic gearing—two key principles in machine theory that helped design the gear trains of the Industrial Revolution—and are still widely used today in fields such as robotics and automotive engineering. The Alexandrian engineers mark a significant transition from simple machines to the more complex ones.

The Great Wall of China that was constructed around 200 BC is considered an achievement of ancient civil engineering. The Chinese have been credited with the development of the wheelbarrow, the rotary fan, and the sternpost rudder that guided their bamboo rafts and, later, their junks. They also began making paper from vegetable fibers and gunpowder (Aggelikki 2011).

Then the Romans came. They used the ideas of the Ancient Greeks to implement their own engineering plans. The Romans, the improvers and adapters, built fortifications, roads, water distribution systems, and public buildings across the territories and cities they controlled. They put a great deal of effort into engineering. Roman engineering led to the building of some incredible engineering accomplishments that have survived to this day throughout Europe. The Roman engineer Vitruvius described watermills, and by the end of the Roman era many were in operation.

After the Roman Empire was divided by barbarians (around AD 476), the so-called Dark Ages (roughly, AD 500–1500) that followed still produced some things that were ingenious. For example, there was the development of the mechanical clock and the art of printing. Ibn al-Haytham (965–1040 AD), known for his work on optics understood that experimentation and measurement are essential to discovering new knowledge. Al-Biruni realized that measuring instruments were prone to research bias, so proposed that experiments needed replication, many times, before a common sense was possible (Shuttleworth 2014). There was the technique of heavy iron casting that could be applied to products for war. This was followed by the Renaissance of the sixteenth century, which the engineer/artist Leonardo Da Vinci dominated. But this whole period came under the influence of the architect/engineer, who built cathedrals and other large buildings, and the military engineer who built castles and other fortifications.

1.4.3 Engineering as a profession

The history of engineering as a profession, where payment is made in cash or kind for services, began with tool and weapon making over 1000 years ago, indicating that engineering is one of the oldest professions. The professionalization of engineering is illustrated by Imhotep who built the Step Pyramid at Saqqara in 3000 BC and was one of the few commoner mortals to be accorded divine status after his death. Engineering professionalization continued with the development of craft and guild knowledge, and the formalization of associated knowledge and education. Leonardo da Vinci, for example, had the official title of Ingegnere

Generale and his notebooks reveal an increasing engineering interest in how things worked (UNESCO 2010).

Engineering began to be recognized as a profession in the eighteenth century, where professional engineering has been established in response to threats to public safety. The modern professional identity of engineers began with the establishment of the Ecole Polytechnique in France and the foundation of professional engineering societies in England.

Engineers from the Victorian era in Britain, such as Isambard Kingdom Brunel and George Stephenson, enjoyed celebrity status. Brunel was responsible for bridges and dockyards including the construction of the first major British railway, the Great Western Railway; a series of steamships, including the first propeller-driven transatlantic steamship; and important bridges and tunnels. Brunel's designs revolutionized public transport and modern engineering. George Stephenson was an English civil engineer and mechanical engineer who built the first public railway line in the world to use steam locomotives. Electrical engineering can trace its origins to the experiments of Alessandro Volta in the 1800s, the experiments of Michael Faraday, Georg Ohm and others, and the invention of the electric motor in 1872. Chemical engineering, as with mechanical, developed in the nineteenth century during the Industrial Revolution. Large-scale production of chemicals was needed, along with the demand for new materials and new industrial processes (EO 2016). Aeronautic engineering turned the ancient dream of flight into a travel convenience for people. Control engineering accelerated the pace of automation while industrial engineering designed and managed mass production and distribution systems.

Gradually, practical thinking became scientific, as engineers developed mathematical analysis and laboratory experiments. Technical training shifted from apprenticeship to university education. Knowledge and know-how streamed more quickly in organized conferences and journal publications as professional engineering societies emerged where today engineering is a well-recognized profession.

1.4.4 Industrial revolutions

Historically, the term industrial revolution, as a worldwide phenomenon, is more convenient than accurate. It is applied to technological change and/or period of major industrialization that changed the whole of civil society. Industrialization is the process by which an economy is transformed from agricultural to one based on the manufacturing of goods. Industrial revolution is appropriate because history requires division into periods for purposes of understanding. History shows that there were waves of innovation at the turn of the eighteenth, nineteenth, and twentieth centuries to fairly justify the choice of periods. Since the start of the

Industrial Revolution, engineering has played an enormous role in the development of society.

These waves of innovation and industrial development have become known as Kondratiev (Nikolai Kondratiev was a known figure in the early history of the Soviet Union) waves, K-waves, long waves, supercycles, or surges, and relate to cycles in the world economy of around 50-year duration consisting of alternating periods of high and low sectoral growth. The K-wave is the rise and fall of a generation and covers both the social and economic life of the period.

Kondratiev's work was rediscovered in the mid-twentieth century by western economists. Among these was Joseph Schumpeter, who became famous for his work in innovation theory. Schumpeter took Kondratiev's work a step further, linking his long wave business cycles to technological innovation. He noted that each wave was linked to a profoundly influential and disruptive new technology. According to Schumpeter and others, there have been five Kondratiev waves since the 1600s (Faught 2015). Most analysts accept the "Schumpeter-Freeman-Perez" paradigm of five waves of innovation since the first Industrial Revolution, although the precise dates, phases, and causes and effects of these major changes are strongly debated, as is the nature of the sixth wave based on new knowledge production and application in fields such as IT, nanotechnology, biotechnology, and materials beginning around 1980. In this book, the industrial revolutions are categorized into eight phases as shown in Figure 1.3. This arrangement is chosen to align with the development of engineering along centuries. The most crucial periods in the development of engineering were the eighteenth and nineteenth centuries, particularly the iron and steam ages, the second K-wave of innovation (1845–1900) (UNESCO 2010), and successive industrial revolutions.

The Germans in the sixteenth century, the Dutch in the seventeenth century, and the French in the eighteenth century have perhaps the best claims to be regarded as being at the forefront of engineering expertise, each with impressive achievements to chalk up (Armytage 1976). The United Kingdom (UK) was more likely to be importing this talent than exporting it. However, engineering powered the Industrial Revolution that

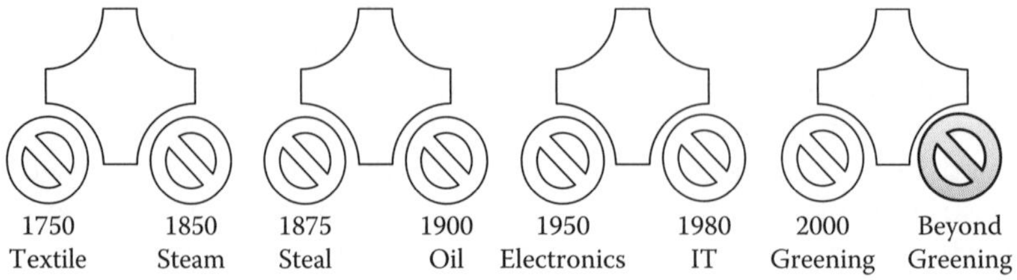

Figure 1.3 Engineering wheeling within the eight phases of industrial revolutions.

really took off in the UK in the eighteenth century, subsequently spreading to Europe, North America, and the world, replacing muscle by machine in a synergistic combination of knowledge and capital.

The first Industrial Revolution took place from 1750 to 1850 in Western Europe. It was important for the inventions of spinning and weaving machines operated by water power. These were eventually replaced by steam. It dominated the evolution of engineering and focused on the textile industry. Innovations in carding and spinning assisted by developments in cast iron technology resulted in the making of larger spinning mules and water frames. This Industrial Revolution was a fundamental change in the way things were made from human labor to machine. The growth of the textile industry was a key development in Britain's industrialization. It was this industry that initially employed the factory system.

The second Industrial Revolution, symbolized by the advent of electricity and mass production, was driven by many disciplines of engineering. In search of something to burn, the UK turned to coal, a largely abundant fossil fuel. Its demand was so high that it caused greater innovation, the introduction of a tremendously powerful invention—the steam engine—and the railways from 1850 to 1900, where the first steam engine was built in 1698 by mechanical engineer Thomas Savery. This device allowed for the beginnings of mass production, where the skills of engineers were suddenly needed. Initially, it was used to pump water out of coal mines. Then these engines were put to other uses in North American cotton machines, and later in steam locomotives and steamships. George Stephenson was an English civil engineer and mechanical engineer who built the first public railway line in the world to use steam locomotives. Aeronautic engineers turned the ancient dream of flight into a travel convenience for ordinary people. Control engineers accelerated the pace of automation. Industrial engineers designed and managed mass production and distribution systems (Bond et al. 2003).

The third Industrial Revolution was based on steel, electricity, and heavy engineering from 1875. Electrical engineering can trace its origins to the experiments of Alessandro Volta in the 1800s, the experiments of Michael Faraday, Georg Ohm and others, and the invention of the electric motor in 1872. The work of James Maxwell and Heinrich Hertz in the late nineteenth century gave rise to the field of electronics.

The fourth Industrial Revolution was based on oil, the automobile, and mass production, taking place between 1900 and 1950 and onward.

The fifth Industrial Revolution started with the invention of the transistor which further accelerated the development of electronics. Electronics was so critical to modern industry that electrical and electronics engineers today outnumber colleagues in any other engineering discipline.

Since the early 1970s, the dominant business mega-trend has been the age of information and telecommunications, the possible sixth Industrial

Revolution. Today information technology (IT) networks have terabytes of storage and the world has more phones than people.

The likely Industrial Revolution based on sustainable "green" engineering and technology is seen to have begun around 2005. The outstanding issue is how this seventh revolution is likely to unfold. It is indeed all about resource efficiency; put simply, doing more with less—less resources and less waste. We are absolutely inefficient in how we use our finite resources (Prentic 2012). Sustainability is converging with digitalized communication Internet and a digitalized automated transportation to create a super-Internet of Things (IoT). In the IoT era, sensors will be embedded into every device and appliance, allowing them to communicate with each other and Internet users, providing up-to-the moment data on the managing, powering, and moving of economic activity in a smart digital world.

1.4.5 *History is an opportunity*

Facts of the past are a reminder that history is an opportunity. History provides successful evidence that emerging sciences and their associated engineering and management disciplines can provide the basis for economic and social progress. First, history provides a compelling perspective on the process of scientific discovery. Another way to look to the rich history of engineering innovations, both successes and failures, is to learn more about motivations (Zaman 2015). For example, Murmann (2006) describes the rise of chemistry in Germany in the nineteenth century; Bush (1945) foresaw the rise of computing in the USA in the mid-twentieth century; and, finally, the product quality and innovation management movements in Japan provide a more recent example. In each of these cases, the key to success was collaboration among stakeholders, including government, industry, and academic sectors.

At today's point in history, start-up entrepreneurship has become the fastest way of creating value, and, thus, the fastest way to move upward in life. But this opportunity is unlike other opportunities humans had in history (Vital 2013). Table 1.1 shows how human utilized the opportunities and created value and money.

The Internet with lots of free resources has made it possible for people anywhere in the world to connect instantly. The world has become a treasure trove of possible investors, partners, employees, and customers. It is certainly a golden age of start-up funding from angel investors and venture capitalists. The amount of funding available to start-ups has grown considerably during the past decade. Entrepreneurs today have more opportunities, more resources, and more potential to succeed than they ever have before, but those new doors also allow new challenges, new restrictions, and new demands to enter in (Demers 2016). Entrepreneurship

Table 1.1 Opportunities along history

Early history	*Hunter*: First sign of survival and living using stone tools
10000 BC	*Grower*: Farming was the first source of income
AD 500	*Warier*: A way of creating value by taking goods of others produced
AD 1200	*Craftsman*: First signs to *be an entrepreneur*
AD 1500	*Explorer*: Came back with silks, spices, and other things
AD 1550	*Merchant*: Risk taker, the old entrepreneur
AD 1700	*Mechanizer*: Owning a machine became the next big thing
AD 1780	*Industrialist*: Industrialists became the big guys
AD 1900	*Oil Driller*: You are big if you discover oil
AD 1930	*Corporate Executive*: Being an executive was the best thing you could do
AD 1960	*Financier*: Being a banker became the best thing to do in life
AD 2000	*Entrepreneur*: IT lowered the cost of starting a company

today is becoming a driving force for economic growth among advanced economies. It is a vehicle that enables prosperity in firms, regions, and nations. The study of technology entrepreneurship, therefore, serves an important function beyond satisfying intellectual curiosity.

1.4.6 Lessons from history

A sense of history teaches students the all-important value of failure in science and engineering. In 1900, construction began on the Quebec Bridge (Canada), which linked Winnipeg to Moncton on the National Trans-Continental Railway. As construction neared completion, it collapsed under the weight of a locomotive loaded with steel. Seventy-five people were killed and an inquiry showed that the accident was due to an error in judgment by the engineers who designed the bridge. Tragedy struck again during the second attempt to build the bridge in 1916. The center span collapsed while being hoisted into place, killing 10 more people. The bridge was finally completed in 1917. The incident led to the tradition of the Iron Ring to symbolize the humility and fallibility of engineers. Today, the Ring signifies the pride in the engineering profession, while reminding engineers of unity and responsibility (Sy 1999).

Integrating case histories such as the Quebec Bridge and others in engineering education would promote a positive professional identity and a sense of tradition. Case histories would also point out a variety of ways that social systems or technical infrastructures can compromise the success of a seemingly appropriate engineering approach. Studying the successes and failures of innovative engineers could help students understand the roots of inspiration and innovation in the profession.

Providing future engineers with exposure to the history of their profession will give them the basis for honing their judgment and critical thinking skills and enhance their professional self-awareness. Successful engineering is defensive engineering, in which solution analysis is proactive and anticipatory. Engineers must consider past lessons and continue to ask questions of other engineers and nonengineering professionals as knowledge expands exponentially (NAE 2004).

1.5 History of engineering education

> If you are thinking a year ahead, sow a seed. If you are thinking 10 years ahead, plant a tree. If you are thinking 100 years ahead, educate the people.
>
> **Chinese Tao patriarch Kuan Tzu (500 BC)**

1.5.1 Early development

An important theme to emerge from the analysis of history of engineering is the growing importance of education. Historically, there exist simple forms of engineering education in ancient societies such as craft training which were developed into vocational technical schools of different types in the Middle Ages, particularly during the Renaissance and later during the scientific revolution of the sixteenth and seventeenth centuries. Initially, engineering was taught unofficially as skills handed down from practicing engineers. Afterward, it was integrated into the curriculum at academies to train engineers to meet the economic need. Figure 1.4 shows the major eras in engineering education development.

French and German academies led in providing of such instruction, while Britain trailed somewhat in the 19th century, owing to its long and highly successful tradition of apprenticeship (Buchanan 2017). Early interest in the development of engineering education took place in Germany in the mining industry, with the creation of a School of Mining and Metallurgy in Freiberg in 1702. Another one of the oldest technical universities is the Czech Technical University in Prague, founded in

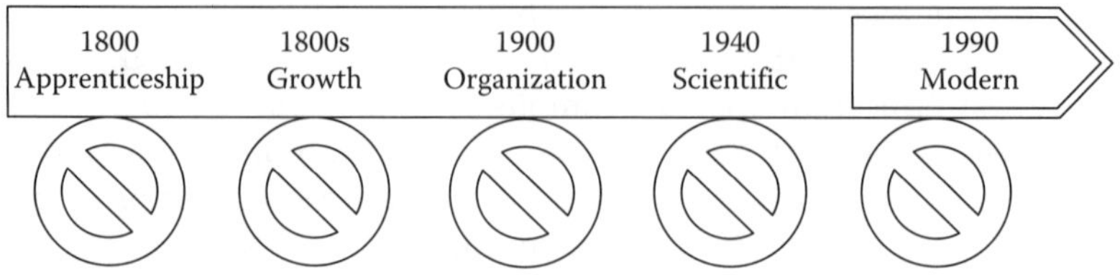

Figure 1.4 Major eras in engineering education development.

1707. In France, engineering education was developed with the creation of the École Nationale des Ponts et Chaussées in 1747 and the École des Mines in 1783. The first formal engineering program in the USA was at the Military Academy at West Point, in 1802 (Emmerson 1973). Around the 1850s, some schools started following the French model, the "polytechnics." Engineering was still separate from the university. This paradigm changed with the Morrill Act of 1862, where engineering became part of the university. The last basic format for offering engineering was the creation of independent scientific schools attached to a conventional university. Rich entrepreneurs funded most of the scientific schools; they realized the demand for scientific degrees. Engineering then became one of the more concrete paths within these schools (Issapour and Sheppard 2015).

At early stage, most American engineers started as apprentices on canal and railroad projects such as the Erie Canal and the Transcontinental Railroad. In Britain, however, engineering education was primarily based on a system of apprenticeship with a working engineer following the early years of the Industrial Revolution when many engineers had little proper training. Men such as Arkwright, Hargreaves, Crompton, and Newcomen, followed by Telford, George and Robert Stephenson, and Maudslay, all were insufficiently educated in engineering but developed technologies that powered the Industrial Revolution and changed the world.

In many fields, practical activity preceded scientific understanding; steam engines exist before thermodynamics, and rocket science is more about engineering than science. Britain tried to retain this lead by prohibiting the export of engineering goods and services in the early 1800s, which is why countries in continental Europe developed their own engineering education systems based on French and German models with foundations in science and math. Through the nineteenth and into the twentieth centuries however, engineering education in Britain also changed toward a science- and university-based system and the rise of the "engineering sciences," partly in recognition of the increasingly close connection between engineering, science, and math, and partly due to fears that Britain was lagging behind the European model in terms of international competition (UNESCO 2010).

1.5.2 Professional engineering education

The first phase of modern engineering occurred in the scientific revolution. Galileo Galilei (1564–1642) developed the scientific approach to the understanding of the natural world and analysis of practical problems. Isaac Newton (1643–1727) made the greatest contribution to the history of the scientific method that requires both deduction and induction. This phase of engineering continued through the first Industrial Revolution, when machines, increasingly powered by steam engines, started to replace muscles in most production. This led to the emergence of the professional identity of engineers in the early eighteenth century with the establishment of the Ecole Polytechnique in France and the foundation of professional engineering societies in England. The scientific schools established a model for transitioning the higher education from a classical model to one that incorporated both the sciences and the liberal arts. They operated independently of their associated parent universities, and faculty and students did not mix. They were looked down upon by the elites and had lower entrance requirements in terms of Latin and other languages. After World War II, most scientific schools steadily became fully integrated with their universities and Europeans professors brought their ideas on education to the USA. Civil and mechanical were the two most dominant engineering fields and engineering programs were closed to women until after World War II. Although the theoretical approach gained momentum in the late 1930s, the US engineering research integrated with science and mathematics curriculum did not become widely accepted until after World War II.

By the 1940s the war had created new opportunities for academic engineering research and the focus of such research in the postwar era was concerned with cutting-edge technologies such as computers and electronics, nuclear power, jet propulsion, rockets, and special materials. Engineer scientists were much more suitable candidates to conduct such research than practically trained engineers and as such they received priority in funding.

The current way of educating engineers, including the structure of the curriculum, was already established by the early twentieth century, but the course content has, of course, changed significantly since then. The last major shift in engineering education in the USA goes back over half a century when the role of science in the educational program increased significantly (Grinter 1955). By the 1950s, the engineering curriculum included much more science than practical applications as part of the curriculum. Engineering education in the 1960s and 1970s was also dominated by science. By the 1980s, hands-on skills dropped tremendously.

In the 1980s and 1990s, concerns about the role of science and math in society began to surface. Scientists, mathematicians, and educationalists began to openly discuss issues such as the contribution of their subjects to solving important real-world problems and a mismatch between what scientists and mathematicians actually do and what gets taught in school (Lucas and Hanson 2014).

1.5.3 Modern engineering education

Deficiencies in engineering education have been exhaustively enumerated in recent years. Engineering schools and professors have been told by countless panels and blue-ribbon commissions and, in the USA, by the Accreditation Board for Engineering and Technology (ABET) that we must strengthen our coverage of fundamentals; teach more about "real-world" engineering design and operations, including quality management; cover more material in frontier areas of engineering; offer more and better instruction in both oral and written communication skills and teamwork skills; provide training in critical and creative thinking skills and problem-solving methods; produce graduates who are conversant with engineering ethics and the connections between technology and society; and reduce the number of hours in the engineering curriculum so that the average student can complete it in four years (Felder et al. 2000).

In the late 1990s, a move to reintroduce more practical aspects of engineering into the curriculum had begun. Some efforts made to re-emphasize design in engineering schools and developed a better balance with engineering science. This move can be referred to as reinventing the wheel, going back toward the post land-grant engineering education. As a result, the accreditation agencies for engineering programs have started to work on changing the content of engineering curriculum (Sheppard 2015).

With the emergence of the Internet, knowledge has been communalized. Everybody has easy access to information; perhaps equally, knowledge is no longer owned by the experts. This change has already transformed industries and raised questions about authorship and ownership of information and scholarly works. Computers have also enabled the average person to create products that previously required large organizations with substantial resources. The same transformation is likely to result in the creation of new trends in engineering education, although the time frame may be rather longer. A significant role of the new trend in engineering education is facilitating experiential learning experiences and transdisciplinary design practices.

1.6 *Engineers*

> A scientist can discover a new star, but he cannot make one. He would have to ask an engineer to do that.
>
> **Gordon Glegg**
> *British Engineer (1969)*

1.6.1 *Who is an engineer?*

> Dreamer, innovator, researcher, problem solver, inventor, creator.
>
> *Tryengineering.org*

All the above are terms that aptly describe the characteristics of an engineer. In addition, an engineer is a mediator between the philosopher and the working mechanic, and, like an interpreter between two foreigners, must understand the language of both, hence the absolute necessity of possessing both practical and theoretical knowledge (Armytage 1976). Being an engineer is part of a profession that makes life better for humanity. An engineer applies the principles of science and math to develop economical solutions to technical problems. Being an engineer is finding the answers to the challenges that confront society. Engineer's work is often hidden in the details of everyday life, invisible because it works like air. Electricity, transportation, water treatment, wireless communication, and data networking are just a few examples of continuous, creative innovation and development by engineers that support lifestyle, drive the economy, sustain safety, and maintain social interactions.

An engineer is also a problem solver, organizer, communicator, planner, innovator, and designer. An individual who is qualified in or practices engineering is designated as engineer, and may be licensed and formally designated as professional, chartered or incorporated (Ciampi and Brito 2011).

Being an engineer is about making a difference and having interest that tends toward innovation, how to take the findings of the sciences and use them to develop useful products for human needs.

1.6.2 The four-dimensional engineer

> The ideal engineer is a composite .... He is not a scientist, he is not a mathematician, he is not a sociologist or a writer; but he may use the knowledge and techniques of any or all of these disciplines in solving engineering problems.
>
> **Nathan W. Dougherty**
> *American civil engineer*

Engineers of the future, like engineers of yesterday and today, should possess strong analytical skills. In this evolving world, however, other skills and knowledge are required to survive in the changing global market. It is crucial to manage technological changes, be creative, take risks, manage stress, think conceptually, and recognize and respect people's diversity and individual differences. de Figueiredo (2008) has tried to show the collective influence of basic sciences, human sciences, design, and the crafts as four dimensions of engineering (see Figure 1.5). This leads us to think of an engineer as a professional who combines in variable proportions the qualities of a scientist, a socialist, a designer, and a doer.

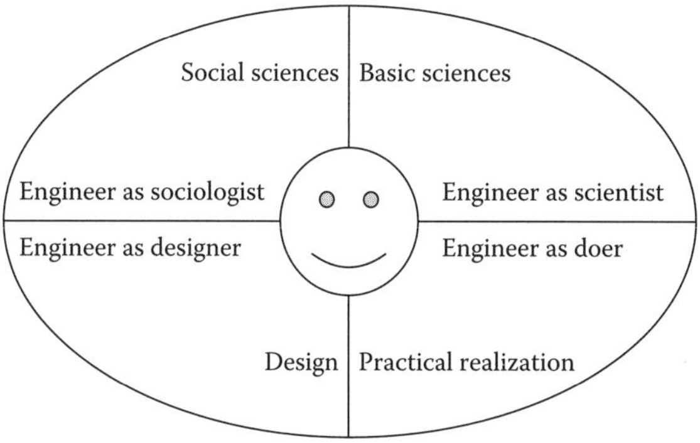

Figure 1.5 The four-dimensional engineer.

The dimension inspired by basic sciences views an engineer as a scientist who applies the basics of natural and exact sciences, stresses the value of logic and rigor, and sees knowledge as produced through analysis and experimentation. The sociologist dimension of an engineer sees engineers not just as technologists, but also as social experts, in their abilities to recognize the social nature of the world they act upon. The designer dimension sees engineering as the art of design. It values ST much more than analytical thinking that characterizes traditional science. The fourth dimension sees engineers as doers, the art of getting things done, valuing the ability to change the world and overcoming complexity. In this dimension, the completed job, which stands before the world, leads to higher recognition. Engineers are often interested in the conceptual development of new products, while technologists are often interested in improving existing technologies and building and refining products.

Engineering develops itself through the interaction of various professions, engaged in a social context. It is undeniable that in engineering, as well as in any other field of professional activity, the actions and decisions taken by its members are steeped in individual perceptions and also by thinking of the professional community. Engineering is, therefore, a social unit, with significant decision-making power in the process of society development. The engineer is, above all, an opinion leader who builds his ideas from the benchmark set by the professional community (Ciampi and Brito 2011).

1.6.3 *The new engineer*

In this evolving world, a new kind of engineer is needed, one who can think broadly across disciplines and consider the human dimensions that are at the heart of every design challenge (Grasso and Martinelli 2007). It is clear that engineering must go beyond pure technology and address matters that are imbedded in the social and economic fabric of society (Akay 2003). Yet, still significant revision is required in engineering curricula if the "new engineer" is to be thoroughly prepared for the challenges that lie ahead.

The "new engineer" will need not only awareness, but also an embedded ethical philosophy that stems from the foundation of their engineering learning. This broad social context of understanding should address the following generic attributes (Vere et al. 2009):

- Understanding of the social, cultural, global, and environmental responsibilities of the professional engineer, and the need for SD
- Understanding of the principles of sustainable design and development
- Understanding of and commitment to professional and ethical responsibilities

Engineers have a major role to meet sustainability. They should work to enhance the well-being, health, and safety, with the minimal use of natural resources and paying attention with regard to the environment and the sustainability of resources. Their work is induced by the opportunities and challenges that bring the sustainability and the structures in which it participates. In its domain, it is necessary to understand the contributions of various fields, which would underline the academic, professional performance, and social environment within which the individuals carry out their activities.

The awareness of social and environmental impacts of engineering profession is one step forward in this direction. Engineers must know how and when to incorporate social elements into a comprehensive systems analysis of their work (NAE 2004). Therefore, engineers have a responsibility to maximize the value of their profession toward building a sustainable world.

This new breed of engineer will not only be truly comprehensive problem solver, but also a problem definer, leading interdisciplinary teams. This is an admirable aspiration, but significant reform of engineering curricula will be required to prepare engineering graduates for their new responsibilities (Grasso and Martinelli 2007). Besides the above talents, one major quality is leadership, and one way to learn how to lead is to have creative leaders. Engineers should also be inquisitive, analytical, and detail oriented.

1.6.4 The entrepreneurial engineer

Engineers have all the skills necessary to be successful entrepreneurs. The technical skills and innovation required for future engineers are vital to convert innovative ideas into reality for common use. The engineer of the twenty-first century should know everything possible, find information about anything quickly, and know how to evaluate and use the information to achieve that goal.

An entrepreneurially minded engineer places product benefits before design features and leverages technology to fill unmet customer needs. The purpose of entrepreneurial engineering is to design value-added products and processes that create demand through innovation, resulting in positive revenue and regenerative profits for the enterprise producing the product (Kriewall and Mekemson 2010).

The development of such entrepreneurial mindset in engineers and scientists requires a significant change of university processes that have to integrate educational and research systems in the final perspective to generate and experiment sustainable innovation processes and practices. However, while academic institutions are central in creating entrepreneurship attitudes, skills, and behaviors, the success also depends on the actions and initiatives undertaken by the other actors outside of the education systems. Entrepreneurship thrives in ecosystems in which multiple stakeholders play key roles (WEF 2011).

The entrepreneurial engineer should have the ability to transform information into knowledge. Such an engineer can do anything possible, can understand engineering basics to quickly assess what needs to be done, and can acquire the tools needed and use these tools proficiently. This engineer should work with anybody, anywhere, and has the communication skills, team skills, and understanding of global and current issues necessary to work effectively with other people. The engineer should have the ability to conceptualize and turn concept to reality. Such an individual requires an entrepreneurial spirit, the imagination, and the managerial skills to identify needs, come up with new solutions, and see them through.

1.7 Disciplines of engineering

> What's nice about having an engineering degree is everybody thinks you are smart.
>
> **Ato Essandoh**

Changes in the engineering profession and engineering education have followed changes in society and its needs. Disciplines were added and curricula were created to meet the critical challenges in society and to provide the workforce required to integrate new developments into society and economy (Duderstadt 2008). For centuries, engineering was focused on war, either building defensive fortifications or the machines to attack these fortifications. In fact, the first nonmilitary engineering discipline is called "civil" engineering to distinguish it from its military counterpart. Today, the broad discipline of engineering includes a range of specialized disciplines or fields of application and particular areas of technology (NAE 2004).

All engineering disciplines derive from military engineering (Augustine 1994), which was formalized in eighteenth-century in France through the creation of technical institutes. Inspired by the French Revolution and the "century of light," the first institute, the Ecole Polytechnique, was established in Paris in 1794 (Bugliarello 1991). The concurrent Industrial Revolution and the second Industrial Revolution associated with the rise of the steel, chemical, and electrical industries (Nybom 2003) were driving forces behind the proliferation of the technical institute/university model that led to the establishment of a host of polytechnics in Europe, the Technische Hochshule in Germany, and institutes of technology in the USA (Rensselaer Polytechnic Institute 1824; Massachusetts Institute of Technology (MIT) 1861; Stevens Institute of Technology 1870; Georgia Institute of Technology 1885; California Institute of Technology 1891; Carnegie Mellon University 1900 and elsewhere). These early institutions, which focused mostly on the industrial arts, began by teaching civil engineering and then gradually introduced other engineering disciplines.

Today's engineering is a broad field that is divided into various disciplines and subdisciplines. There is a diverse and increasing range of areas, fields, disciplines, branches, or specialities of engineering. These disciplines are developed from four major branches: civil, mechanical, chemical, and electrical and electronic engineering. As knowledge developed and differentiated, these disciplines have subdivided, merged, and formed new subdisciplines of the major disciplines that offer specialized knowledge in a particular field or combination of many fields. Aerospace, biomedical, computer, and industrial/manufacturing engineering are among the medium disciplines in terms of degrees awarded annually. Among the smaller disciplines are agricultural, architectural, engineering management, engineering physics/engineering science, environmental, materials, mining, nuclear, and petroleum engineering. The emergence of new branches of engineering is usually indicated by the establishment of new university departments, new professional engineering organizations, or new sections in existing organizations. For more information about engineering disciplines you may visit: whatisengineering.com and www.g9toengineering.com.

With the growth of science and technology to new areas of practice that did not exist in the past and beyond existing scopes of practice, other disciplines are emerging. For example, mechatronics engineering (combination of mechanical, electrical engineering, and computing) is an emerging field that focuses on broad aspects of applications.

Innovation tends to emerge at the edges, at the boundaries between disciplines. Langrish (1985) stressed that much creativity consists of a new combination of existing ideas. Where the existing ideas are present in different people, it requires some kind of interaction to produce the combination. Highly collaborative interdisciplinary teams had an edge, because they used the "disciplines of innovation" to help them exploit interdisciplinary communication, transfer, reasoning, and insights.

1.8 Challenges of engineering

> The path to the CEO's office should not be through the CFO's office and it should not be through the marketing department. It needs to be through engineering and design.
>
> **Elon Musk**

1.8.1 Integration of knowledge

The fundamental need for engineering in the new century is to acknowledge and embrace the fact that engineering profession has changed quite a bit over the years. There are probably several major factors: pace, like most professions; competition from around the world and increasing technological advancement; complexity of the engineering task, so being a single-discipline engineer no longer works.

Knowledge is complex, is multidimensional, and can be either explicit (easily communicated) or tacit which is less tangible and more difficult to transfer (Nonaka 1994). Other knowledge dimensions include content (tasks and interactions), spatial (geographic by nature), temporal (frequency, pace, timing, and rarity), and mindfulness. Heterogeneity of experiences (experience variety across dimensions) has been shown to enhance learning (Schilling et al. 2003), a finding that contradicts the intuitive advantages of specialization.

The volume of information that engineers are collectively called upon to know is increasing far more rapidly than the capability of engineering curricula to cover it. Until the early 1980s, for example, most engineering graduates went to work in discipline-related jobs. Today, they are increasingly employed in nontraditional sectors. To be successful across this broad spectrum of employment, graduates should understand concepts that are well beyond the range of the conventional engineering curriculum. At the same time, the work done by any one engineer tends to occupy a relatively narrow band in the total spectrum of engineering knowledge. Unlike their colleagues of past decades, today's engineering students may never be called upon to work with basic elements of the conventional curriculum.

In the past, engineering responded to the increase in knowledge acquisition by constantly developing and reproducing new areas of focus in the various disciplines of engineering. As more of these areas were established, the depth of individual knowledge increased, but the breadth dramatically decreased. This poses a challenge to a future where transdisciplinarity will likely be critical to the solution of real-life problems. A more effective solution may be to switch emphasis away from providing training in an ever-increasing number of specialty areas to providing a core set of science and engineering fundamentals, helping students acquire and integrate knowledge across areas and disciplines, and training them to gain lifelong active learning skills. The focus in engineering education must shift away from the passive acquisition and possession of knowledge toward collaboration in knowledge creation in a competence learning environment, where critical skills are developed.

In the transdisciplinary educational model, students' characteristics, needs, interests, and personal learning processes are central to the learning experience; these objectives are more important than teaching of specific knowledge and skills (Ertas et al. 2000). Although technical knowledge is the characteristic more relevant and prevalent in the profile of an engineer, humanistic knowledge should not be disregarded. The engineer's ability to interact and collaborate with their peers in a different perspective from that, which favors a mechanistic view, is one of the main thrusts of professional success. To understand this assertion, it is necessary to oppose the idea that highly technical knowledge should be analyzed from a narrow viewpoint, in which problems are identified, classified, and answered with solutions purely rational in nature. The technical ability of engineers should be realized as a fusion of factors resulting from technical, economic, psychological, cognitive, and environmental aspects (Ciampi and Brito 2011).

In summary, the emphasis must move from the mastery of knowledge content to a mastery of the learning process itself. This will require a far more structured approach to continuing engineering education, more comparable to those provided for other learned professions such as medicine, characterized by a rapidly evolving knowledge base and profound changes in professional practice (Duderstadt 2008). In addition, engineers and engineering students are required to use new tools and apply ever-increasing knowledge, all while considering societal consequences and limitations within a multifaceted environment of old and new ideas. They will be working with various teams of engineers and nonengineers to create solutions to yet unknown problems.

1.8.2 ST skills

ST is a way of thinking used to tackle complex and unsure real-world problems. It recognizes that the world is a set of highly interconnected technical and social entities which are hierarchically organized. ST, which is an essential skill for engineers, provides a key intellectual underpinning for engineering. Systems science, systems dynamics, and systems engineering all lay claims to definitions of ST. Standard definitions of ST differ, defining the skill from the use of one's abilities to apply in a given setting, to the application of different kinds of thinking. ST refers to the ability to explore and understand the relationships between a system (health, transportation, energy, weather, heating, etc.) and its component parts to recognize the network of relationships among system components. ST skills are critical for future's engineers as they face twenty-first-century challenges such as meeting society's energy needs, dealing with climate change, and providing healthcare.

ST skills provide a thorough way of integrating people, purpose, process, and performance. Vesilind (1988) says that the most lasting effect of education on students is the maturation of their values and ethical sense. The failure of the engineering curricula to address attitudes and values systematically has had unfortunate consequences. Engineers often make decisions without feeling a need to take into account any of the social, ethical, and moral consequences of those decisions, believing that those considerations are in someone else's purview.

ST is nothing new today. Systems have played an important role in engineering design at least from the nineteenth century and earlier in design thinking (Buchanan 2001). Today's societal problems are getting broader and deeper and are transdisciplinary in nature. An integrative ST approach is needed to solve them and currently engineering education is not adequately preparing engineers for this challenge. Many advocate the development of ST skills, the ability to see the world as a complex system, in which we understand that "we cannot just do one thing" because "everything is connected to everything else." Figure 1.6 portraits ST as shadowing engineering disciplines and specializations.

ST within the engineering community is concerned with the system as whole and elucidating patterns of behavior and interactions, but engineers go beyond observation to actively manipulate technology and manage systems with ill-understood cause-and-effect relationships. Because these systems do not exist until engineers build them, and are therefore not observable, ST within engineering is based on the application of past experience to new situations. The engineering definitions of ST, therefore, place a greater emphasis on interactions and interfaces because these contribute to emergence (Davidz 2006).

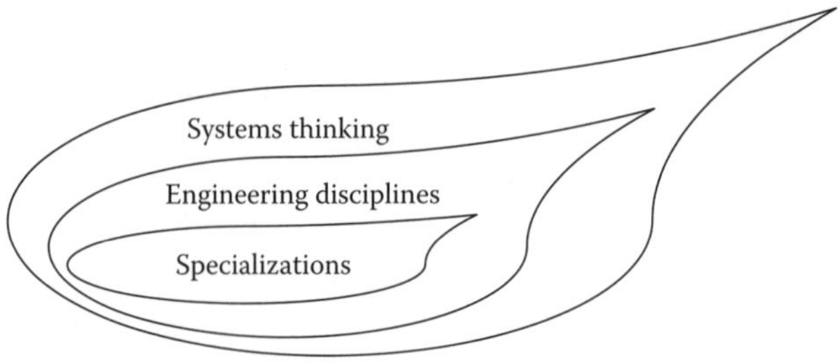

Figure 1.6 ST shadowing engineering disciplines and specializations.

1.8.3 *Sustainability thinking*

Responsibilities of engineers often involve designing and implementing technological solutions to existing problems. It is their responsibility to be aware of the impacts of their designs and products on society in general. They must be aware of subsequent effects upon the environment. They need to be able also to satisfy the often competing priorities as well as constraints specific to the technical challenges at hand. Responding to these competing forces is the essence of the emerging fields of green engineering and sustainability. These two major related areas incorporate many concepts, facts, and tools, all of which are critical for engineers and students to know and realize.

Sustainability is characterized by four dimensions (Penzenstadler et al. 2013) as shown in Figure 1.7. Individual/group sustainability refers to maintaining human capital (e.g., health, education, skills, knowledge, employment, culture, religion, leadership, privacy, security, and access to services). Social sustainability aims at preserving the societal communities in their social capital, including explicit requirements for strengthening safe and caring community, local development, health, poverty reduction, quality of life, sense of place, and urban and rural welfare. Economic sustainability aims at maintaining capital and added value including budget constraints and costs as well as market requirements and long-term business

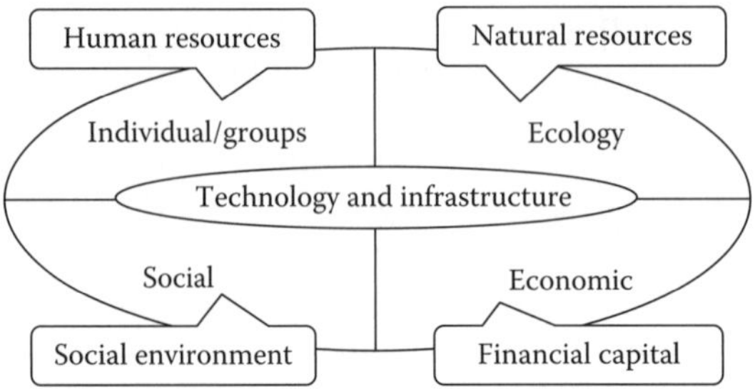

Figure 1.7 Dimensions of sustainability.

objectives. Environmental sustainability refers to improving human welfare by protecting the natural resources: water, land, air, minerals, and ecosystem services. The central technological sustainability refers to longevity of systems and infrastructure and their adequate evolution with changing surrounding conditions. Furthermore, efficiency, especially energy efficiency and hardware sufficiency (WCED 1987; Hilty et al. 2011), is part of the technical sustainability requirements.

Individuals and society may play different roles in the pursuit of sustainability. Barry (1999) thinks that we are not an undifferentiated humanity facing an equally undifferentiated nature. He proposes a citizen–environment perspective, as opposed to the classical society–environment relation, as the most appropriate standpoint from which to judge politically the normative standing of the nonhuman world. Merging individuals and society into one single dimension might fail to capture the complexity of human behavior and the relevance of personal relationships for sustainability. Individuals, who play a fundamental role in the generation, shaping, and maintenance of culture, are in consequence partly responsible for the construction of a culture-dependent notion of nature.

1.9 *The broader engineering education*

> Education is what remains after one has forgotten what one has learned in school.
>
> **Albert Einstein**

1.9.1 *Reengineering of engineering education*

> Education is the acquisition of the art of the utilization of knowledge.
>
> **Alfred North Whitehead**

As late as the 1870s, most engineers entered the field after serving an apprenticeship in the field or in a machine shop. But by the 1880s, what one historian has called the "shop culture" slowly began to give way to the "school culture." After World War II, American engineering colleges completely embraced engineering science as the foundation of engineering education. That decision led to sharp reductions in the time and coursework devoted to practical skills such as drafting, surveying, and other traditional features of engineering curricula (Reynolds and Seely 1999).

Sciences have become the essential component of the engineering curriculum, giving a higher status to analytical courses than intuitive and practical-oriented courses. This predominance of sciences in engineering seems to be a barrier to developing the new set of skills that new engineers need, taking into consideration that industry has become one of the main employers and an important supporter of education. A further challenge resulting from this shift is that globalization has generated an open market of engineers causing the creation of multicultural engineering workplaces around the world, requiring a new set of professional competencies (Ferguson 1992; Freeman 2010).

The modern economy is characterized by the rise in importance of the flexible, innovative small firms that can react quickly to market change and is equipped with people who have the skills and knowledge to generate new ideas and to get those ideas to market relatively quickly (Galloway et al. 2006). On the other hand, the rapid changes occurring in the world, coupled with changes in engineering education already taking place, are likely to result in an extensive reengineering of engineering education.

To be competitive, engineers must provide high value by being fast, innovative, integrative, conceptual, transdisciplinary, and entrepreneurial. Figure 1.8 reflects the components of the reengineering process and shows a model of how engineering education system could transform ideas into real-world innovations.

1.9.2 Transition to deep learner-centered environment pedagogy

> Children needed to be engaged in self-directed activities and the teacher was to serve as the guide.
>
> **Friedrich Froebel**

> Learning through doing.
>
> **Rudolf Steiner**

> Teachers need to prepare the space and then step back and facilitate.
>
> **Maria Montessori**

While pedagogic practices vary greatly between universities and schools, in general, teachers tend to have a highly positive view of the importance of fostering and valuing creativity and innovation. Such educational incentives are not necessarily dependent on teaching tools and facilities but on the creation of a positive educational environment. Technology in this regard is persistent and is used to find and deliver content knowledge and to enable deep learning goals of creating and using new knowledge to solve real-life problems. Most instructional elements of the new pedagogy are not new; however, the active learning

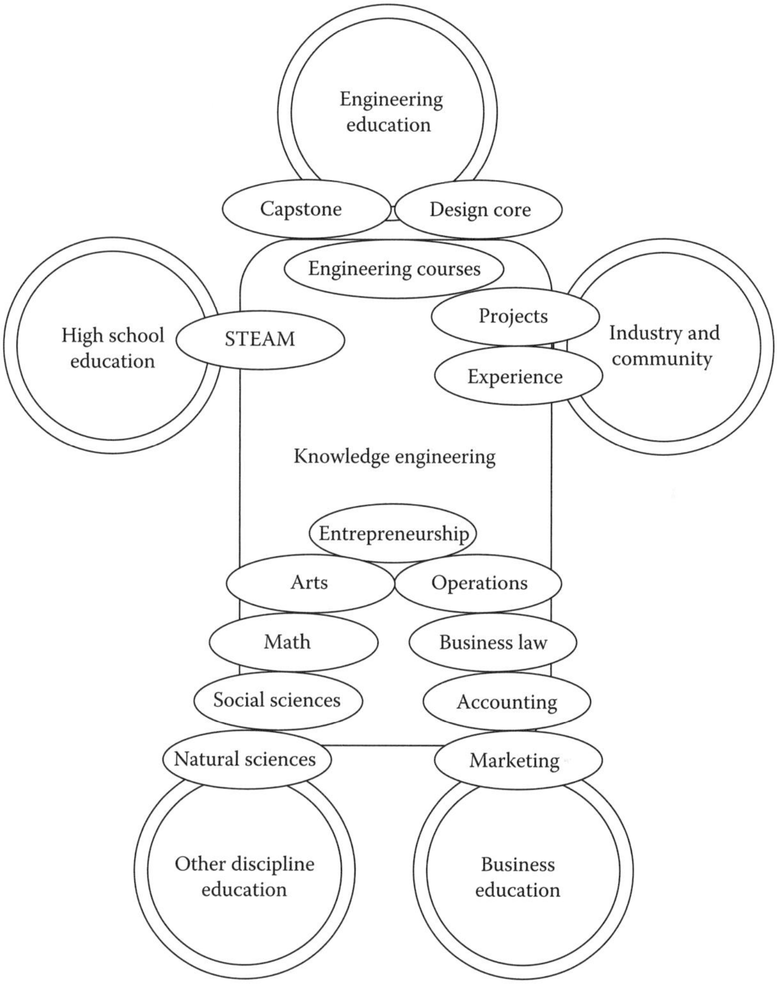

Figure 1.8 The reengineered education landscape.

partnerships with students are new. Figure 1.9 shows the transition from old pedagogy, where a teacher is the only source of knowledge to learning environment pedagogy where the teacher plays the role of guide or facilitator.

Many of the teaching strategies that have been advocated for at least a century by the likes of the German Friedrich Froebel (1782–1852), Austrian Rudolf Steiner (Waldorf) (1861–1925), American philosopher John Dewey (1859–1952), Swiss psychologist Jean Piaget (1896–1980), Italian physician Maria Montessori (1870–1952), and Russian sociologist Lev Vygotsky (1896–1934) are beginning to emerge and be embraced. Previously, the

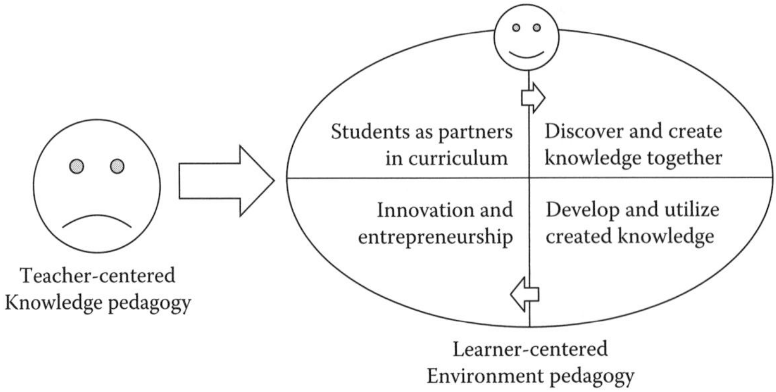

Figure 1.9 Transition from old pedagogy to new pedagogy.

conditions for these ideas to take hold and flourish did not exist. Today, there are signs that this is changing. Crucially, the new ideas, compared to the past, have potentially greater precision, specificity, clarity, and above all greater learning power. A form of positive contagion is emerging as these powerful teaching strategies begin to take hold in regular schools and in fairly traditional public education systems. They are emerging almost as a natural consequence of student and teacher alienation on the one hand and growing digital access on the other hand. These developments have profound implications for curriculum, learning design, and assessment (Fullan and Langworthy 2014; Murphy 2016).

Dewey, Piaget, Montessori, and Vygotsky believed that education should focus on the development of the individual, to nurture children's intellects in an effort to form a better society. Since children were seen as the catalyst for change, all four believed that educational models should center on the child rather than on subject matter. They all believed that children learn by doing.

Montessori developed teaching tools that encouraged learners to explore their environments through self-directed and cooperative learning activities. At the time, this was an innovative and modern approach. As a student of Montessori's work, Piaget believed that learning takes place when a child interacts with his or her environment, and by interacting with the environment a child constructs his or her own knowledge by giving meaning to the people, places, and things in their world (Mooney 2000). Piaget believed that children should be given every opportunity to do things on their own so they could learn from those experiences. Piaget also valued free play, as it helps support cognitive development of "preoperational children." By allowing large blocks of free play, children can develop cognitive skills through real-world experiences and open-ended activities (Meagor 2014).

The learner-centered pedagogies can be defined briefly as a new model of learning partnerships between and among students and teachers, aiming toward deep learning goals and enabled by universal digital access. Most instructional elements of the above pedagogies are not new teaching strategies, although the active learning partnerships with students are new. Many of the teaching strategies that have been advocated by Dewey, Piaget, Montessori, and Vygotsky are beginning to emerge and be embraced. The new pedagogies require students to create new knowledge and connect it to the world by using the power of digital tools. With high level uses of technology, students and teachers can develop creative multimedia presentations, simulations, and animations that enhance deep learning. Previously, the conditions for these ideas to take hold and flourish did not exist. Today, there are signs that this is happening (Fullan and Langworthy 2014).

1.9.3 Greening education and embracing sustainability

Owing to the multidimensional nature of sustainability, based on complex social, economic, and ecological theories, policies, and practice, the concept of sustainability and design can be difficult for students and engineering professionals to fully comprehend and understand. These topics will require critical analysis by academic leaders, teaching and learning pedagogues, and university lecturers and teachers (Coyle and Rebow 2009).

As outlined in Figure 1.10, this book may offer some commentary and attempt to answer some of the above questions by exploring the principles of SD and sustainable design, in the context of engineering profession, design, and education with particular emphasis on sustainable strategies and tools.

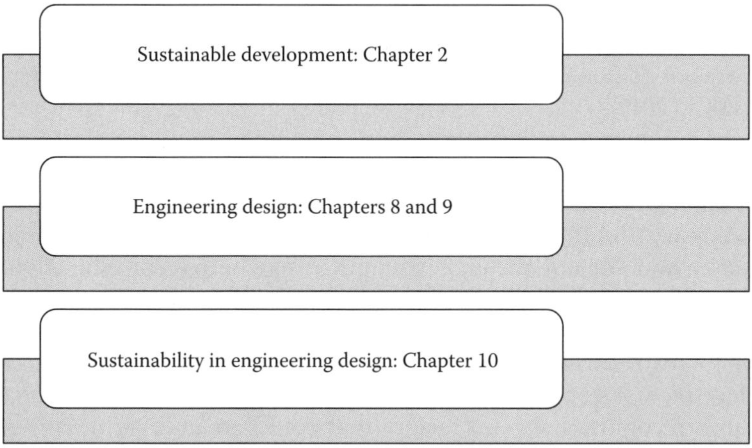

Figure 1.10 Steps toward engineering education for SD and design.

In discussing engineering education for SD and sustainable design, educators should offer clarity in definition and understanding of these concepts. Students must be tutored in the use of multi-, inter-, and transdisciplinary approaches and encouraged to work together in teams comprising people from different disciplinary, social, and cultural backgrounds.

Today's engineering design process requires engagement by many participants, including engineers, politicians, governmental agencies, managers, clients, anticipated customers, and the general public. Defining and measuring the qualities in engineering designs that need to be preserved is a major challenge if we are to fully embrace and understand sustainability. How can engineers measure the quality of engineering systems? This is in the light of taking on board goals, requirements, and constraints of all concerned parties, and at the same time ensuring minimal negative effect on the environment (Coyle and Rebow 2009).

1.9.4 *Education for innovation and entrepreneurship*

One of the key success factors for entrepreneurship education is an effective development of the entrepreneurial ecosystem, in which multiple stakeholders play a role in facilitating entrepreneurship. It includes business (large and small firms, as well as entrepreneurs), policymakers (at international, national, regional, and local levels), and formal (primary, secondary, and higher education) and informal educational institutions. To effectively implement entrepreneurship education, both top-down and bottom-up approaches are necessary. Top-down approaches require the commitment of the most senior policymakers. Building this commitment is not easy, often entrepreneurship education has to be framed within the context of its contribution to economic and social growth to attract proper attention. Bottom-up approaches require champions at the local or regional level who can help drive initiatives on the ground (UNESCO 2010).

The importance of creativity and innovation in addressing the economic, environmental, and social needs has been recognized in policy discussion worldwide. Such policies call for the reinforcement of innovative capacity and the development of a creative and knowledge-intensive economy and society through strengthening the role of education and training in the knowledge landscape and focusing school curricula on creativity, innovation, and entrepreneurship. A common understanding of what creativity is for education and what it entails is, therefore, envisaged as the first step toward creative and innovative education. Moreover, research recognizes several factors that could create a stimulating and creative environment. Teachers, for instance, are key figures in constructing a creative climate, but they need support from both policymakers and institutions. In particular, curricula and assessment are key areas that must be addressed if creativity is to be allowed in the classroom (Cachia et al. 2010). In order for these three elements, innovation, technology, and entrepreneurship, to produce the synergies necessary to considerably transform education, it is needed to build a collaborative platform that allows for the productive integration of careful study, design, and invention, and action at scale.

Innovations in engineering curricula, teaching approaches, and pedagogical activities both inside and outside the classroom are aimed at contributing to a more holistic education that will provide engineering students with a wide range of opportunities to acquire, develop, and practice these professional abilities (Grasso 2002). It is argued that creativity, in the educational context, should be hypothesized as a transversal and cross-curricular skill, which everyone can develop. Therefore, it can be fostered and inhibited. No profession unleashes the spirit of innovation like engineering. Engineers play an integral role in everyday life. They are at the forefront of modern technology and also are at the pinnacle of innovation. Innovation, the route of inventing real-world applications, something new, desirable, useful, and sustainable, happens at the intersection of technology, business, human factors, and complexity. Figure 1.11 shows the essence of innovation in engineering education. It provides an idea of how engineers make things that work or make things work better, doing so in particular ways.

Engineers think and act in distinctive ways. These specific ways of thinking and acting are known as habits of mind (HOM) and engineering habits of mind (EHOM), which have emerged through an iterative process of study and conversations with engineers and educators. To accomplish the goal, creativity and innovation should be embedded in the thinking behind and approach to education policies and national visions and they

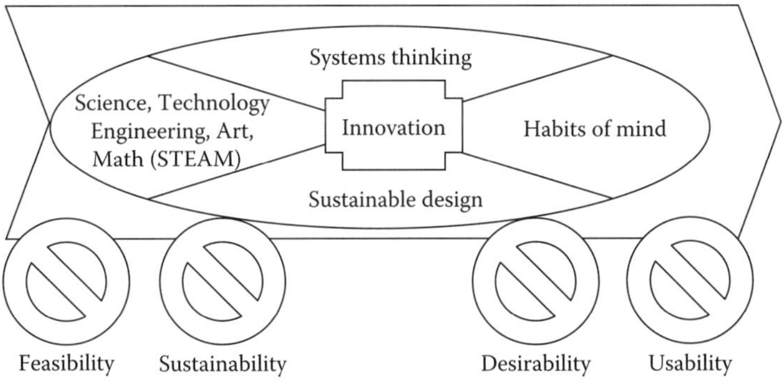

Figure 1.11 Determinants of sustainable innovation.

should be promoted in all curricular areas and across different subjects. Engineering students as well as students from the humanities, arts, social sciences, and business must all realize they are partners in enriching the innovation challenge.

1.9.5 Breadth and depth of knowledge and skills

It has become evident that there is a crucial need for change in the practice of engineering and the education of future engineers. Specifically, engineering disciplines need to be broadened and enriched to better prepare graduates for working in a constantly changing economy, driven by the explosion of knowledge, globalization, and a myriad of other factors described aptly in Thomas L. Friedman's book, *The World is Flat* (Friedman 2005).

Engineers are no longer singularly required to have expertise in a specific technical skill area. Technologies are changing and the boundaries in and between science and technology disciplines are less defined. Future engineers will have to be able to transcend disciplinary limitations, work in different fields, and master communication and intercultural collaboration skills (Haase et al. 2013).

In the past, engineering responded to the explosion in knowledge by continually developing and spawning new areas of focus in the various engineering disciplines. As more of these areas arise, the depth of individual knowledge increases, but the breadth can dramatically decrease. This poses a challenge to an engineering future, where interdisciplinarity will likely be critical to the solution of complex problems (NAE 2004).

Engineering should be realized through the process of active investigation, which is initiative and risk taking, subject to critical thinking, adaptive to life-long learning, and indicative of intellectual development and maturity. As Figure 1.12 shows, the modern professional engineer

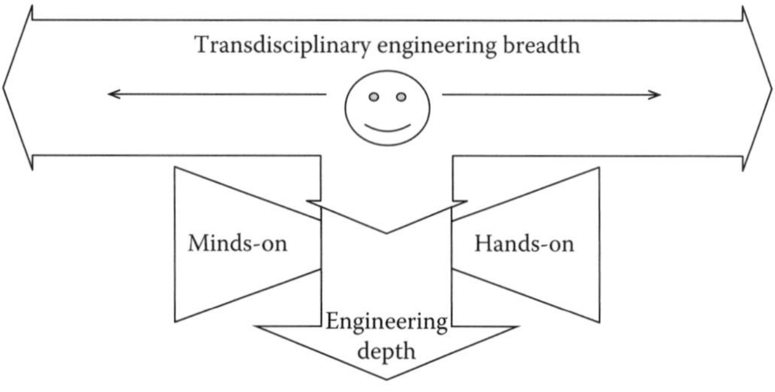

Figure 1.12 Knowledge and skill domain of a modern professional engineer.

must have depth in an engineering discipline with transdisciplinary engineering breadth and a balance between theory and practice. Engineering faculties should ensure that breadth of learning, beyond the technical aspects of the specialist engineering discipline, is a major drive in engineering education. For engineers, this means not only possessing deep, technical skills, but also having broader attributes, such as critical thinking skills and entrepreneurial mindset, compassion, written and verbal communication skills, team building, client interaction skills, and the ability to collaborate in addition to other subjects such as finance and marketing (Doyle 2014).

Today's engineer must span an array of fields, just as modern technology, systems, and processes do. To address this need, many universities have taken the initiative to draft new programs that implement changes required in the education of future engineers. The broader impact of these programs is strengthening the knowledge and proficiency of the future engineering workforce. These programs target toward enhancing and enriching engineering education to better address the challenges of the future. Such programs should use an educational approach that gives engineering students a broader background that goes beyond technical aspects and introduces them to critical issues which include leadership, innovation, and entrepreneurship. This more holistic approach to engineering emphasizes the importance for engineers to comprehend interactions between engineering and nonengineering aspects of a system (Hawken et al. 1999; Hargroves and Smith 2005).

The most important and fundamental role for engineering faculties is to prepare young people to work in various capacities in an evolving world, providing them with an education which is technically focused and has adequate breadth. The industry is moving toward such education, but universities are slow to do the same with their curricula. Society requires that engineering graduates be broadly educated, that they be knowledgeable about the society in which they live and work, and that they be sensitive to the economic, social, political, environmental, cultural, and ethical dimensions of their work (CAE 1999).

Other professions such as medicine and law are currently accepting the idea of broadening their admissions requirements to allow enrollment of students from undergraduate majors in the social sciences and humanities. They seek more well-rounded students who can be molded into caring and compassionate physicians and who better understand the broader context of medical decisions and patient treatment. Furthermore, the recent development of multiple course sequences to provide a concentration or minor in engineering programs for students in other fields provides yet another path for broadly educated undergraduates to consider engineering careers after further graduate studies (Duderstadt 2008).

1.10 *Teaching and learning styles*

> The only way to do great work is to love what you do. If you haven't found it yet, keep looking. Don't settle.
>
> **Steve Jobs**

When we walk into an arbitrarily chosen engineering classroom what do we see? Too often the same thing we would have seen in 1970 or 1940. The professor stands at the front of the room, copying a derivation from his notes onto the board and repeating aloud what he writes. The students sit passively, copying from the board, reading, working on homework from another class, or daydreaming. Once in a while the professor asks a question: the student in the front row who feels compelled to answer almost every question may respond and the others simply avoid eye contact with the professor until the awkward moment passes. At the end of the class, students are assigned several problems that require them to do something similar to what the professor just did or simply to solve the derived formula for some variable from given values of other variables. The next class is the same, and so is the next one, and the one after that...

(Rugarcia et al. 2000)

1.10.1 *Conceptions of learning*

Students learn in many ways—by seeing and hearing; reflecting and acting; reasoning logically and intuitively; memorizing and visualizing; drawing analogies and building mathematical models; steadily and in fits and starts. Teaching methods also vary. Some instructors lecture, others demonstrate or discuss; some focus on principles and others on applications; some emphasize memory and others understanding (Felder and Silverman 1988).

Sfard (1998) labeled the "acquisition" metaphor for learning. That is, learning is seen as something which results in the personal acquisition of knowledge and skills. Rather, less attention has been paid to other conceptions of learning, such as "learning as participation" or "learning as knowledge creation." For example, if learning is seen as a matter of acquiring knowledge, then tools which are set in place for students are there to help them acquire that knowledge: they are solely a means to that end. But if learning is seen as primarily a matter of participating in a social practice, tools are there to be mastered, as instruments of that social practice.

If learning is seen as a matter of collaborating in knowledge creation, then new tools are designed and created by students, as a legitimate outcome of their work (Paavola et al. 2004; Moen et al. 2012; Goodyear 2015).

1.10.2 Deductive versus inductive reasoning approaches

Typically, two solution approaches are followed to investigate and solve scientific problems, the deductive approach and the inductive approach. Inductive approach is sometimes called the "bottom-up" approach, which depends on creative insight into observed phenomena, and may be more applicable to creating new solutions or analysis methods related to sustainable design (McIsaac and Morey 1998). Two Nobel laureates, Barbara McClintock and Albert Einstein, employed inductive and collaborative approaches using research questions to narrow the scope of their qualitative research study. Furthermore, both researchers stressed the need for a connection to living ecosystems. Figure 1.13 shows the difference between deductive (water-fall) and inductive (hill-climbing) reasoning approaches. Induction is usually described as moving from the specific to the general, while deduction begins with the general and ends with the specific. Arguments based on laws, rules, and accepted principles are generally used for deductive reasoning; however, observations tend to be used for inductive arguments.

Many engineers are comfortable with deductive approaches, which will help in quantifying the results of sustainable and green design. In a

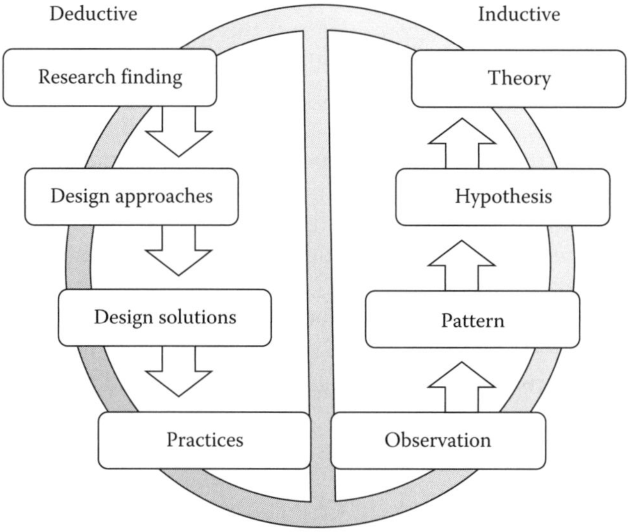

Figure 1.13 Deductive and inductive reasoning approaches.

sense, one can think of both approaches, deductive and inductive logic, as a cycle. Engineers can use the inductive approach to grasp the problems and solutions and then employ the deductive approach to determine a method to quantify and evaluate the results (Bilec et al. 2007).

To be effective, engineering education should adopt both approaches, rather than directing itself primarily to a single one. Induction is the natural human learning style. Most of what we learn on our own (as opposed to in class) originates in a real situation or problem that needs to be addressed and solved, not in a general principle; deduction may be part of the solution process, but it is never the entire process. On the other hand, deduction is the natural human teaching style, at least for technical subjects at the university level (Felder and Silverman 1988).

While induction and deduction are indeed different learning approaches, the effective approach of teaching and learning at least if carried out well is a blended approach of both deduction and induction. An effective way to reach all types of learners is to follow the scientific method in classroom presentations: first induction, then deduction. The inductive approach is enriched by problem/project/product-based learning (PBL), discovery and inquiry learning, or some variation on those themes. Inductive learners need motivation for learning. They do not feel comfortable with the typical teacher's saying: "This material will be useful to you some day (Felder and Silverman 1988)."

PBL begins with an assignment, to one or more students, to carry out specific tasks that would eventually lead to the arrival of a final product—a design, a model, a device, and/or a computer simulation. The end result is normally a written and/or an oral report highlighting the main steps undertaken to produce the product and the outcome (Akili 2015). PBL is an instructional approach that challenges students to learn by working cooperatively in teams seeking solutions to real-life problems. It prepares students to think critically and analytically, as well as to find and use appropriate learning resources. Importantly, PBL encourages students to be in charge of their own education. It underlines critical thinking skills, learning how to learn, and cooperating with others.

Few engineering instructors would have to modify what they usually do in order to present engineering courses: lectures still accomplish this task. What must generally be added to accommodate more students is project and case studies that incorporate analysis, design, visual material, pictures, diagrams, videos, and prototypes (Felder and Silverman 1988).

1.10.3 Teacher-centered and student-centered instruction

In teacher-centered education (TCE), control is of primary importance and "authority is transmitted hierarchically" (Dollard and Christensen 1996), meaning the teacher exerts control over the students. Critics of teacher centeredness argue that in these classrooms, compliance is valued over initiative and passive learners over active learners (Freiberg 1999). Conventional TCE, frontal teaching, and chalk and talk can still prevail in creating the right environment to enhance creativity and innovation. In such pedagogy model, the substance of teacher quality is a teacher's pedagogical capacity and talent to coach strategies rather than delivering knowledge and the ability to develop partnership with students in realizing the real process of learning.

In contrast, a constructivist teacher is interested primarily in helping students engage problems and issues, search below the surface, try out various possible solutions or explanations, and finally construct their own meaning (Ryan and Cooper 2001). In these classrooms, teaching methods or strategies include reflective thinking, inquiry, exploratory discussions, role-playing, demonstrations, projects, and simulation exercises (Edwards 2004).

Learner-centered education (LCE), as the term implies, centers on learners. It builds around the learners' needs and interests. Here, education is seen from the perspective of learners rather than teachers. Learners remain at the center of all educational activities. Students learn through active engagement and interaction. The role of a teacher is only of a facilitator.

According to Halperin (1994), most activities today in a majority of classrooms in higher education continue to reflect an "old" style of instruction wherein "students sit quietly, passively receiving words of wisdom being professed by the lone instructor standing in front of the class." Learning, however, rarely if ever occurs passively. Instruction is most effective, according to cognitive psychologists and educators, when students are encouraged to become actively involved in their own learning. Additionally, an allowance of time must be made for meaningful, open interactions between teacher and student and groups of students that nurture the student's natural curiosity.

The importance of integrating and nurturing creativity is described by Caine and Caine (1991) who have written the following: educators can generate much of the excitement and energy they desire by introducing creativity into the lives of their students… a student's desire to know more about a subject is more important than a measure of performance at any point in time.

One procedure used is a prize award for performance in project work and the other procedure is a Socratic style of teaching in the classroom. Socratic teaching is an old but still a powerful teaching tactic for fostering dialogue. It is a form of cooperative argumentative debate between individuals, based on asking and answering questions to stimulate critical thinking. In Socratic teaching, focus is on giving students questions, not answers to model an inquiring, probing mind by continually probing into the subject with questions.

1.10.4 Convergent (closed-ended) and divergent (open-ended) problems

Convergent thinking is the process of finding a single best solution to a problem that we are trying to solve (Hajesfandiari et al. 2014). Many tests that are used in schools, such as multiple-choice tests, math quizzes, and standardized tests, are measures of convergent thinking.

On the other hand, divergent thinking is the process to create several unique solutions intending to solve a problem. The process of divergent thinking is spontaneous and free flowing, unlike convergent thinking, which is systematic and logical. When using convergent thinking, logical steps are used in order to choose the single best solution. Convergent thinking stands firmly on logic and less on creativity, while divergent thinking is mostly based on creativity. Divergent thinking is mostly used in open-ended problems that creativity is a fundamental part (Hajesfandiari et al. 2014).

The types of problems encountered by engineers have been classified in a number of different ways. One of the most basic classifications has been between well-structured (closed-ended problem, CEP) and ill-structured problems (open-ended problem, OEP). CEPs are those that are simple, concrete and have a single solution, while OEPs are complex, abstract and have multiple possible solutions (Jonassen 1997). OEP solving is a skill that is central to engineering practice and, as a consequence, it is imperative for engineering students to develop skills for solving such problems (Denayer et al. 2003). Courses or projects in systems design should not only teach the logical sequence of design steps that professional engineers take but also stimulate creative thinking with OEP formulations and questions.

OEPs are challenging for students when they are confronted with the fact that there is no unique answer for them to achieve. The uncertainty of no one right answer is somehow intimidating to students. Design based on OEP is a difficult concept for faculty to teach and for students to learn because addressing OEP requires an integrative approach that is not taught in analytic courses. In this regard, engineering design is not teachable without letting students explore the design process.

1.10.5 System-based versus subject-based learning

System-based learning is an approach that is common in medical education and practice. It is one of the most challenging competencies to define, incorporate into training and practice, and evaluate. System-based learning can be thought of as an analytic tool, as well as a way of viewing the world, both of which can make change efforts more successful. The focus is on understanding the interdependencies of a system or series of systems. It emphasizes broad interdisciplinary topics rather than single-subject classes. Instead of teaching single topic, for example, educators could blend aspects of various topics in one system or theme (Johnson et al. 2008).

Bertalanffy (1968), the founder of the scientific, mathematical "Theory of Systems," defined a system as a set of interacting, interrelated, or interdependent elements that work together in a particular environment to perform the functions that are required to achieve the system's aim. The importance of understanding systems as interrelated parts of a whole cannot be overstated. Systems can be continually improved, but one must consider how its products are created, why they are created, and how they can be improved. Comprehending the assembly of the system as a whole can inform the work of those who are trying to create successful, interdependent systems (Batalden and Mohr 1997). Learning to see interrelationships, rather than linear cause-and-effect chains, and grasping the phenomenon of change as a process, rather than as a snapshot, are essential for understanding systems (Senge 1990).

System-based learning is different from the common subject-based teaching, where the latter means that the sequences are based on broad subjects such as electronics, mechanics, thermodynamics, control, etc. There is less potential for integration between the subjects and it is good for students who like unloading information on exams and then forgetting everything. However, the sequence in system-based learning depends on system under consideration, for example, data acquisition system, satellite control system, energy management system, etc. During each sequence, students learn about electronics, computing, control, sensing, and other subjects in an integrated approach.

ST is the cornerstone of how learning organizations think about their world (Senge 1990). Learning organizations are those that measure outcomes and strive for improvement. Many fields including health care, education, telecommunications, and aviation use systems theory to better serve their clients, understand applicable research, improve outcomes, and ensure quality and safety. Recognizing feedback from the system and using that feedback for design and redesign of services is an inherent element of ST (Johnson et al. 2008).

1.10.6 Mastery learning

Mastery learning (or as it was originally called "learning for mastery") is an instructional strategy and educational philosophy, first proposed by Dr. Benjamin Bloom in 1968. It maintains that students must achieve a level of mastery (e.g., 90% on a knowledge test) in prerequisite knowledge before moving forward to learn subsequent information. If students do not achieve mastery on the test, they will be given additional support in learning and then get tested again. In using this strategy, teachers organize the important concepts and skills they want students to acquire into learning units, each requiring about a week or two of instructional time (Bloom 1971).

The mastery learning approach implies that the instruction should be based on the time needed for different students to learn the same subjects and reach the same level of mastery. This is quite different from the classic models of teaching, which focus more on differences in students' ability to learn where all students are given the same time to learn and the same sets of instructions. Therefore, in mastery learning there is a shift in responsibilities so that student's failure is more due to the instruction and not necessary the student's ability to succeed.

1.11 Bridging curriculum through training and education

> Education is not preparation for life; education is life itself.
>
> **John Dewey**

1.11.1 Knowledge engineering

Developing a curriculum that effectively addresses the diverse requirements of engineering is a demanding task. Therefore, it may require building multiskilled teams by connecting teachers of engineering substance, experts on pedagogy, and external experts on the process.

Engineering curricula in educational institutions is recognized as having two distinct functions which must be carried on concurrently: training and education. Training is defined as the inculcation of methods of procedure, the development of adequate vocabularies and skill in communication and in manipulation of mathematical processes, followed by typical exercises with definite solutions or measures of performance. Education, on the other hand, is defined as the broader development of the mind and personality, a guided enlargement of creative ability and understanding. Education develops the ability to meet new situations with confidence and with a degree of wisdom limited only by the inherent capabilities of the individual (Everitt 1962). The central commitment of engineering education must be the welfare of students. There is an old saying that the purpose of education should not be to prepare a student for their first job but to instead prepare them for their last job. This will often require stressing for the far greater long-term value of a truly transdisciplinary education.

A key area to be addressed in to accommodate transdisciplinarity, design, innovation, and entrepreneurship education is curriculum that is tailored to the local environment, by leveraging existing resources and by creating new courses, activities, projects, case studies, and examples of role models that students can relate to. Engineering curriculum should be built around developing and creating knowledge and skills and not around delivering acquiring knowledge only. The process of acquiring knowledge from resources and building a knowledge base is called knowledge engineering. The activity of knowledge engineering is defined in the pioneering work of Feigenbaum and McCorduck (1983) as the art of bringing the principles and tools of research to bear on difficult applications problems requiring the knowledge of experts for their solutions. Knowledge engineering can be viewed from two perspectives: narrow and broad. According to the narrow perspective, knowledge engineering deals with knowledge acquisition, representation, validation, inferencing, explanation, and maintenance. Alternatively, according to the broad perspective, the term describes the entire process of developing and maintaining intelligent systems. In this book, we use the narrow definition. The knowledge possessed by human experts is often unstructured and not explicitly expressed. A major goal of knowledge engineering is to help experts articulate what they know and document the knowledge in a reusable form. The knowledge development process from high school to industry throughout the university is reflected in Figure 1.14. Questions and activities at the end of every chapter in this book are based on the above classification. The above arrangement should be supported by the fact that people with different backgrounds and different working environments tend to understand concepts differently.

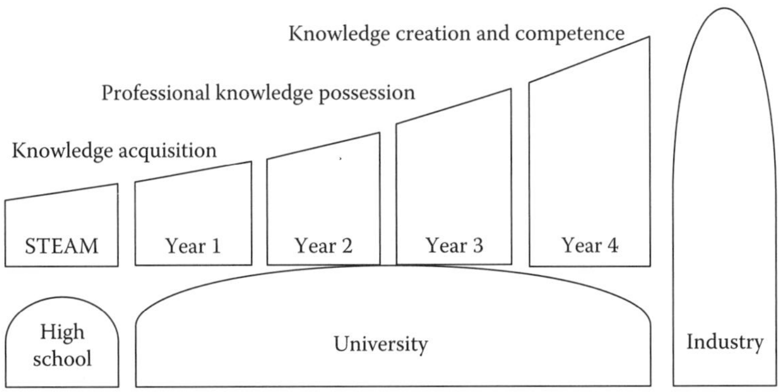

Figure 1.14 Knowledge development process from high school to industry.

1.11.2 STEAM in K-12 education

The art challenges the technology, and the technology inspires the art.

John Lasseter

The development of knowledge and skills, together with problem solving, teamwork and cooperation, creativity, innovation, and learning from failure, is something that begins in early childhood and continues throughout life. Presently, science and math "SM" are fully integrated in the K-12 education almost worldwide while engineering and technology or "ET" component is missing from the STEM core curricula for K-12 courses. In fact, another pre-existing component in K-12 education, art "A," should also be strongly integrated into STEM curricula for enrichment and elevation to STEAM education.

Georgette Yakman, who developed STEAM in 2006, describes the concept as "providing a avenue for formally teaching the inter-relationships of how subjects relate in real-life. The STEAM framework and developing model of education can be paraphrased in the following definition: "science and technology, interpreted through engineering and the arts, all based in a language of mathematics." Letter "A" stands for all liberal art studies including social studies, language, physical, musical, fine, and performing.

The idea of adding the arts to STEM education has been gaining momentum. STEAM is not about simply "adding" arts to the equation or utilizing certain elements of art (visual art and design) in a lesson. Art, including social studies and business, can be seen as a way of offering more diverse learning opportunities and greater access to STEM for all types of learners. Art also provides diverse opportunities for creative thinking, representation, emotion, communication, expression, and leadership. Art education is often project based and a closer representation of real life. The rationale for STEAM should not be so much to teach art but to apply art in real situations. In fact, the idea of STEAM is a mindset, and that is one that expands with ideas and opportunity.

The best way to develop STEAM curricula and create innovative practitioners is through context: placing actual transdisciplinary problems in front of students early on. Through industrial interaction the students will be able to know the types of problems engineers, scientists, and mathematicians face. They will be able to observe the concepts, processes, and tools used to solve those problems, and develop the personal and professional attributes essential to being leaders. Students need to be shown the difference between studying engineering, science, and math and becoming engineer, scientist, or mathematician with hands-on,

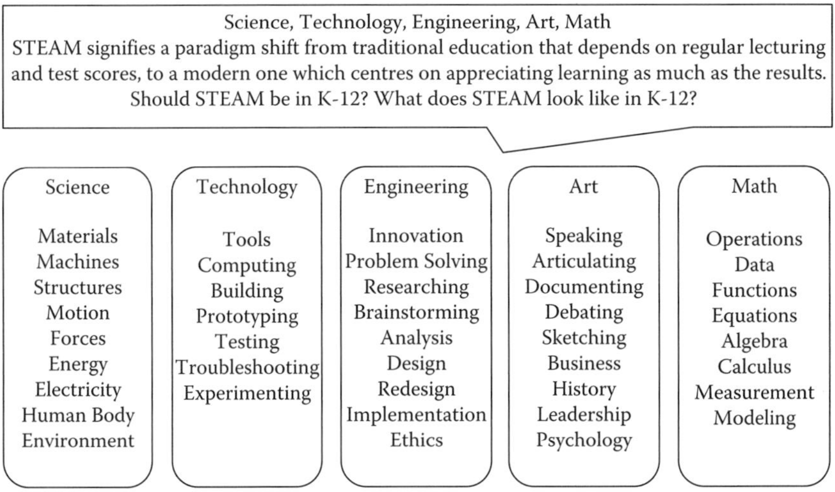

Figure 1.15 STEAM education challenges and components.

minds-on experiences that happen early and often during their education. Figure 1.15 defines STEAM and shows its components.

1.11.3 *First year*

Deep-rooted motivation and desire are the main reason for most students to have chosen their fields of study. Regardless of external influences, these students are primarily affected by internal motives. However, many of these students do not see the relevance of the beginning of engineering courses. First-year students arriving fresh from high schools come with certain expectations, but those students lack technical knowledge, and many are disillusioned about the engineering profession often as they begin a highly demanding curriculum that includes demanding theory of science and math that is based mainly on textbooks. Such courses are not what they expected from an education in engineering, and this leads to disappointment. Many of them, even good students, drop out of the program. It is a serious problem and the retention of these students is difficult.

To overcome this problem, engage students to retain them by reassuring that their coursework will not be devoted solely to textbooks and theory. It is necessary to get them excited early on and help them see what engineering is *really* about. To accomplish this goal, two major components should be integrated in the curriculum. First is the integration and development of communication skills within the framework of classroom. This objective is accomplished by emphasizing the value of technical writing, communication skills, and verbal communication. Special emphasis should be placed on relating the importance of these elements to the engineering profession. Wherever appropriate, students may be encouraged to use resources such as the university writing centers, bookstores, and libraries. Activities like engineering communication tasks and debates can be integrated as assessment elements within the classes.

The second important component is an introductory course to engineering design. An ideal introductory design course should motivate students with little or no technical background and should have connection to most first-year courses; and the course content should be challenging but interesting. This should not be a highly demanding class, with respect to science or problem-solving. Students should be introduced to a completely typical engineering endeavor that is product or service design. At this level, students should be able to differentiate PBL project from entrepreneurial project.

A model curriculum has been proposed (Eder and Hosnedl 2007), where the first-year presentation should start to introduce the concepts of engineering design science, but concentrate on designing in routine situations, especially on redesigning existing items of engineering products. According to the author of this book, the first-year design-oriented course (or design studio) is similar to many capstone design courses, but it differs markedly in its tendency to focus more heavily on conceptual design methods and less on discipline-specific artifacts. The course should be based on PBL approach. PBL is potentially effective as it emphasizes students' participation and involvement in the learning process and helps to develop a sense of creativity and innovative thinking. It is a powerful classroom process and a strategy that promotes lifelong habits of learning. With PBL, students, working in teams, should identify what they know, and, importantly, what they do not know. The role of the teacher is only to facilitate the process, not providing easy answers.

The course should involve conceptual design learning based on the engineering design including preparation of proposal, generation of design alternatives, thought of constraints and criteria, selection of design alternative, testing and verification of design by prototyping, and preparation and presentation of design report. The projects should be intended as purposes/goals in a problem-based approach. Students may make use of transdisciplinary approach to describe the solution. Proposed design solutions may be sustainable: affordable (economic) and address the needs of society (social, health, and safety), use local materials in an environmentally conscious manner (environmental) that can be built from a small number of components (manufacturability).

This design-oriented course or studio seeks to develop active, dynamic drivers of innovation, and strive to uncover, and get rid of, overt and barriers to creativity within each student. This is best achieved by providing students a supportive environment in which to practice. The course is regarded as a student's design "toolkit" to learn skills in drawing, sketching, modeling, problem exploration and formulation, and presentation of their ideas. It may be more appropriate if relevant simulation tools are introduced to students. Software such as Solidworks, MATLAB, LabVIEW, and Multisim are good examples.

In summary, there is a strong belief that first-year design-oriented courses make students quickly learn that generating a great idea is the first step in the process of innovation throughout the following (Akili 2015):

- Enhance students' interest in engineering as well increase students' retention in engineering programs.
- Motivate learning in upper division engineering science courses.
- Enhance performance in design courses, in general, and in capstone design courses, in particular.
- The PBL project may motivate a number of students to get involved into entrepreneurial projects, which require producing new things, innovating, to produce products, systems, services, or events.

1.11.4 Second and third year

Second- and third-year engineering courses with a balance between theory and practice are necessary to maintain engineering breadth during those years as students pursue disciplinary study, and also to prepare students for the senior capstone design experience and eventual engineering practice. A prior PBL little experience from first year may have a great impact on student motivation. Instructors may assign projects to be conceived and developed by student groups to reinforce fundamental concepts and to generate innovating projects. Students will be allowed to develop their own research-based, open-ended projects and propose solutions. By developing projects, students can create solutions outside the box and develop skills to reinforce the students' teamwork and leadership abilities.

Projects should be embedded into a student's courses. Blending business and design approaches with experimental learning in these projects will allow much deeper connections compared with simple lecture-style learning. One of the most important ways in which projects from second and third years may progress through four-year capstone design course is to introduce projects that are socially, as well as technically, more demanding.

The author gives an example of a two-semester project on mechatronics offered to mechanical engineering students studying two electrical engineering courses, namely electric circuits and machines, and electronics. Students conceive the idea of the project and carry out design work during the first course (second year) and then they build the prototype, test it, and operate it during the second course (third year). This interdisciplinary engineering design project deals with the integrated and optimal design of a mechatronics system, including sensors, actuators, and electronic components, and its embedded digital control system. The main objective of this project is to make students think like engineers. The integration is respect to both hardware components and information processing. Performance, reliability, low cost, robustness, efficiency, and sustainability are absolutely essential (Habash et al. 2011). Examples of student projects are exhibited at www.g9toengineering.com.

1.11.5 Final year

The purpose of the fourth year is to strengthen students' entrepreneurial mindsets and behaviors, and to give tangible tools for those who want to become entrepreneurs. During this year, the students will gain systematic business knowledge in an integrated manner using real-life cases. At this stage, design content varies according to the choice of discipline, and transdisciplinary opportunities are very limited. If design content is not taught and implemented during the first three years of study, it will be hard for students to effectively engage in real-life applications that are required by employers after graduation.

The fourth-year curriculum may be enriched further by adopting a feasibility or case study approach, where students are introduced to a broad variety of topics, including societal and environmental issues, business, ethics, health, safety and liability, equity, learning techniques, and creative complex problem solving. Cases may help bridge the gap between theory and practice. They provide the students with a real engineering scenario requiring application of a certain technical discipline while exemplifying the often-critical nontechnical aspects of a problem. Cases often involve circumstances that do not have right or wrong answers. This may help students understand and develop a tolerance for uncertainty.

Common topics to be mastered during the process include business planning and divergent thinking, problem-solving skills (especially analysis, design methodologies, and engineering calculations), professional skills (technical writing, communications, standards and codes, and engineering ethics), and computer simulation skills. Frequently, an appreciation for the engineering profession and professional interpersonal skills are developed through this approach. A major advantage of engaging students with case study is to integrate business aspects into substance teaching rather than teaching business as a separate topic.

Engineers who want to bring their innovative ideas to the market should be taught business at the senior level of their undergraduate design program, since some students are already attempting to launch their ideas at this stage. Students who have at least some business education will be able to utilize both knowledge bases to make new connections and full decisions, which will ultimately increase the chance of success of their business. Figure 1.16 outlines the design and entrepreneurial activities along the four years of study.

1.11.6 Student partnership in curriculum design

> At its roots partnership is about investing students with the power to co-create, not just knowledge or learning, but the higher education itself.
>
> **NUS (2012)**

Across university programs worldwide, students are usually engaged in course evaluations and in teacher–student committees, but it is very rare for institutions to go beyond the student voice and engage students as partners in designing the curriculum and giving pedagogic advice and consultancy. These ways of engaging students as partners complement those discussed in the previous subsections in which student activities and projects are well developed in many institutions of higher education. These ways represent the higher levels of engagement in Arnstein's (1969) ladder of participation as shown in Figure 1.17.

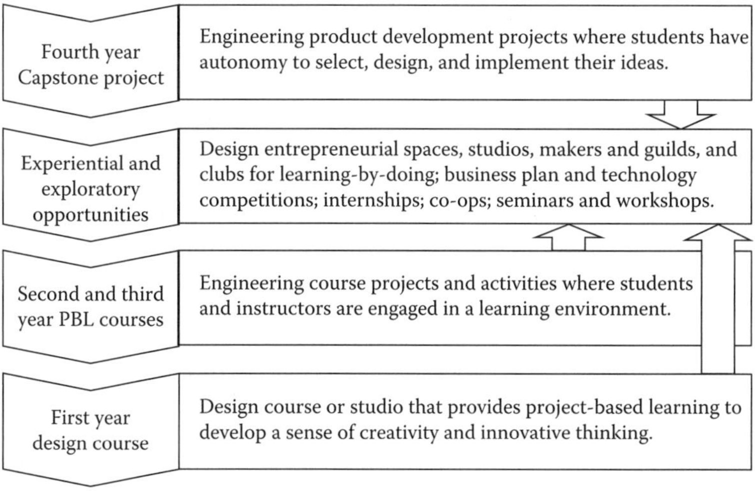

Figure 1.16 Design and entrepreneurial activities along the four years of study.

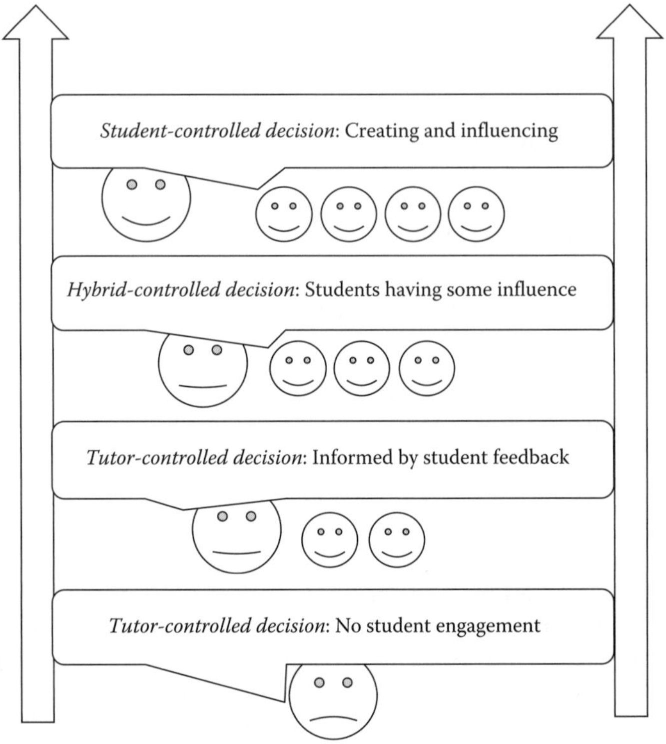

Figure 1.17 Ladder of student engagement and influence.

New learning partnerships between teachers and students are the essential foundations for effective new pedagogies. Partnership is realized as a relationship in which all involved including students, teaching staff, and supporting staff are actively engaged in and stand to grow in the process of learning to promote student learning. To work within the model given in Figure 1.17, designing a curriculum allows students to take ownership of their learning way, from induction through studying and teaching, academic support, and hands-on learning.

The key concept is that students themselves take responsibility for commanding change, based on their own research on aspects of learning and teaching. The approach enables students to gain pedagogical information (understanding how learning happens, and a nature and capacity to shape individual learning). It also helps the students to actively engage with the means of change, often taking on a leadership role. The students are also given the opportunity to teach and mentor each other. Student-to-student support and interaction is particularly important for learning to complement teaching support mechanisms. To enhance the mechanism, students may use discussion forums to ask questions, or develop discussion and support groups on social media.

Students and teaching staff have different expertise to bring to the process, and there will be times when staff may appropriately have more voice, and other times when students may appropriately have more voice. Co-creation is not about giving students complete control, nor about staff maintaining complete control over curriculum design decisions. The relative levels of control over decision-making and appropriate levels of partnership are likely to depend upon the context, the level of study, the relative experience levels of the students and the staff, the attitudes of students and staff, what is being discussed, and the level of influence of professional bodies over the curriculum (Bovill 2013). It is important to see the ladder as a useful tool for exploring practice, not a measure of the quality of engagement. There are many factors that affect the ability to offer greater participation and the ability of staff and students to engage, and students do not always have to be engaged at the top rung of the ladder. In partnership, control and ownership of the curriculum is shared, with different partners taking the lead as appropriate (NUS 2012).

In conclusion, this approach of curriculum development involves a shift in thinking about the design of a course content and a focus on the development, presentation, and revision of engineering curriculum materials that will be interesting, useful, engaging, motivating, and inclusive of most voices in the process.

1.11.7 Peer mentorship in group projects

Peer mentoring is a process through which a more experienced individual encourages and assists a less experienced individual to develop potential within a shared area of interest. The resulting relationship is a reciprocal one in that both individuals in the partnership have an opportunity for growth and development (Gillman 2006).

Problem-based and experiential learning activities where students at different years of study can work together and mentor each other in real tasks with a possibility of being mentored by expert from the industry whenever possible are desirable. Such activities which can be organized with the environment of undergraduate classes can enhance future employability skills—generated by this problem-based approach in solving real-world problems—of all students.

Developing the student for such mentoring experience requires a fundamental shift in how to structure and imagine the whole undergraduate experience. It requires the enhancement of the learning paradigm from the introductory course through the final learning experience. The contrast between subject-domain knowledge subjects and employability skills can be minimized by enabling the students to solve real-life problems so the knowledge (content), technical skills (context), and learning/employability skills (process) can be developed in an integrated manner.

A level of understanding and trust is also required for highest team performance in projects. Team members need to be informed of the personal traits, values, and needs of their colleagues. Team leaders must also provide an environment for trust and assure fairness. One way to do this is through being a team player. Students are analytical by nature and will be quick to react to any effort to layer team-building exercises. A crucial part of managing people is getting them to be as productive as possible, while working in teams. Table 1.2 answers this fact and shows the 17 indisputable laws of teamwork highlighting a clear character profile of the ideal team player (Maxwell 2002).

Table 1.2 Essential qualities of team players

Quality	Definition
Adaptable	Be creative and resourceful when faced with a challenge. Look for unconventional solutions to fit changed circumstances
Collaborative	Joint effort to win
Committed	Not giving up; however, commitment involves risk
Communicative	Open minded and engage in active listening
Competent	Skills and abilities to achieve the desired objective
Dependable	Align personal priorities with those of your team
Disciplined	Do not overreact emotionally and think rationally
Enlarging	Believe in others before they believe in you
Enthusiastic	Strive for excellence
Intentional	Commit yourself to long-term achievement
Mission conscious	Find ways to keep the mission in mind
Prepared	Do more research
Relational	Focus on others instead of yourself
Self-improving	Become highly teachable
Selfless	Promote someone other than yourself
Solution oriented	Refuse to give up
Tenacious	Work harder or smarter

Source: Maxwell, J.C., *The 17 Essential Qualities of a Team Player: Becoming the Kind of Person Every Team Wants*, Thomas Nelson Publishers, Nashville, TN, 2002.

1.12 *Reach out case: Learning by doing*

> I hear and I forget, I see and I remember, I do and I understand.
>
> **John Dewey**

The reflective learning presented in this case shows how "learning by doing in real world" is becoming an integrated part of the learning process in academic institutions by linking theory and practice through engaging in a sequence of activities and the acquisition of competence.

1.12.1 *Experiential learning*

> Tell me and I forget, teach me and I remember, involve me and I will learn.
>
> **Benjamin Franklin**

"Learning by doing" and the term "experiential learning" are commonly used to refer to several different aspects of learning. The initial theories of experiential learning arose in the mid-nineteenth century as attempts to move away from traditional formal education, where teachers simply presented students with abstract concepts, and toward an immersive method of instruction. Students would "learn by doing," applying knowledge to experience in order to develop skills or new ways of thinking (Lewis and Williams 1994). A crucial feature of experiential learning is the structure developed by the teacher within which learning takes place (Gibbs 2013).

Keeton and Tate (1978) offered this definition for experiential learning: "Learning in which the learner is directly in touch with the realities being studied. It is contrasted with the learner who only reads about, hears about, talks about, or writes about these realities but never comes into contact with them as part of the learning process." Many educational institutions offer experiential education programs such as internships, practicums, cooperative education, field projects, and classroom experiential learning exercises including role-playing, games, case studies, simulations to replicate real-life situations, presentations, and group work that add a direct experience component to their traditional academic studies.

Experiential learning is built upon a foundation of interdisciplinary and constructivist learning. Experiential methodology does not treat each subject as being walled off in its own room, unconnected to any other subjects. Compartmentalized learning does not reflect the real world, while the experiential classroom works to create an interdisciplinary learning experience that mimics real-world learning (Wurdinger 2005).

Experiential learning can be divided into two major categories: field-based experiences and classroom-based learning. Field-based learning includes internships, practicums, cooperative education, and service learning. Classroom-based experiential learning can take a multitude of forms, including role-playing, games, case studies, simulations, presentations, and various types of group work (Lewis and Williams 1994).

Learners have a preference for certain learning modes of grasping and transforming experience into understanding which he defines as "learning style." This can be correlated with career choices; for example, learners with a diverging style are often interested in the arts while convergent learners tend to be specialists in technical fields. Assimilative learners are usually interested in theory and abstract problem solving while accommodative learners gravitate toward action-oriented careers such as marketing and sales. Learners may also have a balanced or flexible style that allows them to adapt their learning on a situational basis (Kolb 1984; Kolb and Kolb 2005; Sharma and Kolb 2010).

For a concrete experience, students should have an open mind and be immersed in the situation. For reflective observation, they should reflect on the experience from multiple perspectives. In abstract conceptualization, they should turn their observations into concepts and theories. Finally, for active experimentation, they should use the new theories to solve problems or make decisions. Students may enter the cycle at any point, but need to experience all four modes (Kolb 2015).

1.12.2 Student competitions

Today, valuable skill competitions including technology development, business-plan writing, and marketing that lead to networking, attracting financing (such as seed capital), and connecting with local businessess are an excellent way to actively engage faculty and students in the entrepreneurial and design learning process. As a whole, competitions are geared toward teaching students how to think outside the classroom, fostering collaborations across disciplines, and increasing access to industry and businesses.

Competitions provide an exciting platform for students to learn practical skills, such as how to write a business plan, access venture funding, pitch ideas, build, test, and operate prototypes. Universities should further enhance this, and start transitioning further from single monetary rewards for competitions recognizing milestone achievements with a multitude of prizes, including nonmonetary resources such as incubator spaces and mentorships. Expanding student team competitions to include faculty and alumni, and increasing the scope and size of resources through collaboration with industry and local partners are required.

Finally and importantly, competitions represent the essence of "learning by doing" and should be integrated in the grading system, which can be based on how well teammates perform in a fellow student's presence and on fellow students' evaluation of their peer's performance.

1.12.3 Design entrepreneurial spaces

Developing an entrepreneurial culture within the university includes recognition and reward for developing experiential learning programs, knowledge engagement, lab experiences through establishing incubator-style collaborative spaces; creation of science and technology parks; development of the role of intermediaries such as industrial liaison and technology transfer offices; support entrepreneurs on campus and help launching of new venture programs; link research with real problems; organization of spin off activity; and work with industrial and community partners. This helps to create a flexible infrastructure to support and encourage innovation in teaching and learning. These spaces foster student engagement in developing innovative ideas and beginning businesses. Students who want to pursue innovation with the industry and individuals that can provide financing such as angel investors, start-up accelerators, and incubators may use these facilities.

Most engineering students who wish to become design entrepreneurs do not need to enroll in lengthy, graduate-level programs. Considerably, many of them only need to acquire basic business knowledge and skills to launch a new idea and to succeed in an entrepreneurial endeavor. Engineering students should begin obtaining this knowledge base and skill set at the undergraduate level. Since many undergraduate programs lack the integration of a formal design and business program, professors and students alike must actively work to create these opportunities within their existing programs, and identify outside resources to support the acquisition of that knowledge and skill set.

Entrepreneurial and design spaces provide students with access to learning and networking opportunities with local entrepreneurs and innovators. In addition, such facilities formulate design entrepreneur clubs to understand the needs of the design entrepreneur and to advise students on bringing their ideas to market and develop connections in the business community. Currently, some universities are embracing the entrepreneurial dorm, whereas others are expanding this notion to boost entrepreneurial clusters, within the university and sometimes extending into local communities.

1.12.4 Facilitating university–industry collaboration

To facilitate greater collaboration and innovation, universities should open up their facilities to businesses in order to develop greater economic value. This means developing networks to stimulate partnerships and create an environment of cooperation in the local ecosystem and beyond. Businesses and industries benefit greatly from university research and innovation. For example, creating a university-run consultancy would allow the outside community industry, government agencies, and others to access experts on campus by hiring them as consultants. Universities may put greater emphasis on supporting start-up companies, while continuing to engage established companies that have traditionally been their licensing partners.

As faculty become more interested in commercialization activities, universities should provide additional resources to encourage collaboration with local communities and industries. Universities need to hire individuals, or create teams, to connect faculty with similar interests and research goals—often reaching across academic departments—to share information and experience on creating start-ups, licensing technology, and collaborating with industry. This transdisciplinary effort helps share information on best practices and spurs new ideas for developing and commercializing new products. Community leaders and local entrepreneurs should be invited to become more involved in the development of technology and start-up companies. Programs to link experienced entrepreneurs with faculty to assist in the start-up process, development, and longevity should be developed. Entrepreneurs should also serve in a mentoring role, helping faculty to identify and further develop commercialization opportunities.

1.12.5 *Rewarding faculty innovation and entrepreneurship*

In order to create an infrastructure and culture that encourages, supports, recognizes, and rewards achievements among faculty, universities and colleges celebrate faculty achievements in innovation and entrepreneurship. These recognitions include prizes and award ceremonies that bring the faculty community together to distinguish and know about the activities of their peers across the campus. Awards such as "Innovator of the Year" and "Faculty Entrepreneur of the Year" are popular as they reward faculty for achievements that reach beyond traditional teaching and research accomplishments. Universities and colleges may update tenure and sabbatical leave guidelines to encourage faculty to pursue collaborative and entrepreneurial endeavors, such as launching a start-up company. In addition, entrepreneurial and design thinking and activity needs to be valued and seen as an optional part of a professor's duty, not as an extra-curricular activity. For this change to happen, aspects of entrepreneurship, design, patents, and commercialization should be rewarded in tenure and promotion processes.

1.12.6 Case research questions

- What are the main challenges to traditional university models?
- Does using a PBL approach increase or decrease the breadth of learning that may be achieved?
- Is there really urgency for universities to become more entrepreneurial?
- How can the entrepreneurial potential of a university be developed?
- What might be the model of the future entrepreneurial university?
- What are the main features that should be present in order to define a learning activity or process as experiential? Who benefits from experiential learning?
- Describe any situations where your class, faculty, or university took initiative and/or displayed an entrepreneurial approach.

1.13 Knowledge acquisition

Attempting to answer the following questions involves acquisition of knowledge from this book and other books, documents, and the Internet.

- Define engineering, the types of questions usually engineers ask, and the tools they use to answer those questions.
- What thinking historically means for engineering? What can engineers learn from the past?
- What distinguishes engineers from other professionals?
- What is a discipline? Why are disciplinary studies useful? What are their strengths and limitations?
- What is the difference between interdisciplinary and transdisciplinary education?
- How can looking at the same subject from different perspectives pave the way for progress?
- What strengths make a good engineer?
- What are the main challenges engineering ever faced?
- What do you get out of engineering that you cannot get from any other kind of work?
- What is an entrepreneurial engineer?
- What are the key requirements for engineering sustainability?
- How engineering students think about sustainability across the life cycle?
- What is the difference between teacher-centered and student-centered instructions?
- What is the difference between convergent (closed-ended) and divergent (open-ended) problems?

- What are the current challenges that the engineering curriculum faces today?
- What is STEM? Why is STEM important for students?
- What should STEAM look like in the K-12 classroom?
- Why should engineering educators include history in their lesson plans?

1.14 *Knowledge possession*

Attempting to answer the following open-ended "not explicitly expressed" questions may require research and investigation beyond the scope of this book, mostly by engaging in conversation, class discussion, and Internet-based research.

- Engage in historical analysis using the practices from multiple disciplines, toward an integrated, transdisciplinary understanding of the Industrial Revolution.
- Based on deductive and inductive learning approaches, should engineering courses stress on fundamentals or on applications, or should the two be integrated within courses?
- Should the flow of knowledge within a course or curriculum generally proceed from fundamentals to applications (deductive presentation and expository teaching) or from applications to fundamentals (inductive presentation, discovery learning, and problem-based learning)?
- How to teach engineering in the twenty-first century? Are we currently educating our engineering graduates to deal with the key issue of the twenty-first century?

1.15 *Knowledge creation*

Collaborate with peers on learning or you may work with others outside the class to narrow down the objectives of each activity. You may access online resources, and analyze data and information to create new ideas and balanced solutions. High-level digital tools may be used to develop multimedia presentations, simulations or animations, videos and visual displays, digital portfolios (ePortfolios), reflective practice (online publishing and blogging), or well-researched and up-to date reports.

1.15.1 *Campaign for future engineering*

More than 50 years ago, Dean William L. Everitt wrote a visionary essay (Everitt 1962) about educating engineers "in the future." His future was

2012. His essay asserted that educating engineers means fostering innovative minds—the ability to create and navigate a world that, at any given time, we are only beginning to imagine (Chartoff 2014).

In this task, propose a vision for the future engineering based on understanding of the past. Your vision may be reflected in a form of logo, poster, video, simulation, animation, or any sort of innovative art.

1.15.2 *Debate on transdisciplinary education*

Objective	**Introducing an open-ended debate in the classroom to help students understand argument on the concept of transdisciplinary education**
Time	15 min for debate and 15 min for review
Format	For and against
Learning outcomes	Make an argument about a particular opinion, evaluate the arguments of peers, and understand the concept of counterarguments
Capabilities demonstrated	Developing skills on public speaking, research, teamwork, critical thinking, communication, and professional judgment
Arrangement	Students are organized into two position groups of three speaking in a specific order. Three argue for an opinion and three argue against. One or two students might work on the opening and closing statements while the group is investigating the subject; however, the entire group should revise the statements. Each group should read an opening and closing statement for the debate.
Idea for the topic	Investigate arguments that are for and against transdisciplinary education.
Assessment	Indicate what you consider the best arguments in favor of the topic. How were they substantiated? Identify arguments that are based on poor facts, not ethical and/or sustainable or not well substantiated?

1.15.3 *Portfolio on designing a smart city*

For a smart city to be successful in its endeavors, human resources skills need to be available to ensure that all the different facets of the city are adequately and efficiently addressed. A significant proportion of smart infrastructure creation and maintenance jobs require a good foundation in STEAM education.

In this task, develop a five-day STEAM-based learning program for Grade 11–12 students on the topic of smart cities. The program should include five teaching modules. Each module is designed to total about 5 h of instruction. Through its demonstrations and projects, the program should introduce students to new ideas in science, engineering, and technology, as well as to the scientific method and research practices. The curriculum also explores urban planning, sustainability, and healthy urban living. The program should connect students through teamwork and mentorship while they learn how to build and design more livable, efficient, sustainable, and resilient cities.

1.15.4 Partnership course development portfolio

Thinking "out of the box" to develop a joint course or program with an industrial partner. This could be an idea for an open-ended innovation task based on student's understandings, ideas, and competencies. To narrow down the search you may consider developing an outline for a full credited course (actual or virtual) by your university and *Shopify* as an industrial partner. You may consider another industrial partner based on the nature of the course, company, and proximity.

> Shopify is a company based in the Canadian city of Ottawa. It is the leading cloud-based, commerce platform designed for small- and medium-sized businesses. Merchants can use Shopify software to design, set up, and manage their stores across various sales channels, including web, mobile, social media, marketplaces, brick-and-mortar locations, and pop-up shops. Shopify currently powers over 300, 000 businesses in approximately 150 countries and is trusted by leading companies such as Tesla Motors, Budweiser, Red Bull, the New York Stock Exchange, and others.

- Identify the nature of the course, actual or virtual.
- Name the course.
- Develop a course outline including course objectives, elements, description, learning outcome, materials and reading, evaluation process, components and weights, and course policies.
- Propose a feasible approach where Shopify (or another industrial partner of your choice) can contribute to the course.
- Use a proper course outline template.

Analyze information and data to create a digital portfolio that may be formatted as prescribed by the instructor and would be a key component in assessing the competencies.

1.15.5 Portfolio on design studio for sustainability

Design studios provide a hands-on environment for working on design projects (Carlson and Sullivan 1999). How does an engineering school create a design studio for sustainability in which students understand and practice design from year 1 to year 4? The design studio learning often moves learners from dependent and instructional learning formats to more independent projects and inquiries, building self-knowledge and entrepreneurship. A range of exercises may be introduced to encourage relational and creative complex problem-solving techniques that would help in reinforcing their design solutions. The following represents a guideline to develop the task.

- What is learning through making?
- What are the key components of the proposed design studio?
- What are the best strategies and practices for designing group projects?
- What are the best strategies and practices to engage students in deep design learning?
- As a year 1 task: propose an approach and/or a model based on conceptual design scenario that reflects a technical transport problem of unsustainable (inefficient) car.
- As a year 2 task: propose an approach and/or a model based on conceptual design scenario that reflects a technical transport problem of sustainable (efficient) car.
- As a year 3 task: propose an approach and/or a model based on detail design that reflects a technical transport problem of sustainable (efficient) car.
- As a year 4 task: propose an approach and/or a model based on implementation and operation that reflects a technical transport problem of sustainable (efficient) car.

In the above task, narrow down the objectives of the task. You may focus on one transport model such as the car.

1.15.6 Montessori-based engineering learning module

The Montessori teacher works as a guide and facilitator while students pose a central role as the engineers of the future. The teacher has the specific role of creating a well-prepared environment and an atmosphere of learning and inquisitiveness with the purpose of increasing the participation of the group to the learning activities. Montessori identified two main categories of fundamental needs: physical and spiritual. The physical needs consist of food, shelter, clothing, transportation, and defense. The spiritual needs consist of art, communication, love, and a belief system.

Based on the above fundamental needs, build an outline for a 1-h learning module on the history of transportation. The module should involve creation of images that reflect the engagement of innovation and engineering design.

1.15.7 Entrepreneurial think-tank poster on student engagement

Both teacher and students should participate in the learning process, but the teacher must involve students in the knowledge-building process and encourage collaboration and togetherness in the classroom. This approach is more effective when learners are autonomous, self-directed, and willing to construct their own learning experience and valuable competencies (Rambocas and Sastry 2017). For this task, a team of three to four students are asked to develop three-week entrepreneurial learning activity to be embedded in typical engineering course. The activity should involve open-ended engaging questions and adopt a student-centered approach to teaching and learning. An evaluation criterion as part of the course assessment should also be included. Each team may reflect the outline of the activity in a digital poster format.

1.15.8 Video contest on what Montessori can do for engineering

Montessori methods underscore the importance of three main elements that current engineering education usually includes, but may not have articulated as clearly as the Montessori system. These are the power of the story to engage and need for a meaningful context, the role of the sensorial while teaching (hands-on activities), and the learning spiral in evaluation (evaluating gradual building of knowledge) (NAE 2008; Frances and Ng 2011).

Based on the above three elements, develop a 3-min video for organizing a 1-h lesson, to be offered to primary school students. This engineering-based lesson should leverage Montessori's knowledge of moving from whole to parts, concrete to abstract, and known to unknown. Select one engineering topic and determine its activities that would lead to the next level of engineering in the spiral. Frame the activities in a way that would excite the students as they are re-exposed to the lesson in the future.

References

Aggelikki, K. 2011. History of civil engineering. http://www.brighthubengineering.com/building-construction-design/41552-history-of-civil-engineering/.

Akay, A. 2003. The renaissance engineer: Educating engineers in a post-9/11 world. *European Journal of Engineering Education* 28(2): 145–150.

Akili, W. 2015. Perspectives on engineering design learning: Realities, challenges, and recommendations. *IEEE Frontiers in Education Conference (FIE)*, October 21–24, Washington, DC, USA.

Armytage, W. H. G. 1976. *A Social History of Engineering*. Fourth edition. London: Faber and Faber.

Arnstein, S. R. 1969. A ladder of citizen participation. *Journal of the American Institute of Planners* 35(4): 216–224.

Ashford, N. A. 2004. Major challenges to engineering education for sustainable development: What has to change to make it creative, effective, and acceptable to the established disciplines? *International Journal of Sustainability in Higher Education* 5(3): 239–250.

Augustine, N. 1994. Engineering the next century. *The Bridge* 24(2): 3–6.

Barry, J. 1999. *Rethinking Green Politics: Nature, Virtue and Progress*. London: Sage Publications.

Batalden, P., and J. Mohr. 1997. Building a knowledge of health care as a system. *Quality Management Health Care* 5: 1–12.

Beder, S. 1999. Beyond technicalities: Expanding engineering thinking. *Journal of Professional Issues in Engineering* 125(1): 12–18.

Bertalanffy, L. V. 1968. *General System Theory: Foundations, Development, Applications*. New York: George Braziller.

Bilec, M., R. Ries, and H. S. Matthews. 2007. Sustainable development and green design—Who is leading the green initiative? *Journal of Professional Issues in Engineering Education and Practice* 133(4): 265–269.

Bloom, B. S. 1971. Mastery learning. In Block, J. H. (Ed.), *Mastery Learning: Theory and Practice*. New York: Holt, Rinehard and Winston, pp. 47–63.

Bond, E., S. Gingerich, O. Archer-Antonsen, L. Purcell, and E. Macklem. 2003. The Industrial Revolution. http://industrialrevolution.sea.ca/.

Bovill, C. 2013. Students and staff co-creating curricula: An example of good practice in higher education? In Dunne, E., and D. Owen (Eds.), *The Student Engagement Handbook: Practice in Higher Education*. Bingley, UK: Emerald, pp. 461–475.

Buchanan, R. 2001. Design research and the new learning. *Design Issues* 17(4): 3–23.

Buchanan, R. A. 2017. History of technology. https://www.britannica.com/technology/history-of-technology#toc1350805main.

Bugliarello, G. 1991. The university-and particularly the technological university: Pragmatism and beyond. In Zinberg, D. S. (Ed.), *The Changing University: How Increased Demand for Scientists and Technology Is Transforming Education Institutions Internationally*. Dordrecht, The Netherlands: Kluwer Academic Publishing, pp. 31–37.

Bush, V. 1945. As we may think. *The Atlantic Monthly* 176(1): 101–108.

Cachia, R., A. Ferrari, K. Ala-Mutka, and Y. Punie. 2010. Creative learning and innovative teaching final report on the study on creativity and innovation in education in the EU member states. Institute for Prospective Technological Studies.

CAE. 1999. Evolution of engineering education in Canada. New challenges for Canadian Universities. Engineering Issues, *Canadian Academy of Engineering* 8: 1–2.

Caine, R., and G. Caine. 1991. *Making Connections: Teaching and the Human Brain*. Alexandria, VA: Associates for Supervision and Curriculum Development, p. 134.

Carlson, L., and J. Sullivan. 1999. Hands-on engineering: Learning by doing in the integrated teaching and learning program. *International Journal of Engineering Education* 15(1): 2–31.

Chartoff, R. 2014. Engineering Education @ Illinois: Past, Present and Future. Academy for Excellence in Engineering Education. http://ae3.engineering.illinois.edu/2014/08/29/engineering-education-illinois-past-present-and-future/.

Christensen, C. M. 1997. *The Innovator's Dilemma: When New Technologies Cause Great Firms to Fall*. Boston, MA: Harvard Business School Press.

CIA. 2001. Long-term global demographic trends: Reshaping the geopolitical landscape. Central Intelligence Agency, Washington DC. http://www.odci.gov/cia/reports/ Demo_Trends_For_Web.pdf.

Ciampi, M. M., and C. da R. Brito. 2010. Implementation of an engineering educator graduation program for the formation of new skilled engineering teachers. *IEEE EDUCON Annual Conference. The Future of Global Learning in Engineering Education*. April 14–16, Madrid, Spain: IEEE.

Ciampi, M. M., and C. da R. Brito. 2011. Awareness of social impact of engineering: The task for engineering schools? *American Society for Engineering Education (ASEE) Annual Conference and Exposition*, Vancouver, BC, June 26–29.

Corbett, A. C. 2005. Experiential learning within the process of opportunity identification and exploitation. *Entrepreneurship Theory and Practice* 29(4): 473–491.

Coyle, E., and M. Rebow. 2009. Chapter 16: Sustainable Design: A Case Study in Energy Systems. In Christensen, S. H., B. Delahousse, and M. Meganck (Eds.), *Engineering in Context*. London: Academica.

Davidz, H. 2006. Enabling systems thinking to accelerate the development of senior systems engineers. PhD thesis, Massachusetts Institute of Technology, Cambridge, MA.

Demers, J. 2016. Are we truly in a "golden age" of entrepreneurship? https://www.entrepreneur.com/article/271059.

Denayer, I., K. Thaels, J. V. Sloten, and R. Gobin. 2003. Teaching a structured approach to the design process for undergraduate engineering students by problem-based education. *European Journal of Engineering Education* 28(2): 203–214.

Diamond, J. 2005. *Collapse: How Societies Choose to Fail or Succeed*. London, NY: Penguin Books.

Dollard, N., and L. Christensen. 1996. Constructive classroom management. *Focus on Exceptional Children* 29(2): 1–24.

Doyle, M. 2014. Why engineers need to develop T-shaped skills? Product Lifecycle Report. http://blogs.ptc.com/2014/12/03/why-engineers-need-to-develop-t-shaped-skills/.

Duderstadt, J. J. 2008. Engineering for a changing world: A roadmap to the future of engineering pactice, research, and education. The Millennium Project, Duderstadt Center, The University of Michigan. http://milproj.dc.umich.edu/pdfs/2008/EngFlex%20Report.pdf.

Eder, W. E., and S. Hosnedl. 2007. *Design Engineering: A Manual for Enhanced Creativity*. Boca Raton, FL: CRC Press.

Edwards, C. 2004. *Classroom Management and Discipline*. Fourth edition. New York: John Wiley and Sons.

Elder, L., M. Gorzycki, and R. Paul. 2012. Student guide to historical thinking. *Thinker's Guide Library*. https://www.criticalthinking.org/store/get_file.php?inventories_id=453&inventories_files_id=358.

Emmerson, G. S. 1973. *Engineering Education: A Social History*. New York: David and Charles.
EO. 2016. Engineering: A short history. Engineering Opportunities. http://engopps.com/engineering-in-school/engineering-a-short-history.
Ertas, A., T. Maxwell, V. P. Rainey, and M. M. Tanik. 2003. Transformation of higher education. The transdisciplinary approach to engineering. *IEEE Transactions on Education* 46(2): 289–295.
Ertas, A., M. M. Tanik, and T. T. Maxwell. 2000. Transdisciplinary engineering education and research model. *Society for Design and Process Science* 4(4): 1–11.
Everitt, W. L. 1962. Engineering education-circa 2012 A.D. *Communication and Electronics* May, pp. 571–572.
Faught, D. 2015. Kondratiev waves and sustainability. Converstation in sustainability. https://sustainableworldbd.wordpress.com/2015/07/22/kondratiev-waves-and-sustainability/
Feigenbaum, E., and P. McCorduck. 1983. *The Fifth Generation*. Reading, MA: Addison-Wesley.
Felder, R. M., and L. K. Silverman. 1988. Learning and teaching styles in engineering education. *Journal of Engineering Education* 78(7): 674–681.
Felder, R. M., D. R. Woods, J. E. Stice, and A. Rugarcia. 2000. The future of engineering education. II. Teaching methods that work. *Chemical Engineering Education* 34(1): 26–39.
Ferguson, E. S. 1992. *Engineering and the Mind's Eye*. Cambridge, MA: MIT Press.
de Figueiredo, A. D. 2008. Toward an epistemology of engineering. Workshop on Philosophy and Engineering, The Royal Academy of Engineering, London, November 10–12.
Frances, C., and Y. Ng. 2011. *Engineering in a Montessori Environment*. American Society for Engineering Education.
Freeman, R. 2010. Globalization of scientific and engineering talent: International mobility of students, workers, and ideas and the world economy. *Economics of Innovation and New Technology* 19(5): 393–406.
Freiberg, H. J. (Ed.). 1999. *Beyond Behaviorism: Changing the Classroom Management Paradigm*. Boston, MA: Allyn and Bacon.
Friedman, T. L. 2005. *The World Is Flat: A Brief History of the Twenty-First Century*. New York: Farrar, Strauss, and Giroux.
Fullan, M., and M. Langworthy. 2014. *A Rich Seam: How New Pedagogies Find Deep Learning?* London: Pearson. http://www.michaelfullan.ca/wp-content/uploads/2014/01/3897.Rich_Seam_web.pdf.
Galloway, L., M. Anderson, and W. Brown. 2006. Are engineers becoming more enterprising? A study of the potentials of entrepreneurship education. *International Journal of Continuing Engineering Education and Lifelong Learning* 16(5): 355–365.
Gibbs, G. 2013. *Learning by Doing*. Oxford: Oxford Brooks University. https://thoughtsmostlyaboutlearning.files.wordpress.com/2015/12/learning-by-doing-graham-gibbs.pdf.
Gillman, D. 2006. *The Power of Peer Mentoring*. Waisman Center University of Wisconsin–Madison University Center for Excellence in Developmental Disabilities.
Grasso, D. 2002. Engineering a liberal education, Prism. *American Society of Engineering Education* 12(3): 76.

Grasso, D., and D. Martinelli. 2007. Holistic engineering. *The Chronical Review* 53(28): B8.

Grinter, L. E. 1955. Report of the committee on evaluation of engineering education. American Society for Engineering Education, Washington, DC.

Haase, S. M., H. L. Chen, S. Sheppard, A. Kolmos, and M. M. Mejlcaard. 2013. What does it take to become a good engineer? Identifying cross-national engineering student profiles according to perceived importance of skills. *International Journal of Engineering Education* 29(3): 698–713.

Habash, R. W. Y., C. Suurtamm, and D. Necsulescu. 2011. Mechatronics learning studio: From "play and learn" to industry-inspired green energy applications. *IEEE Transactions on Education* 45(4): 667–674.

Hajesfandiari, B., A. G. Mehrdad, and L. Karimi. 2014. Comparing the effects of convergent and divergent Teaching methods on using articles by Iranian EFL learners. *International Journal of Educational Investigations* 1(1): 313–327.

Halperin, D. 1994. *Changing College Classrooms*. San Francisco, CA: Jossey-Bass Publications, pp. 11–12.

Hargroves, K., and M. H. Smith. 2005. *The Natural Advantage of Nations: Business Opportunities, Innovation and Governance in the 21st Century*. London: Earthscan.

Hawken, P., A. B. Lovins, and L. H. Lovins. 1999. *Natural Capitalism: Creating the Next Industrial Revolution*. Boston, MA: Little, Brown and Co.

Heywood, J., and M. Mina. 2015. The role of transdisciplinary studies in the reform of the engineering curriculum: A case study. *122 ASEE Annual Conference and Exposition*, Seattle, WA, June 14–17.

Hilty, L., W. Lohmann, and E. M. Huang. 2011. Sustainability and ICT—An overview of the field. *Notizie di Politeia* 27(104): 13–28.

Issapour, M. K., and K. Sheppard. 2015. Evolution of American engineering education. *Conference for Industry and Education Collaboration, American Society for Engineering Education*, Sheraton Seattle, Seattle, WA, June 13.

Johnson, J. K., S. H. Miller, and S. D. Horowitz. 2008. Systems-based practice: Improving the safety and quality of patient care by recognizing and improving the systems in which we work. In Henriksen, K., J. B. Battles, and M. A. Keyes et al. (Eds.), *Advances in Patient Safety: New Directions and Alternative Approaches* (Vol. 2: Culture and Redesign). Rockville, MD: Agency for Healthcare Research and Quality.

Jonassen, D. H. 1997. Instructional design models for well-structured and III-structured problem-solving learning outcomes. *Educational Technology Research and Development* 45(1): 65–94.

Kagan, D., S. Ozment, and F. M. Turner. 2001. *The Western Heritage*. Seventh edition. Englewood Cliffs, NJ: Prentice Hall.

Kamp, A. 2014. *Engineering Education in the Rapidly Changing World: Rethinking the Mission and Vision on Engineering Education*. TU Delft, Delft University of Technology Faculty of Aerospace Engineering. Delft, The Netherlands: Kluyverweg.

Keeton, M.T., and P. J. Tate. eds. 1978. *Learning by experience–what, why, how?* San Francisco, CA: Jossey-Bass.

Kolb, D. 1984. *Experiential learning: Experience as the source of learning and development*. Englewood Cliffs, NJ: Prentice Hall.

Kolb, D. A. 1984. *Experiential learning*. Upper Saddle River, NJ: Prentice Hall, Inc.

Kolb, D.A. 2015. Experiential learning: experience as the source of learning and development. Second edition. Upper Saddle River, New Jersey: Pearson Education.

Kolb, A. Y. and Kolb, D. A. 2005. *The Kolb learning style Inventory-Version 3.1: 2005, Technical specifications*. Boston, MA: Hay Resources Direct. www.learningfromexperience.com.

Kozmetsky, G. 1997. Generational road maps and commercialization in science and technology. *Journal of Integrated Design Process Science* 1(1): 4–8.

Kriewall, T. J., and K. Mekemson. 2010. Instilling the entrepreneurial mindset into engineering understanding. *The Journal of Engineering Entrepreneurship* 1(1): 5–19.

Langrish, J. 1985. *Innovation Management: The Role of Creativity*. Institute of Advanced Studies, Manchester Polytechnic, Manchester.

Lewis, L. H., and C. J. Williams. 1994. *Experiential Learning: Past and Present. New Directions for Adults and Continuing Education*. San Francisco, CA: Jossey-Bass, pp. 5–16. https://pdfs.semanticscholar.org/e05d/abcc394a580e9710492f440259381ae179c1.pdf.

Lucas, B., and J. Hanson. 2014. Thinking like an engineer implications for the education system. A report for the Royal Academy of Engineering Standing Committee for Education and Training. Summary Report. Royal Academy of Engineering, 2014.

MacLeod, I. 2009. The education of innovative engineers. Engineering. *Construction and Architectural Management* 17(1): 21–34.

Mäkimurto-Koivumaa, S., and P. Belt. 2015. About, for, in or through entrepreneurship in engineering education. *European Journal of Engineering Education* 23: 14–28.

Maxwell, J. C. 2002. *The 17 Essential Qualities of a Team Player: Becoming the Kind of Person Every Team Wants*. Nashville, TN: Thomas Nelson Publishers.

McIsaac, G. F., and N. C. Morey. 1998. Engineers' role in sustainable development: Considering cultural dynamics. *Lithuanian Journal of Physics* 1244: 110–119.

Meagor, K. 2014. Dewey, Piaget, and Vygotsky on "Learner-centered" education. Team Two Blog. http://aasmith17.blogspot.ca/2014/03/dewey-piaget-vygotsky-on-learner.html.

Moen, A., A. Mørch, and S. Paavola (Eds.). 2012. *Collaborative Knowledge Creation: Practices, Tools, Concepts*. Rotterdam, the Netherlands: Sense.

Mooney, C. G. 2000. *Theories of Childhood: An Introduction to Dewey, Montessori, Erikson, Piaget and Vygotsky*. St. Paul, MN: Redleaf Press.

Murmann, J. P. 2006. *Knowledge and Competitive Advantage: The Coevolution of Firms, Technology, and National Institutions*. Cambridge Studies in the Emergence of Global Enterprise. Cambridge: Cambridge University Press.

Murphy, L. 2016. Montessori, Vygotsky and Piaget! OH MY! How what THEY did influences what YOU do. Ooey Gooey Inc. http://www.ooeygooey.com/wp-content/uploads/2010/08/meet-the-masters.pdf.

NAE. 2004. The engineer of 2020: Visions of engineering in the new century. National Academy of Engineering. https://inside.mines.edu/UserFiles/File/Assessment/Engr2020.pdf.

NAE. 2008. *Changing the Conversation: Messages for Improving Public Understanding of Engineering*. Committee on Public Understanding of Engineering Messages, National Academy of Engineers.

Nazzal, D., J. Zabinski, A. Hugar, D. Reinhart, W. Karwowski, and K. Madani. 2015. Introduction of sustainability concepts into industrial engineering education: A modular approach. *Advances in Engineering Education*. http://files.eric.ed.gov/fulltext/EJ1077845.pdf.

Nonaka, I. 1994. A dynamic theory of organizational knowledge creation. *Organization Science* 5(1): 14–37.

NUS. 2012. *A Manifesto for Partnership*. London: National Union of Students. www.nusconnect.org.uk/campaigns/highereducation/partnership/a-manifesto-for-partnerships/.

Nybom, T. 2003. The Humboldt legacy: Reflections on the past, present and future of the European university. *Higher Education Policy* 16(2): 141–159.

Paavola, S., L. Lipponen, and K. Hakkarainen. 2004. Models of innovative knowledge communities and three metaphors of learning. *Review of Educational Research* 74: 557–576.

Penzenstadler, B., H. Femmer, and D. Richardson. 2013. *Who is the advocate? Stakeholders for sustainability. Second International Workshop on Green and Sustainable Software*, May 18–26, San Francisco, USA.

Prentic, J. 2012. Technological revolutions through the ages: Lessons for leading the cleantech revolution. Business 485: Business Leadership and the Liberal Arts.

Rambocas, M., and M. K. S. Sastry. 2017. Teaching business management to engineers: The impact of interactive lectures. *IEEE Transactions on Education* PP(99): 1–10.

Reynolds, T. S., and B. E. Seely. 1999. Striving for balance: A hundred years of the American Society for engineering Education. *Journal of Engineering Education* 82(3): 136–151.

Rugarcia, A., R. M. Felder, D. R. Woods, and J. E. Stice. 2000. The future of engineering education: A vision for a new century. *Chemical Engineering Education* 34(1): 16–25.

Ryan, K., and J. Cooper. 2001. *Those Who Can, Teach*. New York: Houghton Mifflin Company.

Schilling, M. A., P. Vidal, R. E. Ployhart, and A. Marangoni. 2003. Learning by doing something else: Variation, relatedness, and the learning curve. *Management Science* 49(1): 39–56.

Senge, P. 1990. *The Fifth Discipline*. New York: Doubleday.

Sfard, A. 1998. On two metaphors for learning and the dangers of choosing just one. *Educational Researcher* 27(3): 4–13.

Sharma, G. and D. A. Kolb. 2010. The Learning flexibility index: assessing contextual flexibility in learning style. Department of Organizational Behavior, Case Western Reserve University Working Paper Series. https://weatherhead.case.edu/departments/organizational-behavior/workingPapers/WP-10-01.pdf.

Shattleworth, M. 2014. History of the scientific method. https://explorable.com/history-of-the-scientific-method.

Sy, C. 1999. A little history lesson: the iron ring. http://www2.ensc.sfu.ca/undergrad/euss/enscquire/Vol10No2/page9_1.html.

Tanik, M. M., and A. Ertas. 1997. Interdisciplinary design and process science: A discourse on scientific method for the integration age. *Trans SDPS: Journal of Integrated Design and Process Science* 1(1): 76–94.

WCED. 1987. Report of the world commission on environment and development: our common future. World Commission on Environment and Development. New York: Oxford University Press.
UNESCO. 2010. *Engineering: Issues, Challenges, and Opportunities for Development. United Nations Educational Report, Scientific and Cultural Organization*. Paris, France: UNESCO Publishing.
de Vere, I., K. B. Johnson, and C. Thong. 2009. Educating the responsible engineer: Socially responsible design and sustainability in the curriculum. *International Conference on Engineering and Product Design Education*, University of Brighton, Brighton, UK, September 10 and 11.
Vesilind, P. A. 1988. Rules, ethics and morals in engineering education. *Engineering Education* 78: 289–293.
Vital, A. 2013. The history of creating value—How humans made money in different eras. http://fundersandfounders.com/the-history-of-creating-value/.
WEF. 2011. Unlocking entrepreneurial capabilities to meet the global challenges of the 21st century. Final Report on the Entrepreneurship Education Workstream. World Economic Forum Global Education Initiative, June 2011.
Williams, Y. 2003. Convergent thinking: Definition, examples and quiz. *Education Portal* 13(3).
Wurdinger, S. D. 2005. *Using Experiential Learning in the Classroom*. Lanham, MD: Scarecrow Education.
Zaman, M. H. 2015. Why science and engineering need to remind students of forgotten lessons from history. http://www.newspronto.com/news/the-conversation/34741-why-science-and-engineering-need-to-remind-students-of-forgotten-lessons-from-history.

part one

The Sustainability Landscape

chapter two

Engineering for sustainability and sustainable development

Sustainability: Both a journey and a destination.

Linda Giudice (2015)

2.1 Objectives

- Historically, trace the origin and roots of sustainability.
- Understand the concepts of sustainability and sustainable development (SD).
- Discuss the progress timeline of SD.
- Examine the 8 millennium development goals (MDGs).
- Describe the 17 UN SD goals (SDGs).
- Discuss the 12 guiding engineering principles for SD.
- Understand the changes in the use, distribution, and importance of natural resources on human life.
- Introduce sustainability models including the triple bottom line (TBL) and the egg of sustainability.
- Discuss the driving forces for sustainable solutions, where the technology dimension is central to the three main pillars of sustainability: economy, society, and ecology.
- Explain what indicators are, how indicators relate to sustainability, and how to identify good indicators of sustainability.
- Show how ideas of circular economy (CE) have been proposed to change the "take-make-waste" linear economic system in order to lower resource use and waste of natural capital.
- Describe the stages of sustainability planning.
- Discuss the requirements for sustainability approaches in engineering and the unique role of engineers in SD.
- Investigate various pathways of exploring SD.
- Discuss urban transformation and determinants of sustainable and smarter cities.
- Explain extensively the topic of energy and sustainability and the concept of net energy analysis (NEA).
- Explore the role of engineering education in sustainability literacy and the need for a curriculum reorientation.

- Discuss the role of transdisciplinary research and collaboration on promotion of sustainability.
- Investigate an integrated approach toward meeting SD goals through a case that explores the interaction of three major resources, water, energy, and food, in the state of California.
- Provide three end of chapter pedagogical knowledge strategies, namely knowledge acquisition, knowledge possession, and knowledge creation to help understand the above topics and generate new and alternative ideas and solutions.

2.2 Historical perspective

> The story of civilisation is, in a sense, the story of engineering—that long and arduous struggle to make the forces of nature work for man's good.
>
> **Lyon Sprague De Camp**

2.2.1 Early history

Ancient cultures, those like the aboriginals of North and South America, the Chinese, and the Egyptians, were maintained for thousands of years with primitive tools. This was possible through ideas of sustainability that current cultures are still attempting to adopt. The ancients knew that the main sources of life needed to be in constant supply and that they came directly from nature. Therefore, their primitive agriculture was used in a sustainable manner as they rotated crop areas, kept soil fertile, and made sure to understand when one crop was able to grow best through an understanding of the seasons. Water was also known to be very important to life, so they were careful never to spoil it. Ancient cultures worshiped nature because it provided them with their needs and they tried not to harm it. The ancient Hawaiians were able to create a sustainable fishing society that rivals the fishing regulations of today (Jones 2013).

Exploitation of the environment can be traced back to 3000 BC when settlements began to realize that nature provided an easy and usable source of easy living, such as wood for building and burning. This has caused settlements to become more of a permanent arrangement while natural resources were consumed, generating a growth in population. New techniques of exploiting nature were developed and as a result, there was a steady decrease of the appreciation for nature that the ancients had. As this appreciation decreased, the idea that everything available from nature is for human exploitation was rooted further. Settlements began to develop into cities and nature utilization spread to a level in which a key change had to happen.

The major change that happened around the 1750s was the Industrial Revolution. This gave further way to invention and heavy exploitation of natural resources. Land developers started taking over large tracts of forests and cropping lands. Coal became a huge energy resource and it allowed for an expansion of modern civilization to essentially consume all civilizations around the world that were based on ancient and sustainable cultures. This is how the thousands of years of sustainable cultural practice were lost. Nature was exploited at an extreme rate to create a human civilization that could live easier than ever before, and able to explore further. The implementation of gasoline engines starting in the early 1900s was the beginning for the exploitation of environment by human beings and misuse of natural resources.

However, this does not mean that there was a complete absence of people who care about sustainability through these times. Many economists in the mid- to late 1700s developed the "theory of limits," where continued population growth and resource exploitation were eventually going to reach a limit which is most usefully defined in terms of the point or range of conditions beyond which the benefits derived from a natural resources are judged unacceptable or inadequate. The Earth only has a defined amount of space and resources, and once that is gone, it cannot return (Jones 2013).

2.2.2 *Origin of the concept*

Sustainability is based on a simple factual premise: everything that humans require for their survival and well-being depends, directly or indirectly, on the natural environment (Marsh 1864). This concept is used by many different communities. Its Latin origin *sustinere* was used both as endure and as uphold, furnish (something) with means of support. In modern English, sustainability refers to the "capacity" of a system to endure (Oxford Dictionary of English 2010). But these definitions merely raise further questions (Venters 2014). Tainter (2006) points out we need to ask: Sustain what? For whom? How long? At what cost?

The concept of sustainability takes its roots in German forestry and lumber businesses at the beginning of the eighteenth century. At that time, the German lumber business community agreed on a limitation of cutting wood to an amount that will be compensated by afforesting every year. Hans Carl von Carlowitz (1645–1714) called this principle Nuchhaltige Entwicklung, which translates in English to sustainable development (SD).

Sustainable means "capable of being sustained," which links to the capacity of durability, stability, permanence, or even eternalness. This adjective has a kind connotation of immobility or perpetuity. Development means the act of improving by expanding, enlarging, or refining. This includes both qualitative and quantitative features. The word itself induces the thought of movement as the way of improving. So, dynamics is clearly included in the definition (Garcia-Serna 2007).

By applying the SD principle, a smart balance was reached that secured long-term business economic development without the depletion of natural resources. Furthermore, Hans Carl von Carlowitz also took notice of the ethical and esthetical values of the forest. He insisted on the protection and prevalence of these values for future generations. This concept of balance was again reflected in the famous book *The Limits of Growth* by the Club of Rome in 1972 (Meadows et al. 1972). Over the years, this principle was developed into today's idea of SD.

The environmental revolution of the 1960s and the 1970s was a major stepping stone. Rachel Carson's *Silent Spring* was published in 1962. The book brings together research on toxicology, ecology, and epidemiology to suggest that agricultural pesticides are building to catastrophic levels, linked to damaging animal species and human health. Many consider the book's release a turning point in our understanding of the interconnections among the environment, the economy, and social well-being. Since then, many milestones have marked the journey toward SD.

2.2.3 SD timeline

Warnings about the deterioration of the environment have been sounded around the world since the 1960s. Partly because of these warnings, numerous proposals have been made from the 1980s onward for a worldwide approach to existing and predicted environmental problems. The book *Silent Spring* made an important scientific contribution to public concern about environmental pollution, caused mainly by the use of pesticides. The book describes and documents the harmful effects of pesticides on the environment (Carson 1962). In 1972, the publication *Limits to Growth* by the Club of Rome focuses attention on topics including the problem of depletion of the earth's resources. It suggests that if the present growth trends continue, the earth will no longer be able to meet demands for natural resources by around 2100 (Meadows et al. 1972).

The World Conservation Strategy by the International Union for Conservation of Nature in 1980 and the Brundtland Report by the World Commission on Environment and Development (WCED) in 1987 pushed SD steadily forward. Both reports advocate a departure from nonsustainable consumption and production in favor of SD. Since then, awareness of the global environmental problem has clearly increased (Van de Westerlo 2011). Figure 2.1 shows a chronological overview of the most important lines of thinking in relation to SD.

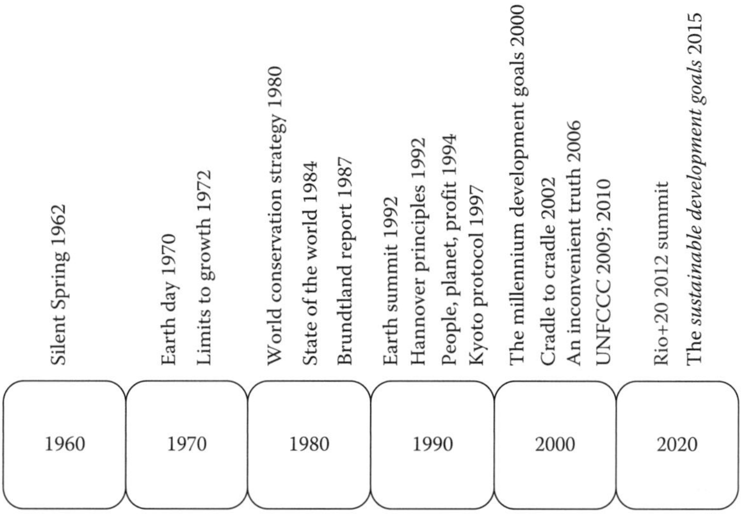

Figure 2.1 Chronological timeline of the most important lines of thinking in relation to SD.

2.2.4 *SD framework*

SD is the overarching paradigm of the United Nations (UN). The theoretical framework for SD evolved between 1972 and 1992 through a series of international conferences and initiatives. An important step in its growth as a mainstream concept and practice was the 1972 UN Conference on the Human Environment, in Stockholm, Sweden. This conference brought the industrialized and developing nations together to delineate the rights of the human family to a healthy and productive environment. A series of such meetings followed—for example, on the rights of people to adequate food, to sound housing, to safe water, and to access to means of family planning. The recognition to revitalize humanity's connection with nature led to the creation of global institutions within the UN system. This marks a transition from a national focus to an international one (Stofleth 2016). This led to the development of the common use of the word "sustainability" as is known today.

In 1978, the Organisation for Economic Co-operation and Development (OECD) Directorate of the Environment relaunched research on environmental and economic linkages. The work built the foundation for the 1987 report, "Our Common Future." In 1980, the Global 2000 report was released. It recognized biodiversity for the first time as critical to the proper functioning of the planetary ecosystem (IISD 2010). It asserts that the robust nature of ecosystems is weakened by species extinction. In 1981, the World Health Assembly unanimously adopted the "Global

Strategy for Health for All by the Year 2000," which affirmed that the major social goal of governments should be for all peoples to attain a level of health that would permit them to lead socially and economically productive lives.

In 1983, the UN convened the WCED, chaired by Norwegian Prime Minister Gro Harlem Brundtland. Comprised of representatives from both developed and developing countries, the Commission was created to address growing concern over the accelerating deterioration of the human environment and natural resources and the consequences of that deterioration for economic and social development. The conceptual definition of the Brundtland Commission contains two key concepts (Mebrato 1998):

- The concept of "needs" in particular the essential needs of the world's poor, to which overriding priority should be given
- The idea of limitations imposed by the state of technology and social organization on the environment's ability to meet present and future needs

In 1987, the WCED, sponsored by the UN, published a report called "Our Common Future." The so-called Brundtland Report of the World Commission on Environment and Development defines SD as "development that meets the needs of the present without compromising the ability of future generations to meet their own needs" (WCED 1987). This report has been taken as a starting point for most current discussions on the concept of SD.

In 1990, the International Institute for Sustainable Development (IISD) was established in Canada and started publishing the *Earth Negotiations Bulletin* in 1992. Also in 1990, the Regional Environmental Centre for Central and Eastern Europe was established to address environmental challenges across the region, with an emphasis on the engagement of business as well as governments and civil society (IISD 2010).

A ground-breaking step came in 1992 with the first UN Conference on Environment and Development in Rio de Janeiro. At this Earth Summit, an agenda called Agenda 21 was adopted, which recognized each nation's right to pursue social and economic progress and assigned to states the responsibility of adopting a model of SD (Stofleth 2016). Even though environmental issues have gained more importance since, the environment has often been seen by policymakers as an ancillary goal to other more important concerns. Development aims have been particularly driven by economic growth; goals for environmental sustainability have often been interpreted as precautions external to or constraining economic performance, rather than something integral to it.

Another notable international protocol designed to guide the international community toward SD, in this case particularly environmental, was the Kyoto Climate Agreement in 1997. Its goal was to reduce the emissions of its signatories, with more emphasis placed on those developed countries which were responsible for most of the air pollution and its subsequent consequences.

2.2.5 Millennium development goals

In September 2000, 189 countries signed the UN Millennium Declaration [A/RES/55/2], committing themselves to eradicating extreme poverty in all its forms by 2015. To help track progress toward these commitments, a set of time-bound and quantified goals and targets, called the millennium development goals (MDGs), was developed for combating poverty in its many dimensions including reducing income poverty, hunger, disease, environmental degradation, and gender discrimination. The MDGs include 8 goals, 21 targets, and 60 indicators for measuring progress between 1990 and 2015, when the goals are expected to be met. The eight MDGs that range from halving extreme poverty to halting the spread of HIV/AIDS and providing universal primary education, all by the target date of 2015, form a blueprint agreed to by all the world's countries and all the world's leading development institutions. They have galvanized unprecedented efforts to meet the needs of the world's poorest countries. Table 2.1 shows the eight MDGs (UNICEF 2014).

The MDG indicators are quantified and time bound, and encourage quick-win initiatives, where environmental problems can be addressed while alleviating poverty. The goal serves to mobilize political commitment and to generate popular awareness around consensus development objectives, as guidelines for coordinated action (Jolly 2010).

The MDGs have proven to be a powerful tool for international efforts to eradicate poverty and focus action toward meeting education, public health, and the environmental goals. However, the MDGs were criticized for lack of careful analysis and reasoning behind their objectives.

The MDGs were obviously adopted for developing countries in order to fulfill certain basic needs. Their success has been recognized, but they

Table 2.1 The millennium development goals

Goal 1	To eradicate extreme poverty and hunger
Goal 2	To achieve universal primary education
Goal 3	To promote gender equality and empower women
Goal 4	To reduce child mortality
Goal 5	To improve maternal health
Goal 6	To combat HIV/AIDS, malaria, and other diseases
Goal 7	To ensure environmental sustainability
Goal 8	To develop a global partnership for development

represent an unfinished work which cannot be abandoned, since the eight objectives are still valid, even if put under a different, broader, framework (Giovannini 2013).

2.2.6 *Toward 2015 SD agenda*

The point of integrating development and environment goals is to take the malicious cycle of environmental degradation and diminished development and make it worthy, leading to both greater short-term efficiencies and long-term sustainability in improving human well-being. For SD to succeed on any time prospect, good integration of the environment in development efforts must therefore be the guiding principle.

Merging the development and environment agendas is very essential. First, the environment is a foundation for development. Deteriorating ecosystem services and degrading natural resources limit the ability to reduce poverty and secure economic development. Natural ecosystems such as oceans, forests, lakes, and rivers provide food, raw materials, and livelihoods for billions of people, and loss of these ecosystems in recent years is already costing billions to communities and economies. Moreover, the degradation of natural ecosystems hits the poorest the hardest. For example, it has been estimated that ecosystem services account for at least half of the gross domestic product (GDP) of the poor, a huge share of the sources of livelihood of poor households worldwide (TEEB 2010). Sustaining the health and resilience of the environment is thus fundamental to meaningful progress toward any human development goals. Second, the environmental concerns of improving development outcomes today affect the ability to do so in the future. Depending on how to pursue development goals, it can either threaten or enhance natural resources and ecosystem services, and similarly determine long-term potential for improving and sustaining human livelihoods. For example, improving access to electricity in certain regions is generally seen as a necessary development outcome. But achieving such a goal by mining and burning coal contributes to global warming that will eventually put people in those very places at risk of catastrophic extremes in climate, environmental risks, and natural resource shortages.

The 2012 UN Conference on Sustainable Development (Rio+20) has laid out some new and inspiring pathways for transitioning toward a green economy. It also opens a political space to resolve the apparent tension between poverty goal and the sustainability of the planet. Herein lies an opportunity to strike a radical shift toward more sustainable patterns of consumption and production and resource use but couched in the reality of poverty eradication and SD. Another policy innovation from Rio+20 is the proposal to develop the Sustainable development goals (SDGs) as a part of the Post-2015 Development Agenda. Any new or revised goals for the environment ideally will embrace broader notions of wealth encompassing natural capital, address environmental challenges directly, and enhance livelihoods and resilience of the poor (Hezri 2013). The Rio+20 concluded: "We adopt the 10-Year Framework of Programmes (10YFP) on sustainable consumption and production. We invite the UN General Assembly to take any necessary steps to fully operationalize the framework." After nearly a decade of moving ahead without formal agreement by all UN countries, the Marrakech Process 10-Year Framework is finally adopted as one of the few successes of a controversial Rio+20 Summit.

Progress has been made on SD metrics. Since the Brundtland Report and the Rio Summit, researchers in universities, environmental organizations, think tanks, and national governments have furthered the measurement of progress on SD.

While important development has been made toward achieving the MDGs by 2015, achievements fall short of our ambitions. At the same time, ever more formidable and crucial environmental challenges now challenge civilization: an increasing biodiversity crisis, climate trajectories often exceeding worst-case forecasts, and shocks to food and water supplies. This realization has prompted a reflection on the nature of SD.

In the following section, the new SD goals and broader sustainability agenda will be discussed. The aim is to complete what the MDGs did not achieve, and move much further, addressing the root causes of poverty and inequality and the universal need for development that works for all people. The SD 2015 website (sustainabledevelopment2015.org) was developed to provide the latest news, information, and expert analysis around the global decision-making process to define a new set of global goals to eradicate poverty through SD, known as the Post-2015 Development Agenda.

In 2015, the antipoverty targets and indicators that made up the eight UN MDGs expired. The 2012 UN Task Report "Realizing the Future We Want for All" assessed the progress made toward achieving the MDGs, although challenges remain regarding achieving certain goals in some countries. The report also identifies several conceptual shortcomings of the MDGs, most notably their failure to address the environment in an integrated and cross-sectoral manner; the need for some goals to deepen their impact; and the challenge of building a partnership for development that does not divide the world into aid recipients and donors, but outlines common but differentiated responsibilities for all. This reflected fact called for a need of expanded goals to succeed the MDGs (Neureuther 2013).

For SD to succeed on any time horizon, reliable integration of the environment in development efforts must be a steering principle. Linking environment and development in one SD agenda will widen the area for sustaining earth's natural capital to secure continuing human well-being.

2.3 *UN SD goals*

A little less conversation, a little more action.

Erna Solberg
Prime Minister of Norway

2.3.1 *The 17 goals*

The SDGs were announced late in 2015, but the difference is noticeable. Unlike the MDGs, the SDGs are universal and every country is required to say how it will meet the goals. The SDGs refer to the 2030 agenda for SD adopted by the 193-Member UN General Assembly at the Sustainable Development Summit held in New York on September 25–27, 2015. The new SDGs include 17 goals and 169 targets. The agenda serves as a launch pad for renewed cooperation over the next 15 years to end poverty in all its forms, promote shared prosperity, and support SD for everyone.

The SDGs are a set of global goals that governments are expected to adopt. When they sign up to them, they will look to society and business, in particular, for help to achieve them. Governments will want to measure and monitor progress and manage the effectiveness of their interventions. In turn, businesses will need to assess its impact on the SDGs and review its strategy accordingly. The SDGs were developed with the guidance and input of people from all over the globe to ensure they represent the needs of its entire population (Preston 2015). The goals are shown in Table 2.2.

A key principle of the SDGs is universality, the goals will be relevant to all countries, and all will contribute to achieving them, but with differentiated targets and actions (Nilsson et al. 2013; Van der Heijden et al. 2014).

2.3.2 *SD perspectives*

It is important that educators, leaders, and citizens recognize that SD is an evolving concept and that the list of sustainability perspectives can therefore grow and change. Accompanying SDGs are perspectives that have become part of the global sustainability dialog, such as the following (UNESCO 2012):

- A systems thinking (ST) approach, rather than an approach that looks at problems in isolation, should be used. Sustainability issues are linked and part of a "whole"
- Understanding local issues in a global context and recognizing that solutions to local problems can have global consequences

Table 2.2 The UN SD goals

Goal 1	End poverty in all its forms everywhere
Goal 2	End hunger, achieve food security and improved nutrition, and promote sustainable agriculture
Goal 3	Ensure healthy lives and promote well-being for all at all ages
Goal 4	Ensure inclusive and quality education for all and promote lifelong learning
Goal 5	Achieve gender equality and empower all women and girls
Goal 6	Ensure access to water and sanitation for all
Goal 7	Ensure access to affordable, reliable, sustainable, and modern energy for all
Goal 8	Promote inclusive and sustainable economic growth, employment, and decent work for all
Goal 9	Build resilient infrastructure, promote sustainable industrialization, and foster innovation
Goal 10	Reduce inequality within and among countries
Goal 11	Make cities inclusive, safe, resilient, and sustainable
Goal 12	Ensure sustainable consumption and production patterns
Goal 13	Take urgent action to combat climate change and its impacts
Goal 14	Conserve and sustainably use the oceans, seas, and marine resources
Goal 15	Sustainably manage forests, combat desertification, halt and reverse land degradation, and halt biodiversity loss
Goal 16	Promote just, peaceful and inclusive societies
Goal 17	Revitalize the global partnership for SD

- Realizing that individual consumer decisions affect and give rise to resource extraction and manufacturing in distant places
- Considering differing views before reaching a decision or judgment
- Recognizing that economic values, religious values, and societal values compete for importance as people with different interests and backgrounds interact
- Seeing all humans as having universal attributes
- Knowing that technology and science alone cannot solve all of our problems
- Emphasizing the role of public participation in community and governmental decision-making. People whose lives will be affected by decisions must be involved in the process leading to the decision.
- Calling for greater transparency and accountability in governmental decision-making
- Employing the precautionary principle: taking action to avoid the possibility of serious or irreversible environmental or social harm even when scientific knowledge is incomplete or inconclusive

2.4 Guiding engineering principles for SD

> Engineers ... are not mere technicians and should not approve or lend their name to any project that does not promise to be beneficent to man and the advancement of civilization.
>
> **John Fowler**

2.4.1 The principles

The idea of satisfying the needs of the present without limiting the ability of future generations to satisfy their own needs is easy to understand but difficult to translate into actions, where engineering is a main player in realizing these actions (ACEC 2014). Engineering for SD is a wide ranging topic and, as such, may be considered to mean different things to different people. For this reason, the broader definitions of SD provide a good starting point. The practice of SD is beginning to provide engineers with indications of techniques and processes that when used on real life lead to better outcomes.

Guiding engineering principles are outlined in 12 principles which are based on Royal Academy of Engineering report: "Engineering for Sustainable Development: Guiding Principles" (Dodds and Venables 2005). The report brings together much of UK engineering's current thinking on applying SD in the real world. These principles are designed to help both engineering students and educators in adopting a sustainability-driven approach. It provides answers to questions of both what action to take and what can be done differently, and has great potential to provide the market stimulus identified earlier. These principles are listed in Table 2.3 (Anastas and Warner 1998; Anastas and Zimmerman 2003; Dodds and Venables 2005).

Principle 1 asks engineers to identify the potential positive and negative impacts of the proposed actions, not only locally but also outside the immediate local environment, organization, and context, as well as into the future.

Principle 2 calls the SD approach to be creative, innovative, and broad, and thus does not mean following a specific set of rules. It requires an approach to decision-making that strikes a balance between environmental, social, and economic factors.

Principle 3 seeks to deliver economic, social, and environmental success all at the same time, and so seeks to avoid any product, process, or project that yields an unbalanced solution.

Principle 4 calls for engagement of stakeholders to bring their different views, perceptions, knowledge, and skills to bear on the challenge being addressed. For example, professional engineers should participate

Table 2.3 Engineering principles for SD

Principle 1	Look beyond your own locality and the immediate future
Principle 2	Innovate and be creative
Principle 3	Seek a balanced solution
Principle 4	Seek engagement from all stakeholders
Principle 5	Make sure you know the needs and wants
Principle 6	Plan and manage effectively
Principle 7	Give sustainability the benefit of any doubt
Principle 8	If polluters must pollute ... then they must pay as well
Principle 9	Adopt a holistic, "cradle-to-grave" approach
Principle 10	Do things right, having decided on the right thing to do
Principle 11	Beware cost reductions that masquerade as value engineering
Principle 12	Practice what you preach

actively in the decision-making process as citizens as well as in their professional roles.

Principle 5 indicates that effective decision-making in engineering for SD is only possible when we know what is needed or wanted. This should be recognized as clearly as possible, including identifying any legal requirements and restrictions.

Principle 6 highlights the fact that when planning engineering projects, objectives should be expressed in sufficiently open-ended terms so as not to preclude the potential for innovative solutions as the project develops.

Principle 7 captures the "precautionary principle" and addresses the future impacts of today's decisions. This principle encourages engineers to demonstrate that improved sustainability will result from the actions they propose.

Principle 8 highlights the fact that the environment belongs to all and its free use for absorption of wastes or its released exploitation is not sustainable. The adverse, polluting effects of any decision should, in some way, be paid for or compensated for by the promoter of an engineering project, scheme, or development.

To deliver Principle 9, the effects on sustainability throughout the whole life cycle of a product should be systematically evaluated. It is also crucial to ensure that the design is maintainable and that the materials are adaptable for reuse or recycling.

Adhering to the principles explained so far should ensure that correct decisions from a sustainability point of view have been made in relation to the circumstances that apply. To deliver principle 10, the sustainability focus on the intended outcome needs to be retained through to the implementation of the solution.

Principle 11 highlights the fact that it is unlikely to arrive at the best decisions the first time, every time. So it is needed to challenge ourselves and refine those decisions, while remaining focused on the intended outcome.

Principle 12 demonstrates how many of the above principles are challenging. This principle indicates that daily practices should not be at odds with what is being asked of others. It also states that engineers must be accountable for their design and engineering decisions. In brief, it states that change should start from oneself (Dodds and Venables 2005).

2.4.2 Applications of the principles

Two central concepts that engineers should attempt to integrate at every opportunity when designing within the principles framework are life cycle considerations and the first principle of green engineering. The materials and energy that are used in each life cycle stage of every product and process have their own life cycle. If a product is environmentally benign but is made using hazardous or nonrenewable substances, the impacts have easily been shifted to another part of the overall life cycle. Figure 2.2 exhibits the sustainable guiding principles now and in the future. It is clear that sustainability is still at the orientation phase. For example, clean technology is not yet economically viable.

The principles of engineering for SD, together with the guidance on its application in practice, should assist all involved in engineering to make their vital and urgent contribution to society to drive down the adverse environmental and social aspects of engineered products, services, and infrastructure; dramatically improve their environmental performance; improve the contribution of engineering products,

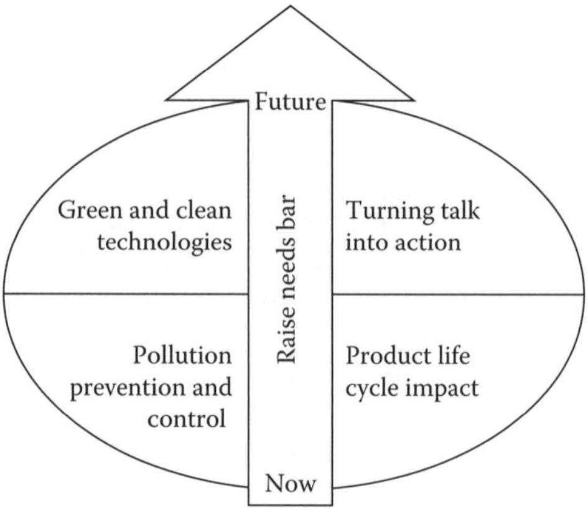

Figure 2.2 Sustainable guiding principles: now and in the future.

services, and infrastructure to a high quality of life; help society to move toward a significantly more-sustainable lifestyle; and ensure products, services, and infrastructure meeting these criteria are competitive in their marketplace and, ideally, the most competitive (Dodds and Venables 2005).

The application of the principles across scales and across disciplines has been documented with case studies from a variety of sectors (Zimmerman et al. 2003; Zimmerman and Anastas 2005). By illustrating how the framework of principles has worked in the past, case studies provide a blueprint for how these guidelines can be applied in future designs for improving quality of life and ultimately advancing sustainability.

2.5 *Sustainability taxonomy*

> Sustainable development requires human ingenuity. People are the most important resource.
>
> **Dan Shechtman**

2.5.1 *Terminology*

2.5.1.1 *Sustainability*

Within the scientific community, the definition of sustainability and what should be sustained (e.g., what might constitute critical natural capital) is by no means agreed on and is subject to value judgments (Bond et al. 2011), up to be interpreted as a shared ethical belief (Seager et al. 2004). Four main interpretations of the concept of sustainability may be identified: ecological, economic, thermodynamic and ecological economic, and public policy and planning theory (Patterson 2010; Sala et al. 2012).

The need to realign the current path of development on a sustainable trajectory was already understood more than two decades ago. This modern concept of sustainability focuses on the establishment of political and economic framework conditions for a well-balanced environmental, economic, and societal development on a global scale.

Sustainability is often thought of as composed of three overlapping, mutually dependent goals: (a) to live in a way that is environmentally sustainable, or viable over the very long term; (b) to live in a way that is economically sustainable, maintaining living standards over the long term; and (c) to live in a way that is socially sustainable, now and in the future. The social dimension of sustainability should be understood as both (1) the processes that generate social health and well-being now and in the future and (2) those social institutions that facilitate environmental and economic sustainability now and in the future (Dillard et al. 2008).

According to Ayres (2008), sustainability is a normative concept about how humans should act in relation to ecology and how they are responsible for each other and future generations. In this context, it is noted that sustainability is conducive to economic growth based on social justice and the efficient use of natural resources (Lozano 2012). The ecological interpretation focuses on a vision of the socioeconomic system embedded in the global biophysical system, the economic interpretation emphasizes the idea of social welfare, the thermodynamic interpretation poses ecological sustainability in the context of the entropic nature of economic–environmental interactions, and the public policy and planning interpretation seeks to achieve a balance of the different aforementioned factors. Each of these interpretations implies a different scientific domain, with some knowledge areas overlapping and others diverging or overlooked.

2.5.1.2 Sustainable development

With the publication of *Our Common Future* (WCED 1987), SD emerged as a publicly recognized and well-defined concept as "meeting the needs of the present without compromising the ability of future generations to meet their own needs." This report prompted numerous actions, which called on governments, local authorities, businesses, and consumers to define and adopt strategies for SD.

SD is a complex concept, normative, subjective, entailing inter- and intrageneration aspects, and can neither be unequivocally described nor simply applied. Pfaff and Stavins (1999) define SD as a process of change in which the exploitation of resources, the direction of investments, the orientation of technical development, and institutional change are all in harmony and enhance both current and future potential to meet human needs and aspirations.

The IISD (2012) presented the following definition of SD for the business community in 2002. For business, SD means adopting business strategies and activities that meet the needs of the enterprise and its stakeholders today, while protecting, sustaining, and enhancing the human and natural resources that will be needed in the future.

One of the prominent characteristics of the term SD is that it means different things to so different people and organizations. The literature is widespread with different efforts to define the term (Ness et al. 2007), and debates have erupted between those who prefer the three-pillar approach (Elkington 1997) and those who prefer a more organic vision focusing more on interrelationships between the economic and environmental dimensions. However, Robinson (2004) argues that it makes sense for definitions, perhaps many of them, to emerge from attempts at implementing SD, rather than having definitional rigor imposed from the outset, so this lack of definitional precision is not a serious problem.

2.5.1.3 *Sustainability science*

The dynamic evolution and the complexity of the challenges posed by sustainability are hardly manageable in the context of classical disciplines and science (Hasna 2010; Bettencourt and Kaur 2011). Therefore, sustainability science begun as a revolutionary concept that, in the Kuhnian sense (Kuhn 1970), is aimed at providing a response to the crisis of present normal sciences, enabling science to contribute more effectively to SD through a holistic approach, able to capitalize, and integrate sectoral knowledge toward the definition of new solutions. SS has emerged as a new discipline, aimed to provide a response to the crisis of present normal sciences, enabling science to contribute more effectively to SD through a holistic approach. This area has become a scientific possibility for transcending reductionist analyses of classical sciences, by means of systemic comprehension of contemporary phenomena within the environmental, ecological, economic, social, and political domains. SS is a discipline that aims at exploring the dynamic interactions between human activities on the earth's life support systems, and between nature and society, to design a path toward SD (Sala et al. 2012).

2.5.1.4 *Sustainability assessment*

Sustainability assessment (SA) is a tool that can help decision-makers and policymakers decide what actions they should take and should not take in an attempt to make society more sustainable (Devuyst et al. 2001). The aim of SA is to ensure that plans and activities make an optimal contribution to SD (Verheem 2002).

SA is one of the most complex types of appraisal methodologies. This entails not only multidisciplinary aspects (environmental, economic, and social) but also cultural and value-based elements. It is usually conducted for supporting decision-making and policy development in a broad context. It is increasingly becoming common practice in product, policy, and institutional appraisals. Concepts such as integrated assessment and SA are introduced to offer new perspectives to impact assessment geared toward planning and decision-making on SD (Hacking and Guthrie 2008).

SA is a process that directs decision-making toward sustainability. It is being increasingly viewed as an important tool that can help decision-makers and policymakers to aid in the shift toward sustainability by which the implications of an initiative or policy are evaluated. This involves tools for informed decision-making and systematic steps (problem definition, policy options, and mitigation efforts) to assess the effects of decisions before they are taken in order to make society more sustainable. SA should assess not only whether an initiative is sustainable and but also assess the direction or target.

2.5.2 Sustainability models

There have been many ways of representing SD in a model that depicts this extremely complex concept and a new way of thinking. This section is an attempt to briefly capture the two popular of these models. Both models clearly emphasize the need for interdisciplinary and transdisciplinary approaches to understanding sustainability. However, the transdisciplinary approach emerges as the most appropriate, which is different from interdisciplinary where it is characterized by unintegrated application of more than one disciplinary methodology to analyze a topic from different perspectives.

2.5.2.1 The TBL

The TBL is to date the most popular framework model used to base the analysis of sustainability on. Developed by John Elkington in 1994, and later expanded in his book *Cannibals with Forks: The Triple Bottom Line of 21st Century Business* (Elkington 1997), the TBL method has been used for a number of years to categorize the different types of sustainability. The three key pillars of Elkington's TBL sustainability are economic (profit), environmental (planet), and social (people). The three-pillar concept is illustrated in Figure 2.3, where SD is supported by even pillars of economic, environmental, and social development. Historically, the values associated with each sustainability pillar were evaluated as capitals: natural, social, and economic.

2.5.2.2 The egg of sustainability

The "egg of sustainability" model is an alternative to the TBL model and was designed in 1994 by the International Union for the Conservation of Nature (Guijt and Moiseev 2001). An illustration of the "egg" concept is shown in Figure 2.4. It illustrates the relationship between people and

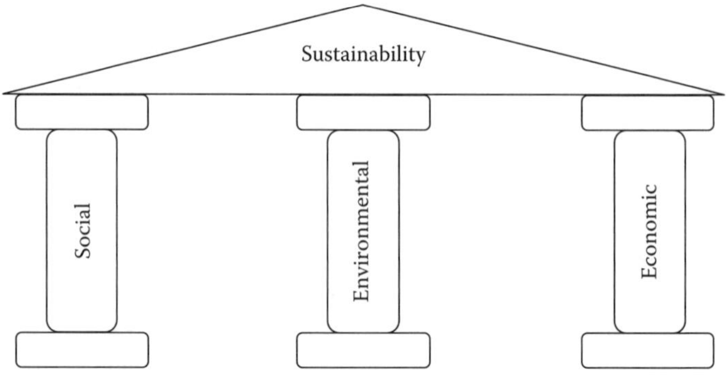

Figure 2.3 The three pillars of sustainability.

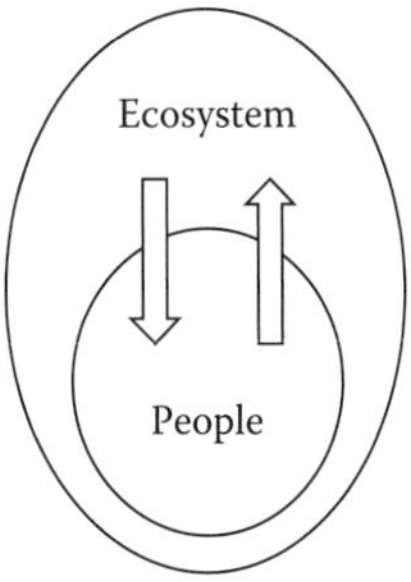

Figure 2.4 Egg of sustainability model illustrating the relationship between people and the ecosystem.

ecosystem as one circle inside another, like the yolk of an egg. Just as an egg is good only if both the white and yolk are good, so a society is well and sustainable only if both, people and the ecosystem, are well. This implies that people are within the ecosystem, and that ultimately one is entirely dependent upon the other. Social and economic development can only take place if the environment offers the necessary resources: raw materials, space for new production sites and jobs, and constitutional qualities (recreation, health, etc.). According to this model,

$$\text{SD} = \text{Human wellbeing} + \text{Ecosystem wellbeing} \quad (2.1)$$

2.5.3 *Interactive zone for sustainability*

Although there is still much confusion and conflict surrounding a precise meaning of SD, many agree that SD is about satisfying social, environmental, and economic goals. In 1994, Holmberg suggests that the ecological, economic, and social systems are independent and may be treated independently (reductionist). The interactive zone where the three different systems interact is the solution area of integration where sustainability is achieved, whereas the area outside the interactive zone is assumed to be an area of contradiction (Bivalent) (Mebrato 1998).

Figure 2.5 shows how driving forces in the interactive zone for sustainable solutions are visualized. In this zone, the science, technology, innovation, policy dimension is central to the three main pillars of sustainability: economy, society, and ecology.

Economy-centric concerns represent the ability of the planet to sustain people, by providing both material and energy resources. To economists, access to resources is a form of capital or wealth that ranges from stocks of raw materials to finished products and factories (Maxwell 2014).

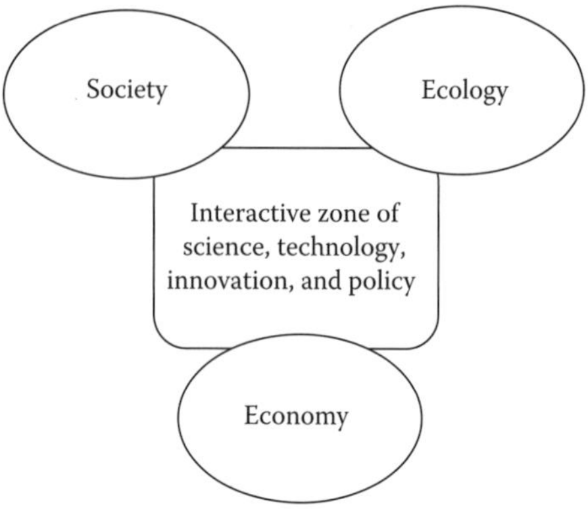

***Figure* 2.5** Interactive zone driving forces for sustainable solutions.

The sociocentric concerns represent social and moral capital. Social capital is investments and services that make the basic framework for society. It is measured by the number of connections among people and social groups (networking). On the other hand, moral capital is made by people choice of moral actions which are beneficial to society. Increasing moral capital causes an internal sense of wellbeing.

The ecology-centric concerns seek to improve human welfare by protecting natural capital, that is, protecting the sources of raw materials used for human needs, and ensuring that sink capacities for recycling human wastes are not exceeded (Maxwell 2014).

The technocentric concerns represent human skills and ingenuity—the skills that engineers must continue to deploy—and the economic system within which we deploy them. While considering SD, interconnectivity must be addressed, as it is decisions taken during projects which have an effect on each pillar. It is not possible to blindly chase the ideal environmental solution, while ignoring the economical aspect (Ugwu and Haupt 2007). This concept of substitution of one type of sustainability capital for another is known as either strong sustainability (where no substitution is allowed) or weak sustainability (where some substitution is allowed).

Although Figure 2.5 is simplistic, it is a reminder that sustainability entails living within all three types of long-term constraint: technology cannot be deployed as though it has no environmental or societal implications. Engineers must therefore be key players in SD and have an obligation as citizens not just to act as isolated technical experts.

In 1994, a study group of the World Bank developed the so-called Capital Stock Model with the basic idea being: If we live only off the interest and not the capital, the basis of prosperity is maintained; however, if we consume the substance, our means of existence is endangered in the long term. The definition of ecological capital for the planning process includes biodiversity, landscape, mineral resources, clean air, and healthy water. Human and social capital equates to health, social security, social cohesion, freedom, justice, equality of opportunity, and peace. The equation is simple and shown in Figure 2.6.

In 2002, nine international banks and the International Finance Corporation (an arm of the World Bank) agreed to voluntarily develop a banking industry framework to address environmental and social risk in project financing that could be applied globally across all industry sectors. It was called the Equator Principles, and the current version applies to all project investment in excess of US$10 million from the 77 member financial institutions. In the developing world, almost all international project finance is affected by these rules, which impose developed country standards wherever in the world the project is located (ACEC 2014).

The phenomenon of globalization has instigated in scientists, educators, and some politicians the search for SD with social promotion of individuals and society. It is the key for the survival of humankind on earth. It is not only a matter of environmental issues that need to be solved but also the social aspects of the mutant world that contemporary society is living in the twenty-first century. It is the application of science to help society to reach the goal of achieving the same level of development as the technological (Ciampi 2011).

Finally, achieving sustainability through SD will require some significant shifts in behavior and patterns in consumption. Often it will be, and should be, engineers who lead processes of making decisions about the use of material, energy and water resources, the development of infrastructure, the design of new products, and so on. One implication is that engineers must recognize and exercise their responsibility to society as a whole, which may sometimes conflict with their responsibility to the immediate client or customer (Dodds and Venables 2005).

Capital Stock of Sustainable Development (CSSD)
=
Capital Stock of the Environment (CSE)
+
Capital Stock of the Economy (CSE)
+
Capital Stock of the Society (SCS)

Figure 2.6 Capital stock of SD.

2.5.4 Sustainability indicators

An indicator is a tool that assists us to measure and understand the condition and progress of a system. Indicators perform many functions. They can lead to improved decisions and effective actions by simplifying, clarifying, and making collected information available to policymakers. A good indicator shows quantified information that leads to direction and alerts to a problem before it gets bad and helps recognize what needs to be done to fix the problem. Indicators of sustainability are different from traditional indicators which are independent of each other such as stockholder profits, asthma rates, and water quality which. In sustainability, indicators point to areas where the links between the economy, the environment, and the society are interconnected. They can provide an early warning to prevent economic, social, and environmental problems. They can help to identify and prioritize sustainability activities. Some examples of indicators include report cards, wind speed and direction, credit card debt, blood pressure, and gas gauge in cars.

The most commonly used technique to analyze the sustainability of a business operation is to use an appropriate set of numerical indicators. Indicator is a parameter, or a value derived from a parameter, which points to, provides information about, or describes the state of phenomenon, with a significance extending beyond that directly associated with its direct value (OECD 2003). The parameter could be quantitative, semi-quantitative, or qualitative derived from a model, often through a tool (Sala et al. 2012).

Indicators act as a guide to the direction of travel, which means the choice of which indicators to use is critical in monitoring and directing progress toward sustainability (Bell and Morse 2008; Singh et al. 2008; Tahir and Darton 2010). Sustainability indicators are increasingly recognized as a useful tool for policymaking and public communication in conveying information on countries and corporate performance in fields such as environment, economy, society, or technological improvement. By visualizing phenomena and highlighting trends, sustainability indicators simplify, quantify, analyze, and communicate otherwise complex and complicated information (Maxwell 2014).

Effective indicators should be relevant, easy to understand, reliable, and based on accessible data. While developing a framework and selecting SD indicators, Spohn (2004) identifies the two main distinctive approaches:

- The top-down approach, which enables experts and researchers to define the overall structure for achieving the sustainability and subsequently is broken down into set of indicators

- The bottom-up approach, which requires systematic participation of various stakeholders to understand the framework as well as the key SD indicators

2.5.5 Circular economy

The CE concept is a response to the aspiration for sustainable growth in the context of the growing pressure of production and consumption on the world's resources and environment. Until now, the economy has mainly operated on a "take-make-dispose" model—a linear model where every product is bound to reach its "end of life" (EC 2015). The CE is best understood by looking into natural, living systems that function optimally because each of their components fits into the whole. Products are intentionally designed to fit into material cycles, and as a result, materials flow in a way that keeps the value added for as long as possible and residual waste is close to zero.

The starting point for the ideas on CE has been to change the linear economic system of "take-make-waste" in order to lower resource use and waste of natural capital. It builds on the notion of cycles in nature fueled by solar energy, where nothing is wasted but just goes around in loops (Berndtsson 2015). Figure 2.7 illustrates the difference between a linear economy and a CE.

Each product produced in a CE should be designed so that the biological and technical components could be easily separated and recirculated in the system in accordance with cradle-to-cradle principles and focus on effectiveness rather than efficiency. It also builds on the ideas of performance economy with new business models that focus on selling services instead of products to lower the resource use (Wijkman and Rockström 2012).

CE emphasizes the importance of closing the loops of material flows. This regards both extracted materials and substances produced by society. By circulating nutrients in biological and technological cycles, keeping the materials separate, the need for virgin materials could be minimized

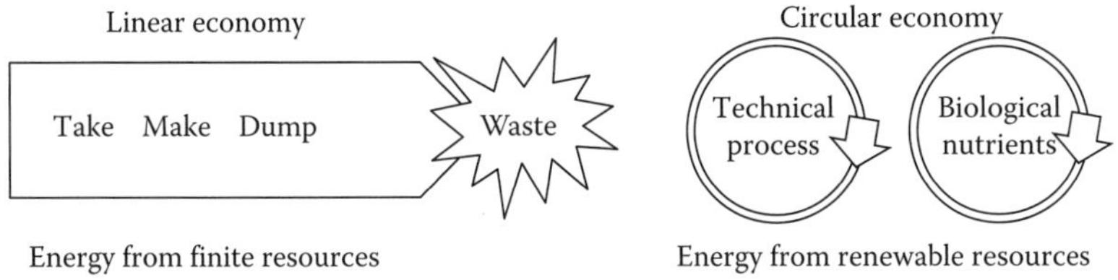

Figure 2.7 Linear economy and CE.

and thus implying less extraction, resource depletion, and degradation of nature. CE also advocates that the materials cycled should be nontoxic to humans and other life forms.

The CE principle of using solar energy is also a clear injunction not to rely on fossil fuel. The transition to a renewable energy system does however imply an increasing need for resources in producing, for instance, solar panels and wind mills. This transition can thus, in short term, be contradicting to the sustainability conditions (Berndtsson 2015).

CE and its underlying theories put a focus on environmental and economic sustainability and do not have any clear principles addressing social sustainability. Wijkman and Skånberg (2015) showed that CE can, at least in the short term, have an accelerating force on job creation. This is one part that could help people to meet their needs.

2.5.6 *Sustainability planning*

Sustainability planning is the process by which involved stakeholders and partners create a road map for decision-making in regard to what to sustain, why, and how. Creating a sustainability plan strengthens buy-in and stakeholders' understanding of the efforts needed to keep the work operating and improving. The integration of people, place, and economy into a single plan over a long-term perspective is a critical process for achieving sustainable community development.

In order for sustainability to become a reality, coalitions need support from key decision-makers as well as community volunteers; sufficient leadership, funding, and channels of communications; and procedures in place to monitor policy results through enforcement and compliance, and to modify strategies accordingly.

A sustainability plan can be used to share the progress of the early learning work with potential funders and partners. It can also be used as a guide to support ongoing management of the work. Communication and stakeholder engagement are critical to successful sustainability planning, including the development and execution of the sustainability plan. The SD stages are shown in Figure 2.8.

The initial stages of plan development are not to start creating plans or policies but to identify the processes and critical stakeholders that will inform the process. This may consist of a preplanning group that is representative of the sectors, key stakeholders, and government departments that need to be involved. This preplanning stage is about identifying the people that need to be bought together, and creating an atmosphere of inclusivity and institutional support. It is not about presupposing what the vision and goals of the community are going to be (Ling et al. 2007).

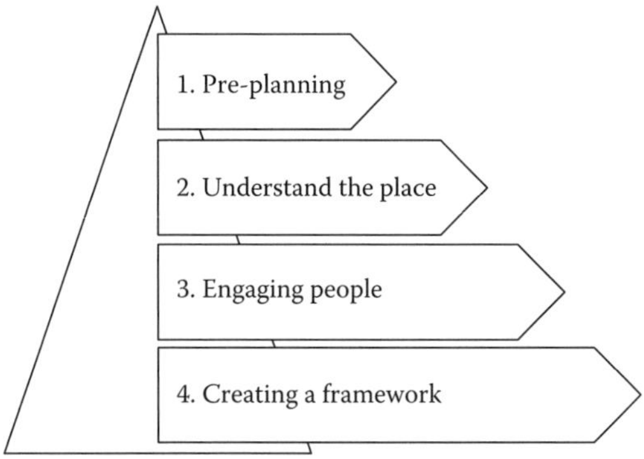

Figure 2.8 Stages of sustainability planning.

The feasibility of proceeding with the project is investigated, and on the acceptance of the proposal, it moves to the next design phase. The criterion for including the indicators in this phase of the project is based on the data defined and fixed in feasibility stage once a case/idea has been generated. A comprehensive review of the ecological, social, and economic capital baseline provides a baseline against which to measure progress, and provides invaluable information to the community. Communities, regardless of scale, are combination of interest, values, and sectors. A planned process will identify and engage key stakeholders from business, community organizations, conservation groups, developers, and government agencies within the community. The plan should reflect the desired nature of the community. This integrated vision reflects the values the community places on things such as diversity, self-sufficiency, accessibility to services, livability, the nature of its development, and the nature of its development.

2.6 Sustainability approaches in engineering

In 500 BC, Chinese Tao patriarch Kuan Tzu was attributed with the following quote: "If you are thinking a year ahead, sow a seed. If you are thinking 10 years ahead, plant a tree. If you are thinking 100 years ahead, educate the people."

2.6.1 Typical and sustainable engineering

Engineering is the application of scientific and mathematical principles for practical purposes such as the design, manufacture, and operation of products and processes, while accounting for constraints invoked by economics, the environment, and other sociological factors. Many technical advances are brought about through engineering. Engineering activities are significant contributors to economic development, standards of living, and well-being of a society, and impact its cultural development and environment. Engineering is continually evolving as a profession (NAE 2004), and engineering education is correspondingly continually changing (Rosen 2012).

Engineering and technology makes the interactive zone that integrates ecology, economy, and society. Engineering can offer varieties of solutions to issues related to the above three aspects of sustainability. Engineers can also play an integral role in not only improving the day-to-day life of people but also putting in place schemes which will continue to enhance the quality of life. Through careful engineering design and manufacturing, the impact on the environment can be managed, resulting in a neutral or positive effect on the consumption that technology uses.

The performance of engineering tasks has been traditionally judged on the basis of suitability for their intended purpose, minimization of cost, and delivery to a fixed schedule. This performance model has been extended over the past 40 years to include considerations of environmental safety. As SD moves into the mainstream of political and business thought, there are signs of a further fundamental shift in the way engineering performance is judged, bringing broad resource and ecological and social issues into the mainstream of engineering design (Gagnon et al. 2008).

In some ways, the concept of engineering sustainability is simply the application of the general definitions of sustainability to engineering. In other ways, engineering sustainability is more complex and involved. Engineering sustainability is taken to involve the sustainable application of engineering in systems. Such systems include processes and technologies for harvesting resources, converting them to useful forms, transportation and storage, and the utilization of engineering products and processes to provide useful services such as operating computers, providing health care, or sheltering people. Therefore, engineering sustainability goes beyond the search for sustainable resources, and implies sustainable engineering systems, for example, systems that use sustainable resources, and that process, store, transport, and utilize those resources sustainably (Rosen 2012). Figure 2.9 shows some of the aspects that differentiate the typical and sustainable approaches in engineering.

2.6.2 Requirements for engineering sustainability

There are several distinct components to the manner in which engineering can be practised sustainably in society, each of which is a requirement for engineering sustainability (Rosen 2012):

- Sustainable resources
- Sustainable processes
- Increased efficiency
- Reduced environmental impact
- Other aspects of sustainability

Most engineering activities utilize resources that are derived from nature. Such resources include water, materials, and energy. The degree to which resources are sustainable depends on many factors, including their scarcity and importance to ecosystems.

Resources are used in engineering processes and operations to yield products and/or services. An important requirement of sustainable engineering is the use of sustainable processes. This implies that the engineering processes utilized must exhibit sustainable characteristics in terms of the operations and steps they involve, and the energy and materials they utilize (Rosen and Kishawy 2012).

High efficiency allows the greatest benefits, in terms of products or services, to be attained from resources, and thus aid efforts to achieve engineering sustainability. Efficiency improvements taken broadly efforts include direct measures to increase the efficiency of processes, devices, and systems.

Numerous environmental impacts associated with engineering processes are of concern and must be addressed in efforts to attain engineering sustainability. These include impacts to the atmosphere, the lithosphere and the hydrosphere, and can be exhibited in many forms.

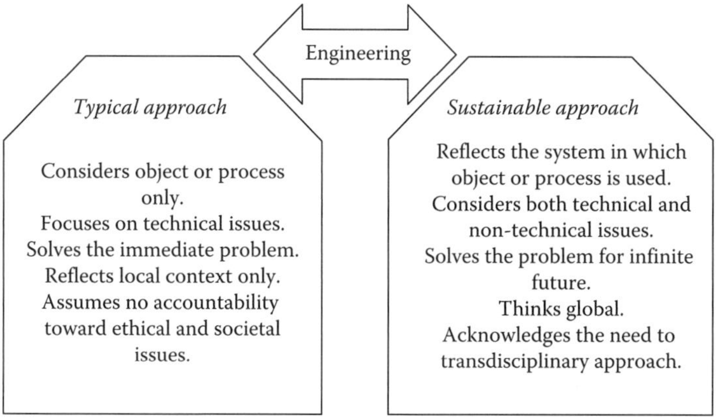

Figure 2.9 Aspects that differentiate traditional and sustainable approaches in engineering.

Many other sustainability factors relate to engineering processes, consequently, need to be considered in the quest for engineering sustainability.

2.6.3 Role of engineers in SD

Technology is the foundation to SD, and engineers provide the interface between its determinants, including economy, society, and environment. To become sustainable, engineer must act as leaders who recognize the world's needs and act accordingly. Globalization offers important opportunities for engineers to promote change through sharing experience and good practice.

Engineers work to help communities access to better services such as water, cleaner energy, health care, transportation, and communication technologies. Engineers can accomplish all that with the minimal use of natural resources and by paying attention to environment and social values. This requires engineers to make determined efforts in discovering all of the relevant facts concerning the design, development, deployment, and every possible outcome of the choices available to them that may positively and negatively affect society and the citizenry. Society is fully dependent on its designed products and services that should be robust, safe, reliable, and economically and environmentally feasible and sustainable.

Engineers have a responsibility to maximize the value of their activity in order to build a sustainable world. This requires the realization of society's needs, the achievement of goals, and the recognition of changes over time. They should understand the potential role for engineering in terms of environmental limits and finite resources; reduce the demand of resources by using less in the first place; reduce waste production by being effective with the resources that are used; use systems and products that reduce embedded carbon, energy and water use, waste, and pollution; adopt full life cycle assessment (LCA) as normal practice including the supply chain; adopt strategies such as reuse, recycling, decommissioning, and disposal of components and materials; minimize any adverse impacts on sustainability at the design stage; harness their skills to minimize the damage to people or the environment from engineering processes and products; undertake a comprehensive risk assessment and finally, undertake a comprehensive risk assessment beyond the life span of an engineered product (Masud et al. 2011).

2.7 Pathways to sustainability

> Sustainable development is the pathway to the future we want for all. It offers a framework to

> generate economic growth, achieve social justice, exercise environmental stewardship and strengthen governance.
>
> **Ban Ki-moon**

2.7.1 Energy and resource efficiency

The current energy system is mainly based on fossil fuels. This trend is unsustainable for a number of reasons: threats of man-made climate change by greenhouse gas (GHG) emissions, the rapid depletion of fossil fuels, rising energy prices due to increasing demand, geopolitical uncertainty, and threat of instability in oil-rich countries. Solutions will be found in massive energy efficiency; development of renewable energy based on sun, wind, biomass, and tides; and improvements in energy storage technologies, such as batteries and flywheels. Carbon capture and storage is not yet proven feasible but could help to mitigate increasing CO_2 emissions (Stephens and Zwaan 2005). A shift from fossil fuels to renewable energy in the energy supply can contribute to achieving ambitious emissions reduction targets, together with significant improvements in energy efficiency. To reduce emissions to a level that would keep the concentration of GHGs at 450 ppm in 2050, the IEA projects that renewable energy would need to account for 27% of the required CO_2 reductions, while the remaining part would result primarily from energy efficiency and alternative mitigation options such as carbon capture and sequestration (IEA 2010).

Greening the energy sector will also require improvements in energy efficiency and a much greater supply of energy services from nuclear and renewable sources, both of which will lead to reducing GHGs and other types of pollution. In most instances, improvement in energy efficiency has net economic benefits. Global energy demand is still likely to grow in order to meet development needs, in the context of growing populations and income levels (UNEP 2011).

In addition to high indirect costs associated with pollution arising from the combustion of fossil and traditional fuels, the use of fossil and traditional energy sources in both developed and developing countries also impacts global biodiversity and ecosystems through deforestation, decreased water quality and availability, acidification of water bodies, and increased introduction of hazardous substances into the biosphere (UNEP 2010). These impacts also reduce the natural capabilities of the planet to respond to climate change.

Renewable energy technologies are not without negative impacts; careful planning to address possible environmental and social impacts is essential. Production of biofuels, for example, can have negative effects on biodiversity and ecosystems, while the environmental and social impacts of large-scale hydroelectric power plant can be significant. However, expanding access to energy is a central challenge for developing countries. Reliable and modern energy services are needed to facilitate poverty reduction, education, and health improvements, as reflected in a number of studies (Modi et al. 2006; GNESD 2007, 2010) identifying access to energy services as crucial for the achievement of most of the SDGs.

Sustainability depends on the evolution of energy technologies. In the short term, dependence on fossil fuels is unavoidable. There are various technological options available to address the energy poverty challenge described previously. In addition to the impact of engineering on providing proper solutions, implementing most of these options requires additional, publicly financed investment, including development assistance, since the commercial market potential is likely to remain limited in some cases.

In terms of technologies for electricity delivery, there are potentially three broad options for expanding access: (1) existing centralized grids can be expanded to nonserved areas, potentially based on new renewable sources of energy; (2) decentralized minigrids can be installed to link a community to a small generating plant; and (3) off-grid access can be facilitated by producing electricity for a single point of demand. The optimal mix of these options for any given country is determined by the availability of energy resources, the regulatory and policy environment, the institutional and technical capacity, geographic considerations, and relative costs (AGECC 2010). Proper planning should allow for flexibility to integrate these systems as countries develop.

Increasing investment in renewable energy, as part of a green economy strategy spanning all major sectors, can contribute to reducing health and environmental impacts from energy production and use, while ensuring the basis for long-term economic growth. Such a strategy is based on the substitution of fossil fuel energy with renewable energy, savings from energy efficiency in manufacturing, buildings and construction, transport, and behavioral change. Such an integrated strategy can increase national energy security and reduce carbon emissions while providing new employment opportunities that may, in global terms, more than compensate for jobs that disappear.

2.7.2 Transport

Transport has a major sustainability dimension and is a main driver for SD. There are a number of SDGs that are directly linked to transport. Six of those targets directly involve transport, and attaining at least another six will significantly depend on it, for example, SDG 3 on road safety and air pollution, SDG 7 on energy, SDG 8 on decent work and economic growth, SDG 9 on resilient infrastructure, SDG 11 on sustainable cities, SDG 12 on sustainable consumption and production, and SDG 14 on oceans, seas, and marine resources. In addition, sustainable transport will enable the implementation of nearly all the SDGs through interlinkage effects including SDG 2 on agricultural productivity and SDG 6 on safe drinking water.

Transport is a huge consumer of energy and it has obvious environmental impacts, yet modern lifestyles depend on innovative transport systems. Therefore, enhanced energy efficiency in transportation systems is of central importance throughout all SDGs. Underway worldwide is the development of improved fuel-efficient transportation systems, integrated urban mass transit planning, energy storage and propulsion including fuel cell and hydrogen systems.

Transport-related pollution, greenhouse effect, noise, and vibration can pose serious threats to human health and well-being. Local air pollution is caused by exhaust emissions produced by traffic, mostly in the form of sulfur oxides (SO_x), nitrogen oxides (NO_x), carbon monoxide (CO), hydrocarbon, volatile organic compounds, toxic metals (TM), and lead particles including black carbon. These emissions represent a large proportion of pollutants, especially in developing cities (UNEP 2011).

A fundamental shift in investment and strategy patterns is needed, based on the principles of avoiding or reducing trips through the integration of land use and transportation planning; localized production and consumption; and shifting to more environmentally efficient modes such as public transport and nonmotorized transport and to rail and water transport; as well as adopting green transport technologies by improving fuels and vehicles through the introduction of cleaner, more efficient fuels and vehicles.

In the near future, cities will witness the beginning of a dramatic shift from petroleum-fueled internal combustion cars toward electric cars. The health and environmental benefits of a shift to electric vehicles (EVs) could be wide ranging.

2.7.3 Water

The water (SDG 6) and the many water-related targets in the other goals well reflect the complexity of water challenges: from access to safe drinking water and adequate sanitation, water quality, water efficiency and sustainability of water use, water governance, protection of water-related ecosystems, water-related disasters, health impacts from water-borne disease and water pollution, and capacity building and stakeholder involvement (Bach 2015).

Water is at the core of SD. It is an integral part of a society's development and a support for human health and dignity, livelihoods, and poverty reduction. Water resources, and the range of services they provide, reinforce sustainability. Many of the potential gains will be achieved simply by deciding to invest in the provision of water and sanitation services.

The use of technologies that encourage efficient forms of water recycling, and reuse should be encouraged. For water-intensive industries, minimizing consumption will become a necessity, and it will be a key factor in determining the market compatibility of industrial products. For the agricultural sector, engineers should investigate new technologies for irrigation that will be needed to minimize water consumption and prevent unsustainable groundwater extraction.

Water should not be treated as an independent sector, but as a cross-linked issue given its crosscutting nature and essential roles in social and economic development and environmental integrity. For example, the interdependence of water and energy demands requires careful attention as arrangements are put in place for a transition to a green economy. There are at least two dimensions to this relationship: (1) Water plays an important role in energy generation, notably as a coolant in power stations, and (2) the water supply and sanitation sector is a large consumer of energy. Relative to its value, water is heavy, and in terms of energy both expensive to pump over long distances and expensive to lift. In developed countries, the relatively high energy costs of pumping and treating water for household, industrial, or mining purposes are broadly accepted. In developing countries, great care must be taken to ensure that water treatment and distribution systems remain affordable (UNEP 2011).

2.7.4 *Agriculture and food*

Agriculture is characterized by farming practices that rely on the use of external farming inputs. Most large-scale industrial farming is considered energy intensive (using 10 calories of energy for every calorie of food produced), whose high productivity relies on the extensive use of chemical fertilizers, herbicides, pesticides, fuel, water, and continuous new investment (e.g., in advanced seed varieties and machinery).

In addition to environmental and socioeconomic aspects, sustainable agriculture is about the efficient production of safe and high-quality products in a way that protects and improves the environment, the conditions of farmers and local communities, and maintains the health and welfare of all farmed species.

Green agriculture could nutritiously feed the global population up to 2050, if worldwide transition efforts are immediately initiated and this transition is carefully managed. This transformation should particularly focus on improving farm productivity of smallholder and family farms in regions where increasing population and food insecurity conditions are most severe. Rural job creation would accompany a green agriculture transition, as organic and other environmentally sustainable farming often generate more returns on labor than on conventional agriculture (UNEP 2011).

2.7.5 *Infrastructure*

Investment in public infrastructure and innovation are crucial drivers of economic growth and development. The infrastructure is essential to the efficient functioning of society and its ability to achieve SD. These include water resource and supply systems, power systems, bridges, roads, as well as communications and transportation facilities. To a large extent, their technologies are well developed. The essential challenge lies in the diffusion and use of such technologies to developing nations, where they are most needed (NAP 1995).

To address the SD agenda, SDG 9 calls for increasing investment in resilient infrastructure but, implicitly, infrastructure development will also play an important role in many other SDGs. It is recognized that growth in productivity and incomes, and improvements in health and education services require substantial investment in infrastructure. For example, energy-related infrastructure and an expansion of the electricity's grid are necessary to provide energy access to urban and rural areas. Transportation infrastructure such as roads, railways, ports, and airports is a key for people's mobility from home to work, and for connecting rural areas to domestic and regional markets, contributing to a country's economic development. Sustainable water infrastructure will improve people's lives by providing access to water and help managing scarce resources in a sustainable manner.

There are various ways for governments to deliver public infrastructure. They range from traditional procurement methods to a range of public–private partnerships (PPPs or P3s). In a PPP model, the different risks and responsibilities are distributed among the public and private partners. This partnership has the potential of delivering more efficient and effective infrastructure, and ultimately value for money for taxpayers (Casier 2015). The future of sustainable infrastructure will increasingly blur boundaries between energy, transportation, water, and waste systems to implement complementary strategies that benefit more than one system. Saving energy saves water, as does switching from fossil fuels to renewable energy sources.

2.7.6 Materials

SDG 12 calls to ensure sustainable consumption and production patterns, the fact that requires minimizing the natural resources and toxic materials used, and the waste and pollutants generated, throughout the entire production and consumption process: (Dodds and Venables 2005). Sustainable materials are materials and products that are natural (not petroleum based), are rapidly renewable or contain largely reclaimed components, have technical attributes that give them valuable properties beyond the aesthetics they provide, and are products that do not off-gas toxic components, which are potentially detrimental to either indoor air quality or personal health.

Sustainable materials are evaluated by the impact they have on the environment and on occupants over life of the material and generally incorporate characteristics that contribute to resources efficiency, indoor air quality, energy efficiency, water conservation, and affordability.

The materials revolution that is now underway has profound implications for the environment. Traditional materials, such as steel, concrete, and plastic, are undergoing significant changes that reduce the environmental impact of their manufacture and use (NAP 1995).

2.7.7 Production and manufacturing

As indicated on SDG 12, the transition to sustainable consumption and production of goods and services is important to decrease the negative impact on the climate and the environment, and on people's health. This involves using resources efficiently, taking account of ecosystem services that are keys to making a living, and reducing the effect of hazardous chemicals.

Production and manufacturing have a key role to play in SD because of their significant share in economic development activities. Contrary to common myths that addressing sustainability issues may have a curbing effect on economic development, findings suggest that companies proactively adopting sustainable manufacturing practices grow more innovative and competitive, resulting in substantial economic benefits among others (Rusinko 2007).

Early works in sustainable manufacturing focused on environmentally conscious manufacturing processes as well as on topics including energy conservation, design for environment, and remanufacturing and recycling. However, the scope of sustainable manufacturing has been enlarged to include concepts and technologies such as lean manufacturing, waste avoidance, innovative manufacturing processes, and green supply chains (Despeisse et al. 2012).

2.7.8 Information technology

The broad-based technological revolution can be both an obstacle in the path to a sustainable future and the bearer of the promise for a new world. New technologies also offer tremendous opportunities to deliver public services, including health care, education, and basic infrastructure, to more people at a much lower cost and with a much lower use of primary resources. Technology now happening is made possible by information technology (IT), which has the potential to alter how and where people work and live, and thus the nature of urban areas of the future. It is changing the way enterprises are operating and managed. It is improving the efficiency of air-, land-, and water-based transportation systems among other sectors of the economy. By considerably reducing the cost, IT services have had tremendous economic impact.

The concept of appropriate technology (AT) is well established in the history of SD, emerging robustly at about the same time as the concept of SD was itself taking shape. Indeed, it can be said that the original arguments for an AT were the arguments for an SD.

From transportation to nuclear power plants, from chemical processing to mineral extraction, ITs allow accurate control of industrial processes, improving the ability to minimize pollution and improve energy efficiency. Energy can also be saved through cloud computing, namely, the principle of outsourcing the programs and functions of personal computers to service providers over the Internet. This also means sharing storage capacity with others.

The connections between IT and development, positive as well as negative, need to be better understood to determine which issues can be effectively addressed using IT, how IT tools can be used, and in what context should the technology be applied. Many of the good ideas connected with green IT and sustainability, for example, involve the combined use of purely virtual space. Digital sharing would be another way of putting it. Even more can be done in this area: energy can be saved by using a laptop instead of a large desktop computer at workplaces or in home offices. This is because laptop components are energy optimized, mainly to ensure that the battery lasts longer (Schäfer 2013).

Another aspect of green IT is the material used for electronic devices, the manufacture of which requires several TMs. A great deal of water is also used during production. The reserves of some of the elements, such as the rare earth elements (a group of metals), are being depleted globally. Recycling is therefore of increasing importance. These issues are being followed by the climate protection organization Germanwatch through an awareness-raising campaign called "makeITfair." Diverse publications on green IT, recycling, and sustainable electronics can be found at germanwatch.org/en.

2.8 Sustainable cities

> Adding lanes to solve traffic congestion is like loosening your belt to solve obesity.
>
> **Glen Hemistra**

2.8.1 Urban transformation

> A smart city has to be a learning city.
>
> **Josep Pique**
> *Barcelona City Council*

Cities and towns have become the primary human living space. They can provide a wide range of socioeconomic benefits to people and communities. By concentrating people, investment, and resources (a process known as agglomeration), cities can reinforce the potential for economic development, innovation, and social interaction. However, these cities play a significant role in consuming energy, materials, and water, and they also generate a high output of carbon oxide, waste, and sewer water when it comes to their construction and exploitation.

By 2050, the number of people living in cities will have nearly doubled, from 3.6 billion in 2011 to more than 6 billion. Yet the world's urban areas are already overcrowded and, particularly in developing countries, suffer from shortages of clean water, electricity, and other resources essential to the support of their surging populations and fragile economies (Macomber 2013).

Sometimes, in comparison with national governments, cities can be more straightforward in taking decisive action, often with more immediate results. Compared to their regionally and nationally elected counterparts, local official might also feel more directly accountable to their constituents, the urban dwellers, for their decisions (Junghans and Grimm 2016).

Urban transformation requires thorough planning that incorporates analysis of information, creating an awareness of challenges and risks, and identification of objectives. The work of engineers alone is certainly not enough to accomplish sustainability goals without other efforts that fall directly under the control or influence of the city through its regulations, bylaws, and decisions about implementing projects.

2.8.2 Sustainable and smarter cities

Cities emit more than 70% of global emissions and consume around the same proportion of the world's primary energy. At the same time, they generate 80% of the global GDP, which depends on healthy urban residents and functioning urban structures and services, many of which are increasingly vulnerable to the impacts of climate change such as heat waves, storm surges, and rising sea levels (GCEC 2016). With more than 70% of today's cities already experiencing the effects of climate change, many of them have initiated policy processes around the issue and engaged in climate action planning in a practical sense (C40 2016).

Sustainable cities are viewed as those that meet human needs for healthy and diverse habitats while preserving nonrenewable resources for future generations and staying within the limits of local, regional, and global ecosystems. On the other hand, smart cities are ones that use information and communication technologies to make the infrastructure components and services of cities. Smart infrastructure provides the foundation for all of the key themes related to a smart city, including people, mobility, economy, living, governance, and environment (UN 2016).

Currently, there is increasing recognition that urban sustainability is tied directly to the quality of life of the population, which is improved by convenient, efficient, and accessible public transportation; recreation areas; convenient shopping; and suitable educational and health services. There is also growing interest in small communities within the urban landscape where all of these services are located in close proximity to one another, where residents can live and work without having to travel long distances. Sustainability issues for such cities tend to be focused on the delivery of these communal needs.

Cities are not systems that can be easily controlled or driven toward narrowly defined objectives, but they are complex ecosystems of independent people, communities, and businesses, each pursuing its own goals using the resources available to them. Accordingly, the objective is not to make the city smarter but to create an environment within which smart ideas are likely to flourish and succeed, wherever they occur (Robinson 2014). Population density provides the opportunity for economical provision of collective solutions that lower the average impact of human activities on the planet. Services that were traditionally offered in cities through individual buildings such as water and wastewater, energy, and waste collection may be more sustainable if provided through district facilities, where population density makes that viable. Heating and cooling at a district level makes the use of nonconventional energy sources more attractive, and there is interest in diversifying water supply beyond the conventional use of potable water for all purposes, to include gray water and collective initial handling of waste water. Waste disposal services have undergone a transformation to include waste separation and material recycling, and composting including the collection and use of methane gas, produced by the decomposition process, as a fuel (ACEC 2014).

Macomber (2013) provides a framework for identifying and pursuing sustainable cities opportunities. The framework rests on three pillars: new business models that generate profits by optimizing the use of resources, financial engineering that encourages investments in efficiency, and careful selection of markets. Although any given company's approach will depend on its capabilities, objectives, and the market it is entering, the broad strategies provided here are relevant to both obvious players such as infrastructure companies and vendors of turbines, trains, and other equipment, and to companies in larger sectors such as IT, financial services, and building products.

2.8.3 Dimensions of urban sustainability

Achieving the sustainability of cities can be conceived as entailing the integration of four pillars: social development, economic development, environmental management, and urban governance. Yet, the ways in which a city is able to build sustainability will reflect its capacity to adapt, within the context of its particular history, to the policy priorities and goals defined by each pillar (UN 2013). Figure 2.10 presents the four pillars for achieving urban sustainability encompassing their balanced accomplishment.

Social sustainability refers to the fairness, inclusiveness, and cultural adequacy of an intervention to promote equal rights over the natural, physical, and economic capital that supports the livelihoods and lives of local communities. Economic sustainability is understood as the capacity and ability of a practice to be able to put local/regional resources to productive use for the long-term benefit of the

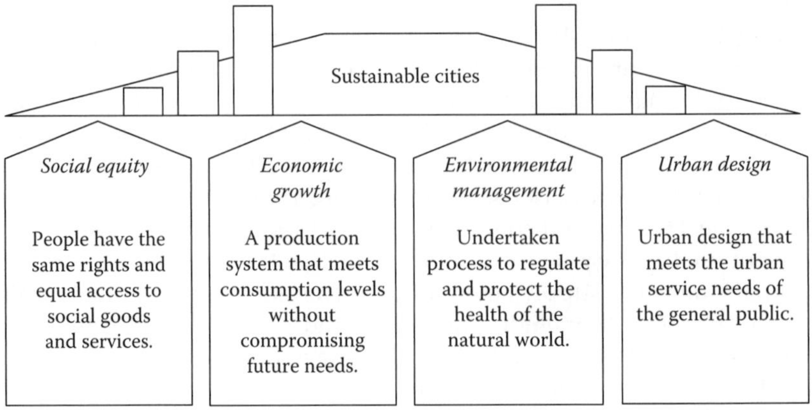

Figure 2.10 Pillars of achieving sustainability in cities.

community, without damaging or depleting the natural resource base on which it depends and without increasing the city's ecological footprint. Ecological sustainability pertains to the impact of urban production and consumption on the integrity and health of the city region and global carrying capacity. The sustainability of the built environment concerns the capacity of an intervention to enhance the livability of buildings and urban infrastructures for city dwellers without damaging or disrupting the urban region environment. Finally, political sustainability is concerned with the quality of governance systems guiding the relationship and actions of different actors among the previous four dimensions (Allen 2009).

2.8.4 Open innovation

Open innovation is the latest buzz word describing how to address the increased volume and complexity of challenges for cities and governments in general; however, what does open innovation mean? Traditionally, public services were designed and implemented by a group of public officials. Open innovation allows the design of these services to be undertaken by multiple actors, including those who stand to benefit from the services, resulting in more targeted and better tailored services, often implemented through partnership with these stakeholders. Open innovation allows cities to be more productive in providing services while addressing increased demand and higher complexity of services to be delivered. To date, a number of cities across the world have already successfully raised financial resources locally through innovative means. New York, Barcelona, Amsterdam, Copenhagen, and many other cities have been experimenting with this concept, introducing challenges for entrepreneurs to address common problems or inviting stakeholders to cocreate new services (Mulas and Barroca 2016).

As an example, the City of Calgary and the City of Vancouver in Canada both have plans to improve their sustainability over the current period to 2020. Both plans discuss diverting waste from landfills, improving air and water quality, better transit, and reduction of GHG emissions. Calgary also specifically mentions the control of storm water discharge, distinctive complete communities with mixed housing, a resilient economy, and jobs for a high-quality workforce. Vancouver is focused on communal heating, increases in population density, reduction in fossil fuel use, carbon neutral buildings, access to nature, and improvements to urban food systems involving reduced transportation.

Another example, Copenhagen is a typical western European city but differs from many others with regard to its ambitious climate change and sustainability targets. Driven by its objective of becoming carbon neutral by 2025, the city is in the process of an integral transformation, transitioning from an industrial and harbor city to a contemporary and sustainable metropolis (City of Copenhagen 2010). The city is determined to combine economic growth with the enhancement of quality of life for its population and to address climate change through various emission reduction and adaptation measures in a sustainable manner. In order to achieve the pioneering position of being the first carbon neutral city, its local government focuses on four pillars of intervention: energy consumption, energy production, green mobility, and city administration (City of Copenhagen 2014).

2.9 Energy and sustainability

> Energy is the ability to do work. It exists in different forms including motion (kinetic), heat (thermal), light (radiant), electrical, chemical, nuclear, etc. Energy sources exist as renewable (easily replenished) or nonrenewable (not easily replenished).
>
> **William A. Smith**

2.9.1 Energy impact

Energy is the ability to do work. It comes in various forms including heat (thermal), light (radiant), motion (kinetic), electrical, chemical, nuclear, and gravitational. There are two types of energy: stored (potential) energy and working (kinetic) energy. Energy sources can be categorized as renewable (can be easily replenished) or nonrenewable (cannot be easily replenished).

Energy plays a central role in every aspect of life. It is a foundation stone of the modern economy. Energy is needed for transport and mobility, to heat and cool homes, and to keep factories, farms, and offices running. The full challenge facing humanity can be related to the global demand for energy which has increased by 40% over the past 20 years, with more than 85% provided by fossil fuels. The number of people described by the World Bank as being middle class doubled from 1.5 billion to 3 billion over that period. The definition of middle class is minimal; such people live in a home with electricity and running water, having a refrigerator for food storage and access to modern communications to warn of any dangers, with no mention of car ownership or access to aviation, the province of the rich. Such people use 3.5 times as much energy as those described as poor (Mackay 2009; BP 2015). Over the next 20 years, the World Bank estimates that the middle class will rise from 3B to 5B, on the basis of which BP estimates a further increase in global energy demand of 40% still to be met in the main by fossil fuels

(Kelly 2016). So, humanity is owed a serious investigation of how the world has gone so far with the decarbonization project without a serious challenge in terms of engineering reality.

2.9.2 Net energy analysis

The origin of NEA might go back to an idea by Nobel Prize winner Sir Frederick Soddy, suggesting that energy is a more fundamental unit of account than money. This idea was not well received, but the idea of analyzing the economy in terms of energy was revived in the 1970s. There were diverse origins, from the idea that dollars and energy flow along the same paths but in opposite directions, to the study of the energy inputs to copper and aluminum production, recycling, and, most notably, nuclear power (IAEA 1994).

The 1973 energy crisis led to studies on a close appraisal of how much energy is used in the production of the various goods and services in our economic systems. In this context, since 1974, various studies on NEA of nuclear power have been carried out, which evaluated each phase of the fuel cycle, including resource extraction, equipment manufacturing, facility construction, facility operation, decommissioning, and waste management. The results of these studies have, in principle, shown that nuclear power has a positive energy balance, with a short tune for payback of the energy initially invested. However, opponents of nuclear power still claim that, in the buildup phase of a nuclear power program, nuclear power requires more energy for its construction and operation than it produces (IAEA 1994).

NEA is defined, in general, as the computation and measurement of energy flows in society, and, in particular, as the quantification of the volume of energy resources sequestered, directly and indirectly, in various commodities. Its main goal is to inform energy policy on the energy performance of various systems. NEA has been applied to a number of important sectors of the economy. NEA seeks to understand how effective a system is at exploiting primary energy sources and upgrading environmental stocks and flows into usable energy carriers (Dale 2013).

The areas which have attracted an unusual amount of interest from energy analysts are food production, transportation, and energy conversion systems. For example, when we consume anything, we consume energy. It takes energy to manufacture, deliver, and sell all types of goods and services. It is possible to add up the energy required at each step of the production process to determine the total "energy cost" of particular goods and services (Bullard et al. 1978). NEA is not equipped to say anything about the long-term sustainability of an energy technology, since the actual amounts of primary energy stocks and flows that are directly extracted, delivered, and transformed into the returned energy carriers are not included in the calculation of the energy return of investment (EROI). Also, NEA does not differentiate between renewable and nonrenewable primary energy sources (Raugei et al. 2012).

2.9.3 *Energy return on investment*

In the field of NEA, the EROI is a commonly used calculation of how much energy is needed to locate, extract, and refine an output of energy. The concept was initially derived in animal ecology and has been moved to analyze human industrial society: a cheetah must get more energy from consuming his prey than expended on catching it; otherwise, it will die. Mathematically, EROI measures the ratio of the net usable energy in a given amount of the extracted and delivered fuel to the total primary energy (e.g., the energy that is directly or indirectly required to extract, refine, and deliver the fuel). It is assumed that EROI > 5–7 is required for modern society to function. This marks the edge of the net energy cliff. Fossil fuels remain comfortably away from the cliff edge but much closer to it for every year that passes.

The ratio decreases when energy becomes scarcer and more difficult to extract or produce. In general, the higher the EROI value, the better the process. When the ratio comes to 1:1, it is no longer cost effective to pursue. As typical energy resources become scarcer, they become more difficult and expensive to obtain, and EROI will continue to approach closer to 1:1. If EROI continues to drop, there will be significant social and economic implications for the society, the fact that is acknowledged by sustainability.

To calculate the energy cost of energy, or any good or service, one must be able to quantify in energy terms the fuel, capital, materials, and labor used in the extraction and processing of the energy in question (Cleveland and Costanza 2008; Haas et al. 2012). Figure 2.11 describes EROI in terms

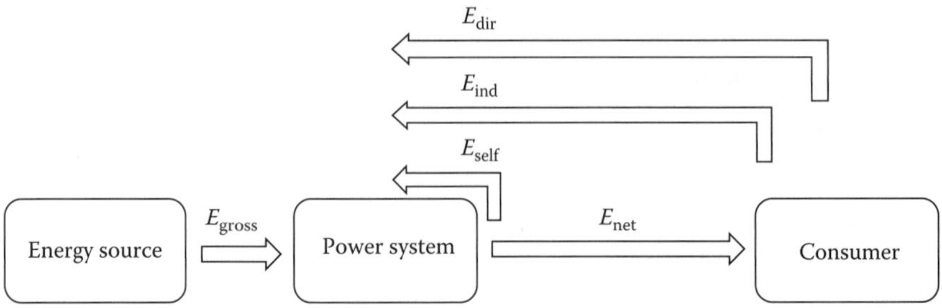

Figure 2.11 Energy return on investment.

of gross energy (E_{gross}), net usable energy (E_{net}), primary energy (E_p), self-consumption energy (E_{self}), direct consumption energy (fuel and electricity) (E_{dir}), and indirect consumption energy (material and capital) (E_{ind}) (Cleveland and Costanza 2008).

$$\text{EROI} = \frac{E_{net}}{E_p} \tag{2.2}$$

$$\text{Energy surplus} = E_{net} - E_p \tag{2.3}$$

$$E_p = E_{self} + E_{dir} + E_{ind} \tag{2.4}$$

E_{gross} is the central concept in the utilization of energy. This utilization is bound to a physical process, which transforms energy from the source to energy defined as the usable work inside a system. In the case of power plants, many technical processes are involved which are categorized as construction, decommissioning, maintenance, and fuel supply. All these technical processes are characterized by their efficiencies which determine the expended primary energy in its different forms (Ayres et al. 1998).

E_p is also called the cumulated energy demand or the embodied energy. The greatest part of LCA studies is devoted to a precise evaluation of E_p, based on material databases. It has a fixed part for construction and deconstruction, and a part that increases with time (e.g., maintenance and fuel provisioning, if required).

EROI calculations are market determined to the degree that they depend on the technology, industry structure, discount rate, and prices that exist at the time. Changes in any of those factors will alter the energy costs of goods, and therefore alter the results of NEA. It is an analytic tool for the evaluation of energy systems that seeks to compare the amount of energy delivered to economy by a technology to the total energy required to find, extract, process, deliver, and otherwise upgrade that energy to a socially useful form.

Within the field of LCA, a different set of metrics may be reported, including the cumulative energy demand, defined as the amount of primary energy consumed during the life cycle of a product or a service (Amor et al. 2010), and the energy or GHG intensity, defined as the ratio of the primary energy consumed to CO_2 emitted for the construction, operation, and decommissioning, per unit of output of electrical energy over the lifetime of the device (Lenzen and Munksgaard 2002).

Currently, ecological economists debate that NEA does not provide a theory of value, but it has several advantages over standard economic analysis including the following. (1) It assesses the change in the physical scarcity of energy resources and shows the increasing energy costs of obtaining energy; (2) because goods and services are produced from the conversion of energy into useful work, net energy is a measure of the potential to do useful work in economic systems; and (3) EROI can be used to rank alternative energy supply technologies according to their potential abilities to do useful work in the economy (Haas et al. 2012).

2.9.4 NEA in power generation

NEA in power generation has been introduced as a feasible and practical additional method for evaluating the engineering, economic, and environmental aspects of power generation systems. It compares total direct and indirect energy investment in construction and operation of power plants with their lifetime energy output. NEA in power generation is a way of evaluating the relation between input energy and output energy, where input energy is an aggregation of all necessary energies for power generation activities including different stages such as construction of a power plant, fabrication and storage of fuels, and transportation of materials, and output energy is an aggregation of energies produced during the life of the power plant. Here, the net energy requirement does not include the energy content of the original source of energy.

Power systems provide electricity to consumers. For the energy output, although the term "available" is easy to implement by defining the connection point to the network or to the consumer (Figure 2.11), the term "usable" is more complicated. It implies that the consumer has an actual need for the energy at the moment it is available. There are certain possibilities to make the energy output fit the demand: (1) by ignoring output peaks and installing multiple times of the necessary capacity as a backup to overcome weak output periods and (2) by installing storage capacities to store the peaks.

The electricity production of any sort of power plant during a period T is shown in generalized form:

$$E = P \times T \times C \tag{2.5}$$

where P is the rated power for the plant, C is the capacity factor, and the units used for E are multiples of kWh. Many analyses convert kilowatt-hours (kWh) to kilojoules (kJ), or vice versa, in which assumptions must be made about the thermal efficiency of the electricity production.

Analyzing this energy balance between inputs and outputs, however, is complex because the inputs are diverse, and it is not always clear how far back they should be taken in any analysis.

2.9.5 NEA as a policy

Most economic activities consume more energy than they produce. Consider steel production: factories consume energy to turn iron ore into useful material products. In contrast, primary energy processes must supply much more energy than they consume. For example, the oil industry historically has output tens to hundreds of times more energy than it consumes in extracting and refining oil (Hall et al. 1986; Carbajales-Dale et al. 2014) or, over its lifetime, a modern wind turbine produces about 80 times more electrical energy than consumed in manufacture and installation, whereas solar photovoltaic systems produce about 10 times more (Barnhart et al. 2013). Shifting the mix of energy supplies between traditional fossil fuels and renewables will affect the energy needed to transform and sustain our energy system. Tracking these levels of productivity is the domain of NEA, which combines analysis of primary energy resources with engineering analysis of device efficiencies, as well as efficiencies and transformations in the broader technological system. NEA supplements traditional economic analyses by systematically accounting for the energy consumed, directly and indirectly, by the energy sectors during the life cycle of energy production (Figure 2.11). NEA can complement traditional energy planning, which focuses primarily on minimizing the financial cost of energy production. For example, using NEA, the success of policies to promote photovoltaics can be judged on cost reductions and installed capacity, as well as on net energy provided to society and net emissions avoided. For photovoltaics, this perspective would prioritize photovoltaics with high efficiency and low energetic inputs for manufacturing. NEA would also favor manufacturing photovoltaic panels in locations with low emissions and high-efficiency energy production, and favor deployment in locations with higher solar irradiation and where the photovoltaic electricity produced can offset electricity with a high carbon footprint (Dale 2013).

Policymakers should conduct NEA when evaluating the long-term sustainability of energy technologies. NEA provides a quantitative way to compare the amount of energy a technology produces over its lifetime with the energy required to build and maintain it. The technique can complement conventional energy planning, which often focuses on minimizing the financial cost of energy production (Carbajales-Dale et al. 2014).

2.9.6 Environmental impact

Human energy consumption diverts energy stocks and flows from nature to society, and deposits waste products into the environment. Fossil fuels provide 85% of current primary energy supply and contribute some 60% of total GHG emissions (IPCC 2007). Climate impacts of renewable resources are much smaller, but renewable energy production can have land and ecosystem impacts. Because impacts from primary energy extraction scale with total energy consumption, energy production pathways with high net energy returns help reduce environmental impacts. In essence, every unit of energy consumed within the energy sector to supply our needs acts as a multiplier that increases environmental impacts associated with our energy use. The Canadian oil sands provide a pertinent example. These resources require more energy for their extraction and processing than conventional oil (Dale et al. 2011). This is due fundamentally to the challenging physical properties of the resource: the bituminous oil sands are viscous and difficult to extract. In addition, the resulting product must be more intensively processed to produce useful fuels for consumers. The oil sands industry supplies about five times more energy to society than consumed from outside sources (Brandt et al. 2013a). This can be compared with traditional oil resources, which supply ten to twenty times the energy consumed in the production process (Dale et al. 2011). This increased energy intensity results in larger climate impacts per unit of energy supplied from the oil sands (El-Houjeiri et al. 2013; Carbajales-Dale et al. 2014).

2.10 Education as a promotor of sustainability

All the SDGs come down to education.

Malala Yousafzai

2.10.1 Sustainability literacy

The subject of SD for engineers and engineering students is a key issue currently facing the higher education sector (Davis 2006). Education for SD does not belong to a single discipline; however, education is held to be central to sustainability. Its components are indivisibly linked, but the distinction between education as we know it and education for sustainability remains undefined.

Education is essential to SD. It is crucial to enhancing the ability of the leaner to create solutions and find new paths to a better, more sustainable future. Engineering graduates will need to have a deeper knowledge and understanding of SD and the overall education needs to ensure context, particularly in the social aspects of SD. It seems likely that many universities will try to address SD broadly so engineering students should benefit. However, engineering faculties may use this opportunity to ensure that the engineering aspects of SD are addressed for all students.

A number of goals—e.g., Goal 6, water and sanitation; Goal 7, sustainable energy; Goal 9, resilient infrastructure; and Goal 11, resilient and sustainable cities—are heavily dependent on engineering. However, an examination of the targets for such heavily engineering-based goals should make it clear that achieving SD as defined in the SDGs will require a truly transdisciplinary education approach.

Expressed at the highest level, a sustainability literate person would be expected to understand the need for change to a sustainable way of doing things, individually and collectively; have sufficient knowledge and skills to decide and act in a way that favors SD; and be able to recognize and reward other people's decisions and actions that favor SD. One of the barriers to integrate SD into education provision and make students knowledgeable about the above expectations is that many teachers feel alone and unsure about the meaning of SD (Parkin et al. 2004).

2.10.2 Challenges

There are numerous challenges that are needed to be addressed or overcome when incorporating sustainability issues in engineering education in general. The concept of sustainability is vast and far more complex than conventional engineering topics. It should be scientifically and socially addressed based on its transdisciplinary ST scenario. As Albert Einstein famously said, "we cannot solve our problems with the same thinking we used when we created them." In other words, the engineering education where sustainability is to be addressed should find a way to promote innovation and creativity, which has not necessarily been common preparation.

It is encouraging that there is an emerging development toward integrating sustainability into curricula of both technical and nontechnical studies. However, there is often a preference in engineering education on incorporating environmental issues on the level of engineering tools and methods, while neglecting the holistic nature of SD, its social component, and the equity principle. Sustainability requires more than putting a social science course into an engineering curriculum (Mulder 2004), as it also requires changes in existing engineering paradigms, a broadening of mental frameworks, and changes in values and basic assumptions.

Significant efforts are being made worldwide to bring sustainability into engineering education. These efforts may be of a pedagogical nature by individual faculty or may be policy decisions at a university or national level that enhance the implementation of sustainability into engineering education. At present and in spite of the progress which has been made in several countries, there are still enormous barriers in the reorientation of engineering education to sustainability (Downey 2004).

Faced with the constant pressure to innovate, the twenty-first century brought to engineering schools the need to keep up the quality of education in order to provide the community that it serves the best engineering professional possible. There is a push coming from enterprises and society to accomplish its mission of forming professionals aware of sustainability and social aspects of engineering (Longo and Telles 1998).

Mulder (2004) sees sustainability as a tool for opening the windows of engineering institutions to shift the paradigm away from modernity. Such a shift might help to bridge or narrow the gap between technology and society. This shift also concerns teachers, as they were educated in the old paradigm. Therefore, a shift toward sustainability in engineering education should include substantial learning by teachers, especially on the level of paradigms and mental framework before they will be fully capable of integrating sustainability in their teaching.

Continuous effort has always been taken in most industrialized countries to keep curricula and courses up to date and in line with societal demands, though there is often a focus on technical capabilities and the demands of business. Interestingly, a recent trend in engineering education in most industrialized countries is to go beyond technical capabilities and to extend curricula with teaching of nontechnical (social and management) skills and the incorporation of ethics and social aspects of technology into course programs. De Graaff and Ravesteijn (2001) have called this method "training complete engineers." In fact, many engineering subjects, regardless of the disciplines, are taught in isolation with a minimal global awareness and exposure, and are usually taught on their own with no reference to the economic, environmental, political, cultural, technological, ethical, and global aspects (McKeown and Hopkins 2003).

Universities have a unique opportunity to operate as a societal test bed for sustainability, transforming the campus into a sort of sandbox in which it is possible to conceive, design, implement, and test sustainable solutions; teach and investigate the involving processes; and work with partners to learn and benefit together. Such an agenda helps the university contribute directly to the solution of societal problems and offer significant opportunities for learning, recruitment, fund-raising, and partnership.

2.10.3 Reorienting curriculum

In order to make a transition to engineering education for SD, it is critical to systematically address a number of common key orientation elements within typical curriculum renewal processes. Reorienting education involves selecting appropriate knowledge, issues, skills, and values for the environmental, social, and economic spheres of sustainability. Driving factors for undertaking a curriculum renewal process include the need to address industry and government demand for engineering graduates who are literate and competent in addressing various aspects of SD, the need

Table 2.4 Educational key reorientation elements for SD

Key element	Description
Awareness	Facilitate opportunities for learners to become aware of the current context of sustainability, through activities that promote ST or a whole-systems approach problem solving rather than looking at problems in isolation is used. This may be accomplished by class discussions, issue analysis, storytelling, keynote lectures, lunchtime seminars, media articles, and profiling of existing sustainability-related initiatives and/or competitions within the university
Integrated approach	Develop and embed sustainability content in case study or project format, across early year courses. This requires commitment from teachers to embed such materials
Flagship approach	Develop a common introductory course for early year students to start the reorientation process. This may comprise either the replacement of a previous course or the development of an existing course
Outreach	Use courses as outreach and bridging material for students considering studies in one of the fields of engineering. Materials from first-year courses could be promoted to high schools as an accelerated K-12 course content

to meet changing student expectations on course content, and the need to respond in a critical timeframe (UNESCO 2012). Table 2.4 summarizes the key reorientation elements for SD.

2.11 Research for sustainability

> Look! Look! Look deep into nature and you will understand everything.
>
> **Albert Einstein**

2.11.1 Transdisciplinary research

Critical to transforming higher education is making sustainability a major research focus. Currently, sustainability-oriented research is gradually funded in the sciences, but plans are also underway to bring the social sciences and humanities into the research domain. The challenges of ensuring food, health, water, and energy security while mitigating environmental change require the involvement of a range of disciplines and stakeholders (Lyon 2014). Disciplines are good at providing essential knowledge, techniques, and tools to address the above challenges. However, disciplinary approaches tend not to have the capability to handle the above complex challenges that demand cross-disciplinary collaboration. These challenges require research that cuts across traditional boundaries which is termed interdisciplinary research.

Transdisciplinarity is a broad, reflexive, and highly contested research approach that addresses societal problems by means of interdisciplinary collaboration as well as the collaboration between researchers and extra-scientific actors; its aim is to enable mutual learning processes between science and society; integration is the main cognitive challenge of the research process (Jahn et al. 2012).

Transdisciplinary approaches draw on a range of paradigms and emerging processes shaping the generation of knowledge and the concern with sustainability. Transdisciplinarity draws on ideas of coproduction and the generation of knowledge from a range of stakeholders (Lemos and Morehouse 2005). Such approaches do not prioritize the role of academia in this knowledge production process, but rather see a range of actors collaborating. Transdisciplinarity as a research approach requires practices that use the simplest language possible and produce results which are widely understandable (Brandt et al. 2013b). Transdisciplinary processes require openness to a choice of methods and steps to correspond with the specific research problem being addressed as well as the skills, backgrounds, and competencies of the group of people participating. The steps for transdisciplinary research include the following:

- Identify a real problem that may create sizeable harm or damage.
- Establish a methodological framework to enable the reintegration of knowledge.
- Produce solution-oriented and moveable knowledge which requires incorporating various knowledge bodies by collaboration among various disciplines as well as between researchers, community, and business.
- Integrate and deploy produced knowledge into societal and scientific practice.

2.11.2 Collaborative framework for SD

The main objective of research for sustainability is to realize the interactions between socioeconomic and technical systems on one side, and environmental systems on the other. What research is expected to do in this context is to provide information as to whether and how the global and local changes in various social, economic, and environmental systems are related to use practices, and whether and how changes in the above systems affect each other. Information and knowledge are also required on how practices can be improved and how their implementation can be regulated. Figure 2.12 presents a collaborative framework to illustrate the interrelationships that exist among the various efforts needed to effectively implement SD, and the ways in which an evolving research agenda can be integrated into this framework.

The fields that sustainability issues traditionally draw from include engineering and environmental science, but sustainability issues transcend most discipline divisions to include areas such as education, philosophy, business, and law (Bezbatchenko 2010). Applying science to solve environmental problems in ways acceptable to society requires negotiating the goals of research, policy options, and public acceptability. A growing body of research seeks to investigate complex environmental challenges from a transdisciplinary perspective. This is a more deliberative form of science that requires knowledge of multiple disciplines. Embedded within transdisciplinary research is the attention to the complexity of working across multiple disciplinary perspectives and scales, as well as moving across the divides between academic science and professional knowledge (Lyon 2014).

One key aspect of sustainability is the involvement of actors from outside academia into the research process in order to integrate the best available knowledge, reconcile values and preferences, as well as to create ownership for problems and solution options. Transdisciplinary, community-based, interactive, or participatory research approaches are often suggested as appropriate means to meet both the requirements posed by real-world problems and the goals of sustainability science as a transformational scientific field (Lang et al. 2012).

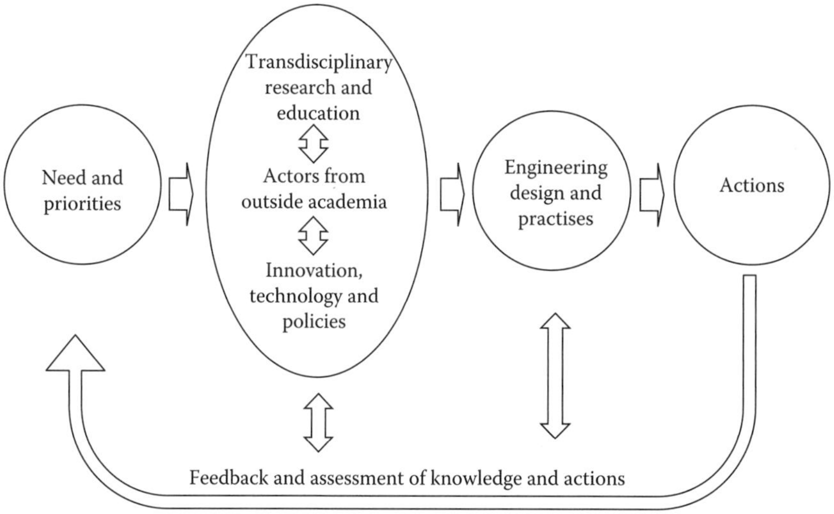

Figure 2.12 A collaborative framework for SD.

Finally, while much research is reflected on the transdisciplinary process from the perspective of academia, there is less research which presents the voice of other participants although there are some emerging examples (Maynard 2013) which include a focus on those whose contribution makes interdisciplinary research transdisciplinary, such as business, civil society, policymakers, and the wider public. While these actors make valuable contributions to the transdisciplinary research projects, their views should also be captured so they are able to make valuable contributions to the wider debates about the benefits, pitfalls, and future directions of transdisciplinary research.

2.12 California case: Integrated approach to water, energy, and food

> We never know the worth of water till the well is dry.
>
> **Thomas Fuller**

2.12.1 Exploring interaction

A growing number of scientists and policy analysts in recent years have emphasized linkages between water, energy, and food, and encouraged an integrated approach to those sectors. The approach emerges from a long history of system analysis and is backed by a robust body of scientific evidence, but is only beginning to take hold in policymaking and planning. The guiding principles of the integrated approach are to promote sustainable and efficient resource use, doing more with less; to ensure access to resources for the most vulnerable, especially the poor; and to maintain healthy and productive ecosystems (Hoff 2011).

It is essential to examine how the SDGs interact, and where there are potential trade-offs and synergies among them. It is needed to identify possible "nexus targets" that would be positioned between sectors, aiming to maximize overall efficiency, within the three sectors under consideration (water, fuel, and food) and beyond. Table 2.5 shows the interaction between water, energy, and food.

As is evident from Table 2.5, some of the targets focus on ensuring access to resources, some on efficiency, and some on long-term sustainability. The three are linked: efforts to ensure access must be combined with efficient management and protection of the resource base and ecosystems in order for the outcome to be sustainable (e.g., if we expand access to agricultural irrigation, over-abstraction must be avoided in order to ensure that the resulting productivity gains can be sustained in the long term).

Table 2.5 Target interaction between water, energy, and food

Water	Energy	Food
Ensuring access to water	Ensuring access to energy services	Making food and agricultural systems sustainable
Improving water use efficiency	Improving energy efficiency	Sustainably improving agricultural yields
Ensuring good management of water resources	Increasing the share of energy that comes from renewable sources	Addressing land conversion and climate impact for/of agriculture

Clearly there are many connections between water, energy, and food targets, but in order to be able to address them effectively in the SDGs, we need to understand the nature of those interactions. In our second analysis, we looked at natural resources as enablers of development: for example, food production requires water, land, and energy. It is important to note that although the focus here is only on a narrow set of targets and resources, there are many other enablers of development, such as health, education, governance, access to technology and knowledge, equality, peace, and security (Weitz 2014).

2.12.2 California's major drought

Water security is a growing concern for California—the most populous U.S. state and producer of more than one-third of America's vegetables and two-thirds of its fruit and nuts, most through intensive, irrigated farming. Only northern California has plentiful water, whereas the rest of the state relies on major aqueducts, most notably the state water project (SWP), which delivers water to local agencies serving 25 million people and more than 3000 km^2 of irrigated farmland in Central and Southern California. Allocations from the SWP are reduced when water is scarce, and in January 2014, with reservoirs at historic lows due to severe drought, the SWP cut off allocations entirely for the first time in its 54-year history (Lovett 2014).

In California, where large volumes of water are transported over long distances, the water sector consumes 19% of the state's electricity and 30% of its natural gas (Klein et al. 2005). Much of the state's surface water comes from winter precipitation and spring snowmelt, and over the past 30 years, winters have been getting warmer, the snowpack has declined, and spring stream flow timing has changed.

Part of the solution to California's demand for water in the face of the state's crippling drought may lie 10,000 ft beneath the surface of the state's Central Valley. Research published in the *Proceedings of the National*

Academy of Sciences (*PNAS*) suggests that the region's aquifers, areas deep underground where water can collect, have three times the usable groundwater as previously estimated (Worland 2016).

Farmers have already started looking to underground springs to feed their crops, so the windfall may be welcome news. The discovery is not a panacea, however. Drilling for water so deep is expensive and can be hazardous. It may add to the gradual sinking of the land already taking place in the Central Valley, according to a statement from Stanford. Groundwater pumping from shallow aquifers has already caused some regions to drop by tens of feet, the researchers said. Those deep groundwater resources are also vulnerable to contamination from oil and gas. Oil and gas drilling occurs in up to 30% of the sites where deep groundwater is located as well as from other human activities, like hydraulic fracturing, according to the study (Berlinger 2016).

The ongoing California drought has led state officials to enact dramatic water conservation measures in hopes of preserving enough water to fuel the state's agriculture sector and provide for its nearly 40 million residents. While the new study offers a potential solution, extracting water from such depths would not be easy. Deep drilling is costly and much of the water would likely be salty and require the installation of desalination facilities (Worland 2016).

Scientists from Stanford University used data from 360 oil and/or natural gas fields, including 35,000 wells, in order to refine California's groundwater estimates. Despite reserves existing several kilometers beneath the surface, little information is available on groundwater deeper than 300 m. The team assessed shallow and deep groundwater sources in terms of quantity and quality. The latter was assessed through salinity and total dissolved solids (TDSs). In California, water with a TDS of less than 3000 ppm can be considered fresh. Findings published in the *PNAS* showed that the amount of usable groundwater was 2700 km^3, which is three times the current estimate. Most was found between 300 and 1000 m underground (Osborne 2016).

One of the most significant drought impacts to the state's energy systems is on traditional hydropower systems. Throughout California, hydroelectric production has declined by 60% over the past 3 years. Historically, the state had been the third largest producer of hydroelectric power nationally (Larson 2015).

In Southern California, several utilities have been investigating seawater desalination as a way to improve water security. But desalination is very energy-intensive, and California's water sector already uses 19% of the state's electricity, including energy for pumping, transporting, and treating water, and energy-intensive residential, commercial, and agricultural water end uses. The SWP is California's single largest power consumer, with a net energy use of two million MWh per year. A nexus perspective thus raises the question: how would desalination affect energy use and efficiency in the state?

An Stockholm Environment Institute (SEI) analysis by Mehta and Yates (2012) linked two modeling tools, the water evaluation and planning system and the long-range energy alternatives planning system, to examine the implications of meeting roughly 5% of Southern California's current urban water demand with desalinated seawater through 2049. It found that desalination could reduce the need for water imports by about 300 million m^3 per year, on average, but it would also increase the water sector's electricity use by about 3 TWh per year; producing that additional energy, in turn, would increase the GHG emissions of the energy sector, contributing to future climate change.

2.12.3 Water and energy: Smart solution

Solar and wind each require little or no water to operate. Both California and Arizona have been among the states with the largest gains in solar market share over the past 5 years, with California also in this category for its growing wind sector (Larson 2015). Renewable energy, in general, and wind energy, in particular, are the perfect technologies that integrate water and energy. Understanding the implications of renewable energy plans on water use will be essential in states most severely impacted by drought.

Wind energy saved 2.5 billion gallons of water in California in 2014 by displacing water consumption at the state's thirsty fossil-fired power plants, playing a valuable role in alleviating the state's record drought. Wind energy's annual water savings work out to be around 65 gallons per person in the state (200 gallons per household), or the equivalent of 20 billion bottles of water. One of wind energy's most overlooked benefits is that it requires virtually no water to produce electricity, while almost all other electricity sources evaporate tremendous amounts of water. In 2008, the nation's thermal power plants withdrew 22–62 trillion gallons of freshwater from rivers, lakes, streams, and aquifers, and consumed 1–2 trillion gallons. By displacing generation from these conventional power plants, U.S. wind energy currently saves around 35 billion gallons of water per year, the equivalent of 120 gallons per person or 285 billion bottles of water (Goggin 2015).

California is currently facing a record drought, and unfortunately, scientists say it is being worsened by climate change. In addition to saving valuable water, wind energy is helping to guard against threats the drought poses to electric reliability as well. This is an even less frequently discussed aspect of the complex relationship in the energy–water interconnection.

The drought has taken a toll on California's hydroelectric generation, but wind energy is helping to pick up the slack. Last year, California's hydroelectric generation was down to 7366 GWh from its 2013 levels.

California-based wind generation is more than made up for that shortfall, providing 13,776 GWh in 2014.

2.12.4 EROI values

As discussed in Section 2.9, the EROI indicator is an expression of energy payback: a numerical quantification of the benefit that the user gets out of the exploitation of an energy source, in terms of how much energy is gained from an energy production process compared to how much of that energy (or its equivalent from some other source) is required to extract, grow, etc., a new unit of the energy in question (Murphy and Hall 2011).

Using the data from several studies, the EROIs for the most commonly used energy techniques are obtained, each calculated for an unbuffered and buffered scenario. The power plant's lifetime should be carefully considered since the EROI scales directly with it. It is dominated by the lifetime of the most energy-intense parts. Whereas wind- and solar-based techniques have estimated lifetimes from 20 to 30 years (limited to turbine rotor or silicon degradation), fossil-fueled power plants can reach 35 years and even more than 60 years (new and refurbished nuclear plants). Lifetimes for refurbished steam turbines often exceed 50 years (Leyzerovich 2007). These longer lifetimes are often ignored in LCA studies. Hydropower has a lifetime of more than 100 years. On the other hand, no statistically relevant experience exists for the lifetime of solar cells. Aging test procedures are still being developed (Lütke et al. 2009) with the goal of a lifetime guarantee of 25–30 years. The given EROIs have uncertainties due to material inventory and maintenance assumptions, which cannot be determined in detail here because LCA database material was not available. These errors affect all techniques roughly equal, so the EROI's relative error is assumed to be about 10%.

Weißbach et al. (2013) found EROI values of 4 for solar and 16 for wind, without storage, or 1.6 and 3.9, respectively, with storage. That is to say, they found that for every unit of energy used to build solar panels, society ultimately gets back 4 units of energy. Solar panels generate four times as much energy as it takes to manufacture them over their lifetimes. The authors assume that half of all solar power is thrown away.

For wind energy, it is possible to get 16 times more energy out of the wind turbine than investing in it during manufacturing, installation, operation, and dismantling (Weißbach et al. 2013). The EROI of wind turbines depends on the invested energy in the turbine, the produced energy, and the life span of a turbine. In the scientific literature, EROIs normally vary between 20 and 50 (Zimmermann 2013).

On the other hand, EROI values for hydro and nuclear energy are 49 and 75, respectively. But these power plants require large amounts of water during the process of electricity generation.

2.12.5 Case research questions

- What are the most important approaches to follow that make cities more sustainable in their use of energy? Can cities be self-sufficient in terms of energy; if possible, how?
- Under what conditions can natural resource extraction and exploitation provide joint social and environmental benefits?
- Apply California's case to another region of your choice with similar challenges to outline a proper agenda for a smart solution.

2.13 Knowledge acquisition

Attempting to answer the following questions involves acquisition of knowledge from this book and other books, documents, and the Internet.

- What is sustainability?
- Illustrate the three-pillar model of sustainability.
- What are the sustainability topics of interest to you?
- What is the difference between sustainability and SD?
- What are the measures for SD?
- What is SA?
- Do you think that SDGs can be achieved by 2030?
- What is sustainable engineering?
- What are key requirements for sustainable engineering?
- What are the primary goals of SD that are directly related to engineering?
- What is the difference between sustainable engineering and environmental engineering?
- What roles can engineers play in SD?
- What is the difference between a "green" firm and a "sustainable" firm?
- What is an indicator of sustainability? What are the features of an effective sustainability indicator?
- How is possible to make farming more sustainable in our generation?
- How can agriculture become an appealing entrepreneurial undertaking, decent, and satisfying work?
- How is business transforming markets so the SDGs are met?
- The GHG effect is a term used frequently in relation to climate change. What is the major cause of this GHG effect, and why is it of concern to environmental scientists?
- What is sustainability planning? What are some core elements for sustainability planning success?
- How can development initiatives best incorporate arts and culture?
- How can businesses be encouraged to better understand and deliver on SD?

- In what means can CE contribute to SD?
- Differentiate between conventional and nonconventional energy sources. Which will you support? Why?
- What makes a sustainable city? Why it is important?
- What role will new technology play in sustainable cities?
- Universities are usually focused on education, research, and service. Do universities really need to add sustainability to list of important matters to address?
- How to engage young and adult learners in SD?
- What is education for SD?
- What are the implications of SDGs for engineering education and practice?

2.14 *Knowledge possession*

Attempting to answer the following open-ended "not explicitly expressed" questions may require research and investigation beyond the scope of this book, mostly by engaging in conversation, class discussion, and Internet-based research.

- Describe the acceleration in world population, technology, science, communication and IT, and transportation. Explain how they have benefited and threatened society.
- What have been the most important approaches of thinking in SD during the past 50 years?
- What are the scientific challenges to the realization of efficient and effective SD in the built environment?
- Which methods, techniques, and policies are currently available in relation to the realization of SD in the built environment?
- Engineering and technology can help to move the products, processes, and systems developed by society toward sustainability. Substantiate this statement.
- How can reliable, universal, sustainable, and affordable electricity services best be financed to rural regions?
- Consider a green building. How it differs from a conventional building? Compare it in several aspects?
- How can architecture, urban design, and planning address social sustainability effectively?
- What are the technological and social driving forces that shape the process of designing smart cities?
- How can educational systems be adapted and developed to maximize youth's capacities for sustainable livelihoods through employment and/or entrepreneurship?

- How does the transformation of higher education influence development pathways?
- What is the current state of research on sustainability within higher education institutions, and how might the research be enhanced?

2.15 *Knowledge creation*

Collaborate with peers on learning or you may work with others outside the class to narrow down the objectives of each activity. You may access class and online resources, and analyze data and information to create new ideas and balanced solutions. High-level digital tools may be used to develop multimedia presentations, simulations or animations, videos and visual displays, digital portfolios (eportfolios), reflective practice (online publishing and blogging), or well-researched and up-to-date reports.

2.15.1 *Writing tasks on greening thinking*

This contest exercise discusses different assignments and methods used to introduce writing curriculum with sustainability principles. It inspires students to learn more about sustainability and to use their class assignment to search for ways business, the campus, and the local community can become more sustainable. This contest requires teams of three to four students to carry out each of the following activities.

2.15.1.1 *Online quiz*

Take the online carbon footprint quiz (http://myfootprint.org/subscription.php), and then write about your response to the results, including what steps you could take to lessen footprint and to live more sustainably.

2.15.1.2 *Public speaker*

Prepare a speech about sustainability asking stakeholder including government, people, businesses, nongovernmental organizations, and individual the following questions.

What are you doing to protect the environment and contribute to SD?

2.15.1.3 *Business plan*

Write a sustainability business plan for one of the following topics. Organize your plan into the following four components: research, solving a need, being different, and starting small.

- Pesticide-free floral business
- Home energy retrofit plan
- Residential solar energy system

- Redesigned furniture out of recycled materials
- Green gym (energy from exercise)
- Ethanol conversion kit for a car
- Digital notebooks to replace textbooks
- Biodiesel from dormitory dining commons cooking oil

2.15.1.4 Project proposal

Write a sustainability project proposal for one of the following topics. Organize your concept with three sections: mission statement, statement of need, and project narrative.

- Small wind turbine in a farm
- Energy incentives for home owners
- Creating a solar rooftop plant on a campus building
- Wind solar hybrid LED streetlight for community

2.15.2 Reflective practice on path to sustainability

Investigate your university path to sustainability by tracing year by year accomplished activities. Develop a sustainability timeline that tells the story of becoming a sustainable university, if it is the case. You may include activities and programs such as Leadership in Energy and Environmental Design (LEED)-certified building projects, recycling programs, hybrid vehicles and EVs, renewable energy projects, lighting retrofit projects, heating, ventilation and air conditioning (HVAC) optimization projects, energy monitoring systems, energy and water conservation programs, fossil fuels divestment campaign, rain garden projects, climate change initiatives, temperature setting policy, sustainability leadership awards, sustainability curriculum activities, sustainability research/teaching programs, sustainability design labs, energy education campaigns, and earth day activities.

2.15.3 Survey task on SD among engineering students

One of the most popular definitions is that contained within "Our Common Future," the 1987 Report of the UN World Commission on Environment and Development (WCED 1987), which reads as follows: "Humanity has the ability to make development sustainable, to ensure that it meets the needs of the present without compromising the ability of future generations to meet their own needs".

The objective of this task is to develop digital knowledge content for a survey that involves a brief two-page, tick-box style questionnaire to be delivered to 100 engineering students across several disciplines and at different stages of their courses. The questionnaire may be divided into several parts, starting with information about students, their level of knowledge and understanding of SD, the perceived importance of SD by the students, and previous sustainability education.

2.15.4 *Writing tasks on routes to urban agriculture*

Thinking of growing food! While the rural landscape might be what immediately comes to mind, cities, too, can be places of significant food production. Urban agriculture is a term that describes many activities, all of which are connected with the growing, processing, and circulation of food and food-related products in and around cities. Urban agriculture is growing in acceptance and plays a vital role in the resilience and sustainability of cities. More people around the world are considering urban farming, which offers to make food as local as possible. By growing fresh produce close to a living, the food miles associated with long-distance transportation is decreased. Another benefit of urban agriculture is that it can enhance greenery of cities, reducing harmful runoff, increasing shading, and neutralizing the unlikable heat island impact.

Conduct a case by creating an urban agriculture program at your city or campus. In the context of the case, you may consider the following.

- Where does our current food come from?
- What could be the array of activities that exist under the umbrella of urban agriculture?
- What are the benefits of urban agriculture?
- What is the role of city planners in promoting urban agriculture?
- Propose policies to support and implement urban agriculture.
- Investigate examples of cities promoting urban agriculture: leading by example.
- Investigate at least two innovative entrepreneurial activities and inventive business models in the field of urban agriculture.
- Propose a model on entrepreneurial backyard garden.
- What are therapeutic gardens?
- Identify city-owned park and/or public space suitable for urban gardening and community greenhouses.
- How to promote and strengthen entrepreneurial and apprenticeship programs that teach entrepreneurial and business planning skills?
- You may develop a social networking site that integrates all resources, organizations, and initiatives to inspire collaboration and realize the full benefits of urban agriculture.

2.15.5 *Feasibility study of sustainable distributed generation*

Infrastructure projects have a major influence on the attainment of SD. Energy generation systems, and solutions, usually come with detailed feasibility studies, taking into consideration the technical and operational details, investment involved, any incentives available, and the running costs entailed. The concept of the feasibility study is to investigate which sustainable energy technologies could be appropriately used. In this task, develop a five-page preliminary feasibility study in support of the development of a state-of-the-art sustainable energy project for a remote community of 5000 people with thinly industries including agriculture, hunting, fishing, and a small-scale wastewater treatment plant. The primary services include government office, health care, and schools. Unemployment is 30% of population. The community is served by a single 69-kV transmission line and a radial 34-kV distribution line. The project should focus on energy efficiency and renewable energy (solar and/or wind). Data collected from the community site indicate wind and solar potential. To proceed, identify a location within your region to approximately match this case, collect data and information, and proceed with the required procedure.

2.15.6 Piece of art on the engineering principles for SD

Read the 12 principles of sustainable engineering given in Table 2.3. Create 12 logos or pieces of art that correspond to the 12 principles and arrange them in a table format.

2.15.7 Debate on energy and sustainability

Objective	Introducing an open-ended debate in the classroom to help students understand argument on the concepts of energy and sustainability
Time	15 min for debate and 15 min for review
Format	For and against
Learning outcomes	Make an argument about a particular opinion, evaluate the arguments of peers, and understand the concept of counterarguments
Capabilities demonstrated	Develope skills on public speaking, research, teamwork, critical thinking, communication, and professional judgment
Arrangement	Students are organized into two position groups of three speaking in a specific order. Three argue for an opinion and three argue against. One or two students might each work on the opening and closing statements while the group is investigating the subject; however, the entire group should revise the statements. Each group should read an opening and closing statement for the debate.

(*Continued*)

Idea for the topic	It is possible to create a sustainable, cleaner, and safer world by making wiser energy choices. The scenario of 100% supply by renewables can be argued strongly. How to justify the above notion in terms of energy efficiency and lifestyle modernization? What requirements are imposed on energy system in case of a high penetration of renewable resources? What are the consequences in terms of sustainability? Evaluate the importance of above and discuss on the various "for and against" issues.
Assessment	Indicate what you consider the best arguments in favor of the topic. How were they substantiated? Identify arguments that are based on poor facts, not ethical and/or sustainable or not well substantiated.

2.15.8 *Video contest on life cycle emission of a vehicle*

In the production process of vehicles there is a hierarchy of production layers, and each one of them needs inputs like materials and energy. The raw materials and parts to manufacture the car will be purchased from a range of specialized industries upstream. It is likely that they themselves obtained materials from other industries and so on. The parts of the vehicle are transported downstream to factories in order to put the car together and deliver it to retailers. All these steps use up energy resources and emit pollution in the process, pollution and resource use that should be accounted for when calculating the emissions associated with purchasing a vehicle (Skot 2009).

In this task, collaborate with peers or you may work with others outside the class to narrow down the objectives of the task. You may access online resources to create a 3-min video contest that highlights various sources of emissions in the timeline of car manufacturing, operation, and end of life (disposal). Each stage is linked with carbon dioxide and other GHG emissions, but those emissions differ between gas-powered cars and EVs. You may compare the life cycle of a gasoline car with that of an electric car accounting for pollution from battery manufacturing.

In the process of evaluation and comparison, you may address the following two key questions (Nealer et al. 2015):

- What are the global warming emissions from operating an EV today?
- How much does the manufacturing of EVs affect their total global warming emissions benefits?

References

ACEC. 2014. Sustainable development for Canadian consulting engineers. Association of Consulting Engineering Companies, Ottawa, ON, Canada.

AGECC. 2010. Energy for a sustainable future, summary report and recommendations. The Secretary-General's advisory group on energy and climate change (AGECC). http://www.unenergy.org/publications/558-agecc-report-energy-for-asustainable-future.

Allen, A. 2009. Sustainable cities or sustainable urbanisation? White Paper. https://www.ucl.ac.uk/sustainable-cities/results/gcsc-reports/allen.pdf.

Amor, M. B., P. Lesage, P. Pineau, and R. Samson. 2010. Can distributed generation offer substantial benefits in a northeastern American context? A case study of small-scale renewable technologies using a life cycle methodology. *Renewable and Sustainable Energy Review* 14: 2885–2895.

Anastas, P., and J. Warner. 1998. *Green Chemistry: Theory and Practice*. London: Oxford University Press.

Anastas, P., and J. Zimmerman. 2003. Design through the twelve principles of green engineering. *Environmental Science and Technology* 37: 94A–101A.

Anders, W., and R. Johan. 2012. *Bankrupting Nature: Denying our Planetary Boundaries. A Report to the Club of Rome.* Abingdon, UK: Routledge.

Anders, W., and S. Kristian, 2015. The circular economy and benefits for society. Club of Rome. https://www.clubofrome.org/wp-content/uploads/2016/03/The-Circular-Economy-and-Benefits-for-Society.pdf.

Ayres, R. U. 2008. Sustainability economics: Where do we stand? *Ecological Economics* 67(2): 281–310.

Ayres, R. U., Leslie, W. A., and K. Martins. 1998. Exergy, waste accounting, and life-cycle analysis. *Energy* 23: 355–363.

Bach, C. F. 2015. Water is key to sustainable development. https://www.unece.org/info/media/blog/previous-blogs/water-is-key-to-sustainable-development.html.

Barnhart, C. J., M. Dale, A. R. Brandt, and S. M. Benson. 2013. The energetic implications of curtailing versus storing solar- and wind-generated electricity. *Energy and Environmental Sciences* 6: 2804–2810.

Bell, S., and S. Morse. 2008. *Sustainability Indicators: Measuring the Immeasurable*. London: Earthscan Publishing, p. 192.

Berlinger, J. 2016. Is California sitting on the solution to its drought? *CNN*. http://www.cnn.com/2016/06/28/us/california-drought-underground-water/index.html.

Berndtsson, M. 2015. Circular economy and sustainable development. Master thesis, Department of Earth Sciences, Uppsala University, Uppsala, Sweden.

Bettencourt, L. M. A., and J. Kaur. 2011. The evolution and structure of sustainability science. *Proceedings of the National Academy of Science* 108: 19540–19545.

Bezbatchenko, A. W. 2010. Sustainability in colleges and universities: Toward institutional culture shifts. *Journal of Student Affairs at New York University* VI: 1–11.

Bond, A. J., T. Dockerty, A. Lovett, A. B. Riche, A. J. Haughton, D. A. Bohan, R. B. Sage, I. F. Shield, J. W. Finch, M. M. Turner, and A. Karp. 2011. Learning how to deal with values, frames and governance in sustainability appraisal. *Regional Studies* 45(8): 1157–1170.

BP. 2015. The BP energy outlook. http://www.bp.com/content/dam/bp/pdf/energy-economics/energy-outlook-2015/bp-energy-outlook-2035-booklet.pdf.

Brandt, A. R., J. Englander, and S. Bharadwaj. 2013a. The energy efficiency of oil sands extraction: Energy return ratios from 1970 to 2010. *Energy* 55: 693–702.

Brandt, P., A. Ernst, F. Gralla, and H. Wehrden. 2013b. A review of transdisciplinary research in sustainability science. *Ecological Economics* 92: 1–15.

Bullard, C. W., P. S. Penner, and D. A. Pilati. 1978. Net energy analysis: Handbook for combining process and input-output analysis. *Resources and Energy* 1: 267–313.
C40. 2016. Why cities? Ending climate change begins in the city. Cities Climate Leadership Group. http://www.c40.org/ending-climate-change-begins-in-the-city.
Carbajales-Dale, M, C. J. Barnhart, A. R. Brandt, and S. M. Benson. 2014. A better currency for investing in a sustainable future. *Nature Climate Change* 4: 524–527.
Carson, R. 1962. *Silent Spring*. New York: Houghton Mufflin.
Casier, L. 2015. Why infrastructure is key to the success of the SDGs? https://www.iisd.org/blog/why-infrastructure-key-success-sdgs.
Ciampi, N. N. 2011. Human side of engineering: Dealing with complex and ethical challenges. *41st ASEE/IEEE Frontiers in Education Conference*, Rapid City, SD, USA, October, 12–15, 2011.
City of Copenhagen. 2010. *Copenhagen, City of Architecture*. Technical and Environmental Administation, Copenhagen, Denmark, p. 14. Accessed August 11, 2016, http://kk.sites.itera.dk/apps/kk_pub2/pdf/961_BLdBaJTGzV.pdf[].
City of Copenhagen. 2014. *Copenhagen: Solutions for Sustainable Cities*. Copenhagen, Denmark, pp. 6–9, 30–33 and 46–47, 2014. http://kk.sites.itera.dk/apps/kk_pub2/pdf/1353_58936BnEKE.pdf.
Cleveland, C., and R. Costanza. 2008. Net energy analysis in: *Encyclopedia of Earth*. http://www.eoearth.org/article/Net_energy_analysis.
Dale, M. 2013. A comparative analysis of energy costs of photovoltaic, solar thermal, and wind electricity generation technologies. *Applied Science* 3: 325–337.
Dale, M., and S. M. Benson. 2013. Energy balance of the global photovoltaic (PV) industry-is the PV industry a net electricity producer? *Environmental Science and Technology* 47(7): 3482–3489.
Dale, M., S. Krumdieck, and P. Bodger. 2011. Net energy yield from production of conventional oil. *Energy Policy* 39(11): 7095–7102.
Davis, G. U. 2006. The role of case studies for the integration of sustainable development into the education of engineers. *World Transactions on Engineering and Technology Education* 5(1): 159–162.
De Graaff, E., and W. Ravesteijn. 2001. Training complete engineers: Global enterprise and engineering education. *European Journal of Engineering Education* 26(4): 419–427.
Despeisse, M., F. Mbaye, P. D. Ball, and A. Levers. 2012. The emergence of sustainable manufacturing practices. *Production Planning and Control* 23(5): 354–376.
Devuyst, D., W. De Lannoy, and L. Hens (Eds.). 2001. *How Green Is the City? Sustainability Assessment and the Management of Urban Environments*. New York: Columbia University Press, pp. 129–156.
Dillard, J., V. Dujon, and M. King. 2008. *Understanding the Social Dimension of Sustainability*. New York: Routledge.
Dodds, R., and R. Venables (Eds.). 2005. Engineering for sustainable development: Guiding principles. The Royal Academy of Engineering. http://www.raeng.org.uk/publications/reports/engineering-for-sustainable-development
Downey, P. R. 2004. Sustainability takes time. *International Journal of Sustainability in Higher Education* 5(1): 81–90.

EC. 2015. The circular economy: Connecting, creating and conserving value. European Commission. http://www.eesc.europa.eu/resources/docs/the-circular-economy.pdf.
El-Houjeiri, H. M., A. R. Brandt, and J. E. Duffy. 2013. Open-source LCA tool for estimating greenhouse gas emissions from crude oil production using field characteristics. *Environmental Science and Technology* 47(11): 5998–6006.
Elkington, J. 1997. *Cannibals with Forks: The Triple Bottom Line of 21st Century Business*. Mankato, MN: Capstone Publishing.
IAEA. 1994. *Net Energy Analysis of Different Electricity Generation Systems.* International Atomic Energy Agency, Vienna, Austria.
IEA. 2010. *Energy Technology Perspectives.* International Energy Agency. Paris: OECD Publishing.
IISD. 2010. The sustainable development timeline. http://www.iisd.org/pdf/2009/sd_timeline_2009.pdf.
IISD. 2012. International Institute for Sustainable Development. http://www.iisd.org/.
IPCC. 2007: Climate change 2007: Synthesis report. In Pachauri, R. K., and A. Reisinger (Eds.), *Contribution of Working Groups I, II and III to the Fourth Assessment Report of the Intergovernmental Panel on Climate Change, Core Writing Team*. Geneva, Switzerland: IPCC, 104 pp.
Jahn, T., M. Bergmann, and F. Keil. 2012. Transdisciplinarity: Between mainstreaming and marginalization. *Ecological Economics* 79: 1–10.
Jolly, R. 2010. The MDGs in historical perspective. *IDS Bulletin* 41: 48–50.
Jones, O. 2013. Sustainability: A brief history. New Leaf Sustainability Consulting. http://newleaf-llc.com/2013/01/sustainability-a-brief-history/.
Junghans, L., and J. Grimm. 2016. Hand in hand for urban transformation roles and responsibilities at the local, national and international level to enable cities' climate leadership. https://germanwatch.org/en/12659.
Gagnon, B., R. Leduc, and L. Savard. 2008. From a conventional to a sustainable engineering design process: Different shades of sustainability. Working Paper, Groupe de Recherche en Économie et Développement International, University of Sherbrooke, Sherbrooke, QC.
Garcia-Serna, J., L. Perez-Barrig, and M. J. Cocero. 2007. New trends for design towards sustainability in chemical engineering: Green engineering. *Chemical Engineering Journal* 133(1): 7–30.
GCEC. 2016. Seizing the global opportunity. Partner-ships for better growth and better climate. The 2015 New Climate Economy Report. The Global Commission on the Economy and Climate. Washington, DC. http://2015.newclimateeconomy.report/wp-content/uploads/2014/08/NCE-2015_Seizing-the-Global-Opportunity_web.pdf.
Giovannini, E. 2013. Using "equitable and sustainable wellbeing" to build the post-MDGs framework. *IDS Bulletin* 44(5–6): 89–96.
Giudice, L. 2015. Sustainability: Both a journey and a destination. http://www.sustrana.com/blog/2015/8/17/sustainability-both-a-journey-and-a-destination.
GNESD. 2007. Reaching the millennium development goals and beyond-access to modern forms of energy as a prerequisite. Global Network on Energy for Sustainable Development, Roskilde, Denmark.
GNESD. 2010. Energy, climate change and poverty alleviation, global network on energy for sustainable development (GNESD) policy paper prepared by AFREPREN, ENDA-TM and Fundacion Bariloche.

Goggin, M. 2015. Wind energy saves 2.5 billion gallons of water annually in drought-parched California. http://www.aweablog.org/wind-energy-saves-2-5-billion-gallons-of-water-annually-in-drought-parched-california.

Guijt, I., and A. Moiseev. 2001. *Resource Kit for Sustainability Assessment*. Gland, Switzerland and Cambridge: IUCN.

Haas, W., Singh, S. J., and A. Musel. 2012. Energy return on energy input/invested. Environmental justice organizations. Liabilities and trades. http://www.ejolt.org/2012/12/eroi-energy-return-on-energy-inputinvested.

Hacking, T., and P. Guthrie. 2008. A framework for clarifying the meaning of triple bottom-line, integrated and sustainability assessment. *Environmental Impact Assessment Review* 28(2–3): 73–89.

Hall, C. A., C. J. Cleveland, and R. Kaufmann. 1986. *Energy and Resource Quality: The Ecology of the Economic Process*. Hoboken, NJ: John Wiley and Sons.

Hasna, A. M. 2010. Sustainability classifications in engineering: Discipline and approach. *International Journal on Sustainability Engineering* 3(4): 258–276.

Hezri, A. A. 2013. Broadening the environmental agenda in the post-2015 development agenda. *IDS Bulletin* 44(5–6): 81–88.

Hoff, H. 2011. Understanding the nexus. Background paper for the Bonn2011 Conference: The Water, Energy and Food Security Nexus, Stockholm Environment Institute, Stockholm, Sweden. http://sei-international.org/publications?pid=1977.

Kelly, M. J. 2016. Lesson from technology development for energy and sustainability. *MRS Energy and Sustainability* 3 (E3 (13 pages)). http://dx.doi.org/10.1557/mre.2016.3.

Klein, G., M. Krebs, V. Hall, T. O'Brien, and B. B. Blevins. 2005. California's water-energy relationship. Californian Energy Commission. *Ecosystems and Human Well-Being: Synthesis*. Washington, DC: Island Press.

Kuhn, T. 1970. *The Structure of Scientific Revolution*. Chicago, IL: University of Chicago Press.

Lang, D. J., A. Wiek, M. Bergmann, M. Stauffacher, P. Martens, P. Moll, M. Swilling, and C. J. Thomas. 2012. Transdisciplinary research in sustainability science: Practice, principles, and challenges. *Sustainability Science* 7(1): 25–43.

Larson, K. N. 2015. Blue-green intersections: Impacts of water shortages on western state energy policies. http://lexingtoninstitute.org/wp-content/uploads/2015/09/Blue-Green-Intersections.pdf.

Lemos, M. C., and B. J. Morehouse. 2005. The co-production of science and policy in integrated climate assessments. *Global Environmental Change* 15: 57–68.

Lenzen, M., and J. Munksgaard. 2002. Energy and CO_2 life-cycle analyses of wind turbines—Review and applications. *Renewable Energy* 26: 339–362.

Leyzerovich, A. S. 2007. *Steam Turbines for Modern Fossil-Fuel Power Plants*. Boca Raton: CRC Press. http://www.steamshed.com/pdf/040SteamTurbines2.pdf.

Ling, C., A. Dale, and K. Hanna. 2007. Integrated community sustainability planning tool. Royal Roads University. https://crcresearch.org/sites/default/files/icsp-planning-tool.pdf.

Longo, W. P., and M. H. C. Telles. 1998. Programa de desenvolvimento das Engenharias: Situação Atual. *Revista de Ensino de Engenharia* 18: 74–82. Rio de Janeiro: ABENGE.

Lovett, I. 2014. Parched, California cuts off tap to agencies. *The New York Times*. http://www.nytimes.com/2014/02/01/us/amid-drought-california-agency-will-withhold-water-deliveries.html.

Lozano, R. 2012. Towards better embedding sustainability into companies' systems: An analysis of voluntary corporate initiatives. *Journal of Cleaner Production* 25: 14–26.

Lütke, D., H. H. Kuehnlein, H. Verbunt, and S. Irsen. 2009. Accelerated ageing tests of solar cells. G.I.T. *Imaging and Microscopy* 1(April): 42–43.

Lyon, F. 2004. Transdisciplinary environmental research: A review of approaches to knowledge co-production. Nexus Network Think Piece Series, Paper 002.

MacKay, D. J. C. 2009. Sustainable energy: Without the hot air, UIT Cambridge Ltd. http://www.withouthotair.com.

Macomber, J. D. 2013. Building sustainable cities. *Harvard Business Review*, July/August.

Marsh, G. P. 1864. *Man and Nature: Or, Physical Geography as Modified by Human Action*. Cambridge, MA: Belknap Press of Harvard University.

Masud, M. M., S. M. Akhtaruzzaman, B. Sadakatul, and A. Farhat. 2011. Engineers' obligation toward sustainable environment. *Second International Conference on Professional Ethics and Education in Engineering*, Kuala Lumpur, Malaysia, pp. 85–92.

Maxwell, S. 2014. Integrating sustainability into the engineering design process using the global reporting initiative indicators. MSc Thesis, University of Iceland Reykjavik.

Maynard, C. M. 2013. How public participation in river management improvements is affected by scale. *Area* 45(2): 230–238.

McKeown, R., and C. Hopkins. 2003. EE does not equal ESD: Defusing the worry. *Environmental Education Research* 9(1): 118–128.

Meadows, D. H., D. L. Meadows, J. Randers, and W. W. Behrens III. 1972. *The Limits to Growth: A Report of the Club of Rome's Project on the Predicament of Mankind*. New York: Universe Books.

Mebrato, D. 1998. Sustainability and sustainable development: Historical and conceptual review *Environmental Impact Assessment Review* 18: 493–520.

Mehta, V., and D. Yates. 2012. Integrated water-energy-emissions analysis: Applying LEAP and WEAP together in California. SEI policy brief. Stockholm Environment Institute, Davis, CA. http://www. sei-international.org/publications?pid=2149.

Modi, V., S. McDade, D. Lallement, and J. Saghir. 2006. Energy and the millennium development goals. Energy Sector Management Assistance Programme, United Nations Development Programme, UN Millennium Project, and World Bank, New York. http://www.unmillenniumproject.org/documents/MP_Energy_Low_Res.pdf.

Mulas, V., and J. P. G Barroca. 2016. How to implement open innovation in city government? *The World Bank*. http://blogs.worldbank.org/sustainablecities/how-implement-open-innovation-city-government.

Mulder, K. 2004. Engineering education in sustainable development: Sustainability as a tool to open up the windows of engineering institutions. *Business Strategy and the Environment* 13: 275–285.

Murphy, D. J., and C. A. S. Hall. 2011. Energy return on investment, peak oil, and the end of economic growth. *Annals of the New York Academy of Sciences* 1219: 52–72.

NAE. 2004. The engineer of 2020: Visions of engineering in the new century. National Academy of Engineering. https://inside.mines.edu/UserFiles/File/Assessment/Engr2020.pdf.

NAP. 1995. The role of technology in environmentally sustainable development. The National Academies Press. https://www.nap.edu/read/9236/chapter/5.

Nealer, R., D. Reichmuth, and D. Anair. 2015. Cleaner cars from cradle to grave: How electric cars beat gasoline cars on lifetime global warming emissions. The Union of Concerned Scientists. http://www.ucsusa.org/sites/default/files/attach/2015/11/Cleaner-Cars-from-Cradle-to-Grave-full-report.pdf.

Ness, B., E. Urbel-Piirsalu, S. Anderberg, and L. Olsson. 2007. Categorising tools for sustainability assessment. *Ecological Indicators* 60(3): 498–508.

Neureuther, A.-K. 2013. Five questions about sustainable development goals and the potential role of landscapes. Forest News. https://blog.cifor.org/19995/five-questions-about-sustainable-development-goals-and-the-potential-role-of-landscapes?fnl=en.

Nilsson, M., P. Lucas, and T. Yoshida. 2013. Towards an integrated framework for SDGs: Ultimate and enabling goals for the case of energy. *Sustainability* 5(10): 4124–4151.

Patterson, M. 2010. Is there more in common than we think? Convergence of ecological footprinting, energy analysis, life cycle assessment and other methods of sustainability assessment? Proceedings of *ISEE 2010 Conference on Advancing Sustainability in Time of Crisis,* Bremen and Oldenburg, Germany, August 22–25,2010.

OECD. 2003. Environmental indicators-development, measurement and use. Organization for Economic Co-operation and Development. Reference Paper, Paris.

Osborne, H. 2016. California's water windfall: 3 times more groundwater sitting under Central Valley than thought. International Business Times. http://www.ibtimes.co.uk/californias-water-windfall-3-times-more-groundwater-sitting-under-central-valley-thought-1567691.

Oxford Dictionary of English. 2010. Oxford: Oxford University Press.

Parkin, S., A. Johnston, H. Buckland, F. Brookes, and E. White. 2004. Learning and skills for sustainable development developing a sustainability literate society. Guidance for Higher Education Institutions. Higher Education Partnership for Sustainability. https://www.forumforthefuture.org/sites/default/files/project/downloads/learningandskills.pdf.

Pfaff, A. S., and R. N. Stavins. 1999. Readings in the field of natural resource and environmental economics. Faculty Research Working Paper Series, R99-02, Harvard University Press.

Preston, M. 2015. Make it your business: engaging with the sustainable Development Goals. http://www.pwc.com/id/en/Consulting/Asset/S&CC/SDG%20Research.pdf.

Raugei, M., P. P. Fullana, and V. Fthenakis. 2012. The energy return on energy investment (EROI) of photovoltaics: Methodology and comparisons with fossil fuel life cycles. *Energy Policy* 45: 576–582.

Robinson, J. 2004. Squaring the circle? Some thoughts on the idea of sustainable development. *Ecological Economics* 48(4): 369–384.

Robinson, R. 2014. Smart city design patterns. The Urban Technologist. https://theurbantechnologist.com/design-patterns/.

Rosen, M. A. 2012. Engineering sustainability: A technical approach to sustainability. *Sustainability* 4: 2270–2292.

Rosen, M. A., and H. A. Kishawy. 2012. Sustainable manufacturing and design: Concepts, practices and needs. *Sustainability* 4: 154–174.

Rusinko, C. A. 2007. Green manufacturing: An evaluation of environmentally sustainable manufacturing practices and their impact on competitive outcomes. *IEEE Transactions on Engineering Management* 54(3): 445–454.

Sala, S., F. Farioli, and A. Zamagni. 2012. Progress in sustainability science: Lessons learnt from current methodologies for sustainability assessment: Part 1. *International Journal Life Cycle Assessment* 10(9): 1–12.

Schäfer, T. 2013. Green IT: sustainable communications and information. https://www.alumniportal-deutschland.org/en/global-goals/sdg-09-infrastructure/green-it-sustainable-information-technology/ technology.

Seager, T. P., J. Melton, and T. T. Eighmy. 2004. Working towards sustainable science and engineering: Introduction to the special issue on highway infrastructure. *Resources, Conservation and Recycling* 42(3): 205–207.

Singh, R. K., H. R. Murty, S. K. Gupta, and A. K. Dikshit. 2008. An overview of sustainability assessment methodologies. *Ecological Indicators* 15(1): 281–299.

Skot, K. 2009. A literature review on sustainable lifestyles and recommendations for further research. Stockholm Environment Institute (SEI). http://www.unep.fr/scp/marrakech/taskforces/pdf/sei_sustainable_lifestyles_evidence_report.pdf.

Spohn, O. M. 2004. *Sustainable Development Indicators within the German Water Industry—A Case Study*. Gothenburg, Sweden: Chalmers University of Technology.

Stephens, J. C., and B. V. D. Zwaan. 2005. The case for carbon capture and storage. *Issues in Science and Technology*, Fall, XXII(1): 69–76.

Stofleth, D. 2016. A short history of sustainable development, Rethinking Prosperity. http://rethinkingprosperity.org/a-short-history-of-sustainable-development/.

Tahir, A. C., and R. C. Darton. 2010. The process analysis method of selecting indicators to quantify the sustainability performance of a business operation. *Journal of Cleaner Production* 18(16–17): 1598–1607.

Tainter, J. A. 2006. Social complexity and sustainability. *Journal of Ecological Complexity* 3: 91–103.

TEEB. 2010. The economics of ecosystems and biodiversity. Mainstreaming the economics of nature: A synthesis of the approach, conclusions and recommendations of TEEB. The Economics of Ecosystems and Biodiversity, Geneva, Switzerland.

Ugwu, O. O., and T. C. Haupt. 2007. Key performance indicators and assessment methods for infrastructure sustainability—A South African construction industry perspective. *Building and Environment* 42(2): 665–680.

UN. 2013. World economic and social survey 2013: sustainable development challenges. Department of Economic and Social Affairs. United Nations. https://sustainabledevelopment.un.org/content/documents/2843WESS2013.pdf

UN. 2016. Smart cities and infrastructure. Report of the Secretary-General. Commission on Science and Technology for Development. http://unctad.org/meetings/en/SessionalDocuments/ecn162016d2_en.pdf.

UNEP. 2010. Advancing the biodiversity agenda: A UN system wide contribution. United Nations Environment Programme, Nairobi, Kenya.

UNEP. 2011. Towards a green economy: Pathways to sustainable development and poverty eradication. United Nations Environment Program. ISBN: 978-92-807-3143-9. http://www.unep.org/greeneconomy.

UNEP. 2010. *Advancing the Biodiversity Agenda: A UN System Wide Contribution.* United Nations Environment Programme, Nairobi. https://www.cbd.int/doc/strategic-plan/BIODIVERSITY_Corrections_finales_.pdf.

UNEP. 2011. *Towards a Green Economy: Pathways to Sustainable Development and Poverty Eradication.* United Nations Environment Program, Nairobi. https://sustainabledevelopment.un.org/content/documents/126GER_synthesis_en.pdf.

UNESCO. 2012. *Education for Sustainable Development.* ISBN 978-92-3-001063-8. http://unesdoc.unesco.org/images/0021/002163/216383e.pdf.

UNICEF. 2014. Millennium development goals (MDG) monitoring. https://www.unicef.org/statistics/index_24304.html.

Van de Westerlo, B. 2011. Sustainable development and cradle to cradle approach. A literature study of the opportunities to apply the Cradle to Cradle® approach in the built environment. University of Twente, Enschede, the Netherlands.

Van der Heijden, K., O. S. Hoiberg, and A. Scott. 2014. From solidarity to universality: How global interdependence impacts the post-2015 development agenda. Background paper for Towards a Transformative Post-2015 Development Agenda, Tarrytown, NY, February 9–10.

Venters, C. C. 2014. Software sustainability: The modern tower of Babel. *RE4SuSy: Proceedings of the Third International Workshop on RE for Sustainable Systems.* Karlskrona, Sweden: CEUR-WS 1216.

Verheem, R. A. A. 2002. Environmental impact assessment in the Netherlands: Views from the Commission for EIA in 2002. The Netherlands.

WCED. 1987. *Our Common Future. Commission on Environment and Development.* Oxford: Oxford University Press.

Weißbach, D., G. Ruprechta, A. Huke, K. Czerskia, S. Gottlieba, and A. Hussein. 2013. Energy intensities, EROIs, and energy payback times of electricity generating power plants. *Energy* 52: 210–221.

Weitz, N. 2014. Cross-sectoral integration in the sustainable development goals: A nexus approach, Stockholm Environment Institute, Stockholm, Sweden.

Wijkman, A., and J. Rockström. 2012. *Bankrupting Nature: Denying Our Planetary Boundaries.* New York: Routledge.

Wijkman, A. and K. Skånberg. 2015. The circular economy and benefits for society. Jobs and climate clear winners in an economy based on renewable energy and resource efficiency. Study pertaining to Finland, France, the Netherlands, Spain and Sweden. The Club of Rome. www.clubofrome.org/wp-content/uploads/2016/03/The-Circular-Economy-and-Benefitsfor-Society.pd.

Worland, J. 2016. Drought-hit California has a bonanza of water-underground. *TIME.* http://time.com/4384505/california-drought-groundwater/.

Zimmermann, J. B. 2013. Parameterized tool for site specific LCAs of wind energy converters. *The International Journal of Life Cycle Assessment* 18: 49–60.

Zimmerman, J. B., and P. T. Anastas. 2005. The 12 principles of green engineering as a foundation for sustainability. In Abraham, M. (Ed.), *Sustainability Science and Engineering: Principles.* Amsterdam, the Netherlands: Elsevier Science.

Zimmerman, J. B., A. F. Clarens, S. J. Skerlos, and K. F. Hayes. 2003. Design of emulsifier systems for petroleum- and bio-based semi-synthetic metalworking fluid stability under hardwater conditions. *Environmental Science and Technology* 37(23): 5278–5288.

chapter three

Technology and sustainability

> Technology is, of course, a double edged sword. Fire can cook our food but also burn us.
>
> **Jason Silva**

3.1 *Objectives*

- Trace the history of technology.
- Understand the definitions and scope of science and technology.
- Discuss transition models to new technologies.
- Provide a brief history of technological milestones from the beginning of the nineteenth century.
- Realize how science, technology, and society interact with and impact one another.
- Discuss the role of technology in sustainable development (SD).
- Focus on challenges related to leveraging applications of technology for SD.
- Show how science, technology, and innovation (STI) serves as a crucial driver of rising prosperity and improved national competitiveness.
- Explore the impact, attributes, and characteristics of appropriate technologies with regard to development problems.
- Know about the performance analysis of technology using the S-curve model.
- Explore the various forms of green technology (GT) practices and products in various sectors.
- Learn about energy mix and energy efficiency and their impact on sustainability.
- Understand the scope and impact of GT products.
- Discuss the concept of technology transfer (TT) and its determinants.
- Realize TT through university innovation and entrepreneurship culture.
- Learn through a TT case study about process in the area of energy-efficient transportation system.
- Provide three end of chapter pedagogical knowledge strategies, namely knowledge acquisition, knowledge possession, and knowledge creation to help understand the above topics and generate new and alternative ideas and solutions.

3.2 Historical perspective

> Engineering or technology is the making of things that did not previously exist, whereas science is the discovering of things that have long existed.
>
> **David Billington**

The word "technology" a combination of the Greek "technē", meaning art or craft, with "logos", meaning word or speech; meant in Greece a discourse on the arts, both fine and applied (Buchanan 2017). It encompasses essentially three meanings: tools and instruments to enhance human ability to shape nature and solve problems (such as a hammer and nail), knowledge of how to create things or how to solve problems (such as to brew beer or to make an atomic bomb), and culture (our understanding of the world, our value systems). Historically, the emergence of human civilization has been closely connected to the development of tools for hunting, agriculture, irrigation and water management, and navigation. In the second, meaning–knowledge–technology becomes reflexive in that understanding of how to make and use tools and instruments becomes encoded and transmissible as technological knowledge and know-how. The development of modern scientific knowledge is related to this second meaning of technology, based on empirical observations, hypotheses, and generalizations on the natural laws concerning the behavior of materials and the living environment (Vergragt 2006). In the third, sense–culture–technology has permeated society to such an extent that separation between technology and culture is no longer meaningful. All human activities, like housing, nutrition, transportation, work, leisure, even art and imagination, become heavily enmeshed with technology. We "own" products of technology by a process of "cultural appropriation," in which the use of technologies is learned, interpreted, and given meaning in everyday life (Hard and Jamison 2005).

The concept of instruments, or of tools independent of the hand that held them, emerged only during the twelfth century. In a tool-using culture according to Postman (1992), tools were largely invented to do two things: solve specific and urgent problems of physical life and serve the symbolic world of art, politics, myth, ritual, and religion. Additionally, in a tool-using culture, technology is not seen as autonomous, and is subject to the jurisdiction of some binding social or religious system.

De Diversis Artibus, complied by Theophilus Presbyter, is the first handbook on technology in Europe (Dodwell 1961). His carefully annotated drawings from the workrooms of the carpenter, the blacksmith, the tailor, and others, illustrated devices disembedded from the activity of artisans. They thus depict, perhaps, the first classification of tools as such.

Moreover, they typify an epoch that may be called the "Technological Age," one characterized by devices that embody human intentions. For example, the hammer became an instrument or tool when conceptualized as a device intended for hammering. For the centuries that followed, this subordination of tools to human purposes implied that technology was a means to personal and communal ends (Samerski 2003).

During the twentieth century, technology was implicitly redefined as the application of industrial tools to the satisfaction of professionally defined needs. So defined, technology cemented the conceptual conflation of tools, needs, and the professions (Illich et al. 1977). It reinforced the prejudice for industrialization, which was exported worldwide as development (Sachs 1992). Increased interest in technological change can be traced back to the years following the Great Depression, when the bicentennial debate on the role of mechanization or employment reemerged (Pigou 1924). Worldwide industrialization entailed the technological transformation of the world. While technologists engineered the machinery of industrial society, professionals shaped its basic creed: only a consumer can satisfy his needs.

Among the most insightful twentieth-century thinkers on the philosophy of technology are, without a doubt, Jacques Ellul and Ivan Illich. In *The Technological Society* (1964), Ellul analyzed, for the first time, the consequences of a society pervaded by professional technicians. He forcefully emphasized the erosion of ethics brought about by technicians of the professional–industrial complex committed to ever-expanding the means for unexamined goals. In *Tools for Conviviality* (1973), Illich argued that tools, in their technical aspects, can breach thresholds beyond which they become critical to society. When a tool acquires such a critical character, it inevitably and counter-productively affects the distribution of political power, the culture, and the social structure of the community which uses it.

By the last quarter of the twentieth century, these reflections on the appropriateness of tools and institutions had identified three independent dimensions of public choice (Illich et al. 1977; Turner 1978):

- The technical choice between hard (oversized machines) and soft (smaller, local tools).
- The ethical option between heteronomy and autonomy, respectively, exemplified by homo "economicus" who is satisfied by consuming the products of slaves and machines, and "*Homo habilis*" who seeks pleasure from doing and making things for immediate use.
- The political decision between right and left, where right refers to centralized decisions about goals and professional control over means, while left connotes the local definition of ends and the communal reappropriation of means.

3.3 *Science and technology*

The science of today is the technology of tomorrow.

Edward Teller

3.3.1 *Science defined*

Scientists see things as they are and ask, why? Engineers see things as they could be and ask, why not?

Robert F. Kennedy

On the simplest level, science is knowledge of the world of nature (Williams 2017). It may be defined as the study of subjects like chemistry, biology, physics, geology, psychology, sociology, and other fields in order to establish given properties, behaviors and interactions about such things. Science aims to gather and validate knowledge. It investigates current phenomena to acquire knowledge. Science in general does not include concrete expressions about real applications. Its core is the performing experiments. Basically, a theory is made, analysis and experimenting are conducted with the use of several controls, and when a specific, measurable result occurs, and can be proven, the theory then becomes scientific law. In general, science can be divided into basic and applied science.

In contrast to technology, science is seen as an organized search for truth and objective knowledge about reality and the laws of nature. Science can be characterized by a rigorous methodology exemplified by Popper's claim that science is an unending process of conjecture and falsification. In practice, the boundaries between modern science and technology have become blurred; moreover, modern philosophy of science treats scientific knowledge to a certain extent as socially constructed (Vergragt 2006).

Science and philosophy have always worked together to try to uncover truths about the world and the universe. Both are a necessary element for the advancement of knowledge and the development of human society. Scientists design experiments and try to obtain results verifying or disproving a hypothesis, but philosophers are the driving force in determining what factors determine the validity of scientific results. Therefore, science and philosophy are a necessary element for the advancement of knowledge and the development of human society (Popper 1963).

Science and technology were by and large separate worlds in the eighteenth century. In this sense, advances in science did not yet call for advanced technological backup. However, science did contribute in two ways to subsequent developments in technology: first through setting the example of experimental method, the very core of the scientific revolution, as a means of analysis (but without having to communicate any of its explicit findings); second through its focus on instrumentation and measurement (Tunzelmann 1996). Thomas Edison played a significant role in the deepening relationship between science and technology, because the remarkable "trial-and-error" process by which he selected the carbon filament for his electric lightbulb in 1879 resulted in the creation of what may be regarded as the world's first genuine industrial research laboratory. From this success the application of scientific principles to technology grew quickly (Buchanan 2017).

3.3.2 Technology defined

When "technology" first appeared in English in the 17th century, it was used to mean a discussion of the applied arts only, and gradually these arts came to be the object of the designation (Buchanan 2017). The German term *Technik* was used in the nineteenth and twentieth centuries. *Technik* is the whole of processes, tools, machines, and systems employed in the practical arts and engineering. Today, technology underlines civilization and encompasses several meanings: tools and instruments to enhance the human ability to shape nature and solve problems (such as a hammer and nail), knowledge of how to create things or how to solve problems (such as to brew beer or to make an atomic bomb), and culture (our understanding of the world, our value systems) (Vergragt 2006). The development of modern scientific knowledge is related to this meaning of technology, based on empirical observations, hypotheses, and generalizations on the natural laws concerning the behavior of materials and the living environment. Figure 3.1 briefly defines technology.

Another definition of technology, one that some engineers may like, is that in which technology refers generally to items of use, created from applied science. Although this definition offers important insight into the nature of modern technology, it is not applicable to all technology. The medieval Chinese had a highly developed technology, but they had neither the notion of laws of nature nor a concept of controlled experiment. However, the National Academy of Engineering (NAE 2004) defines technology as the outcome of engineering; it is rare that science translates directly into technology, just as it is not true that engineering is just applied science.

Figure 3.1 Possible definition of technology.

Yet another definition, favored by many contemporary scholars, is that technology is best understood as a system composed of physical objects and tools, knowledge, inventors, operators, repair people, managers, government regulators, and others. Notice that this definition views technology as being firmly embedded in a social network. It implies that it is difficult to understand technology without understanding the society of which it is a part, and it carries at least the suggestion that technology both influences and is influenced by society at large (Harris et al. 2009).

Relating to technology, technological innovation is generally understood as bringing a new product, process, or service successfully brought to market. Technological innovation thus goes beyond invention, which depicts the elaboration and prototyping of a new technological principle; it is related to diffusion, which refers to the spread of new technology into the wider society (Vergragt 2006).

Finally, technology, taken in its widest domain, is an integrated knowledge system, only one of whose elements is actual physical technology and equipment. Connecting the above elements, engineering aims to provide tools and resources to solve given tasks, independent of the degrees to which the applied knowledge are understood, especially in its interactions.

3.4 *Transition to new technologies*

> Technology is just a tool. It's a powerful tool, but it's just a tool. Deep human connection is very different. It's not a tool. It's not a means to an end. It is the end – the purpose and the result of a meaningful life.
>
> **Melinda Gates**
> *Philanthropist, Duke University Commencement Address (2013)*

3.4.1 *Transition models*

The transition from technology as tool use to knowledge began around the emergence of the first Industrial Revolution more than two centuries ago. The transition to technology as culture accelerated after the Second World War and is closely related to the rise of information and communication technologies, biotechnology, computers, and the Internet (Vergragt 2006). The impact of new technologies on society affects various interests and domains, and the effective and appropriate use of technology requires

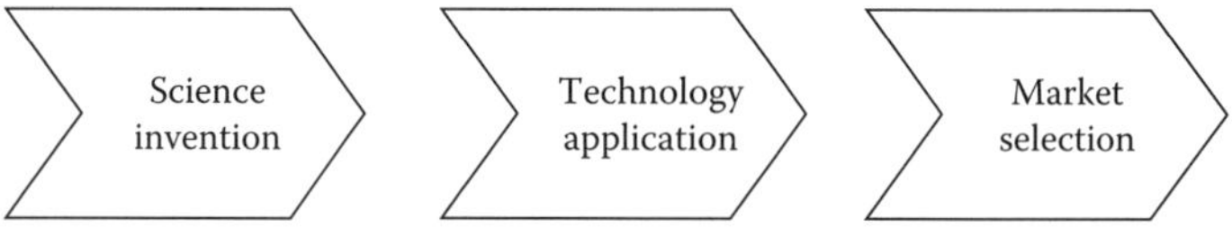

Figure 3.2 The linear model of technological innovation.

consideration of social, cultural, economic, political, and environmental contexts and effects.

The development of science and technology from tools to an encompassing culture obscured questions about their helmsmanship, especially the possibility of democratic decision-making directing them. Such questions were obscured as well by the dominant philosophy and history of science and technology that emerged in the 1930s which posited that scientific invention is driven by innate human curiosity and that scientific discovery eventually leads "automatically" to technological application and commercial deployment. This approach, generally called the "linear model" of technological innovation, in which "science invents," "technology applies," and the "markets select," suggests that some inexorable laws of nature, rather than human choices, are directing this endeavor as shown in Figure 3.2 (Vergragt 2006).

Historically, most technology in the west did not derive directly from science in its modern sense either. The inventors of the seventeenth and eighteenth centuries were not usually well versed in mathematical physics. Instead, they were practical people tinkerers who found solutions to problems by intuition and trial and error. Thomas Edison did his creative work without knowing the electromagnetic theory of James Clark Maxwell. In fact, Edison thought physicists had little to contribute to technology (Harris et al. 2009).

The linear model presented in Figure 3.2, also called "technological determinism," is no longer supported by many academics, but is still widely believed in general society. Research into the processes of scientific and technological discovery has shown that the linear model is not valid, disguising the role of human choice and values in shaping technology, as well as the social and economic interests guiding scientific inventions and technological innovations (Vergragt 2006). It has been replaced by a model like social construction of technology (SCOT) (Pinch and Bijker 1987; Bijker 1995) (Vergragt 2006). This theory includes social actors, problem definitions, and social networks as shown in Figure 3.3. For example, in the SCOT theory, technological innovation is steered by the meaning that "relevant social groups" give to a technological artifact, generating problem definitions that lead in turn to revised technological artifacts, a process highly contingent on its particular context.

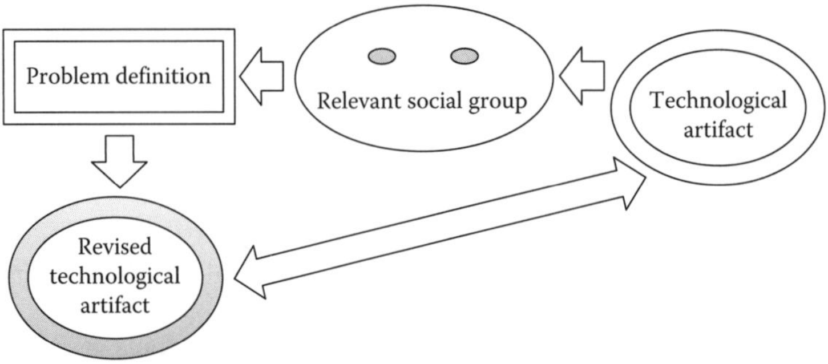

Figure 3.3 Social construction of technology (SCOT).

3.4.2 Technology milestones

The twentieth century without a doubt contributed most significantly to theoretical as well as practical applications of science and technology, which had great global influence on human society. A somewhat detached observation of the development of technology shows that there are certain stages in which several products play a decisive role in the overall level and trends of technology, when compared with other products. Table 3.1 shows the technological milestones and achievements from the beginning of the nineteenth century.

Technology has developed at an increasingly rapid pace in recent years. By taking a look at the latest two decades, we are able to get a good idea of

Table 3.1 Technological milestones from the beginning of the nineteenth century

Duration	Technology
Beginning of 1860	Combustion engine
Mid-1870s	Telephony
Mid-1880s	Calculating machines
Mid-1990s	Wireless telegraphy
Beginning of the twentieth century	Car industry
Beginning of the twentieth century	Building materials
Beginning of the 1920s	Radio broadcast
Beginning of 1930s	Telegraphy
Beginning of the 1930s	Analog computers
End of the 1930s	Airplane industry
Beginning of the 1940s	Television broadcast

(*Continued*)

Table 3.1 (Continued) Technological milestones from the beginning of the nineteenth century

Duration	Technology
Mid-1940s	Electronic general computer
Mid-1940s	Atomic bomb
End of the 1940s	Polaroid land camera
End of the 1940s	Gas turbine
Beginning of the 1950s	Nuclear power plant
Mid-1950s	Transistor and color television
End of the 1950s	Rockets, satellites, and space exploration
End of the 1950s	Microchip
Beginning of the 1960s	Laser and satellite communications
Mid-1960s	Computer operating system
End of the 1960s	Beginning of Internet
Beginning of the 1970s	Genetics and biotechnology
End of the 1970s	Apple II personal computer
Beginning of the 1980s	Cellular communications
Mid-1980s	Personal computers and Microsoft Windows
Mid-1980s	Information technology
Beginning of the 1990s	WWW and Netscape Web browser
Beginning of the 1990s	3D printing; hydrogen fuel cell for automobiles
Mid-1990s	Evolution of Google
End of the 1990s	Interactive television
End of the 1990s	Human genome
Beginning of the 2000s	Wireless Internet
Beginning of the 2000s	High-definition and 3D television
Beginning of the 2000s	GPS
Beginning of the 2000s	Hybrid car and renewable energy development
Beginning of the 2000s	Text messaging
Beginning of the 2000s	Wikipedia as user generated content
Beginning of the 2000s	Digital cameras
Beginning of the 2000s	Nanotechnology
Mid-2000s	iPhone; iPad; and digital music
Mid-2000s	Facebook; YouTube
Mid-2000s	LEED Green Building Rating System; record of efficiency in solar cells
Beginning of the 2010s	Genetically engineered immune cells
Beginning of the 2010s	Smart grid
Mid-2010s	Autonomous driving
Mid-2010s	Robotics
Mid-2010s	High-efficiency and less-expensive solar panels

where things may be headed. Today, the majority of technology has been made portable. In 1992, we saw the first release of the World Wide Web. This was the first time home users got online and began accessing websites. Today, the Internet hosts billions of websites. Around the year 1998, Google started becoming recognized as a pioneer in the field of Internet technology. They opened several work locations, and began growing the business that today impacts almost every single business. Google has become the cornerstone to Internet marketing, regardless of what field you are in. During the 2000s, wireless Internet was incorporated in home computers when Intel added it to their Centrino chip. Today, almost every house is set up with Wi-Fi technology for connectivity. The popularity of mobile phones and text messaging also surged in the 2000s.

Perhaps, the most challenging developments in technology are taking place in the realm of biotechnology. Genetic modification of crops has already made it possible to increase their yield, protect them from insects and pests, and enable them to grow in brackish water, among many other unprecedented alterations. Another fast-emerging technology is nanotechnology, essentially the design of technology at the molecular level. Some current and near-future applications of nanomaterials include catalysts, dry lubrication, coatings, clothing, and materials (Vergragt 2006).

There are major advances in hybrid vehicles and greater interest in future energy development due to global warming and the potential scenario of peak oil. Photovoltaics increased in popularity and decreased in cost as a result of increased public interest and generous government subsidies. Developments in the early 2000s showed upward trends in global renewable energy investment, capacity, and integration across all sectors. For the energy sector, lighting based on nanotechnologies could reduce the energy demand for lighting. In photovoltaics, nanotechnology could raise efficiency and lower costs. Today, the use of renewable energy technologies to provide electricity, heating and cooling, and transportation is now spread across the globe, and recent trends suggest sustained growth worldwide.

3.4.3 *Society interaction domain*

Progress of technology alone cannot improve society. As social systems, communities have dynamic interactions among entities that occur in uncertain ways. Technologies coevolve with societies; technological developments influence society and vice versa. The questions about who makes decisions about the development and direction of new technologies have seldom been asked and even less often answered. "Today, technology confronts civilization with the need to make decisions about how to use the massive power available to society positively rather than

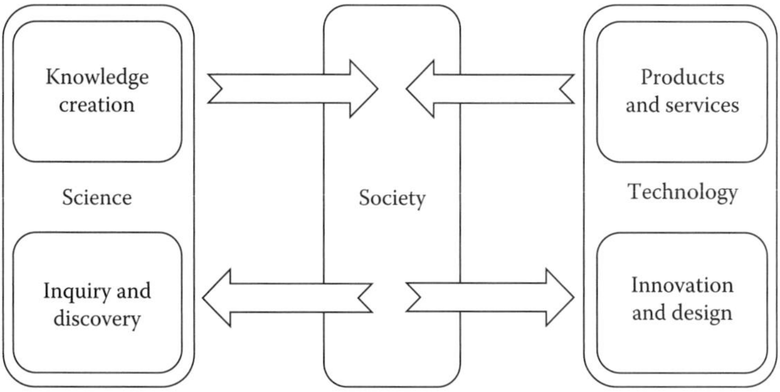

Figure 3.4 Interaction between science, society, and technology.

harmfully by regulating its application to creative social objectives. Importantly, is the ecological problem, whereby the products and wastes of technical processes have polluted the environment and disturbed the environmental balance (Buchanan 2017). This requires a survivable system to pass through unreliable circumstances. Such system, whether in a limited resource setting or a fully industrialized one, will consist of infrastructure and people that will come in contact with the technology. Therefore, technology can be referred to as things which we make, but are developed by applying scientific law and knowing what such things are capable of doing, positive or negative. Figure 3.4 shows society as a central interaction element between the domains of science and technology."

3.5 *Role of technology in sustainable development*

> Cities in the past were built on river-banks. They are now built along highways. But in future, they will be built, based on availability of optical-fibre networks and next-generation infrastructure.
>
> **Shri Narendra Modi**
> *Prime Minister of India*

3.5.1 *The challenge of right technology*

Technologies coevolve with societies (Saviotti 2005). Technological developments influence society and vice versa. Technology is often seen as a path forward to achieve sustainability in all of its aspects: social, environmental, and economic (Tlusty 2015). Technological innovation is at the heart of SD. Innovation itself is one of the Sustainable development goals

(SDGs) (goal 9) and also a means for achieving other goals. It has been identified with a host of problems and solutions that are closely linked with elements and aims of SD. Technology can influence sustainability in a positive way by reducing waste and by increasing efficiency and finding alternatives to inadequate resources. Technology, especially science-based technology, has offered the potential of a healthier and better world through the eradication of disease and material improvements to standards of living. However, resource extraction, emissions of hazardous materials, and air pollution already caused irreversible destruction to the planet.

One of the greatest challenges that society faces in realizing SD is obtaining and implementing the right technologies especially in areas crucial to SD. As discussed in Chapter 2, there are several broad areas, energy, health, communication and information, and agriculture, all of which are crucial for SD and to which technology could contribute significantly.

It is the unintended consequences of an innovation, often brought about by success and overuse of the technology, that ultimately end up challenging sustainability (Tlusty 2015). While access to technology depends to a certain extent on economic resources, it is not only an economic matter. In many instances, legal and institutional bodies impede the development, transfer, and use of technologies for SD. The opportunities and challenges associated with applying new technologies to old problems require guidance from a new breed of professionals and policymakers who can integrate expertise and understanding of their wider implications as a guide to developing strategies for employing technology to tasks of achieving SD.

The persisting contradictions between pursuits of a better life created and supported by technology and the increasing environmental degradation and continuing poverty call for a serious investigation of the nature of technology and its relationship to a sustainable society. In the context of the effort to catalyze a great transition to a sustainable global society, in which deep changes in culture, values, consumption patterns, governance, business, and institutions are envisaged (Raskin et al. 2002), questions about the role of technology become even more pressing. For example, would a great transition society require an intensive use of technology to lessen the environmental degradation of the ecosphere, or might technology play a much more modest role in such a society? Would that society essentially return to the time before the first Industrial Revolution, when technology offered a limited, incremental extension of human capacity to transform nature? In either of these scenarios, the unique challenge society faces is to leverage technological strengths and expand strong entrepreneurial spirit into a broader business innovation culture (Vergragt 2006).

3.5.2 Science, technology, and innovation

STI infuses all aspects of modern life. STI serves as a crucial driver of rising prosperity and improved competitiveness. Science has always functioned on two levels that may be described as curiosity driven and need driven, and they interact in sometimes surprising ways. On the other hand, the frontiers of science are defined by the capabilities of instrumentation, that is, of technology. The needs of pure science are a huge but poorly understood stimulus for technologies that have the capacity to be disruptive precisely because these needs do not arise from the marketplace. Necessity is said to be the mother of invention, but in all human societies, necessity is a mix of culturally conditioned perceptions and the actual physical necessities of life. The concept of need, of what is wanted, is the ultimate driver of markets and an essential dimension of innovation (Vergragt 2006). Typically, science and technology are applied to innovation within a social and economic context.

There are two essential STI issues that need to be tackled simultaneously in the post-2015 SD agenda. First, innovation-driven growth is no longer the prerogative of high-income countries alone; some developing countries have achieved significant economic growth through the creation and deployment of STI capacity. But this has not been the case for all countries. Second, STI policy has often been pursued independently of the broader developmental agenda; it is important that STI be integrated into public policy goals, giving particular focus to the nexus between STI, culture, education, and development (ITU et al. 2015).

What causes innovations to be adopted and integrated into economies depends on their ability to satisfy some perceived need by consumers, and that perception may be an artifact of marketing, or fashion, or cultural inertia, or ignorance. Some most profitable industries in the developed world like entertainment, automobiles, clothing and fashion accessories, and health products depend on perceptions of need that go far beyond the utilitarian and are notoriously difficult to predict. And yet these industries clearly depend on sophisticated and rapidly advancing technologies to compete in the marketplace. Of course, they do not depend only upon technology. Technologies are part of the environment for innovation, or in a popular and very appropriate comparison part of the innovation ecology (Marburger 2011).

Advances in renewable and restorative systems will be mandatory. Until these truly sustainable resources occur, it is needed to embrace additional solutions outside of technology to help meet our sustainability goals, if nothing more than to provide the time to develop the renewable and restorative innovations. However, solutions are not all-or-none phenomena. We do need restorative technologies to reduce our impacts on our planet, but we cannot wait for the single solution so we can take a technologically aided large jump to be sustainable. To more rapidly travel down the path toward sustainability, we need innovation, and more importantly, we need to couple that to simple, nontechnological, and immediate steps (Tlusty 2015).

A well-functioning STI ecosystem needs to include well-functioning institutions; an educated workforce; sound research and education infrastructure and linkages between public and private innovation actors; enterprises committed to R&D; as well as a balanced intellectual property (IP) rights framework (ITU et al. 2015). To quickly join the path toward sustainability, innovation is needed, and more seriously, linkage to simple, nontechnological domains is required.

3.6 Alternative and appropriate technologies

> An expert is a person who has made all the mistakes that can be made in a very narrow field.
>
> **Niels Bohr**

In contrast to the areas of so-called high-tech innovation and fast-emerging technologies which have been considered so far, there is a very different strand of technologies, often called intermediate (Schumacher 1973) or appropriate technology (AT) that are inherently people and environmentally friendly. Through the use of regulation, fees, taxes, or incentives, society can exercise control over which technologies are permitted to enter the marketplace, allowing only those that are very low risk to be implemented. Two categories of technologies that are often described as having these attributes are alternative technology and/or AT.

3.6.1 Alternative technology

The term alternate technology was first used by Peter Harper from the Centre for Alternative Technology in Wales in the 1970s and is still commonly used as a label to describe technologies that are relatively benign. It refers to technologies that are considered friendly to the environment, conserves or renews natural resources, encourage recycling, use renewable resources, and limit the usage of fossil fuels. Solar hot water heating, anaerobic digestions, landfill gas extraction, biofuels, and wind energy generators are examples of alternative technologies. In this regard, these technologies use fewer or no critical elements due to better design or manufacturing procedures.

3.6.2 Appropriate technology

The concept of AT stemmed from the work of British economist Dr. Fritz Schumacher in the 1970s (Schumacher 1973). It emerged in the context of the 1973 energy crisis and 1970s environmental movement. AT has been defined as technology tailored to fit the psychosocial and biophysical context prevailing in a particular location and period (Willoughby 1990). AT is small scale, energy efficient, environmentally sound, labor-intensive, and controlled by the local community. The term is used in two primary contexts: technology that most effectively meets peoples' needs in developing or limited resource settings; and technology that is environmentally and socially acceptable in the developed world. AT best fits with the community it serves because it is created by the people to fulfill a need. Therefore, the communities are placed at the center of decision-making and create technologies that will best serve their communities in the long term. Figure 3.5 shows the criteria for AT (Vergragt 2006).

During the mid-1970s, the AT movement expanded from its initial focus on low-income countries to consider the problems in industrialized high-income countries. Advocates of AT were concerned about social as well as environmental problems. AT was designed not to dominate nature but to be in harmony with it. It includes the concept of alternative technology but, in addition to considering the environmental attributes of a technology, also considers its ethical, cultural, social, and economic aspects. It can refer to technologies that are either the most effective for addressing problems in developing countries or ones that are socially and environmentally responsible in industrial countries.

AT has been advocated as a solution for rural development problems, but has also gained support as a direction for sustainable technologies. However, it has often been identified as "cheap," "second hand," or second best by adherents of massive Western TT to developing countries and by ideologues who believe in modernization by technological innovation (Vergragt 2006). AT is a grass roots approach to technology that builds a strong sense of community and encompasses benefits that span across social, environmental, cultural, economic, and spiritual facets.

Effective: In theory and in practice
Safe: Not easy to use incorrectly
Affordable: Initial and recurrent costs
Acceptable: To all who are affected by it
Sustainable: Can be maintained and repaired

Figure 3.5 Criteria for AT.

One of the best-known early proponents and popularizers of AT was the British economist Schumacher (1973), who talked about "intermediate technology" in his book *Small Is Beautiful: A Study of Economics as if People Mattered*. He was principally concerned with development in low-income countries, and recommended a technology that was aimed at helping the poor in these countries to do what they were already doing in a better way.

Clarke (1974) differentiated between the AT response and the "technological fix" responses to environmental problems. For example, he characterized the technological-fix response to pollution as solving pollution with pollution control technology"; the AT response, instead, would be to invent nonpolluting technologies. Similarly, the technological-fix response to exploitation of natural resources was to use resources more cleverly; the AT response was to design technologies that only used renewable resources.

3.6.3 Attributes and characteristics of technological appropriateness

Attempts to invent and design different types of technology that fit with natural systems are not new. The AT movement which blossomed in the 1970s attempted to do just this. Appropriateness of scale is a fundamental feature of successful technology implementation.

The main strength of ATs exists in its appropriateness. Sometimes, an AT will be appropriate for some applications, others may only be appropriate for one specific application, and even some designed technologies are not appropriate or have limited appropriateness. By using tangible tiers, obstacles in design engineering can be clearly observed and addressed to find solutions; however, the ultimate tier permanently incorporates intangible issues to ensure AT diffusion. In addition, AT needs to be adapted to the human, physical and financial resources available in the environment where they will be used. Defining the attributes and characteristics of ATs need to take place early in the product development cycle to ensure that the technology is adapted to user needs, rather than users having to adapt to the technology. Figure 3.6 shows AT as part of the ecosystem. The component of people incorporates their knowledge, skills, culture, and beliefs.

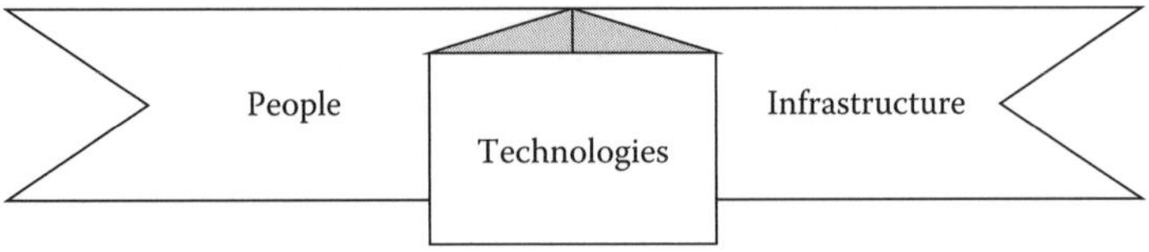

Figure 3.6 AT as part of the ecosystem.

The component of infrastructure includes equipment, facilities, communication systems, finances, policies, and regulations.

3.6.4 Tiers of technological appropriateness

The first tier of technological appropriateness (Figure 3.7) is the level where most ATs are designed. As a technology, an AT is designed based on a set of specifications. The basic understanding of the term "specification" exists in two tangible aspects: technical and economic. They reflect one of the famous terms in engineering, "price-to-performance," in which the significance of a technology is indicated by its economic value for each technical performance unit (Sianipar et al. 2013).

Above the first tier, a basically AT can be further engineered to incorporate the environmental aspect, which will produce an "environmentally appropriate" AT. There has been some evidence of this type of appropriateness, but there are two interpretations of how the environmental aspect should be diffused into the design of an AT. The first interpretation is that an AT incorporates the preservation of environmental conditions into its application (Sianipar et al. 2013).

Although the second tier of appropriateness already incorporates an issue other than the first two basic aspects, it does not incorporate the key to the successful introduction of an AT application: the social aspect. Although considerations in AT design start by involving the social aspect, the whole concept will become closely related to and will automatically incorporate intangible aspects. Sometimes, some parameters in previous tiers can be clearly stated in their own aspect, but this is not the best way

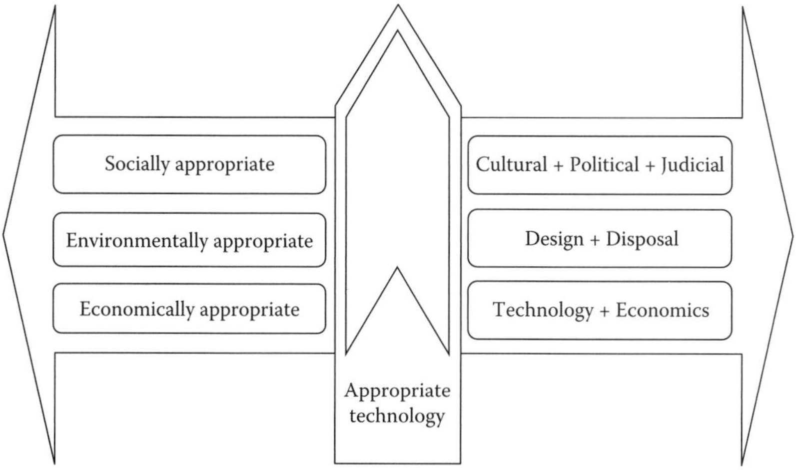

Figure 3.7 Tiers of technological appropriateness.

to truly solve community problems with AT. The most common occurrence is that engineers pick a set of given problems and find a single solution to solve them all.

When engineers require more information, they prefer to obtain it numerically and to treat each problem as a technical, economic, or environmental parameter. In other words, engineers tend to design AT as only a technical artifact (Kroes 2012). That type of engineering approach is not ultimately wrong but, as previously explained, cannot ensure the successful diffusion of an AT into a community's daily routines or further into the community's long-term survival efforts. It is crucial to reach survivability beyond sustainability because local resilience is mostly influenced by survival capabilities during future changes (Kaplinsky 1990).

3.7 Green technology

All our wisdom is stored in the trees.

Santosh Kalwar

3.7.1 GT defined

The history of GT goes back to the pre-Industrial Revolution era in the seventeenth century, when windmills were used to power looms and mills. The situation changed with the entrance of fossil fuel powered engines and since then GT took a back seat. However, the potential in GT to satisfy energy requirements cannot be denied. Today, GT covers a broad area of manufacturing, production, and consumption technology. It combines different methods and technologies for continually evolving new solutions. It involves the use of environmental technologies for monitoring and controlling, pollution prevention, remediation, and restoration. Examples of GT subject areas are energy, green building, green transportation, environmentally preferred purchasing, green chemistry, and green nanotechnology.

GT is a general term often used interchangeably with clean technology. Other terms such as "environmental technologies," "climate-related technologies," and "mitigation and adaptation technologies," essentially refer to the same thing (Clarke 1974). The term which has been adopted in this book is the application of the environmental science and technology for the development and application of products, equipment, and systems to conserve the natural resources and environment as well as to minimize or mitigate the negative impacts on the environment from human activities (Chu 2013).

GT is a link that connects sustainability with applications, where natural resource productivity is efficiently maintained by thoroughly planning the conservation and exploitation of resources such as air, soil, water, plants, and animals. GT is a driver of the future economy that would contribute to overall green growth and SD. The process of transferring existing technology to cost-effective and environmentally friendly applications should be taken simultaneously. The ideal technology should be efficient, practical, cost-effective, and pollution free. GT can be applied in various sectors including energy, building, water and waste management, agriculture, and transportation.

GT's goal is to replace practices and methods that damage or deplete natural resources with alternative practices that are sustainable and efficient. By sharing in alternatives like renewable energy, it is possible to cut down on pollution and avoid depleting natural resources. Main areas of interest in GT are sustainability, reuseability, source reduction, innovation, and viability. GT covers a wide range of services for improving efficiency, preserving natural resources, and reducing emission.

3.7.2 Adoption of GT

The adoption and diffusion of GTs like other technologies depends on attaining a balance between economic profitability and environmental sustainability. Adoption and diffusion of technology are two interconnected concepts expressing the decision to use or not to use and the spread of a given technology among economic units over a period of time. The adoption and diffusion of an innovation within an institution does not ensure its effective integration in other institutions or its sustained use (UNIDO 2014).

Foster (1986) proposed a simple theory to explain technological evolution. He suggested that technological performance on some key dimension, as a function of research, effort evolved along S-shaped curve (Figure 3.8). With continued research, the technological platform crosses a threshold after which it makes rapid progress. After a period of rapid improvement in performance, the new technology reaches a period of maturity, when progress occurs very slowly or reaches a ceiling. In this situation, the variation point in the S-curve is significant because it signals the approaching doom of the old technological platform and the need to focus on a new platform. So, a good strategy is to switch from an old technology on the mature or upper flat of its S-curve to a new technology on the upward or growth trajectory of its S-curve.

The adoption and use of GTs involves the use of environmental technologies for monitoring and assessment, pollution prevention and control, and remediation and restoration. Monitoring and assessment technologies are used to measure and track the condition of the environment, including the release of natural or anthropogenic materials of a harmful nature. Prevention technologies avoid the production of environmentally hazardous substances or alter human activities in ways that

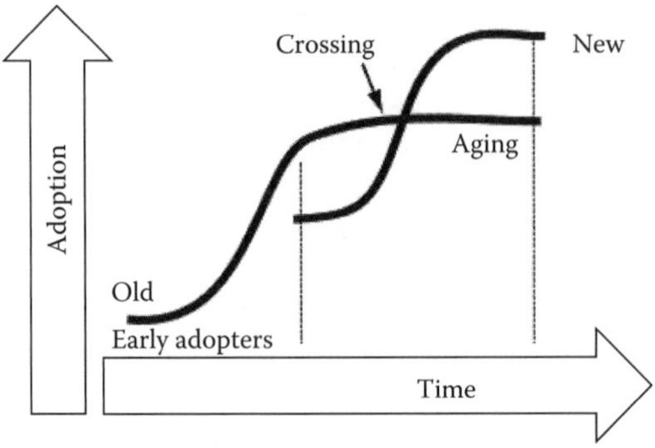

Figure 3.8 Adopters of innovation.

minimize damage to the environment; it encompasses product substitution or the redesign of an entire production process rather than using new pieces of equipment. Control technology renders hazardous substances harmless before they enter the environment. Remediation and restoration technologies embody methods designed to improve the condition of ecosystems, degraded through naturally induced or anthropogenic effects (Soni 2015).

To encourage the adoption of GTs, some governments are providing subsidies to consumers who buy solar panels, wind turbines, electric vehicles, waste management, carbon capture and store, and the like. When governments offer consumer subsidies on GTs, the goal is generally to achieve some critical level of demand that will serve to bring down prices, encourage product improvements, establish wider familiarity, and ensure a continuing stream of future customers. Setting a subsidy (e.g., one year, then second year, and so on or multiyear) program that will elicit the desired response while minimizing government expenditures is not easy. Part of the problem is that policy makers do not know in advance how consumers will respond to a given subsidy (Stauffer 2013).

3.7.3 GT products

GTs are goods and services that improve the quality of air, water, soil, waste and noise-related problems. They vary from very complex and expensive advanced technology (high-tech) to more simple (low-tech) solutions. GT products are items which factor environmental awareness into their design and use. They aim to reduce waste, cut pollution, and diminish fossil fuel use. Some of the major types of GT products include energy creation products, green chemicals, sustainable or recyclable products, and technology that run on alternative energy. Products that help create alternative energy, such as solar panels and thermal heating discs, are some of the most important GT products used in everyday life (Soni 2015).

Generally, GT is more expensive than the technology it aims to replace, because it accounts for the environmental costs that are expressed in numerous production processes. Because it is relatively new, the associated development and training costs can make it even more costly in comparison with typical technologies. The benefits are also dependent on other factors such as related infrastructure, technology readiness, human resource potentials, and geographic features. The adoption and circulation of GTs can be confined by a number of other obstacles. Some may be institutional, such as the lack of an appropriate regulatory structure; others may be technological, financial, political, cultural, or legal in nature.

3.8 *GT practices*

> Don't get me wrong: I love nuclear energy! It's just that I prefer fusion to fission. And it just so happens that there's an enormous fusion reactor safely banked a few million miles from us. It delivers more than we could ever use in just about 8 minutes. And it is wireless!
>
> **William McDonough**
> *Stanford University*

3.8.1 *Energy mix*

This use of fossil fuels (initially mainly lignite and coal; later oil and natural gas) has served humanity well during the historically short time period of about two centuries, having allowed the world population, with its supporting agricultural and industrial productivity, to grow to previously unimaginable numbers while providing an average standard of living that is higher than ever before.

In the context of energy options, "sustainable" implies the ability to provide energy for indefinitely long time periods without depriving future generations and in a way that is environmentally friendly, economically viable, safe and able to be delivered reliably. It should thus be concluded that the term "sustainable" in this context is more restrictive than the term "renewable" that is often applied to energy derived from wind, sunlight, biomass, waves, tides, and geothermal resources, which for certain applications do not meet all the criteria of sustainability (Brook et al. 2014).

Replacement of fossil fuels is needed to sustain society while mitigating environmental impacts, and sustainable forms of renewable and nuclear energy offer a realistic and effective way of achieving this goal.

3.8.1.1 Renewable energy

The energy sources popularly known as "renewables" (such as wind and solar) will be hard pressed to supply the needed quantities of energy sustainably, economically, and reliably. They are inherently intermittent, depending on backup power or on energy storage if they are to be used for delivery of base-load electrical energy to the grid. This backup power has to be flexible and is derived in most cases from combustion of fossil fuels (mainly natural gas). If used in this way, intermittent energy sources do not meet the requirements of sustainability, nor are they economically viable because they require redundant, underutilized investment in capacity both for generation and for transmission (Brook et al. 2014).

Solar technologies extract directly from the vast power of the sun and use that energy to produce power. Solar electric systems consist of three main components: cells that convert sunlight into electricity; inverters that convert electricity into alternating current; and batteries that store extra electricity produced by the system.

Biomass is an organic material with stored sunlight in the form of chemical energy. Biomass technologies break down the organic material to release the stored energy and produce biofuels. Ethanol is the most common biofuel. It is an alcohol made from the fermentation of biomass high in carbohydrates. Biomass from corn, wheat, soybeans, and wood can also be used to produce chemicals and materials that are usually obtained from petroleum.

Wind turbines are highly sophisticated power conversion systems that capture the wind's energy by means of blade and rotor designs and convert it to mechanical energy. Mechanical drive systems, combined with advanced generators and control topologies convert the mechanical energy into electricity.

Not all renewable energy resources come from the sun. Geothermal energy extracts the earth's internal heat for electric power generation, and the heating and cooling of buildings. This heat comes out from the core, heating the neighboring area, which can create underground reservoirs of hot water and steam.

The ocean produces two types of energy: thermal energy from the sun's heat and mechanical energy from the tides and waves. Ocean thermal energy can be used for many applications, including electricity generation. Electricity conversion systems use either the warm surface water or boil the seawater to turn a turbine, which activates a generator.

Some natural resources, such as moving water, wind, and sunshine, are not at risk of depletion from their use for energy production. Biomass, however, is a renewable resource only if its rate of consumption does not exceed its rate of regeneration. Renewable energy technologies turn these fuels into usable forms of energy, most often electricity, but also in the form of heat, chemicals, or mechanical power (NRC 2016).

The recent growth of renewable energy markets has been accompanied by remarkable technology improvements and cost reductions. However, the new technologies must compete with fossil fuel technologies and depend on the overall evolution of demand. A wide range of energy-producing technologies and equipment have been developed over time to take advantage of these natural resources. As a result, usable energy can be produced in the form of electricity, industrial heat, thermal energy for space and water conditioning, and transportation fuels (NRC 2016). Table 3.2 shows several types of renewable energy resources and their corresponding technologies.

The principal impediment to the wider commercialization of renewable energy technologies (e.g., solar PV, wind, biopower, biofuels, concentrating solar power, ocean energy) is their higher cost compared to conventional technologies and, in the case of wind, wave and solar, their short-term variable character. To overcome cost barriers, carbon emissions and related externality costs need to be accounted for to raise the price of conventional fossil fuels relative to renewables (OECD 2007).

3.8.1.2 Nuclear energy

The practical generation of nuclear energy was demonstrated on the second day of December 1942 when the first human-controlled self-sustaining nuclear fission reaction was achieved at the University of Chicago under the guidance of Italian-born physicist Enrico Fermi. This experimental reactor (in those days called an "atomic pile") made use of "slow" (usually called "thermal") neutrons, capable of sustaining a chain reaction in the rare "fissile" uranium isotope U-235 that constitutes only 0.7% of

Table 3.2 Types of renewable energy resources and their corresponding technologies

Type of energy	Associated technologies
Hydro	Turbines and generators
Wind	Wind turbines; generators; energy converters; transformers; cables
Solar	Photovoltaic cells; energy converters; storage systems
Tidal	Underwater tidal turbines and generators
Biomass	Feedstocks; converters; generators
Geothermal	Pumping systems and generators
Solid waste	Combustion, boilers, and generators
Hydrogen	Fuel cells

natural (mined) uranium; the rest (99.3%) being the "fertile" isotope U-238. From this small experimental reactor, an entire industry emerged that has led to hundreds of operating nuclear power reactors as well as numerous research reactors around the world, delivering clean energy and a large number of products and services for use in many human activities, including medical diagnosis/therapy, industry, and agriculture (Brook et al. 2014).

There are two types of nuclear reactions, namely fission and fusion that release energy due to the existence of highly powered atomic bonds between particles within a nucleus. In nuclear fission, an atom is split into smaller subatomic particles. It is the basic principle behind the functioning of controlled atomic bombs. Nuclear fusion, in contrast, is the exact opposite of fission. It occurs when two or smaller atoms combine together, creating a larger and heavier atom and releasing vast amount of energy. Nuclear fusion reactions appear in cosmic bodies like sun and stars. In fact, fusion is what fuels the sun. To carry out an artificially fission reaction, a highly controlled environment must be created.

Nuclear energy from fission of uranium and plutonium is sustainable because it can provide the world with clean, economical, and reliable energy. Such energy (transmuted from U-238) is capable of replacing most of the performed efforts by the combustion of fossil fuels. However, many environmental organizations and governments oppose the application of abundant nuclear energy because it is unsafe and has links to proliferation of nuclear weapons.

Today, scientists are working on controlling nuclear fusion in order to make a fusion reactor to produce electricity. They are encouraged by the idea that fusion creates less radioactive material than fission and has a nearly unlimited fuel supply. Research continues into ways to better harness the power of fusion, but research is in early experimental stages. Many countries take part in fusion research led by the European Union, the United States, Russia, and Japan, with active programs also underway in China, Brazil, Canada, and Korea.

The first and largest machine of its kind is currently under construction at the French scientific research center "Cadarache," which specializes in nuclear power research. It is called ITER, Latin for "The Way." It is one of the most ambitious energy projects in the world today and is expected to usher in a new era of nuclear fusion-powered electricity something scientists and engineers have been working toward for over 40 years (Orwig 2016). The goal of ITER is to operate at 500 MW (for at least 400 s continuously) with less than 50 MW of input power, a tenfold energy gain. No electricity will be generated at ITER. The conceptual design of Demo is expected to be completed by 2017, with construction beginning in around 2024 and the first phase of operation commencing from 2033.

Fusion power offers the prospect of an almost inexhaustible source of energy for future generations, but it also presents so far insurmountable scientific and engineering challenges. This energy is inexpensive, virtually limitless, cleaner with no greenhouse gases (GHGs), and with little or no nuclear waste.

If commercial-scale fusion plants were to become a reality, we would have an unlimited, nearly free, clean source of energy. And if limited energy supply and climate change were our only problems, or, should We say the only geophysical constraints imposing themselves on our way of life, then that would be the happy ending to the story (Siegel 2012). Fusion has much promise, but at its current state of engineering design, it is not more sustainable than fission. Table 3.3 outlines the technologies for fission and fusion power.

3.8.2 Energy efficiency

Energy consumption indicators are an essential element in development. They are used to identify target priority areas in industry. Energy intensity, defined as energy consumption relative to GDP, is an indicator that is commonly used to measure progress in energy efficiency. While increased energy use clearly has many benefits, it has negative impacts as well. These negative impacts are experienced globally and locally in the form of climate change and degradation of local environments in terms of poor air quality, degradation of soil, and resource depletion such as water and noise pollution (IEA 2014).

In general, sources of energy consumption may be divided into four sectors: residential, commercial, industrial, and transportation. All above sectors are growing energy consumers worldwide. Energy efficiency activities are therefore increasingly important, depending not only on the total power consumption but also on the potential for cost-effective improvements.

Energy efficiency sometimes is considered to be an easy solution for achieving instant energy savings. Efficient technical solutions are available today for most applications and uses. Technological developments are offering and will offer a range of technical solutions for improving energy efficiency, but there are barriers: organizational, financial, and behavioral that need to be addressed in a holistic way. There is considerable technical

Table 3.3 Corresponding technologies for fission and fusion power

Energy process	Associated technologies
Fission	Reactors; coolants; controllers; generators
Fusion	Reactors; magnetic confinement fusion; coolants; controllers; generators

potential for energy efficiency improvements along the entire energy value chain: from the extraction of primary energy resources: oil, gas, coal, uranium, and others, to their transformation into heat and electricity, transportation, and distribution of energy, and ultimately to the end use by appliances, equipment, and devices (WEC 2013).

Energy efficiency technologies aim to make such improvements, with some options being well known and proven over many years of application. In the residential and commercial sectors, technology contributes to reducing electricity consumption through advanced controls and loads. Energy management control systems (EMCSs) have already proven to cut energy use with a reasonable rate of payback for businesses and residences. Technology advances can help the industrial sector become more energy efficient by reducing inventories, efficient utilization of space and better power supply systems. The most positive impact technology can have on easing energy use by the transportation sector is in the form of enabling more energy-efficient transportation options, as is discussed in Section 3.12. Table 3.4 shows several types of energy efficiency approaches and their corresponding technologies.

Table 3.4 Examples of energy efficiency approaches and their corresponding technologies

Type of energy efficiency	Associated technologies
Motor system optimization	Reengineering and/or use of software
Isothermal melting process	Immersion heaters in a closed loop multiple bay arrangement
Server virtualization	Software application
Green buildings	Passive building materials; renewable energy; heating and cooling services; EMCSs, energy-efficient lighting; occupancy sensors; operating practices
Energy-efficient devices	Energy Star appliances, compact fluorescent light, LED; motion detection lighting, or programmable thermostats which reduce the establishment's energy consumption
Industry	Standard process; efficient equipment; data collection; control and instrumentation
Transport	Switching modes (e.g., from road to rail); public transport; road maintenance
Power generation	Boilers and furnace control; generating units; cogeneration of electricity and heat
Transmission and distribution systems	Transmission and distribution line upgrading; improved control and operation
Oil and gas	Exploration and production

Improving energy efficiency is the cheapest, fastest, and most environmentally friendly way to meet a significant portion of the world's energy needs. Improved energy efficiency reduces the need for investing in energy supply. Many energy efficiency measures are already cost-effective, and they will pay for themselves over their lifetime through reduced energy costs. There are diverse barriers to greater deployment of energy-efficient options. Consumers are often ill-informed and few are concerned with energy efficiency when buying appliances, homes, or cars. Business management tends to give energy efficiency a low priority in decision-making. There are opportunities for energy efficiency that consumers never see because the manufacturers of refrigerators, televisions, or cars do not always take full advantage of the technologies that exist to make their products more energy efficient (OECD 2007).

3.8.3 GHG emissions reduction

GHGs are a group of compounds that are able to trap heat (longwave radiation) in the atmosphere, keeping the earth's surface warmer than it would be if they were not present. These gases are the fundamental cause of the greenhouse effect. An increase in the amount of GHG in the atmosphere enhances the greenhouse effect which is creating global warming and consequently contributes to climate change (Allison 2010).

GHG allows sunlight (shortwave radiation) to pass through the atmosphere freely, where it is then partially absorbed by the surface of the earth. But some of this energy bounces back out toward space as heat. Of the heat emitted back to space, some is intercepted and absorbed by GHGs in the atmosphere.

Carbon dioxide (CO_2) is the most important GHG. It is produced from the use of fossil fuels to generate electricity (e.g., coal-fired power plants), in manufacturing processes, and to power vehicles. Methane (CH_4) is another effective GHG, but with a shorter lifespan in the atmosphere than CO_2. It comes from a variety of sources. Some sources are natural: methane escapes wetlands and oceans at a significant rate. Other sources are anthropogenic, which means man-made. The extraction, processing, and distribution of oil and natural gas all release methane. Nitrous oxide (N_2O) occurs naturally in the atmosphere as one of the many forms nitrogen can take. However, large amounts of released nitrous oxide contribute significantly to global warming. The main source is the use of synthetic fertilizer in agricultural activities. N_2O is also released during the manufacturing of synthetic fertilizers. Motor vehicles release N_2O when operating with fossil fuels like gasoline or diesel (Beaudry 2016).

Ozone is a naturally occurring gas positioned in the upper part of the atmosphere, protecting earth from much of the damaging sun rays. In the lower parts of the atmosphere, ozone is produced as other chemicals break down (e.g., nitrogen oxides). This ozone is thought a GHG, but it is short-lived and although it can contribute to warming, its impact is usually local rather than global.

GHGs take many years to leave the atmosphere. CO_2, CH_4, N_2O, and the fluorinated gases are all well-mixed gases in the atmosphere (Ramaswamy et al. 2001). They do not react to changes in either temperature or air pressure and thus do not get removed easily like water, which condenses to become rain or snow. Their long atmospheric lifetimes allow them to have a lasting effect on global warming and climate change.

Climate change is caused by increasing levels of GHG emissions in the atmosphere and efforts must focus on reducing those levels to avoid more rapid and severe changes in the climate. This is a major challenge as much of the global economy is powered by fossil fuels that, for example, support transportation and the generation of energy. Table 3.5 shows several types of GHG approaches and their corresponding technologies.

3.8.4 Pollution reduction and removal

Air pollution is a main environmental issue in urban areas. Sometimes, city air holds high levels of pollutants that are unsafe to human health. It may lead to human health problems and reduced visibility, and it may impair the health of plants and wildlife. Table 3.6 shows several types of pollution reduction and removal approaches and their corresponding technologies.

Conventional air pollution management programs focus on controlling the source of air pollutants. This approach successfully reduces the emission of new air pollutants but does not address the pollutants already in the air. Innovative approaches can be adopted to remove existing air pollutants, thereby reducing pollution concentrations to an acceptable level. One way to reach that goal is the use of urban vegetation which can reduce air pollutants through a dry deposition process and microclimate

Table 3.5 Examples of GHG emissions technologies

Process	Associated technologies
Carbon capture and storage	Turbines; particle removers; coolers and condensers
Alternative forms of transportation	Fuel-efficient vehicles, cycling, or mass transit
Methane emissions reduction and/or reuse	Enteric fermentation reduction, landfill gas capture, manure biogas recovery

Table 3.6 Examples of pollution reduction and removal approaches and their corresponding technologies

Process	Associated technologies
Flue gas desulfurization	Methods that do not produce hazardous secondary waste or by-products
Catalytic/thermal destruction of NOX	Basic thermal oxidation equipment to effectively incinerate a wide variety of hazardous industrial wastes
Dioxins reduction	Removal methods and/or practices
Low volatile organic compound paints and sealers	Application techniques and emission controls
Solvent recovery systems	Distillation; liquid–liquid extraction; absorption systems; film evaporation; crystallization; membrane separation

effects. Other techniques that relay on advanced technologies are widely used (Yang et al 2008).

3.8.5 Reducing, reusing, recycling, and recovery

Waste minimization, or dematerialization, can be defined as achieving material and energy reduction per unit of product or service produced. The most successful way to manage waste is to minimize producing it (waste minimization). Making a new product requires a lot of materials and energy—raw materials must be extracted from the earth, and the product must be fabricated then transported to wherever it will be sold. As a result, reduction and reuse are the most effective ways you can conserve natural resources, protect the environment, and save money (Haghseta 2003).

The real cost of waste is not just the cost of discarded materials; it includes ineffective use of raw materials, unnecessary use of energy and water, defective products, waste disposal of by-products, waste treatment, and wasted work.

Engineers are involved in all steps of waste management. They design and use materials to package products, and use recyclable and reusable materials. Chemical engineers in particular develop environmentally recyclable materials. Engineers also investigate ways to accelerate the decomposition process, develop industrial systems that burn trash for energy at power plants, and design innovative landfills that are more economical and reduce pollution.

There are several technologies and practices to reduce or eliminate the creation of waste materials as a result of operations. Table 3.7 shows several types of pollution reduction and removal approaches and their corresponding technologies.

***Table* 3.7** Examples of pollution reduction and removal approaches and their corresponding technologies

Process	Associated technologies
Air pollution control	
Collecting and reusing or recycling waste materials	Natural ingredient materials; packaging reinnovation
Managing and/or recycling wastewater effluent	Reusing treated wastewater; gray water; industrial cooling processes
Composting solid waste	Waste pretreatment; digesters; digestates
Remanufacturing	Rebuilding and refurbishing

3.8.6 *Agriculture*

The objective of agricultural and technology development has been to advance productivity and soil and water conservation; sustain reasonable costs of food and fiber; improve human nutrition, food quality, and safety; and contribute to the economy as a whole.

Agriculture in a sustainable society will provide plentiful food supplies at prices local populations can afford, at a level of quality that promotes health, and without damage to the environment or reduction of biodiversity. Achieving this goal will require a prudent combination of new technologies and ecological sensitivity. Ecological agriculture would be accepted and practiced as standard throughout the world, taking different forms in different places depending on tradition, local circumstances, and specific opportunities (Vergragt 2006).

Agricultural policy issues with respect to SD include R&D priorities; regulations regarding trade, food prices, land ownership, and environmental quality; and food security and employment.

3.8.7 *Natural resource conservation*

The term conservation came into use in the late nineteenth century and referred to the management, mainly for economic reasons, of such valuable natural resources as timber, fish, game, topsoil, pastureland, and minerals, and also to the preservation of forests, wilderness, and watershed areas. In recent years, the science of ecology has clarified the workings of the biosphere, that is, the complex interrelationships among humans, other animals, plants, and the physical environment (Show 2010). At the same time, burgeoning population and industry and the ensuing pollution have demonstrated how easily delicately balanced ecological relationships can be disrupted (air pollution; water pollution; solid waste). Table 3.8 shows several types of approaches to natural resource conservation and their corresponding technologies.

Table 3.8 Examples of natural resource conservation approaches and their corresponding technologies

Process	Associated technology
Managing land resources	Organic agriculture; reforestation, water resource management and ecotourism projects
Water treatment	Coagulation, flocculation, and filtration
Waste water management	Separation, screening, and sedimentation
Managing storm water	Storm pipes and drains; soil and pavement; rain gardens
Conserving soil, water, or wildlife	Wildlife-friendly farming; land sparing
Organic agriculture	Alternatives to use pesticides, fertilizers, genetically modified organisms, antibiotics, and growth hormones
Implementing a paperless office or reducing paper usage and consumption	Digital media

3.8.8 *Green health*

Green health is emerging from the convergence of the global health economy. Scientifically, green health embodies the epidemiological connections between human health and the environment. Culturally, it represents the understanding of nature as a powerful binding force between people, their health, and the world in which they live. Socially, green health occurs at a nexus of morally laden decisions about living in the world as patients, workers, consumers, and citizens (Falcon and Lueck 2009).

Health care is obviously of central importance for every person on the planet, and takes quite different forms in so-called developing and developed countries. In the south, health care will concentrate on the eradication of poverty-related diseases such as malaria, TBC, diarrhea, typhus, and HIV. This can be accomplished through a combination of poverty alleviation, sanitation, safe drinking water, prophylaxis, vaccination, and western and traditional medicines. In the north, health care will concentrate on lifestyle issues, such as achieving balance between work and relaxation, stress reduction by meditation and exercise, healthy nutrition, as well as new drug and medical treatment development (Vergragt 2006).

Sustainable health care is governed by several primary principles. Maximizing human health by efficient medical treatment is the first principle. This requires a comprehensive understanding of new medical technologies. Making medical treatment economically sustainable is the second principle. Third, the focus of medicine also should shift from curing to the prevention of disease.

3.8.9 Lifestyle change

Lifestyle helps to fulfill people needs and aspirations and function as social conversations through which people communicate their social position in society and likes and dislikes to others. Much of this communication is mediated by the products we consume, the services we use, and the possessions we keep. In general, several key challenges may impact lifestyle. These include consuming, living, moving working, and health. Therefore, lifestyle has great impact on the flow of products and services in society and is closely linked to production and consumption patterns (UNEP 2007).

Initiating a sustainable lifestyle takes a lifelong commitment to learning, exploring, experimenting, and committing to increasingly sustainable practices. Incorporating sustainability into individual lifestyles involves an education on the impact of choices in food, products, and energy use; use, reuse, and disposal of products properly; and incorporating physical activity. Moving toward this goal requires fundamental changes in human attitudes and behavior. Progress in this direction is thus critically dependent on education, public awareness, and technological tools.

Technology has been among the primary drivers of consumption. However, it has the potential to significantly affect sustainability by affecting consumption patterns and preferences. One of the most prevalent current examples is teleworking or telecommuting. Technology advances have helped to provide transportation alternatives that are more amenable to sustainable development by reducing the transportation load due to commuting. Teleconferencing also allows employees to conduct meetings without having to travel. In addition, the availability of online stores has reduced the need to go to physical stores for shopping. Finally, transportation needs for the physical delivery of goods is being alleviated through the online delivery of goods, such as music, books, and other information that can be downloaded right to a user's computer without the need of a physical transportation intermediary. These examples help to illustrate how technology has led to the development of new services and have helped shape consumer preferences as a result (Haghseta 2003).

3.9 Technology planning

> Strategy without tactics is the slowest route to victory;
> tactics without strategy is the noise before defeat.
>
> **Sun Tsu**
> *Ancient Chinese Military strategist*

3.9.1 Technology planning process

Technology planning (TP) is the process that results in an actionable plan to leverage new and existing technologies. The plan identifies goals, their respective objectives, and key strategies. It focuses on strategies pivotal for creating a competitive advantage for the organization through the effective use of existing and emerging technologies. The TP provides strategic and tactical knowledge based on a high-level framework. This TP process leads to determining which technologies not yet adopted will have a strategic impact on the market. During the innovative phase, for example, the objective of TP is the formation of a technology plan. During the transition phase, the goal is the establishment of an effective ongoing planning and implementation structure and process. The transition phase of TP sets forth the goals and objectives and the framework for accomplishing these goals and objectives.

Treating technology as a type of knowledge is helpful, as knowledge management through TP can be useful for more effectively managing technology (Nonaka 1991). TP also requires a universal approach that supports planning of unpredictable technology impacts as well and impacts indirectly related to the technology (Marek 2013). As a result, it is imperative for industry to clearly understand what technology is available to them and how these technologies can be used to meet the needs of their target markets.

The TP process begins with the formation of a transdisciplinary team that best represents the functional areas crucial for identifying and developing critical technology. It provides guidance to evolve and mature relevant technologies to address future mission needs, communicate vital information to stakeholders, provide the technical portion of the overall program plan (cost and schedule), and gain strong executive support.

TP processes can be realized in a platform that consists of four key areas of input: technology development, knowledge acquisition, knowledge usage, and product development as shown in Figure 3.9. By considering these four knowledge bases together, industry will be able to place the TP process within a broader business context that helps maximize business success.

3.9.2 The challenge of GT

In regard to GT, a plan envisages main objectives that are expected to be fulfilled in the process of taking actions including mitigation of climate change and energy independence; creating new engines for economic growth; and improvement in quality of life by greening the land, water and building the green transportation infrastructure. Usually, the above

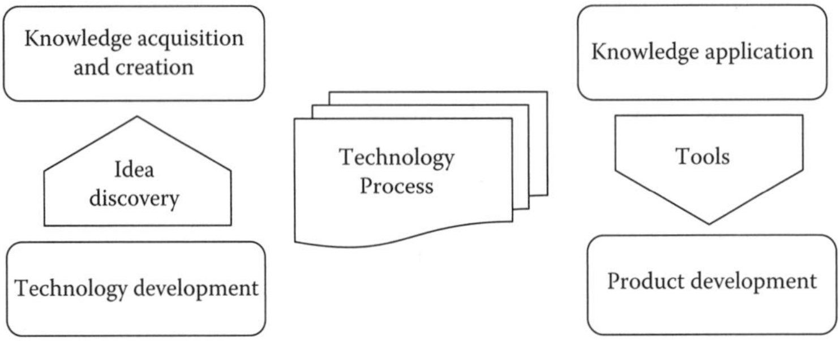

Figure 3.9 Platform for technology planning.

objectives and the related policy directions are based on the consensus among the society, businesses, academia, and government.

The critical factor of technological competitiveness is not a technology innovation process, but to find a new application for an existing technology (Rycroft 2006) and green growth investment. Generally, small and medium enterprises (SMEs) are more competitive in regard to the use of emerging technological inventions and know-how, while larger companies have more capacities for own R&D and planning, but also more cost advantages. SMEs do not usually have formulated an explicit technology strategy, while adapting their technological activities to the general market development (Lee and Lee 2008).

Nurturing green SMEs that are looking to venture into GT through comprehensive and customized training programs and investing is usually a critical component of any green TP. GT investing typically involves the selection of investments in companies with sustainable and environmentally friendly practices (Relander 2015). While some GTs can offer improvements that increase resource productivity and efficiency, others decrease environmental impact. As GT continues to emerge as a growing force, several strong industries have emerged in sectors like water and waste water, energy, advanced materials, agriculture, transportation, energy efficiency, and manufacturing.

Ideally, all GT investments should be considered as good investments; however, as is the case with any other type of investment, there are risks associated with investing in any type of new technology. The challenge of investing in GT is twofold; the objective is to both increase wealth, and to at the same time, make the world a better place through socially responsible investing. Admittedly, this can be a somewhat daunting task, but taking the time to conduct research prior to making an investment can help to select opportunities that will help to protect wealth, as well as the environment (Relander 2015).

3.10 *Technology transfer*

> Patience is a virtue, and I'm learning patience. It's a tough lesson.
>
> **Elon Musk**

3.10.1 TT process

TT means the expedition of technological knowledge, skills, and equipment from an originator's environment to a user's environment. The initial step in the transfer of technology process starts with the recognition of an economic need and accordingly, a business idea. If the transferred technology goes on to be developed and adopted, it becomes an innovation. Figure 3.10 describes the major steps of the TT process and innovation as its outcome.

Initially, a business idea that comes mostly from business should be realized. This process involves vision, intuition, observation, creativity, and insight into a new direction that might turn out to meet desires and/or needs that could not have been known before. The discovery process refers to the process of determining whether the transferred product or method can be developed and, if developed, determining whether it will be adopted. Through internal trials and market tests, a modern economy adds to its knowledge of what can be produced and what methods work. And if a transferred technology goes on to be developed and diffused, it becomes an innovation in the new setting. The first process does not continue until the realization of the idea to the stage of experimental testing in the economy because some technologies might be missing. This process is called first-order discovery. Second, the discovery process is undertaken after the idea has been materialized into a new product or process through TT process and will generate the economic knowledge about the question as to whether the new idea can work economically

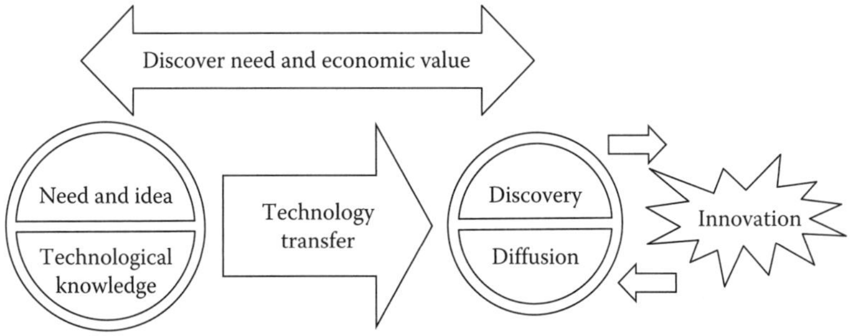

Figure 3.10 Innovation as a TT outcome.

[being developed, implemented, and adopted (UNCTAD 2014)]. This is the second-order discovery process through which the business idea is tested experimentally in the real economy.

The operation of TT process is therefore commenced while all the uncertainties about the economic value of the business ideas remain. The first discovery procedure is a first approach to reduce economic uncertainty; however, it is not sufficient, since the idea cannot be materialized unless the TT has been accomplished, opening the possibility for full economic experimentation (second-order discovery). The operation of TT is therefore characterized by high uncertainty that can be formulated in terms of an intertemporal consistency problem: the fixed cost of the TT is incurred while the knowledge about its economic value is not yet available.

There are two ways to deal with this dilemma. First, it is possible to establish and manage the first-order discovery process in such a way that a significant part of the economic uncertainty will be eliminated through this first process. Second, it is possible to contain the costs of the TT until the full economic knowledge will be produced through the second-order discovery process. A clear advantage of the first-order discovery process to solve the inter-temporal consistency problem is that the earlier economic knowledge is gained, the more possible it is to adjust the channel for TT to what is known about the economic viability of the business idea. For instance, licensing might be the appropriate channel if it is rather clear that the business idea will work economically. On the contrary, public or private partnership and other collaborative networks might be better channels when the first-order discovery process raises a big question mark about the economic value of the idea (UNCTAD 2014).

Many TT operations fail simply because the aim was to find a home for an available and public technology, without worrying about economic needs and business ideas to be tested and experimented in the developing economy: when it comes to TT, the technology-push logic almost inevitably leads to failure (Arora 2007).

3.10.2 TT modes

TT modes have been categorized basically as being passive or active, which refers to the transferor's function in technology application to solve user's problem. According to Mogavero and Shane (1982), TT classification as positive and negative refers to the level of activity in applying the technology in the transfer process. If the TT mechanism presents the technology to the potential user without assistance regarding its application, then the mode is called passive. In the passive mode (Figure 3.11), only the knowledge part of technology is transferred, the skills surrounding the technology are not transferred. These mechanisms can include

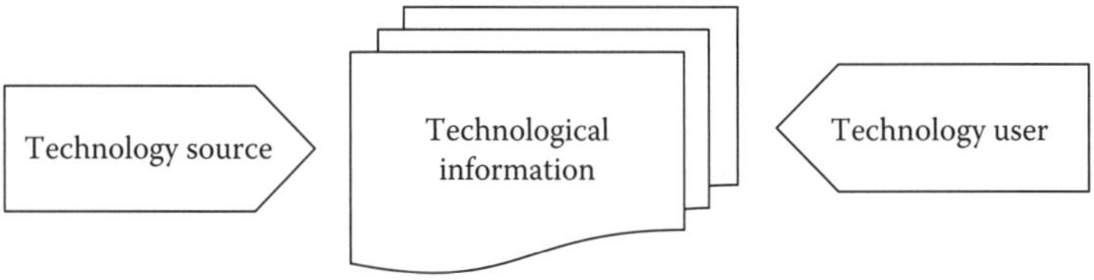

Figure 3.11 Passive mode TT.

presentations in a report. If, on the other hand, the provider of the technology assists with the application of the technology, then the mode is called active (Figure 3.12). These mechanisms include training. The boundaries between passive and active are not very clear and therefore a semiactive mode may sometimes be proposed.

3.10.3 Technology diffusion

Technological diffusion refers to the stage in which technology innovation spreads to the level of use and by which the innovation can be adopted and extended within and across business for application. Diffusion occurrence and the scale at which it occurs is dependent on several factors including the type and quality of the innovation, how information about the innovation is transferred, and the nature of business or environment into which it is transferred. Diffusion is driven by three broad categories including knowledge, institutions, and demand.

New technology brings new production processes, machines, products, and services which typically are not straightforward to implement. A significant part of the cost of adopting new technologies is the cost of figuring out what technology is needed to produce the desired goods or services and how to use it individually or as part of an existing production process. Therefore, any prior knowledge that reduces the magnitude of these costs should foster technology adoption (Comin and Mestieri 2013).

New knowledge is not enough to create innovation for adoption without the development of proper institutions (e.g., political, educational, and financial). While the knowledge needed to address certain needs is scientific and technological, the cause remains in the policy domain.

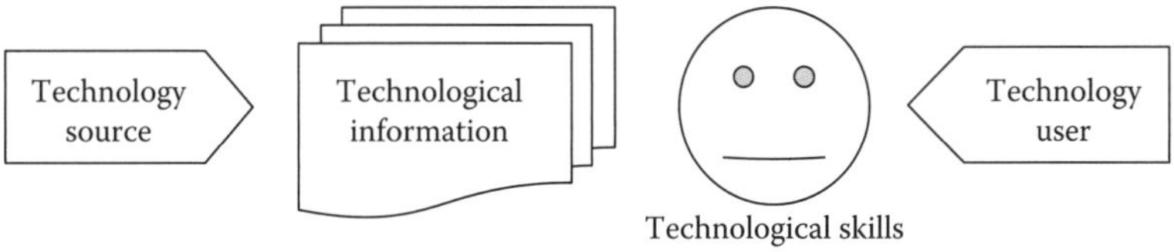

Figure 3.12 Active mode TT.

Technology policies need to be in context with the real conditions of specific social, economic, and educational domain. Technology needs to be realized as an enabler, rather than an applied solution that is imported when needed. Successful TT depends on reliable knowledge flows and the convergence of various areas of science and technology into an innovation system. Policies that combine knowledge sources with TT will improve the product's desirability for investments.

The level of demand is an important determinant of the return to adopting a technology. A higher demand allows adopters to cover the overhead costs of adoption among more buyers of the goods and services produced with the technology, thereby increasing the profitability of the investment (Comin and Mestieri 2013). Schmookler (1966) argued that demand should play a key role both in the amount of innovation activity as well as in the sectors where it is concentrated.

Within this context, adoption refers to the stage within the TT process in which a technology is selected for use. Currently, many approaches of TT that involve diffusion exist. For example, it may be the result of reverse engineering, which is the procedure of realizing the technological principles of a system through the study and analysis of its structure, function, and operation. A second approach might be a description of new products or services which can be realized in publications or patent applications. A third approach of transmission could be a research finding which may be published but requires further development. These different approaches of transmission produce different diffusion patterns with diverse effects on productivity and competitiveness. Figure 3.13 describes the R&D approach of TT. Technology diffusion involves the dissemination of technical information and know-how and the subsequent adoption of innovation by users.

Although a classic approach of technological development implies a simple linear path from basic R&D to technology commercialization and adoption, technology diffusion is more often a complex and iterative process in practice (Edquist and Jacobsson 1991). Technology can diffuse in multiple ways and with significant variations, depending on the particular technology, across time, over space, and between different industries and enterprise types. Moreover, the effective use of diffused technologies

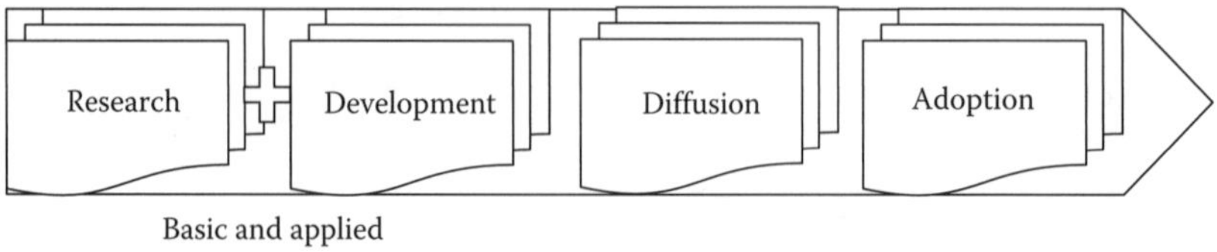

Figure 3.13 R&D approach of TT.

by firms frequently requires organizational, workforce, and follow-up technical changes.

Technology diffusion is especially appropriate for new or emerging technology innovations which suffer from limited sales history. The deployment of such technologies particularly within GTs may improve living standards, create employment, enhance funding, and accelerate growth and investment.

3.10.4 Determinant of TT process

Development and innovation are complex and path dependent. The determinants of successful TT are closely related to the actors involved. In a transfer process, the capacity to absorb and reuse that technology can either enhance or undermine the success of the transfer (Duan et al. 2010).

The first entry path to TT is institutional determinants. These are classified as TT office (TTO) and licensing policies. TTO requires appropriate staffing, clearly articulated mission statements, customer-friendly orientation, clear policies and procedures, and university cultures supportive of technology practice (Hsu and Bernstein 1997). In general, the TTO integrates various TT determinants including organizational, psychological, and university-industry models (Friedman and Silberman 2003; Smilor and Matthews 2004).

The major steps in this process include the disclosure of innovations, patenting the innovation concurrent with the publication of scientific research, and licensing the rights to innovations to industry for commercial development. Sometimes, new technological inventions developed from university research can be the basis for establishing new joint ventures, so-called university spin-off companies in which the university and inventors may both hold shares and therefore benefit from invention (Sukhochev 2011).

The second type of TT determinants is technology related. The most important determinants are idea and innovation technologies, nature and sophistication, technology's significant benefits and advantages when compared to current competing products, technology's sustainable competitive advantages and superiority, the availability of a functioning prototype, the technology's degree of compatibility to other necessary technologies) (Rogers 1995), technology scope or future uses, technology uniqueness and superiority, the barriers to entry, the newness and the nonobviousness in the technology (Nerkar and Shane 2007), the technology's degree of dependability on other necessary technologies, the technology's identifiable and quantifiable technological risks and weaknesses, the technology development time to market, the stage of development of technology, and the technical feasibility (Rahal and Rabelo 2006).

The third type of TT determinants is commercialization-related determinant. Based on a literature review of Rahal and Rabelo (2006), the determinants are classified as the technology's identifiable current and immediate market needs; the absence of a dominant competitor in the technological field; the technology's market growth anticipation; the technology's expected market trend; the time for the technology to reach the target market penetration; market accessibility for the technology; the technology's competitive pricing; the R&D necessary for the technology to reach the product development stage; the technology's expected payoff period, the technology's expected positive return on investment within a specified period; and the technology's financial risk. At this stage, invention becomes value that generates revenue. The commercialization process generally includes negotiating license agreements, building a prototype, establishing a manufacturing process, and marketing the invention. Following production and sales, royalties are distributed according to university policy. Figure 3.14 shows the main determinants of TT.

3.10.5 IP protection

IP is an important way of rewarding the commercialization of innovation which underpins growth and development, as well as promoting the disclosure and dissemination of technological information. It is as such a key element of the ecosystem, but it is not an end in itself (ITU et al. 2015). IP is classified as the complete technology's literature search, and the completed patent search. The patent search must be clear and clean; the confidentiality of the technology must be maintained (no oral or written disclosures); the technology must have no prior claims; the IP must have strength; and the IP must be exclusive (Rahal and Rabelo 2006). IP rights take on a variety of forms including copyrights, patents, trademarks, and trade secrets (nondisclosure). Each IPR has particular economic characteristics, terms,

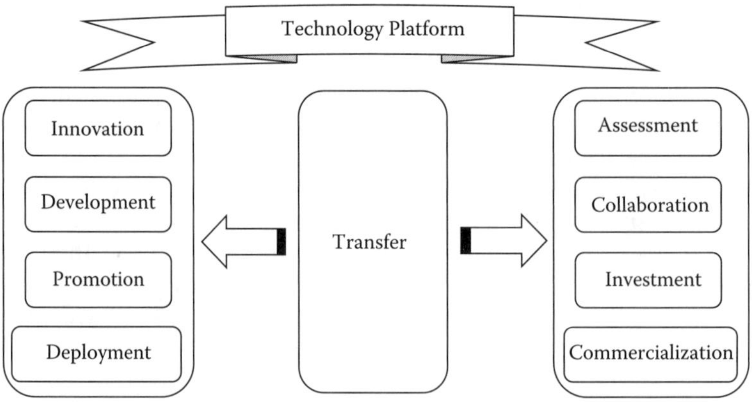

Figure 3.14 Main determinants of TT.

and duration of legal protection and impact on TT, mainly depending on the level of development of the technology recipient (UNCTAD 2014). Usually, every country has an agency that issues patents to inventors and businesses for their inventions, and trademark registration for product and IP identification.

Commitment to the protection of IP through cooperation among states should be coupled with a commitment to ensuring that all countries are able to benefit from the use of IP rights for economic, social, and cultural development. Finding the right balance between accessibility and reward (for creativity and innovation) remains a fundamental challenge in building inclusive and sustainable development paths. Different types of IP are protected in different ways as shown in Table 3.9.

3.10.6 Licensing

Licensing is the only mode of disembodied TT that can be measured. Under licensing agreement, the licensor (technology owner or rights holder) continues to own the technology and gives a defined right to the licensee for the use of the technology. The licensee by the terms of the license is permitted to exploit the IP.

There are two main types of licenses: one which grants an exclusive right to use the technology; another with nonexclusive rights, which implies that the patent owner may transfer the right to use the technology

Table 3.9 Types of intellectual properties

Design patent	A design patent may be granted if the product has a distinct configuration, distinct surface ornamentation or both
Utility patent	A utility patent has a detailed technical disclosure along with drawings (where appropriate) and one or more claims; the claims of a utility patent list the elements of the invention and establish the boundaries of patent coverage
Plant patent	A plant patent is granted by the government to an inventor who has invented or discovered and asexually reproduced a distinct and new variety of plant
Copyright	Copyright deals with the rights of intellectual creators in their creation; it is a form of protection given to the authors or creators of original works of authorship, including literary, dramatic, musical, artistic, and other intellectual works
Trade secret	Trade secrets include any valuable business information that derives its value from the secrecy
Trademark	A trademark is a recognizable sign, design, or expression which identifies products or services of a particular source from those of others, although trademarks used to identify services are usually called service marks

to other companies in the same area. Additionally, the licensing agreement could include a sublicensing clause which permits the licensee to grant to someone else the right to use the technology.

The OTT is the university point of contact for companies who are interested in evaluating and licensing university IP for commercialization and collaborating on research. The signing of a license agreement begins a long-term collaboration between the university and the licensee. These collaborations often lead to funding for additional research, the advancement and dissemination of additional knowledge, other inventions, and the use of inventions for other applications.

3.11 Educating sustainable technology

> I'm most grounded on the role of technology. Ultimately to me it's about the human capital and the human potential and technology empowers humans to do great things. You have to be optimistic about what technology can do in the hands of humans.
>
> **Satya Nadella**
> *Chief Executive Officer, Microsoft Corporation, USA*

3.11.1 The design dimension of technology sustainability

Education is an important means through which SD objectives can be realized. Needs range from education at the local level for self-and quality-of-life improvements to education at the decision-making level to promote knowledgeable planning and policy choices. Developing the right technology depends both on far-sighted educational and entrepreneurial systems and on a deep vision in technological opportunities and societal consequences (Vergragt 2006). None of this is easy or self-evident.

Sustainability education as an academic space for examining nature society interactions should be integrated in the learning system. The objectives are to evolve scientific understanding of human–environment systems; to improve linkages between research communities and relevant policy communities; and to build capacity for linking knowledge with action to promote sustainability.

Today, several educational tools exist that foster a quick understanding of complex issues. For example, computer simulation technologies that present a problem and a variety of possible solution pathways allow us to observe the results of their choices and increase their understanding of the downstream impacts of decisions. Adding sustainable design to the curriculum would heavily require such solutions. It is more important for students to learn how to be a capable designer first, and then learn sustainable technologies later. Students learning design should focus on big ideas or conceptual themes, and how to give those ideas physical form.

3.11.2 The technology entrepreneurial university

Most universities do not yet exist to operate as entrepreneurship drivers in their current form. They exist mainly to teach academic subject matter, a role that they carry out with various degrees of success. However, universities still do not teach entrepreneurship, really do entrepreneurship, or even support entrepreneurship as a priority. There is also confusion between "support for entrepreneurship" and "technology/innovation transfer." There is certainly overlap between these areas; however, technology/innovation transfer tends to be considered more in terms of the IP developed in university labs than as a direct business and selling opportunity (Wells 2012).

The entrepreneurial university does value both innovation and execution, and encourages partnerships between academics and entrepreneurs (Thorp 2010). According to Etzkowitz (2015), there are three stages to the university entrepreneurial transformation process: (1) the university starts to define its priorities and diversify its income sources; (2) the institution starts commercializing the IP that arises from its research activities; and (3) the university takes an active role in participating in its regional innovation environment.

A university can act as business incubator, allowing students and faculty to meet, form teams, and experiment with the idea of bringing technology from research labs to the market. Universities can effectively offer spin-offs an incubation period, in which students and faculty have the freedom to develop technology and form their strategic plans, incrementally reducing the venture's market and technological risk. During their time at the university, students can work on the initial stages of the spin-off without the opportunity cost of foregoing a paid job. This incubation and experimentation can only take place, however, if the university offers programs or opportunities for transdisciplinary teams to meet, and provides resources to help teams develop the technology and plans for the spin-off (Colyvas et al. 2002; Jain and George 2007; Boh et al. 2012).

In general, universities can provide several programs and practices that enhance entrepreneurial efforts for commercializing university technologies. Spin-offs and start-ups offer academic and student entrepreneurs with an unusual pathway for spreading and commercializing research for key motives to undertake these venturing activities including self-realization, reputation, career development, and independence. Figure 3.15 shows the examples of entrepreneurial programs that many universities currently offer.

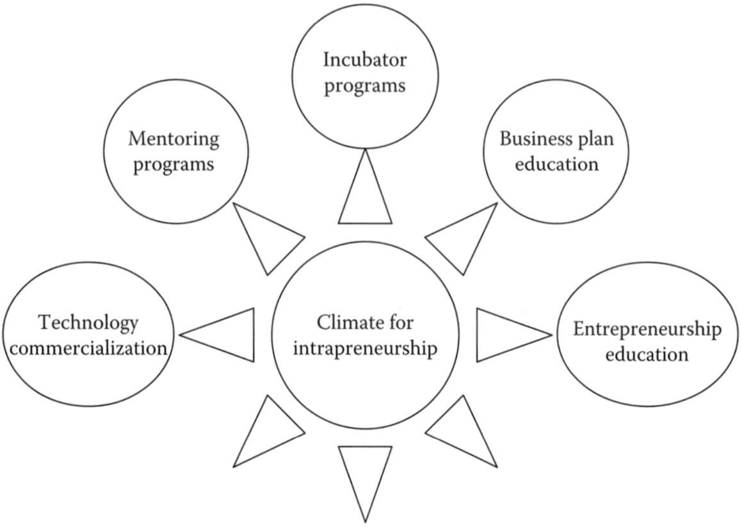

Figure 3.15 Examples of university programs that support technology entrepreneurship.

The benefits of concluding the TT process with a spin-off or start-up company include the potential for the spin-off or start-up to generate a long-term payoff, create jobs, and generate high returns if the firm is taken public (Siegel and Phan 2005). The university connection and proximity is advantageous for the spin-off because the university provides skilled labor, specialized facilities, and topical expertise (Bercovitz and Feldman 2006). Emphasizing spin-offs as a TT strategy can lead to an agglomeration of high-tech firms around the university, eventually resulting in a technology-based cluster (Rogers et al. 2001).

Project-based classes may bring together transdisciplinary teams to work on business plans and create ideas for the commercialization of new technologies. One way that universities can contribute to entrepreneurship is through R&D projects. Universities often offer mentoring services that provide direction and advice to new entrepreneurs, as well as referrals to industry experts, potential customers, licensees, and investors who help founding teams build their networks. Formal accelerator or incubator programs at universities often help start-ups over a period of time, providing mentoring, funding, space, and in some cases, supervision and management. Business plan competitions often play a key role in university spin-off development. Not only do they provide a platform for team establishment, but they also offer potential founding teams the opportunity to develop a business plan and strategic roadmap for the technology.

While successful commercialization of faculty research will always depend, to a certain extent, on the ideas generated in university laboratories and the personalities and talents of the individuals involved in the research, universities can create an environment that fosters new business creation on university campuses. Recognition of the value and potential impact of university technologies for the broader population, of the need for university resources and support, and of the important role students can play in these processes is a critical first step. These efforts also have the potential to inspire future entrepreneurs who will bring continued innovation and growth to our economy (Boh et al. 2012).

3.12 *TT case: Energy efficiency*

> For 300 years, higher education was not disruptable because there was no technological core.
>
> **Clayton Christensen**

3.12.1 TT components

Taking scientific and technological knowledge to commercialization is a frontier mission of universities besides their traditional mission of education. Innovation is one of the core activities of the university along with education and research. To adapt to these challenges, the organizational structure of the university will need to adapt. Universities must transform into environments of innovation and entrepreneurship to meet the current market needs and ties well with the community, industry, and governments (Sukhochev 2011). In this case, a context for TT assuming a collaborating structure is presented (Figure 3.16). The technology provider within the project is a team of one university and two start-up companies, "A" and "B." They include researchers who had set the foundation for many

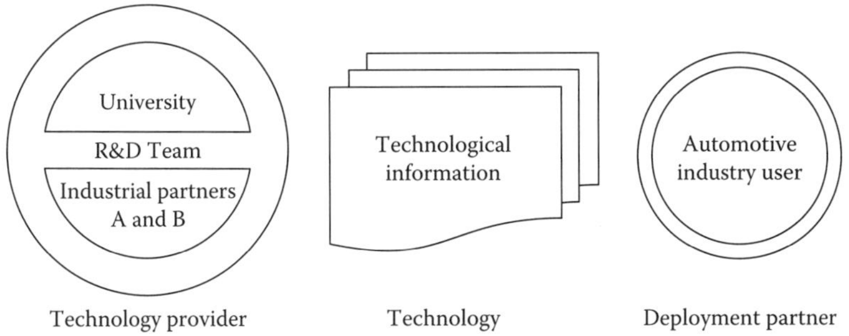

Figure 3.16 Technology development and diffusion scenario.

formal methods and tools, in addition to having been involved in several industrial applications. There is a shared knowledge base among the collaborating partners. The technology user is an automotive supplier who is interested in the technology and plans to commercialize it within its sector.

3.12.2 TT approach

At the start, a couple of TT approaches may be employed: first, some introductory information and learning sessions may be organized for all technology development partners. The intended purpose of these sessions is to introduce the partners to possible customer needs and all required aspects of the relevant existing project methods and tools. Second, several mini-pilots may be introduced by the partners, and the TT process will be carried out through these channels. The mini-pilots act as the bridge between the partners and user by clarifying certain problems and by showing how to apply the techniques taught in the initial phase.

3.12.3 Platform for communication

To simplify the TT process, a platform for communication and for exchanging TT material is set up. The components of this TT platform are as follows:

1. The project development team set up an internal platform (only visible to project members). This includes a shared space for exchanging documents, presentations, and a set of mailing lists, both structured by work packages.
2. The project team set up a public Web platform. This platform includes essential information about the project techniques and tools, including modeling methods, training sessions, and a repository for storing publications and developments related to the project including articles, books, deliverables, conferences, and/or workshop items.

The TT platform allows partners and other contributors to share various types of materials with different audiences. Project-related internal documents can be shared using the internal shared space and discussed via internal e-mail lists. More general material may be made available publicly to a wider audience via the Web platform.

3.12.4 Mini-pilots

After the initial phase of general training, the second phase of TT involves working on the mini-pilot projects. These projects act as a platform for sharing information between the technology provider and the deployment partner. While the technology provider focuses on understanding domain-specific problems, the deployment partner needs to see how the project techniques and tools could help with these. The feedback for TT at the start of project will be detailed. The report contains the views of both deployment and academic partners. One of the main points raised in the report by the deployment partner is that TT should address domain-specific issues that are key to deployment success in different industrial sectors. In other words, TT should be directed by the need to solve domain-specific topics that are required for the deployment of formal methods in industry. TT should be adapted to meet this need and, as a result, there should be different TT materials prepared for diverse audiences in different contexts.

Because of the differences between the domain-specific issues, one should not expect a single method or approach to meet all the needs of the deployment partners, in particular, across different sectors. Instead, there might be the need to adapt existing techniques, create new techniques, and even combine them. The supporting tools should be adapted as the techniques evolve and as different alternatives need to be explored. As a result, the corresponding TT material must also be updated as part of the process. While substantial effort should be devoted to the evolution of techniques and tools, the task of keeping TT material up-to-date is also time-consuming and should not be neglected. Keeping TT material up-to-date increases the effectiveness of techniques and tools, helping transfer them to the relevant audiences more easily. In reality, however, updating TT material is often overlooked.

3.12.5 TT needs

To manage TT at the project level, and to avoid duplicated efforts, a procedure centered around wish list should be designed and maintained within the internal platform for TT. This contains information about what material is available and what material is requested, which partner is responsible and how much time is expected to fulfill the request. The wish list and the accompanying procedure helps ensure that requests related to training and documentation are taken into account and managed accordingly. A limitation of this procedure is that documentation is created or updated on demand, for example, only when there is an explicit request for it. This does not ensure the quality and promptness of the delivered documentation. Documentation quality is a key aspect of TT. Documentation alone does not allow an engineer to start using the tools without significant support from expert. The project can be scheduled for several months, involving representatives from both industrial partners and the academic partner.

3.12.6 Energy-efficient technologies

The project involves dual innovation that engages performance and efficiency optimization, both mechanically and electrically for a hybrid/electric vehicle drive system. Longer range due to the extension of battery life, higher energy efficiency due to geography-aware optimized regenerative breaking, faster acceleration due to the implementation of an ultra-capacitor (UC), and the involvement of an enhanced performance induction machine demonstrates the significance that this innovation can have if implemented in a vehicle.

First, an off-the-shelf three-phase induction machine is modified based on a novel passive technique which demonstrates a significant enhancement of operating performance. The new machine employs an auxiliary winding, which is only magnetically coupled to the stator main winding. The simulation and lab measurement results show that the operating performance of the modified machine has been significantly enhanced in terms of suppressed signal distortion and harmonics, severity of resistive losses and overheating, power factor, and preventing high inrush current at starting (Habash et al. 2012).

In electric and hybrid vehicles, batteries typically do not have a long lifespan, which is dependent on the number of charge/discharge cycles and the depth of state of charge. The improvement of battery performance can be achieved by using a supplemental UC. UCs are robust in terms of repeated charge/discharge cycles, are not temperature dependent (to an extent), and do not rely on a chemical reaction, resulting in an immediate release of energy with minimal loss. These characteristics allow UCs to be used as supplemental energy storage, as well as a source. A parallel configuration for the battery and UC results in an increase in efficiency of the battery, thereby reducing battery drain, and finally increasing the life of the battery bank.

In addition to the mechanical optimization of the drivetrain, optimization can be implemented on the electronics/software side. It is crucial that a high-performance motor has an efficient controller and storage system. To meet this requirement, two controllers are implemented, one that represents the car computer and another one responsible for geography-aware energy management optimization. The application utilizes location services to determine the density of intersections in a given area. Three "scenarios" for the density of intersections: high density (city downtown), medium density (city outskirts), and low density (highway) have been selected. Assuming the driver has this application on the phone; the application monitors the car position in real-time using global positioning satellite (GPS) system, extracts the road scenario using a location service such as Google Maps, and sends a command to the Arduino depending on what scenario the driver is in.

3.12.7 *Case research questions*

- What is energy efficiency?
- Why should energy efficiency be automotive industry priority?
- How do universities provide incentives to companies to develop university technologies into commercial products?
- Is the engagement of educational institutions in TT central to their primary missions of education, research, and public service?

3.13 *Knowledge acquisition*

Attempting to answer the following questions involves acquisition of knowledge from this book and other books, documents, and the Internet.

- What are the lessons taken from technology history?
- How have tools affected the evolution of humankind?
- How much did the Romans and Greeks exploit mechanical power?
- What allowed horses to pull about 6000 kg?
- When did iron-making begin? How was it originally done?
- What counts as science and what counts as technology?
- Who are the scientists and who are engineers?
- How do you define a scientific revolution?
- How do technologies evolve?
- What was the relationship between science and technology during the Industrial Revolution? Did science come before technology?
- What will be the greatest invention in the 2010s?
- Do society control technology, or does technology control society?
- What technologies should one learn to stay ahead in the industry for the next 5 years?
- What would be the most important technology of the future?
- How do technologies evolve?
- How would your life change if all your technological devices were taken away for a day?
- What are the advantages of GT? What are the ways to promote this technology?
- Does technology serve community knowledge and/or relationships?
- What is the difference between fission and fusion energy?
- Why are light-emitting diodes (LEDs) considered a GT?
- What is the best way to promote renewable energy technologies successfully?
- What is TT?
- What is the proper definition of a start-up?
- What is the difference between adoption and diffusion of technology?
- Why patenting something? What kinds of things are patentable?

- What is technology entrepreneurial university?
- What is the relationship between enterprise, entrepreneurship, and innovation in universities?
- What are the most recent technological developments that have appeared in cars?

3.14 Knowledge possession

Attempting to answer the following open-ended "not explicitly expressed" questions may require research and investigation beyond the scope of this book, mostly by engaging in conversation, class discussion, and Internet-based research.

- History of technology: What are some systems we live with today that were designed for a world of the past?
- How technology could contribute to a sustainable world?
- Which new technologies that have emerged over the last 20 years have had the most impact on your lives?
- In an expanding technology firm, what are the most important issues on which an engineering director should spend time?
- Should AT endeavors focus only on low- or high-technology products?
- Would a great transition society require a serious use of technology to lessen the environmental degradation of the ecosphere, or might technology play a much more restrained role in such a society?
- What type of incentives is preferred to engage in an STI educational programs, so that a significant amount of STI skills can be developed?
- How can a country continue to develop, attract, and retain the world's top research talent?

3.15 Knowledge creation

Collaborate with peers on learning or you may work with others outside the class to narrow down the objectives of each activity. You may access class and online resources, and analyze data and information to create new ideas and balanced solutions. High-level digital tools may be used to develop multimedia presentations, simulations or animations, videos and visual displays, digital portfolios (ePortfolios), reflective practice (online publishing and blogging), or well-researched and up-to-date reports.

3.15.1 Reflection practice on a prosumer city

According to Wikipedia, a "prosumer" is a person who consumes and produces media. Cities make up only 2% of the earth's surface; however, they are home to over half of the world population. Technology is a tool cities use to accomplish their full capacity. It facilitates residents, businesses, and government to work more efficiently, interact with each other in better ways, and increase overall quality of life. Currently, technology is converting consumer cities into prosumers cities at a rapid rate. Technology will enable us to live efficient lives. With the emergence of advanced, smarter technology, consumers can now make more informed choices about energy usage, building, agriculture, transportation, and other aspects of life.

For this task, a class may be divided into groups of three students working toward the above goal where every group considers one sector of a prosumer city. Groups may reflect their outcomes in a form of poster, portfolio, or digital art. The work of every group could be uploaded to an online library under the supervision of the instructor.

3.15.2 Communication on primary energy sources and demand

Engineers know that communication, both oral and written, is an essential part of their jobs. The need to communicate with customers, managers, technicians, and other engineers is something they become aware of early in their careers. Therefore, engineering educators must ask themselves if they are preparing students in this area which is as important to their careers as is their technical training.

Primary energy sources include petroleum, coal, natural gas, nuclear energy, and renewable energy. Major demand sectors that rely on the above primary sources include electric power utilities—residential, commercial, and industrial; and transportation. The mix of primary energy sources varies across demand sectors. Energy policies meant to influence the use of a particular primary energy source for environmental, economic, or energy security reasons often focus on sectors that are major users of that energy source. In addition, there is currently focus on the next transition, on the possibility of a substantial change in the energy mix.

3.15.2.1 Poster 1

At the beginning of the twentieth century, coal and wood provided more than 95% of the world's energy requirements. From that point, it took more than half a century for oil, a cleaner and more adaptable alternative to exceed coal as the world's prime energy source. Following, it took several more decades to develop the technologies and infrastructure for natural gas, nuclear energy, and renewable energy to supply of an even cleaner energy to the growing needs (WEF 2013).

In this regard, develop a world energy timeline that starts when human begin using wood as fuel. You may extensively expand the following incomplete table:

100,000s BC	Wood for fuel
1881	First hydropower station; expand
1830	First commercial coal powered steam locomotive; expand
1859	First oil well; expand
...	...
...	...

3.15.2.2 Poster 2

- Draw a block diagram in a poster format that describes the relationship between primary energy sources and demand sectors. You may place the primary sources on one side of the poster sheet and sectors on the opposite side.
- Use arrows to link each primary source to the corresponding demand sector(s).
- Use the Internet to investigate the percent of sources and sectors in certain state or country.
- What are the commonly used units for each primary energy source?
- To compare fuels, a common unit of measure is used. What is that unit?
- Which primary energy source supplies a more diverse range of demand sectors?
- Outline the main technology pathways employed for each primary energy source.
- Outline the main technology pathways employed for each demand sector.
- Do sources and usages of energy change over time? Investigate and give examples.
- What would be the nature of the changing energy mix? What would drive it? How fast could it happen? How long might it take?
- What are the key participants that will largely influence the changes in the global energy mix?
- What is the great energy challenge of the future?

3.15.3 Debate on technology, ecosystem, and culture

Thinking about the progress of science prior to the Age of Enlightenment and the progress since gets students considering what counts as science, what makes a revolution, and if it is possible for a revolution to go on for over 300 years. In this activity, students will assume a position, build an argument in support of it, and defend it in an open-ended debate within class.

Objective	Introducing an open-ended debate in the classroom to help students understand argument on the concepts of technology, ecosystem integrity, and cultural biodiversity
Time	15 min for debate and 15 min for review
Format	For and against
Learning outcomes	Make an argument about a particular opinion; evaluate the arguments of peers; and understand the concept of counterarguments
Capabilities demonstrated	Developing skills on public speaking, research, teamwork, critical thinking, communication, and professional judgment
Arrangement	Students are organized into two position groups of three speaking in a specific order. Three argue for an opinion and three argue against. One or two students might each work on the opening and closing statements while the group is investigating the subject; however, the entire group should revise the statements. Each group should read an opening and closing statement for the debate.
Ideas for the topic	Does technology sustain or reduce ecosystem integrity? Does technology sustain or reduce cultural biodiversity?
Assessment	Indicate what you consider the best arguments in favor of the topic. How were they substantiated? Identify arguments that are based on poor facts, not ethical and/or sustainable or not well substantiated.

3.15.4 Cases on IP rights

Case 1: After many years of investigation in the laboratory, inventor "A" has developed software relating to pay-per-click Web search engine for marketing and advertising. Could "A" patent this invention? Why? What conditions must be passed to allow the invention to be patentable?

Case 2: "B" has discovered a naturally occurring material which can be used as weed control agent. Could "B" patent this discovery? What kind of work would "B" have to do in order to make this qualified? What kind of exclusions to patents might pertain?

3.15.5 Piece of art on green IT

There are many different aspects of green IT. Most students know about IT through the use of their PCs, laptops, IPads, smart phones, and all the social media used with those devices. Thus, energy-efficient IT or green IT is an area that should grab their interest. However, students will need to understand other aspects of green IT, for example, data centers since data centers house the servers that students access when they use Google, Facebook, or any of the many Internet applications used by all users, young and old (Lamb and Marimekala 2015). In this educational task, students are invited to create a digital piece of art that focus on energy and technologies used in green IT. The student would need first to become familiar with the basic electricity concepts and relationship of volts, amps, and watts. They should explore way(s) to promote energy-saving initiatives and/or emerging technologies that contribute to green IT.

3.15.6 *Poster on appropriate engineering for underserved communities*

The idea of AT is to develop products that are perfectly tailored to a specific context in order to enable a positive change in a user's life. In this context, appropriate engineering may be defined as the development of an AT. Engineers are very important where many of their decisions are magnified thousands of times by mass production. This gives them leverage to make real, material changes. Engineering should be more than mechanics and materials and controls: it should include the human component. There is a slow shift in engineering education toward including this holistic focus but it is not occurring rapidly enough (Felser 2011).

Key Point: *The majority of the world's designers focus all their efforts on developing products and services exclusively for the richest 10% of the world's customers. Nothing less than a revolution in design is needed to reach the other 90%!* Dr. Paul Polak, International Development Enterprises.

This poster activity focuses on the role of engineers in development and adoption of ATs to provide innovative solutions to underserved people and communities, both globally and locally. As an outcome, do you think that technology can eliminate poverty?

3.15.7 *Engineering consulting on challenges of the transit-elevated bus*

With the increase in population and the economic growth of China, the congestion of traffic has also increased. One of the most prominent incidents, known as the China National Highway Jam, began to form on August 14, 2010, mostly on China National Highway 110 (G110) and Beijing–Tibet expressway (G6) resulted in thousands of cars being stuck in traffic for about 10 days.

A China-based company called Shenzhen Huashi Future Parking Equipment developed a transit-elevated bus (TEB) that can reduce the country's traffic congestion. In 2010, it was named one of the 50 best inventions of the year by Time Magazine. This straddling bus towers over cars and runs along tracks installed on the road at an average speed of 40 km/h. Its main compartment can hold 1200–1400 individuals and is elevated above the street to leave room for cars to travel beneath it (Cilento 2010).

The vehicle is entirely solar and electrically powered, with solar panels on the roof and at bus stops. It uses relay direct current electrification located on the tracks to supply electric power to the bus (Howkins 2016). The construction of the TEB is also much cheaper than the construction of a subway system. This innovation has the carrying capacity of 40 buses while potentially saving up to 860 tons of fuel and reducing 2640 tons of carbon emissions. In August 2016, a prototype of the TEB was tested in the northern China city of Qinhuangdao. It is 72 ft long and 26 ft wide while being 16 ft above the road (Howkins 2016). Production and testing of the bus are currently in progress.

Many questions and concerns about the vehicle have also been raised such as the following:

- How will the TEB turn corners?
- How will the cars underneath the TEB switch lanes?
- How will larger vehicles such as trucks fit beneath the TEB?
- Is this concept realistic?

3.15.8 Video contest on disruptive technologies in digital age transport

Demand for transport continues to grow steadily as the global population in urban areas reaches its high levels. The current wave of digital technologies have brought planning to computers and phones and provided further access to customer information. In the last few years, disruptive technologies like Uber and AirBnB have emerged and grown to extend to become global enterprises. It is now timely to investigate new trends that provide a wider range of options and choices in transportation (Goodall and Dixon 2015).

In this task, collaborate with peers or you may work with others outside the class to narrow down the objectives of the task. In a 3 min video, can you disrupt the transportation system with ideas that improve the way we travel? Explore new ways of getting around and try out new technologies. Explore possible threats and risks that involve the transformation process. You may finally propose a business model that enables the above disruptive technologies to grow dramatically.

References

Allison, I. 2010. *The Science of Climate Change: Questions and Answers*. Canberra: Australian Academy of Science.

Arora, A. 2007. Intellectual property rights and the international transfer of technology: Setting out an agenda for empirical research in developing countries. International Conference on IP and the Creative Industries, World Intellectual Property Oraganization *WIPO Conference*, October 31, Geneva, Switzerland.

Beaudry, F. 2016. What are greenhouse gases? https://www.thoughtco.com/what-are-greenhouse-gases-1203888.

Bercovitz, J., and M. Feldman. 2006. Entrepreneurial universities and technology transfer: A conceptual framework for understanding knowledge-based economic development. *Journal of Technology Transfer* 31(1): 175–188.

Bijker, W. E. 1995. *Of Bicycles, Bakelites, and Bulbs: Toward a Theory of Socio-technical Change*. Cambridge, MA: MIT Press.

Boh, W. F., U. De-Han, and R. Storm. 2012. *University Technology Transfer through Entrepreneurship: Faculty and Students in Spinoffs*, Ewing Marion Kauffman Foundation. Kansas City, MI: Ewing Marion Kauffman Foundation.

Brook, B. W., A. Alonso, D. A. Meneley, J. Misak, T. Blees, and J. B. van Erp. 2014. Why nuclear energy is sustainable and has to be part of the energy mix. *Sustainable Materials and Technologies* 1(December): 8–16.

Buchanan, R. A. 2017. History of technology. https://www.britannica.com/technology/history-of-technology#toc1350805main

Chu, J. M. W. W. 2013. Developing and diffusing green technologies: The impact of intellectual property rights and their justification. *Journal of Energy Climate and Environment* 53: 53–102. http://scholarlycommons.law.wlu.edu/jece/vol4/iss1/4.

Cilento, K. 2010. Straddling bus/Shenzhen Huashi future parking equipment. http://www.archdaily.com/74140/straddling-bus-shenzhen-huashi-future parking-equipment.

Clarke, R. 1974. Technical dilemmas and social responses. In Cross, N. (Ed.), *Man-made Futures: Readings in Society, Technology and Design*. London: Hutchinson Educational.

Colyvas, J., M. Crow, A. Gelijns, R. Mazzoleni, R. R. Nelson, N. Rosenberg, and B. N. Sampat. 2002. How do university inventions get into practice? *Management Science* 48(1): 61–72.

Comin, D., and M. Mestieri. 2013. Technology diffusion: Measurement, causes and consequences. *Handbook of Economic Growth*. http://www.dartmouth.edu/~dcomin/files/chapter_v8.pdf.

Dodwell, C. R. 1961. *Theophilus, Presbyter: The Various Arts*. New York: Nelson.

Duan, Y., W. Nie, and E. Coakes. 2010. Identifying key factors affecting transnational knowledge transfer. *Information and Management* 47: 356–363. http://dx.doi.org/10.1016/j. im.2010.08.003.

Edquist, C., and S. Jacobsson. 1988. *Flexible Automation: The Global Diffusion of New Technology in the Engineering Industry*. Oxford: Basil Blackwell, p. 227.

Ellul, J. 1964. *The Technological Society*. New York: Knopf.

Etzkowitz, H. 2015. The entrepreneurial university as a technopole platform. A global phenomenon. In Miao J. T., P. Benneworth, and N. A. Phelps (Eds.), *Making 21st Century Knowledge Complexes: Technopoles of the World Revisited*. New York: Routledge, pp. 23–40.

Falcon, R., and M. Lueck. 2009. *The Greening of Health: The Convergence of Health and Sustainability*. Health Horizons Program. Institute for the Future. http://www.iftf.org/uploads/media/SR-1215%20green%20health%20 3.27jr_sm.pdf.

Felser, J. 2011. An engineer's guide to applying appropriate technology. AHS Capstone Projects. Paper 21. http://digitalcommons.olin.edu/ahs_capstone_2011/21.

Foster, R. 1986. *Innovation: The Attacker's Advantage*. New York: Summit Books.

Friedman, J., and J. Silberman. 2003. University technology transfer: Do incentive, management and location matter? *Journal of Technology Transfer* 28(1): 17–30.

Goodall, W., and S. Dixon. 2015. Transport in the digital age: Disruptive trends for smart mobility. https://www2.deloitte.com/content/dam/Deloitte/ch/Documents/public-sector/ch-en-public-sector-transport-digital-age.pdf.

Habash, R.W.Y., Q. Tang, P. Guillemette, and N. Irfan. 2012. Performance evaluation of an induction machine with auxiliary winding for wind turbine power. *International Scholarly Research Network, ISRN Renewable Energy* 2012:1–7. http://downloads.hindawi.com/journals/isrn/2012/167192.pdf.

Haghseta, F. S. 2003. Information technology and sustainable development: Understanding linkages in theory and practice. Master of Science in Technology and Policy Thesis, Massachusetts Institute of Technology, Boston, MA.

Hard, M., and A. Jamison. 2005. *Hubris and Hybrids: A Cultural History of Technology and Science*. New York: Routledge.

Harris, C. E., M. S. Pritchard, and M. J. Rabins. 2009. *Engineering Ethics, Concepts and Cases*. Belmont, CA: Wadsworth Cengage Learning.

Hawkins, A. J. 2010. China finally built an elevated bus that straddles traffic and it's totally bizarre. http://content.time.com/time/specials/packages/0,28757,2029497,00.html.

Howkins, A. J. 2016. China finally built an elevated bus that straddles traffic and it's totally bizarre. https://www.theverge.com/2016/8/2/12360620/china-TEB-elevated-straddling-bus-unveiled.

Hsu, D. H., and T. Bernstein. 1997. Managing the university technology licensing process: Findings from case studies. *Association of University Technology Managers* 9: 1–33.

IEA. 2014. Essentials for policy making and its companion document energy efficiency indicators. https://www.iea.org/publications/freepublications/publication/IEA_EnergyEfficiencyIndicators_EssentialsforPolicyMaking.pdf.

Illich, I., I. K. Zola, and J. Mcknight. 1977. *Disabling Professions*. London: Marion Boyars Publishers.

ITU, OHCHR, UNCTAD, UNEP, UNESCO, UNFCCC, UNIDO, WIPO, WMO. 2015. Science, technology and innovation for sustainable development in the global partnership for development beyond 2015, UN System Task Team on the Post-2015 UN Development Agenda.

Jain, S., and G. George. 2007. Technology transfer offices as institutional entrepreneurs: The case of Wisconsin Alumni Research Foundation and human embryonic stem cells. *Industrial and Corporate Change* 16(4): 535.

Kaplinsky, R. 1990. *The Economies of Small: Appropriate Technology in a Changing World*. London: Intermediate Technology Publications.

Kroes, P. 2012. *Technical Artefacts: Creations of Mind and Matter*. Dordrecht, the Netherland: Springer.

Lamb, J., and S. K. V. Marimekala. 2015. STEM and green IT. *12th International Conference and Expo on Emerging Technologies for a Smarter World (CEWIT)*, Melville, NY, October 19–20.

Lee, Y.-J., and J.-D. Lee. 2008. Technology strategy for enhancing the public-to-private technology transfer: Evidence from the duration of patent. *Applied Economics* 40(2): 229–240.

Marburger, J. H. 2011. Science, technology and innovation in a 21st century context. *Policy Science*. Springer Science+Business Media, LLC. http://science-policy.colorado.edu/publications/special/sip_marburger1.pdf.

Marek, J. 2013. Systemic introduction to technology planning in the context of technology competitiveness. *Journal of Competitiveness* 5(3): 10–13.

Mogavero, L. N., and R. S. Shane. 1982. *What Every Engineer Should Know about Technology Transfer and Innovation*. New York: Marcel Dekker Inc.

NAE. 2004. *The Engineer of 2020*. National Academy of Engineering. Washington, DC: The National Academies Press.

NRC. 2016. About renewable energy. Natural Resources Canada. http://www.nrcan.gc.ca/node/7295

Nerkar, A., and S. Shane. 2007. Determinants of invention commercialization: An empirical examination of academically sourced inventions. *Strategic Management Journal* 28(11): 1155–1166.

Nonaka, I. 1991. The knowledge-creating company. *Harvard Business Review*, November–December, pp. 96–104.

OECD. 2007. Energy and sustainable development. Organization for Economic Cooperation and Development, United Nations, New York.

Orwig, J. 2016. This $14-billion machine is set to usher in a new era of nuclear fusion power. http://www.sciencealert.com/this-14-billion-machine-is-set-to-usher-in-a-new-era-of-nuclear-fusion-power.

Pigou, A. C. 1924. *The Economics of Welfare*. Second edition. London: Macmillan.

Pinch, T. F., and W. E. Bijker. 1987. The social construction of facts and artifacts: Or how the sociology of science and the sociology of technology might benefit each other. In Bijker, W., T. Hughes, and T. Pinch (Eds.), *The Social Construction of Technological Systems*. Cambridge, MA: MIT Press, pp. 11–44.

Popper, K. R. 1963. *Conjectures and Refutations: The Growth of Scientific Knowledge*. London: Routledge and Kegan Paul.

Postman, N. 1992. *Technoply: The Surrender of Culture to Technology*. New York: Knopf.

Rahal, A. D., and L. C. Rabelo. 2006. Assessment framework for the evaluation and prioritization of university inventions for licensing and commercialization. *Engineering Management Journal* 18(4): 28–36.

Ramaswamy, V., O. Boucher, J. Haigh, D. Hauglustaine, J. Haywood, G. Myhre, T. Nakajima, G. Y. Shi, and S. Solomon. 2001. Radiative forcing of climate change. *Climate Change 2001: The Scientific Basis*. Contribution of Working Group I to the Third Assessment Report of the Intergovernmental Panel on Climate Change. Cambridge: Cambridge University Press.

Raskin, P. T., G. Banuri, P. Gallopin, A. Gutman, R. K. Hammond, and R. Swart. 2002. *Great Transition: The Promise and Lure of the Time Ahead*. Boston, MA: Tellus Institute.

Relander, B. 2015. Investing in green technology—The future is now. http://www.investopedia.com/articles/investing/040915/investing-green-technologythe-future-now.asp.

Rogers, E. M. 1995. *Diffusion of Innovations*. Third edition. London: The Free Press.

Rogers, E. M., S. Takegami, and J. Yin. 2001. Lessons learned about technology transfer. *Technovation* 21(4): 253–261.

Rycroft, R. W. 2006. Time and technological innovation: Implications for public policy. *Technology in Society* 28(3): 281–301.
Sachs, W. (Ed.). 1992. *The Development Dictionary. A Guide to Knowledge as Power.* London: Zed Books.
Samerski, S. 2003. Alternative technology. Jean Robert. http://www.pudel.uni-bremen.de.
Saviotti, P. P. 2005. On the co-evolution of technologies and institutions. In Weber, K. M., and J. Hemmelskamp (Eds.), *Towards Environmental Innovation Systems.* Berlin, Germany: Springer, pp. 9–32.
Schmookler, J. 1966. *Invention and Economic Growth.* Boston, MA: Harvard University Press.
Schumacher, E. F. 1973. *Small Is Beautiful: A Study of Economics as if People Mattered.* London: Blond & Briggs.
Show, K. Y. 2010. Green technology. Encyclopedia of life support systems (ELSS). Paris: UNESCO. http://www.eolss.net/sample-chapters/c05/e6-35-55-00.pdf
Sianipar, C. P. M., H. Taufiq, H. R. Estiningtyas, K. Dowaki, A. Adhiutama, and G. Yudoko. 2013. Materials selection in appropriate technology: Four focuses in design thinking. *Advanced Materials Research* 789: 379–382.
Sianipar, C. P. M., and G. Yudoko. 2012. Understanding issue dissemination and arrival patterns on supply-chain using network analysis and social media. *Journal of Asia Pacific Business Innovation and Technology Management* 2(3): 93–110.
Siegel, R. P. 2012. Fusion power: Pros and cons. http://www.triplepundit.com/special/energy-options-pros-and-cons/fusion-power-pros-cons/.
Siegel, D. S., and P. H. Phan. 2005. Analyzing the effectiveness of university technology transfer: Implications for entrepreneurship education. In Liebcap, G. (Ed.), *Advances in the Study of Entrepreneurship, Innovation, and Economic Growth.* Amsterdam, the Netherlands: Elsevier Science/JAI Press, pp. 1–38.
Smilor, R., and J. Matthews. 2004. University venturing: Technology transfer and commercialisation in higher education. *International Journal of Technology Transfer and Commercialisation* 3(1): 111–128.
Soni, G. D. 2015. Advantages of green technology. *Social Issues and Environmental Problems* 3(9:SE): 1–5.
Stauffer, M. W. 2013. Incentives for green technology adoption: Getting government subsidies right. *Energy Futures, the Magazine of the MIT Energy Initiative.* http://energy.mit.edu/news/incentives-for-green-technology-adoption-getting-government-subsidies-right/.
Sukhochev, A. 2011. Technology transfer: Turning knowledge into value. International Research and Exchanges Board, George Mason University, Fairfax, VA.
Thorp, H. 2010. The entrepreneurial university. https://www.insidehighered.com/views/2010/09/27/entrepreneurial-university.
Time. 2010. The 50 best inventions of 2010. http://content.time.com/time/specials/packages/0,28757,2029497,00.html.
Tlusty, M. 2015. Why sustainability requires more than technological advances. http://www.triplepundit.com/2015/05/sustainability-requires-technological-advances.
Tunzelmann, N. V. 1996. Engineering and innovation in the industrial revolutions. Science Policy Research Unit, Mantell Building, University of Sussex, Brighton, UK.

Turner, J. 1978. Housing in three dimensions: Terms of reference for the housing question redefined. In Burguess, R. (Ed.), *World Development*, Vol. 6, no. 9–10. London: Oxford Pergamon Press.

UNCTAD. 2014. Transfer of technology and knowledge sharing for development science, technology and innovation issues for developing countries. United Nations Conference on Trade and Development. http://unctad.org/en/PublicationsLibrary/dtlstict2013d8_en.pdf.

UNEP. 2007. Concept paper for the task force on sustainable lifestyles. https://esa.un.org/MARRAKECHPROCESS/pdf/TFSL_ConceptPaper.pdf.

UNIDO. 2014. Diffusion strategy of green technology and green industry in Africa. United Nations Industrial Development Organization.

Vergragt, R. J. 2006. How technology could contribute to a sustainable world? GTI Paper Series 8. Frontiers of a Great Transition. Tellus Institute, Boston, MA.

WEC. 2013. World energy perspective: energy efficiency technologies. World Energy Council. https://www.worldenergy.org/wp-content/uploads/2013/10/World-Energy-Perspective_-Energy-Efficiency-Technologies_Exec-summary_EMBARGOED.pdf.

WEF. 2013. Energy Vision 2013 Energy transitions: Past and Future. *World Economic Forum*. http://www3.weforum.org/docs/WEF_EN_EnergyVision_Report_2013.pdf.

Wells, J. 2012. The role of technology in technology entrepreneurship. *Technology Innovation Management Review*, April, pp. 35–40.

Willoughby, K. 1990. *Technology Choice: A Critique of the Appropriate Technology Movement*. Boulder, CO: Westview Press.

Yang, J. Q. Yu, and P. Gong. 2008. Quantifying air pollution removal by green roofs in Chicago. Atmospheric Environment 42: 7266 -7273.

chapter four

Engineering ethics and public policy

Engineering is a profession, not just a job.

4.1 *Objectives*

- Provide an opinion about engineering ethics and public policy (PP) and how they relate to each other.
- Provide a historical perspective, from ancient Greece to the end of the twentieth century, on ethical issues and principles commonly associated with engineering in Western civilization.
- Extend knowledge about ethics, ethical theories, and codes of ethics.
- Understand engineering as a profession and the role of ethics in engineering.
- Explore engineering ethics as an area of applied or practical ethics.
- Know about the domains of ethics including research, teaching, and engineering practice.
- Discuss the categories of engineering ethics including technical, social, professional, and entrepreneurial.
- Examine how engineers are required to practice within their field of knowledge and experience.
- Realize the role of engineering profession in the development of PP.
- Recognize how PP development takes into account appropriate technical requirements for success.
- Develop innovative solutions to PP problems based on technology.
- Understand how engineering professionals translate complex technical issues into a publicly accessible discourse.
- Realize the impact of increasing the PP and ethics content on engineering education.
- Communicate clearly and proactively on the contributions of the engineering profession to ethical policies for sustainability.
- Discuss the initiation of PPs and ethics for sustainability and introduce a policy case from the Canadian city of Calgary.
- Develop through a sociotechnical case an intellectual basis for understanding the ethical and policy questions and challenges posed by transitions in energy systems, as well as criteria and approaches for

evaluating the ethical desirability of future energy options including renewable resources and smart grid (SG).
- Provide three end of chapter pedagogical knowledge strategies, namely knowledge acquisition, knowledge possession, and knowledge creation to help understand the above topics and generate new and alternative ideas and solutions.

4.2 Introduction

Both engineering ethics and PP are relatively new fields that have increased in importance and influence over the past few decades. Although each field has a distinctive emphasis, in both cases engineers need to move beyond their traditional disciplinary comfort and merge knowledge and tools from other disciplines. In fact, engineers have been involved in developing both fields through education, research, and the activities of professional engineering societies. In addition, applied ethicists, working on their own or in conjunction with engineers, have made significant contributions to the theory and practice of engineering ethics, as have social scientists in the area of engineering and PP. Although a number of individual engineers, as well as the major professional societies, have been active in both realms, few formal efforts have been made to achieve greater integration and cross-fertilization of the two fields (Herkert 2000a).

Engineering creativity emerges within the constraints of physical laws, commercial considerations, the needs of the client or employer, society, the law and ethics. Constraints provide boundaries within which to explore problems and propose engineering solutions. Ethical considerations in relation to safety and the environment can provide opportunities and inspiration for engineers to devise innovative solutions, directing their creativity to improve the performance of engineering technologies and systems. Ethical concerns about climate change drive engineers to devise creative solutions to the problem of providing reliable, cheap renewable energy. Ethical concerns about global poverty lead engineers to work with local communities to develop new technologies for water supply and sanitation in the developing world. Engineering ethics is a constraint to bad practice and an inspiration to innovation and creativity (Lawlor 2013).

The public role of engineers, which involves the responsibilities of engineers with respect to technology policy at various levels, has been a central focus of effort coming under the banner of engineering and PP. This involves the making of decisions within the public domain that inspire, and sometimes demand, that the engineering bring forth products and services that at least do no harm and ideally contribute to the social good, and the common good of people. A real value in stressing PP is to establish the need for a possible reorientation of engineering toward designing and manufacturing products and services that the end user will find inherently tuned toward serving the common good, products that are humane, sustainable, and risk-free.

There is a need for an integrated approach that evolves from the recognition that the social implications of technology permeate the three spheres in which the engineer operates: personal, professional, and public. Limiting the focus of engineering ethics to the personal and professional spheres and PP to the public sphere (the conventional approaches) can leave the engineers and engineering students with a disjointed sense of his or her role in addressing important issues of ethics and PP. The result can be PP positions by engineers and professional engineering societies that are lacking in ethical foundations.

In order to illustrate these issues, this chapter, following a general discussion of ethics and PP, will discuss several major issues that attract the attention of engineers and professional engineering societies. A major sustainable energy case will be discussed and analyzed from ethical, policy, and risk aspects.

4.3 *Ethics*

The unexamined life is not worth living.

Socrates

The word "ethics" comes from the Greek word *ethos* and is defined as the study of standards of right and wrong; part of science and philosophy dealing with moral conduct, duty, and judgment. Ethics can be defined in different ways: first, ethics refers to well-based standards of right and wrong that prescribe what humans ought to do, usually in terms of rights, obligations, benefits to society, fairness, or specific virtues. Second, it refers to the study of development of one's ethical standards (Velasquez et al. 1987).

The main focus of ethics is to establish right and wrong conduct, both in theory and in given situations. While issues in ethics are often disputed, major ethical imperatives, such as not committing murder, may be codified into law, which allows for a standard of justice. Practically applied, ethics is essential because it gives individuals a basis on which to praise or decry an action and punish or reward it. Without the study of ethics, there can be no government and no law. Without an ethical system in place, all actions are equally acceptable and no one is safe from their neighbor.

4.3.1 Historical perspective

In Western history, much of what is known about moral reasoning mostly began with the ancient Greeks, especially with the philosophers Socrates, Plato, and Aristotle. From the time of ancient Greece up to the current times, there appeared two principal schools of ethics. Some moralists maintained that ethical conceptions are inspired in human from above, and they consequently connected ethics with religion. Other thinkers realized the source of morality in human and endeavored to free ethics from the sanction of religion and to create a realistic morality. Some of these thinkers maintained that the main motivation power of all human actions is found in what is called pleasure and happiness, and all action is toward this end.

The first developments of the profession of engineer coincide with early developments within Western Europe, when, about the year 1000, from the scraps of ancient civilizations, a new world began to be built (Meijknecht 2014). The history of engineers' dedication in the eighteenth century also remains a gray area. Philosopher Thomas Reid compares a system of morals to "laws of motion in the natural world, which, though few and simple, serve to regulate an infinite variety of operations throughout the universe". However, he contrasts a system of morals with a system of geometry: a system of morals is not like a system of geometry, where the subsequent parts derive their evidence from the preceding, and one chain of reasoning is carried on from the beginning; so that, if the arrangement is changed, the chain is broken, and the evidence is lost. It resembles more a system of botany, or mineralogy, where the subsequent parts depend not for their evidence upon the preceding, and the arrangement is made to facilitate apprehension and memory, and not to give evidence (Goldfarb and Pritchard 1999).

In modern history, the work of Darwin was not only limited to biology. Already in 1837, when he had just written a rough outline of his theory of the origin of species, he entered in his notebook this significant remark: "My theory will lead to a new philosophy." And so it did in reality. By introducing the idea of evolution into the study of organic life, he opened a new era in philosophy, and his later sketch of the development of the moral sense turned a new page in ethics. In this sketch, Darwin presented in a new light the true origin of the moral sense, and placed the whole subject on such a firm scientific basis that, although his leading ideas may be considered as a further development of those of Shaftesbury and Hutcheson, he must be, nevertheless, credited with opening a new path for science in the direction faintly indicated by Bacon. He thus became one of the founders of the ethical schools, together with such men as Hume, Hobbes, or Kant (Kropotkin 1922).

The leading ideas of Darwin's ethics can easily be summarized. In the very first sentence of his essay, he states his object in very definite terms. He begins with a praise of the sense of duty, which he characterizes in the well-known poetical words, "Duty! Wondrous thought that works neither by fond insinuation, flattery, nor by any threat... etc." He undertakes to explain this sense of duty, or moral conscience, "exclusively from the viewpoint of natural history" an explanation, he adds, which no English writer had hitherto attempted to give (Kropotkin 1922).

The progress made by the natural sciences in the nineteenth century awakened in modern thinkers the desire to work out a new system of ethics on positive bases. After having established the fundamental principles of a universal philosophy free from postulates of supernatural forces, and at the same time, majestic, poetic, and capable of stimulating in men the highest motives, modern science no longer needs to resort to supernatural inspiration to justify its ideals of moral beauty. The years 1850–1950 are considered the golden age of technology. Every technological innovation in this period was by definition an improvement.

Professional engineering societies in the United States began to be structured in the late nineteenth century, with new societies created as new engineering fields developed. Codes of ethics for engineers were developed along with their corresponding professional societies. As these societies matured, many of them created codes of ethics to guide practicing engineers. The first civilian engineering organization in the United States, the Boston Society of Civil Engineers, was founded in 1848. The American Society of Civil Engineers (ASCE) was founded four years later. Though the early leaders of these organizations often referred to the "high character and integrity" engineers needed to serve the interests others committed to them, the history of codes of ethics really began a half century later. In 1906, the American Institute of Electrical Engineers (AIEE) voted to embody in a code the ideas expressed in an address by its president, Schuyler S. Wheeler. After much debate and many revisions, the AIEE Board of Directors adopted a code in March 1912. The AIEE Code was adopted by the American Society of Mechanical Engineers in 1914. Meanwhile, the American Institute of Consulting Engineering, the American Institute of Chemical Engineers, and ASCE each adopted their own code. By 1915, every major engineering organization in the United States had a code of ethics (Luegenbiehl and Davis 1992; Fleddermann 2012).

In the mid-twentieth century there was a proliferation of important ethics codes that still guide engineering professional behavior and research activities. The first reference to a National Society of Professional Engineers (NSPE) code of ethics is found in the May 1935 issue of

The American Engineer in the form of a suggestion for membership consideration. In 1946, the Board approved the canons of ethics for engineers as prepared by a joint committee sponsored by the Engineers' Council for Professional Development, a coordinating body of technical engineering societies (NSPE 2017).

Currently, these codes are mostly concerned with issues of how to conduct business. Codes also spell out the duties that engineers had toward their employers. Relatively less emphasis was given to issues of service to the public and safety when compared to contemporary codes. This imbalance changed greatly in recent decades as public perceptions and concerns about the safety of engineered products and devices changed. Now, most codes emphasize commitments to safety, public health, and even environmental protection as the most important duties of the engineer (Fleddermann 2012).

4.3.2 Ethical theories

Ethics constitutes an entire branch of philosophy. Ethical theories are a philosophical approach to the moral reasoning of correct action. The fundamental ethical theories rationalize moral principles such as goals, obligations, duties, rights, and social conventions. The notion that practices of organizations, where professional activities are socially constructed through managerial subjectivity, are unlikely to be aligned with the corporate ethics communicated through explicit rhetoric that is not consistently sustained by everyday organizational practice; are ethical and are ethics of the organization is equivocal.

Historically, philosophers have developed theoretical approaches of differentiating right from wrong and for providing guidelines about how to live and work ethically. They have showed an increasing interest in applying moral theories to real-life problems; that is called today, "applied ethics," especially in professions like engineering.

There are a number of different ethical theories of appropriate professional and personal conduct. There are also a number of different perspectives on the role of such theories. Some people consider that they should be used purely to provide guidance and highlight issues in ethical decision-making, whereas others consider that they can be applied to obtain the correct decision (Hersh 2015). One way of categorizing the field of ethics is by distinguishing between its three branches. These are metaethics, normative ethics, and applied ethics.

4.3.2.1 Metaethics

The first type of ethical theory is metaethics, sometimes known as analytic ethics, which talks about the nature of ethics and moral reasoning. This theory addresses the origin and definition of people's ethical principles.

Metaethics deals with whether morality exists. Universal truths, God's will, and how reason plays a role in ethical decisions are all a part of metaethics. Some examples of metaethical questions are as follows:

- What does it mean to say something is ethically good?
- How, if at all, do we know what is right and wrong?
- How do moral attitudes motivate action?
- Are there objective values?

4.3.2.2 *Normative ethics*

The concrete theory of normative ethics defines the moral standards that adjust right and wrong options. Morals refer to mostly accepted standards of right and wrong in society, often learned during childhood, but ethics are learned at the time of the problem. Normative ethics assumes an agreeing answer to the existence question, deals with the reasoned creation of moral principles, and at its top level, decides what the fundamental principle of morality is. It helps find out what things have what moral types, to provide a basis for ethics. This theory is subdivided into three parts: virtue, deontology, and consequentialist.

According to virtue theory, it is ought to possess certain character traits such as courage, generosity, compassion, and these ought to be manifest in actions. Deontology theory concentrates on the act being performed. These acts ought or ought not to be performed, irrespective of the consequences. Deontology approach is based on independent moral rules and duties, which should be defined "objectively" rather than subjectively. Consequentialist theory holds to act in the way that brings about the best consequences. It does not matter what those acts are; the end justifies the means. All that matters for ethics is making the world a better place. It is concerned with the consequences of actions and with the balance between benefits and harms (Ersdal and Aven 2008).

The ends/means problem is a general scenario in ethics. This is the essence of the difference between two central ethical positions: deontology and consequentialism. Deontology explains that whether an act is good or bad depends on some quality of the act itself. Consequentialism, instead, says that whether an action is good or bad depends on the consequence. Most people's ethical beliefs fall into some hybrid version of the above two.

4.3.2.3 *Applied ethics*

Applied ethics is a special category of ethical philosophy. It is the most real of the three categories of the ethics philosophy. Applied ethics deals with tough moral questions and controversial moral issues that people face in their lives including abortion, euthanasia, sex before marriage, death penalty, gay/lesbian rights, etc.

Applied ethics is usually divided into various fields. Business ethics discusses ethical behavior in the corporate world, while professional ethics refers directly to a professional in the field. Biomedical and environmental ethics delve into health, welfare, and the responsibilities we have toward other people and our environment. According to Hersh (2015), another applied human-centered ethics focuses on people first with the organization and technology in the second and third places and on the needs of individuals and groups rather than those of vested interests and power structures. The high degree of uncertainty in evaluating the risk associated with many new technologies also raises ethical questions. Particular examples include nuclear energy and genetically modified foods. Both these examples can be considered social experiments with unknown outcomes and possibly unforeseeable long-term consequences, making the use of the precautionary principle appropriate (Hersh 2015).

The main goal of applied ethics is to verify the appropriate principles of conduct in certain areas to which it relates. There are many areas of applied ethics. Given their situational nature, they are often different from one another. Examples include animal ethics, biomedical ethics, business ethics, environmental ethics, information ethics, law ethics, engineering ethics, etc.

4.3.3 Code of conduct and ethics

A code of ethics is the characteristic of a profession. A code of conduct and ethics is an aspirational system of principles and rules or the guideline that sets satisfactory behaviors for a given group of people or profession. The task of the code of ethics is not to derive obligations, but to make obvious what the public anticipates from the profession. It expresses the rights, duties, and obligations of the members of the profession. Professionally, the code of ethics is a comprehensive guide to professional conduct. The code is usually designed to help practitioners maintain the highest level of ethical conduct, standards of practice and integrity with respect to their professional activities. Every profession has its professional ethics; for example, engineers, lawyers, and physicians typically adhere to a code of ethics. Importantly, a code of ethics is not a legal article, so a professional cannot be arrested for breaching its provisions.

Ethics codes are as old as antiquity. Religious traditions and civic cultures have codes as their foundations. They often capture a vision of excellence, of what individuals and societies should be striving for and what they can achieve. In this sense, codes, which are often mistaken as part of law or general statements of mere aspiration, are some of the most important statements of civic expectation (Gilman 2005). Codes of ethics are drawn up to express the expectations of a group of persons of common vocation with regard to their conduct. These codes serve as a framework for ethical judgment for a professional engineer.

Mostly, a code of ethics provides a framework for ethical judgment for a professional. The key word here is framework. No code can be totally comprehensive and cover all possible ethical situations that a professional engineer is likely to encounter. Rather, codes serve as a starting point for ethical decision-making. A code can also express the commitment to ethical conduct shared by members of a profession. It is important to note that ethical codes do not establish new ethical principles. They simply reiterate principles and standards that are already accepted as responsible engineering practice. A code expresses these principles in a coherent, comprehensive, and accessible manner. Finally, a code defines the roles and responsibilities of professionals (Harris et al. 1995). As an umbrella body, the World Federation of Engineering Organisations (WFEO 2001) has published a model code of ethics. Most national and international professional engineering institutions follow a code of ethics which reads along similar lines. Apparently, the codes of ethics are not thorough enough to cover all possible ethical dilemmas that an engineer might face in career. The codes serve as starting points for making ethical decisions. They reflect the values and beliefs of both professional engineering bodies and individual engineers and are thus themselves present an evolving (microethical: personal conduct, organizational; and macroethical: societal, economic and political structures, and PP) construct. A common theme among all those code of ethics statements is also a narrow focus on the individual agent, to the detriment of a broader context, such as a responsibility to act an agent of cultural or societal change (Byrne et al. 2010).

4.4 *Engineering ethics*

Engineering ethics is part of thinking like an engineer.

Michael Davis

4.4.1 *Engineering ethics defined*

Engineering ethics is an area of applied ethics although other theories of ethics come into play in engineering ethics. It is practical in the sense that its aim is to shed light on ethical concerns related to engineering practice. It is applied in the sense that ethical considerations are directed to practice rather than theory (Pritchard 2005). Engineering ethics is the field that examines and sets standards for an engineer's obligations to the public, clients, employers, and the profession. Ethical issues involve quality, safety, legal compliance, conflict of interest (e.g., bribery and gifts), and treatment of confidential or proprietary information. Engineering is full of ethical decision-making. By the nature of this profession, engineers design, build, and create technologies that disrupt the status quo, enable new ways of behaving, and have unknown consequences on the world (Harris 2008).

The field of engineering ethics emerged in the mid-1970s, when humanists and social scientists joined engineers in addressing issues of moral and social responsibility in engineering (Weil 1984). It is the study of moral issues and decisions confronting individuals and organizations engaged in engineering. It is the field of study that focuses on the ethical aspects of the actions and decisions of engineers, both individually and collectively. A rather broad range of ethical issues are discussed in engineering ethics: professional codes of conduct, whistle-blowing, dealing with safety and risks, liability issues, conflicts of interest, multinational corporations, and privacy (Van Gorp 2005). A key concept in engineering ethics is "professional responsibility," that is, moral responsibility based on an individual's special knowledge (Herkert 2002). As Martin and Schinzinger (1996) note, the goal of responsible engineers is "the creation of useful and safe technological products while respecting the autonomy of clients and the public, especially in matters of risk-taking." In addition to a fundamental commitment to public health, safety, and welfare, engineering ethics is typically concerned with conflicts of interest, the integrity of data, whistle-blowing, loyalty, accountability, giving credit where due, trade secrets, and gift giving and bribes (Wujek and Johnson 1992).

Critiques of engineering ethics have been raised by others, including engineers such as Vanderburg (1995) who draws a distinction between "micro-level" analysis of "individual technologies or practitioners" and "macro-level" analysis of "technology as a whole." Also, Ladd (1980), an ethicist, argues that professional ethics can be delineated as "microethics" or "macroethics" depending on whether the focus is on relationships between individual engineers and their clients, colleagues, and employers or on the collective social responsibility of the profession.

An integral part of engineering progress has been the development of many thousands of technical standards. These standards constitute an engineering triumph, a glory of our civilization, comparable to the development of regulatory laws and agencies. In fact, these two marvelous phenomena are interrelated. Voluntary standards are developed by professional groups, and then government agencies, when they see fit, adopt the standards, and give them the force of law. Without these laws, regulations, codes, and rules, each engineer would be given unwarranted and unwanted powers (and, incidentally, each engineering problem would entail reinventing the wheel) (Florman 2002).

An online resource on ethics for engineering and science maintained by the Center for Engineering Ethics and Society at the National Academy of Engineering (NAE) is available at: www.onlineethics.org. It provides engineers, scientists, faculty, and students with resources for understanding and addressing ethically significant issues that arise in scientific and engineering practice and from the developments of science and engineering; and serve those who promote learning and advance understanding of responsible research and practice in engineering and science.

4.4.2 Scopes and categories of engineering ethics

Consideration of engineering ethics takes place largely in three domains: research, teaching, and engineering practice. A common philosophical approach to engineering ethics is to employ moral theories, such as utilitarianism and duty/rights-based ethical theories, to the solution of moral dilemmas in engineering. Utilitarianism is an ethical system that judges an action to be morally correct if its outcome results in the greatest good for the greatest number of people. Duty and rights approaches to ethics, on the other hand, focus on actions themselves, and on whether or not individuals abide by duties to do good and avoid harm, or act out of respect for the moral rights of other individuals. Although these two types of moral theories often result in the same conclusion regarding a particular act, they might produce conflicting conclusions, as when an engineering project built to benefit the public results in the eviction of individuals without their prior consent. Figure 4.1 shows the scopes of engineering ethics.

McLean (1993), an engineer, uses three categories in discussing engineering ethics. These include technical ethics, covering technical decisions by engineers; professional ethics, dealing with interactions among managers, engineers, and employers; and social ethics, concerning

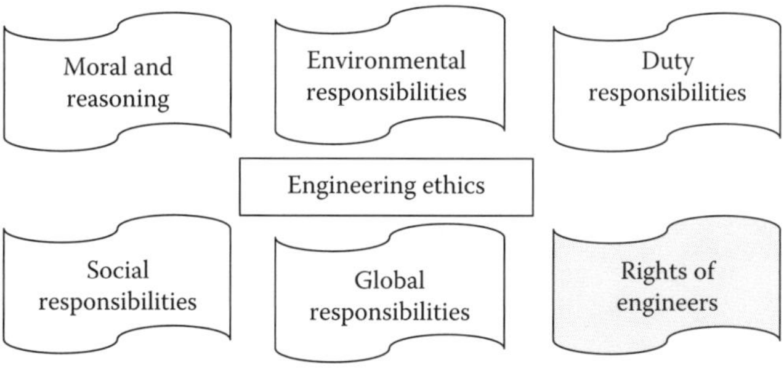

Figure 4.1 Scopes of engineering ethics.

socio-political decisions about technology. Figure 4.2 shows the three categories of engineering ethics.

Langdon Winner, well known for his critical analyses of technological development, is equally critical of traditional approaches to engineering ethics (Winner 1991). Such approaches, Winner argues, focus almost entirely on specific case studies of ethical dilemmas to the exclusion of larger issues relating to the development of technology and to the career choice of engineers: ethical responsibility involves more than leading a decent, honest, truthful life, as important as such lives certainly remain. It involves something much more than making wise choices when such choices suddenly, unexpectedly present themselves. Our moral obligations must include a willingness to engage others in the difficult work of defining the crucial choices that confront technological society and how to confront them intelligently.

While some work in engineering ethics has focused on the public role of engineers, Winner's critique is essentially correct. The primary focus of engineering ethics, whether by ethicists, engineers, or professional engineering societies, has been on the personal and professional roles of engineers. Similarly, engineering codes of ethics, though stressing moral responsibilities to the public, tend to be aimed at the duties and rights of individual engineers in carrying out such responsibilities.

4.4.3 *Professional ethics*

The words "profession" and "professional" have many uses in modern society that go beyond the definition of a job or occupation. Professionalism also involves being honest about level and areas of competence, and never agreeing to work in areas in which you are not competent or not able to easily achieve competency.

Engineering practice can be defined as a "profession," as opposed to an "occupation" or "job." To date, there is no complete agreement about

Figure 4.2 The three categories of engineering ethics.

the definition of the profession. To be a professional involves the recognition of responsibility to the public. According to Hooker (2000), professionals can be defined by three characteristics: they are experts, they use their expertise responsibly, and they mark themselves as professionals. Professionalism best refers to a set of attitudes rather than to a specific job. Professional obligations are usually summed up in a professional code of ethics. A profession renders services based upon advanced education and knowledge, skill, and judgment. It requires sophisticated skills, judgment, and exercise of responsibility (not routine); formal education; being honest and independent and serving with faithfulness the public; membership in special societies; established standards for admission into the profession and conduct of its members; substantial degree of public obligation and positive service results from the practice of the profession. A profession depends on confidence of two kinds for efficient pursuit of work, the personal confidence of the client or employer in the technical competence of the engineer and the confidence of the public at large in the integrity and ethical conduct of the professions as a whole.

Professional ethics encompasses the personal, organizational, and corporate standard of attitude and behavior expected of professionals. Usually, professional attitude and behavior are guided by "codes of conduct," while the operational activities within a profession are guided by "codes of practice." A professional engineer should apply the code of ethics not in passive performance but as a set of principles guiding the professional conduct.

One way to approach engineering ethics is by thinking of it as a profession and then examining ethical issues according to its features. So, for example, given that engineering profession usually have a codified set of principles and/or rules for its professionals, it is possible to articulate, expand, and flesh out such principles. Another way to approach engineering ethics is by starting with certain historical cases and/or hypothetical kind, and then draw out any moral lessons and perhaps principles from them (Martin and Schinzinger 2005).

Engineering ethics and professional responsibility have become more relevant to engineering during the last quarter of the twentieth century. Responsibility can be attributed to causes, role, legal, and moral issues. Responsibility is shared whenever multiple individuals collaborate as a group. As technology and its impacts have become more complex and far-reaching, the importance of responsible engineering decisions toward employers and the public have been emphasized. Engineering work requires sophisticated skills, the use of judgment, and the exercise of discretion. Membership in the profession requires formal education, not simply practical training or apprenticeship. Four years of undergraduate training leading to a bachelor's degree in an engineering program is essential, followed by work under the supervision of an experienced engineer. Many engineering jobs even require advanced degrees beyond the bachelor's degree.

The strength of engineering ethics lies in the strong grounding in professionalism and engineering practice (Herkert 2005). Engineering ethics aims to guide engineers in the profession that do not harm. Engineers should realize the value that is given to their professional judgment. Engineering ethics also evaluates the impact of engineered products and services on the society. Engineering creativity arises within the applications of physical laws, market considerations, client and employer needs, he law and ethics. Constraints provide boundaries within which to explore problems and provide solutions. Ethical considerations in issues of safety and the environment can provide opportunities and inspiration for engineers to plan innovative solutions, guiding their creativity to improve the performance of products and systems.

Some aspects of engineering professionalism, such as (1) sensitivity to risk, (2) awareness of the social context of technology, (3) respect for nature, and (4) commitment to the public good, cannot be adequately accounted for in terms of rules, certainly not negative rules (Harris 2008).

As a guideline, professional ethics, one may see general professional norms and profession-specific guidelines as follows (Murray 2009):

- Conflicts of interest, accepting gifts, bribery, perks, client relationships
- Intellectual property (trade secrets, patents, trademarks, copyrights)
- Publishing, authorship, plagiarism, peer review
- Confidentiality, whistle-blowing
- Fraud, forgery, data fabrication, perjury
- Public relations, media, marketing
- Professional versus personal duties; legal versus moral responsibility
- Public safety, health, and welfare
- Social norms, social pressures, cultural taboos
- Public service, expert testimony, responsibility to inform the public
- Determining risk, liability, accountability, rights, and responsibilities

4.4.4 Technology ethics

Technology ethics is a transdisciplinary area that draws on theories and techniques from several knowledge domains (such as engineering, science, social sciences, technology, applied ethics, and philosophy) to provide insights into ethical dimensions of technological systems and practices for advancing a technological society. It views technology and ethics as socially embedded enterprises and aims on discovering the ethical use of technology, protecting against the misuse of technology, and devising common principles to guide new advances in technological development and application to benefit society (Luppicini 2010). Technology in ethics raises exceptional moral questions because technology has created about historic social, political, and conceptual change. Because technology impacts the way things are accomplished but the way of thinking about them, it challenges some of the basic concepts of moral and philosophy such as property and privacy.

Technologies, particularly revolutionary technology, generate many ethical problems. Sometimes the problems can be treated easily under extant ethical policies. Engineers also have to manage the risks posed by the technologies they develop. Part of this is understanding the immediate safety implications of its development and use, but unforeseen or long-term risks are also of importance to the engineer (Lawlor 2013). According to engineering codes of ethics, the engineer's significant professional responsibility is to ensure the safety, health, and welfare of the public. Although everyone must avoid endangering others, engineers have a special responsibility to ensure the safety of the objects that they produce. Making an ethical decision involves thinking about the lifecycle of a project or product or design, not just assuming everything is fine.

Emerging technologies, like IT, artificial intelligence, genetic technology, neuroscience, nanotechnologies, SG, self-driving cars, drones, and robots, among others, have been progressing for a while, and all offer potential benefits and opportunities. Although the above technologies are not fully developed, it is not unreasonable to expect that they will continue along a revolutionary path and bring with them an increasing cluster of new ethical issues. First, all of the technologies possess an essential feature of revolutionary technology, namely they are propelled in vision and in practice by some important generic capability. All are malleable in some way (Moor 2005). These may pose new legal and ethical issues and risks to the environment or society when fully implemented, where the issues of responsibility and liability are not yet agreed or realized. The introduction of new technologies into society will put people, organizations, and governments in situations where they will need to make choices that are novel to them. A good degree of uncertainty always exists concerning the fast pace of changes in technology and the economy, of which people are part, but with which people have little experience. And for all such emerging technologies, people may have fears or objections about their use that engineers need to seriously take into consideration this relationship between human and technology.

Technological consciousness is a term that describes the above relationship. Technology is seen as an integral component of human consciousness and development. Technology, consciousness, and society are intertwined in a relational process of creation that is a key to human evolution. Technology is rooted in the human mind, and is made manifest in the world in the form of new understandings and artifacts. The process of technological consciousness frames the inquiry into ethical responsibility concerning technology by grounding technology in human life (Luppicini 2010). An ethics of technology must, therefore, concern itself with society's accountable conduct and with technology services and use. Because of the technical way of thinking, there is a kind of consensus on the most current ethical approaches to this matter. In other words, the technological scenario of the world defines contemporary ethics, which may influence engineers as well.

4.4.5 Social ethics

Social ethics deals with human needs and aspirations. It involves a code of conduct created by a society in order to ensure a smooth functioning of the said society. Engineering has enormous role to help provide benefits to society and the idea of social responsibility is very common in engineering ethics. It helps to provide basic needs such as water, food, housing, and energy, and does that on a way crucial for industrial sector to function.

One of the most powerful gifts of engineering is how it improves lives. But, in the wrong hands, it can be just the opposite. The work of engineers is critical to many aspects of economic and social progress of humanity. Contributing in this way requires not only technical competence, but also imagination, persistence, and integrity. At work, engineers and scientists regularly make decisions that have ethical significance or moral relevance at varying scales and aspects of life (Murray 2009).

The practice of engineering does not exist outside the domain of societal interests. That is, the practice has an inherent (and unavoidable) impact on society. Engineering is based upon that relationship with society. An engineer's conduct (as captured in professional codes of conduct) toward other engineers, employers, clients, and toward the public is an essential part of the life of a professional engineer, yet the education process and professional societies pay inadequate attention to the area (Nichols 1997). Therefore, engineers should conduct themselves in a manner which enhances the stature of the professional and its ability to service the public. The work of engineers serves the public good by providing design and operation services in various sectors of life.

The social awareness of engineering is perception of the way in which technology both impacts and is impacted by the larger social environment. Engineers should desire high standard of leadership in the operation and management of technology. They possess a confidential and trusted position in society, and are expected to exhibit that they are seeking to assist the society and to be considerable to public concerns. Technology as a means of social advancement is the common good that engineers should pursue. Modern engineering developed in the nineteenth century, an age when technology was perceived in an almost clearly positive light.

Social responsibility has been identified as the responsibility embodied in the Paramountcy principle, the fundamental and primary ethical principle of engineering included in the professional engineers' code of ethics: "Engineers, in the fulfillment of their professional duties shall hold paramount the safety, health and welfare of the public (NSPE 2016)." The social responsibility of engineers requires that they also focus on the societal impacts of their work, particularly as these impacts affect the safety, health, and welfare of society. That responsibility flows in part from confidential status.

An important notion for the social ethics approach is that design results from human choices. The design process is a humanly organized process. In a less extreme view, technology powerfully influences social institutions and forces, but there is little, if any, causal effect in the other direction. However, the engineer who is sufficiently aware of the social dimension of technology understands that technology both influences and is influenced by the larger social context. On the one hand, technology can be an instrument of the power elite and can be used for such things as the deskilling of labor (Devon and Van de Poel 2004). On the other hand, technology can be utilized by grassroots movements, as protesters did in China and bloggers do in the United States. In any case, engineers are often called on to make design decisions that are not socially neutral. This often requires sensitivities and commitments that cannot be incorporated into rules. We believe that such social awareness is an important aspect of a professional character that will take seriously the obligation to promote public welfare through professional work (Harris et al. 1995).

The challenge of engineers, now and in the future, is to provide infrastructure to rural and semirural communities in the developing world. Also, with increasing urbanization, additional challenges are added including of how to economically provide infrastructure in new urban areas, how to retrofit the existing infrastructure, and how to accomplish all this in a responsible and sustainable manner (Parkinson 2010).

Many engineers and ethicists are critical of the traditional preoccupation of engineering ethics with specific moral dilemmas confronting individuals and call for greater attention to macroethical issues related to the societal implications of technology as a complement to the traditional microethical approach that focuses on individual cases. One response to this critique would be to broaden discussions of engineering ethics so as to include the ethical implications of PP issues relevant to engineering, such as risk and product liability, sustainable development, globalization, health care, and information technology (Herkert 2000b).

4.4.6 *Engineering ethics of entrepreneurship*

> Your reputation is more important than your paycheck, and your integrity is worth more than your career.
>
> **Ryan Freitas**
> *Cofounder of About.me*

There are fundamental reasons to take the ethics of entrepreneurship more seriously than other topics. First, entrepreneurship has emerged as a distinctive area of academic inquiry, with unique problems and questions that can be productively studied in their own right. Second, entrepreneurship is an inescapably ethical activity, whether one views it from the societal, the organizational, or the individual level; entrepreneurial action has powerful ethical dimensions and implications (Dunham 2005).

Given the prominence of entrepreneurship these days, there is also the tendency to extol and even romanticize the entrepreneur. For example, George Gilder (1992) comments that the entrepreneur's "success is the triumph of the spirit of enterprise, a thrust beyond the powers and principalities of the established world to the transcendent sources of creation and truth." And yet there is another side to the story. Though there are many examples of the benefits of entrepreneurship, there are also abundant examples of its misuse. Some employees of Enron engaged in a number of entrepreneurial undertakings that were illegal and unethical. And, on a far less grand scale, there are examples of entrepreneurs who brew whiskey illegally in the hills of Appalachia or who sell prescription drugs without a physician's prescription on the Internet. In these and other cases, entrepreneurs have broken legal and moral rules (Brenkert 2009).

Entrepreneurship is the process of discovering and developing opportunities in order to create value for an existing or new organization (Fisscher et al. 2005). So entrepreneurship is more than being innovative or creative in coming up with new ideas for products or services. Entrepreneurs must also either create an organization or work through an organization (intrapreneurs) to develop the new opportunities and values they envision. This involves not only risks of various sorts, but also obstacles and barriers that may stand in the way of the entrepreneur's efforts (Brenkert 2009). A more objective analysis of entrepreneurship reveals that it can best be characterized as a multifaceted process which includes a vision, a high level of personal commitment and drive, innovation, change, and the creation and building of something of significant value over time. It also involves taking both personal and financial risks, building and motivating a team of people, and mobilizing human, material, and financial resources (Starcher 1997).

Given the tremendous complexities involved in operating in new, untested areas, and at the edge of knowledge, entrepreneurship requires entrepreneurs to make important, and as we have seen, often ethical, decisions under conditions of extreme uncertainty, ambiguity, and ignorance. As a result, the entrepreneurial process places extraordinary hybrid ethical demands on entrepreneurs, whether they realize it and act on it or not, that involves innovation, law, business, technical, and engineering depending on the nature of the product (Dunham 2005).

Entrepreneurs can lay a firm ethical foundation for their enterprise by ensuring a "yes" answer to these four crucial questions (Hagenbuch 2015):

- *Fairness*: Is your business model based on win–win outcomes?
- *Integrity*: Can your business's products/services be promoted with the truth?
- *Decency*: Can you unashamedly tell others what your business does?
- *Sustainability:* Does your business make efficient use of resources?

Engineers are increasingly involved in start-up companies in which they make business decisions as well as engineering decisions. Often, at the same time they are entrepreneurs, managers, and engineers. Even in large firms engineers are often directly involved in the business processes and decisions. The project management which is often associated with the engineering profession in fact brings together management skills and engineering competence with the purpose of producing successful project results. It is evident that engineers must think about ethical issues that were once the provenience of business managers. The firms also have a legal obligation to provide a safe product. The standard of care is defined by generally accepted norms in the engineering profession (Hooker 2000). In such scenario, business ethics deals with the needed principles of business, whereas engineering ethics is involved in determining specific values that characterize a particular profession.

4.4.7 The transdisciplinary ethical engineer

Being an engineer is both an enormous privilege and a very great responsibility. This gives rise to the question of an engineer's responsibilities. At the simplest level, this should require using his or her skills to make a positive difference to people, animals, and/or the environment, as well as drawing attention to abuses, preferably with colleagues to avoid victimization (Hersh 2013).

During the past few decades, engineering ethics has been oriented toward protecting the public from professional misconduct by engineers and from the harmful effects of technology. This "preventive ethics" project has been accomplished primarily by means of the promulgation of negative rules. However, some aspects of engineering professionalism, such as (1) sensitivity to risk, (2) awareness of the social context of technology, (3) respect for nature, and (4) commitment to the public good, cannot be adequately accounted for in terms of rules, certainly not negative rules (Harris 2008).

Several character traits might be a part of such a professional character portrait of an ethical good engineer. The first character trait is professional pride, particularly pride in technical excellence. If engineers want their work as a professional to contribute to public welfare, the first thing they must do is be sure that their professional expertise is at the highest possible level. Professional expertise in engineering includes not only the obvious proficiencies in mathematics, physics, and engineering science but also those capacities and sensitivities that only come with a certain level of experience.

Engineers have a number of ethical duties and responsibilities toward various targets, including the following (Hersh 2015):

- People who are or will be using the technologies they are researching, developing, supplying, or, otherwise, working with
- Society as a whole, including anyone who will be affected, whether positively or negatively, by any technologies they are researching, developing, supplying, or otherwise working with
- Planet in general
- Other species

Engineers are required to practice within their field of knowledge and experience. For example, Engineers Australia (2010)'s code of ethics classifies these practices into four key areas:

- To demonstrate integrity, meaning that members of the engineering profession will act impartially and without favor, even if this means some personal discomfort
- To practice competently, meaning to maintain their professional competence through ongoing personal development, mentoring, and learning from their peers and leaders
- To exercise leadership, where engineer is expected to represent the profession in an honest and trustworthy way

- To promote sustainability, meaning to perform their work in a way that recognizes all of the stakeholders with the needs and requirements of future generations in mind

4.5 *Public policy*

> Our differences are policies; our agreements, principles.
>
> **William McKinley**

4.5.1 *PP defined*

The term "policy" refers to a constructed unity imposed on diverse and disparate measures (Page 2005). A policy often comes in the form of general statements about priorities, written regulations, procedures, and/or standards to be realized. Policy comes from those who have legitimate authority to enforce normative guidelines for action. It is made by elected officials acting in concert with advisors from higher levels of administration. Described simply, policy refers to a distinct path of action which is suitable for the pursuit of anticipated goals within a specific context, directing the decision-making of an organization or individual.

Policy can be formal or informal. A formal policy might take the form of a designed policy that has been analyzed, discussed, documented, reviewed, approved, and published by a recognized policy-making body. An informal policy might be an ad hoc, general, unwritten but widely recognized practice or understanding within an organization where a course of action is to be followed. Policy can also be categorized as reactive or proactive. Reactive policy develops in response to a concern or catastrophe that must be tackled. Proactive policies, by contrast, are initiated and followed through thoughtful choices (Mackay and Shaxton 2016).

The term PP always refers to strategic action led by a public authority and intended to conclude those actions in a given field. PP is the means by which the authority keeps order or addresses the needs of its citizens through actions defined by its constitution. A major aspect of PP is law which includes specific legislation, mandates, or regulations established through a political process. Because PPs are in place to address the needs of people, they are regularly broken down into different categories as they relate to society.

The domain of PP often revolves around politics and interest groups. PP is a course of action that directs a range of related actions in a given field. They rarely undertake one problem, but rather deal with clusters of long-term problems. Usually, PPs provide direction to governments and responsibility links to citizens.

4.5.2 PP making

PPs are the result of efforts made by governments to alter aspects of their own or social behavior in order to carry out some end or purpose and comprise complex arrangements of policy goals and policy means (Capano 2015). Policy-making is essentially a search for the best ideas to solve a problem or realize a public goal. Reliable policy work entails clear thinking, expert knowledge, and profound political judgment on trade-offs and concessions.

The emerging PP challenges require policy-makers to see issues from multiple perspectives, not only from the perspective of the authority, but also from the viewpoint of those whose actions will be essential to the achievement of collective goals. Meeting these challenges comprises complex arrangements of policy goals and policy means, not only of analysis, design, and evaluation, but also of negotiation, conflict resolution, and consensus building. In this view, policy design involves the effort to more or less systematically develop efficient and effective policies through the application of knowledge about policy means gained from experience, and reason, to the development and adoption of courses of action that are likely to succeed in attaining their desired goals or aims within specific policy contexts (Montpetit 2003; Bobrow 2006).

Political scientists often use a model of the policy-making process that focuses on the stages through which ideas and proposals move before becoming PP. Different scholars label the stages differently and place different emphases on them, but the components in Figure 4.3 are common. The development of a PP begins with public recognition that a problem exists which requires definition (the emergence of a problem that requires the attention of the public and decision-makers). The problem should be placed on the authority's agenda in order to find a solution. Various alternatives should be formulated to resolve the problem. This is followed by policy demands and agenda formation.

PP development is an iterative process, rather than a linear one. Following the pre-policy stages, the next major stage in the development of a PP is deliberation and policy adoption. From the policy agenda, decision-makers, with the input of interest groups, policy experts, and constituents, debate and bargain over alternative policy formulations, settling on an alternative or a combination of alternatives to respond to the problem. Decisions are made; policies are formulated; and policy statements are issued, taking such forms as orders, regulations, or laws Also important is the constitutional and statutory structure of the institution that makes the policy decision. Structure often determines which outcomes have a greater chance of success in the political struggle. Policy implementation includes outputs and impacts. Policy outputs are the tangible manifestations of policies, the observable and measurable

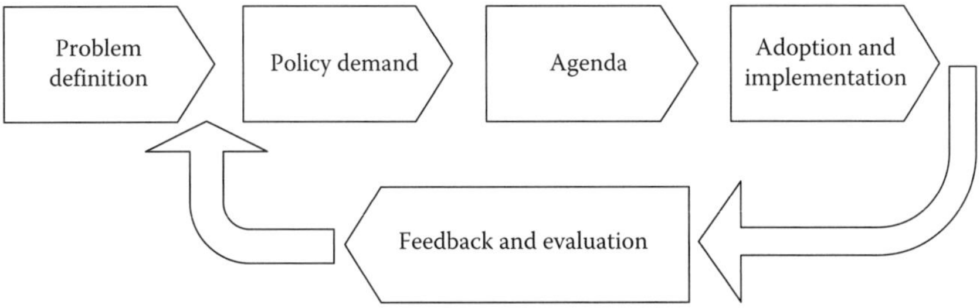

Figure 4.3 Stages of policy development.

results of policy adoption and implementation. Feedback and evaluation will continuously be in the loop coming out from various stakeholders (Cochran et al. 2016).

4.5.3 Key players in PP making

The policy sequence links a variety of key players in the policy-making process through their involvement with the different stages including individuals, institutions, and agencies. Government is often thought of to be the only entity involved in policy-making. Government does have the ultimate decision-making and funding power, but there are many other actors who contribute to PP, often in a network on which government relies for the delivery of complex policy goals (Mackay and Shaxton 2016). Figure 4.4 shows the role of various players in PP making. Government is the power of coercion and the body that makes decisions. Public servants provide technical knowledge and policy advice. They are service providers which include engineers who work to solve societal issues through technological development and innovation. Political parties develop relationships in exchange for political support. Interest groups seek to advance interests of their respective members. Legal systems interpret laws and acts independently. Media reports information to the public and shape public opinion. The public elect government forms opinions, joins interest groups, and coalitions. Often, the public is involved in the various PP making processes including consultation, deliberation, and engagement.

4.6 Engineering and PP

> As engineers, we were going to be in a position to change the world, not just study it.
>
> **Henry Petroski**

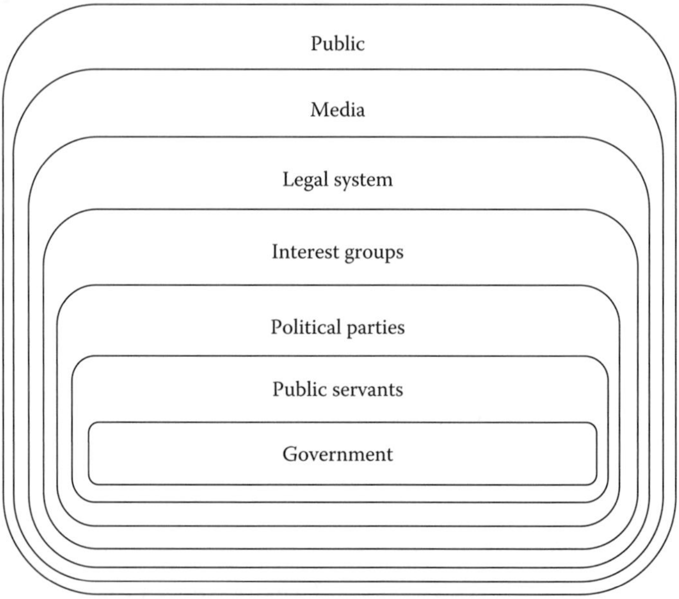

Figure 4.4 Role of various players in PP making.

4.6.1 *Engineering design and PP*

Engineering is a key power that influences the world socially, economically, and environmentally. From the supply of essential services such as power, water, sanitation, transport, communication to health technologies, engineering products and services are threaded through peoples' lives both individually and communally. PP engineering is the application of engineering, computer science, mathematics, or science to solving problems in PP.

The PP process has a lot in common with the engineering design process. Engineering design activities are becoming increasingly entwined with PP considerations. There are a wide variety of practical approaches to incorporating PP considerations into engineering design education (Hyman 2003).

Engineering design PP making is the process of designing a system, component or process to meet desired needs. It is a decision-making process (often iterative), in which the basic social sciences, mathematics law, and engineering science design concepts are applied to convert resources optimally to meet a stated objective. Among the fundamental elements of the design policy process are the establishment of objectives and criteria, synthesis, analysis, construction, testing, and evaluation. Central to the process are the essential and complementary roles of synthesis and

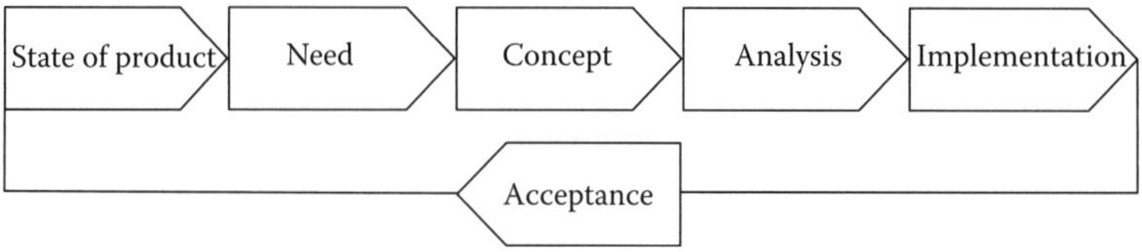

Figure 4.5 Basic model of the engineering design process.

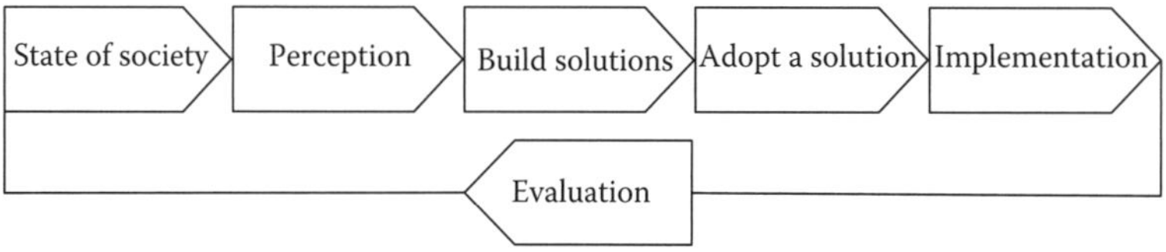

Figure 4.6 Basic model of the PP process.

analysis (Hyman 2003). Figures 4.5 and 4.6 display well-known models of both processes, each extracted from their respective specialized literature (Barke 1986; Dieter 2000). The similarity between the two models is quite remarkable. Upon reflection, what is really significant is that most engineers and policymakers do not recognize this close resemblance.

The crucial task of the engineer is to identify, understand, and integrate the constraints on a design in order to produce a successful result. It is usually not enough to build a technically successful product; it must also meet further requirements. Constraints may include available resources, physical or technical limitations, flexibility for future modifications and additions, and other factors, such as requirements for cost, manufacturability, and serviceability. By understanding the constraints, engineers deduce specifications for the limits within which a viable object or system may be produced and operated (Galloway 2005). These are the very constraints and considerations that are important when considering PP.

Regulations are rules that are made to make people comply and behave in a certain manner. A regulation has the effect of a law and is considered a restriction that is imposed by authorities to make people follow the desired code of conduct. Many engineering designs must satisfy environmental, health and occupational, public, and consumer product safety regulations (Hyman 2003). Examples are emission standards for cars, safety guards for industrial machinery, and impact resistance requirements for bicycle helmets. Figure 4.7 describes seven generic classes of PP activities that have implications on engineering design decisions. This typology is offered as a mechanism for structuring efforts to incorporate social criteria into design education.

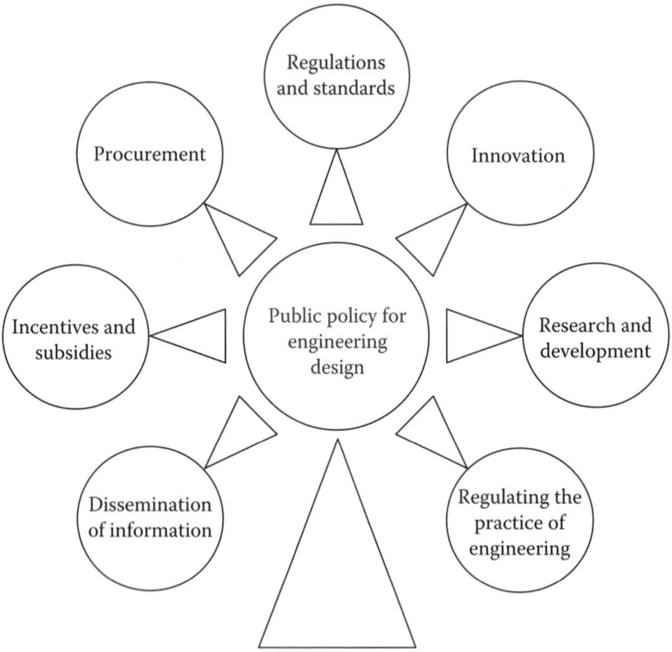

Figure 4.7 Generic classes of PP activities that have implications on engineering design decisions.

Engineers clearly embrace the system as a crucial mechanism for protecting design concepts embodied in patented devices. Acts allow universities and other institutions conducting federally funded research to secure patent rights and retain licensing royalties. The patent system is also a valuable source of design ideas and a tool for reducing the need to "start from scratch" or to "re-invent the wheel" in the early stages of the design process. IP is the foundation of the commercialization of engineering and scientific inventions.

Governmental R&D mission have a significant impact on engineering design by ensuring a continuous flow of new knowledge and technologies for use in design activities. In most countries, the government supplies the bulk of necessary funding for scientists and engineers to conduct research at universities and research centers. For example, in the United States, the National Science Foundation allocates billions of dollars annually, and in Canada the National Science and Engineering Research Corporation.

Many local, state, and federal agencies adopt policies to purchase products that are more environmentally friendly. Because government agencies are large customers in many markets, government specifications for those products have a big effect on their design. Government agencies analyze and disseminate a wide range of technical information. Many engineering design decisions rely on this information.

PP also addresses issues of professional competency and behavior of engineers through professional licensing and regulations (Hyman 2003).

4.6.2 *Engineer's role in PP*

The roles that engineers take on today go beyond the domain of knowledge and technology. Engineers must step up to the challenges of PP especially technology-based and technology-driven ones, local and universal, from waste disposal, road traffic, and air pollution to water shortage, energy supply, and climate change. As technology becomes increasingly deep-rooted into various aspects of life, the convergence of engineering with PP will also strengthen. This will require that engineers develop a deeper sense of how technology and PP interact.

Policies today often involve complicated engineering systems including energy, transportation, information systems, buildings, waste treatment, health and safety, agriculture, and more. To assess the impact of various solutions or options, considerable engineering analysis, design, and simulation are required. Engineering knowledge is invaluable and it enables good policy development and minimizes unintended consequences. Accordingly, policies made without the involvement of engineering input naturally result in expensive unintended outcomes.

PP and professionalism are key areas where engineers ought to be in the forefront. Policies prepared by professional engineering organizations assist legislation and the lawmakers who vote on that legislation. These engineering policies that are prepared by the engineers behind the scenes are actually used by regulators in determining infrastructure for various services and infrastructure. It is these policies upon which codes and standards are developed and promoted for infrastructure projects. However, PP is not just a professional engineering organization program; it goes to the heart of the engineering profession and requires the energies at all levels of government (Galloway 2005).

Accordingly, engineers have a responsibility, but also a great opportunity, to ensure that they have a positive influence on society. Because engineers are such a central part of society, it is important that they engage with the design and development of PPs. Engineers have a major role to play in listening to people, informing them of what engineering can achieve, and focusing their efforts to ensure that engineers can be effective players by developing polices that meet the needs of society (Lawlor 2013). In fact, one of the key ingredients of engineering leadership is the understanding of PP. Engineering needs to be recognized in the framework of its role in society, and the role of engineers has to be realized in the context of work within an organization, and eventually within society.

Engineers have an exceptional opportunity to help plan the enormous infrastructure revitalization that governments around the globe plan to undertake. Engineers embedded in PP discourse can contribute to decisions that have intergenerational implications. Engineers and scientists should be providing sage advice, especially for those policy considerations that have broad impact, including, urban congestion by designing transit and roads for optimal function to reduce atmospheric emissions; achieving a delicate balance between national security without infringing on personal privacy (Krantzberg 2009).

Engineers in whatever discipline have had a long history of involvement in PP. The most significant discipline has been civil engineering where public works has seen it best demonstrated. Other areas may include energy where nuclear engineers help shape policies for nuclear power. Electrical engineers impact the power network of utilities (Hom 2013).

It is both the responsibility of engineers and important to the image of the profession that engineers make a better connection with PP (NAE 2004). Engineers' value to society and authorities is their ability to find engineering solutions that maximize value and minimize unintended consequences. To do this, engineers must get educated in PP and learn how to make compelling arguments with supporting facts. Crucial problems represent challenging opportunities for innovation, motivating professional engineering work and a genuine contribution to society.

4.7 Sustainability and engineering: Ethical and PP implications

> First they ignore you, then they laugh at you, then they fight you, then you win.
>
> **Mahatma Gandhi**

4.7.1 Engineering ethics and sustainability development

From among the published international codes, it is clear that SD/sustainability is envisioned as an area of ethical responsibility for practicing professional engineers. However, rather than the codes of ethics setting sustainability/SD as the very context of engineering practice, whereby as Allen et al. (2008) envisage SD equates with good engineering, that is good engineering in both practical and ethical terms, these concepts instead appear more by way of add-on statements that may accompany terms such as "social," "environmental," "safety," and "health and safety."

Over the last three decades, global attention has focused on the concept of sustainability, and SD has been introduced to address and overcome causes and effects of human activities' increasing negative impacts on environment. Parallel to the global trend there is an increasing demand in both the public and private sectors to understand sustainable design. This demand is driven by the realization of the need for sustainable practices that not only help the environment but that can also improve economic profitability and relationships among many stakeholder groups (Darwish et al. 2009).

Sustainability offers an obvious macroethical dimension to engineering ethics. Historically (beginning with the ancient Greeks) and academically, ethics has been presumed to mean how humans are related, because that is traditionally where the boundaries of community stopped. This has started to change. Environmental ethics, a subdiscipline of ethics as a whole, focuses on human thought and behavior in relation to nature, nonhuman animals, plants, ecosystems, air and water quality, and so on. Since its emergence in the 1970s, environmental ethics has been helpful in shifting the conversation about ethics toward a more inclusive understanding of human relationships to and values regarding others (Keller 2010).

An examination of most professional engineering codes of ethics reveals that there are indeed requirements that mandate engineers to understand and promote the principles of sustainability and/or SD and have due regard for their environmental and social equity, as well as economic obligations. Accordingly, engineers must factor environmental costs into the equation when designing new products and technologies. In the past, pollution control was considered after the fact, at the end of a process. Companies added scrubbers to smokestacks and catalytic converters to cars. Today, however, environmental considerations must be factored into the entire design process. Those concerned realize that pollution control must encompass all stages. Design engineers must invent products that can be manufactured cleanly, maintained easily, and which can ultimately be recycled or salvaged. Materials engineers must design materials that do not include, or do not require in their production, chemicals that harm the environment. Production engineers must design processes to use more environmentally friendly chemicals. Sales engineers must convince clients that they will benefit from purchasing environmentally sound products.

An ethic of sustainability must address a number of specific principles, which help fill out the most important values of sustainability in relation to social, economic, and environmental concerns. Kothari (1990) claims that the shift to SD is primarily an ethical matter. It is not a technological fix, nor a matter of financial investment. It is a shift in values such that nature is valued in itself and for its life support function, not merely for how it can be converted into resources and commodities to feed the engine of economic growth.

The dictum to "Think tomorrow, act today" (Milbrath 1996) has its complement in the well-known recommendation to "Think globally, act locally." Thinking globally involves becoming aware of and responsive to the webs of interdependence that connect us to distant peoples, cultures, and ecosystems. The ethics of sustainability require that we equitably share rights and responsibilities, benefits and burdens with our local and global neighbors. These relationships of shared duties, rights, risks, and opportunities are not dissolved, though they may be attenuated and complicated, by distinctions or divisions arising out of differences in class, race, gender, ethnicity, belief systems, and nationality.

An examination of engineering codes of ethics reveals that there have, indeed, been attempts to integrate environmental and social equity concepts with engineering ethics. Until recently, only two major engineering societies, the ASCE and the IEEE, even mentioned the environment in their codes of ethics. The former ASCE code, adopted in 1977, contained a "Guideline to Practice" committing engineers to "improving the environment." The first article of the current IEEE code, adopted in 1990, pledges engineers "to disclose promptly factors that might endanger the public or the environment." The third canon of the IEEE code of ethics is more relevant for sustainable engineering fields. It reads: to be honest and realistic in stating claims for estimates based on available data. We should not desire to be involved with promoting exaggerated results; we need to be honest and realistic in the benefits of our green engineering projects, for our sustainable engineering solutions. According to Harris et al. (1995), however, these codes provide only limited support for environmental principles beyond the impact of the environment on human health.

One effective way to accomplish the latter goal is to expose engineers to environmental ethics, a branch of applied ethics. Environmental ethics relates to the relationship between humans and the environment and is defined as a system of ethical values, human reasoning, and knowledge of nature which endeavors to forge patterns of right conduct toward environment. These patterns are necessary so that the needs of living beings of the present generation are fulfilled without compromising the ability of the future generation to meet their own needs. Environmental ethics is relationship between humans and the natural environment in which they live. It is the area of study that can teach important knowledge about the responsibility of academia to educate future decision-makers in the area of environmental ethics, so they will become more confident when standing up for their own opinion or resisting outside pressure if needed. Engineers can improve their critical thinking with a background in environmental ethics when decision-making arises (McDonough and

Broungart 2002). Engineering practices can be claimed to be in the benefit of the public only if they are performed in an environmentally responsible manner. Proper manners are those that recognize not only the established norms and ethical values of human societies, but also values of nonhuman nature. Two ethical principles are identified that address the environmental aspects of sustainability.

4.7.2 Policies for sustainability

SD theory emerged from the field of ecological economics. According to this theory, SD involves achieving objectives in three realms: ecological, economic, and social (see Figure 2.1). The ecological objective involves maintaining a sustainable scale of energy and material flows through the environment such that the carrying capacity of the biosphere is not eroded. The economic objective seeks to provide an efficient allocation of resources in conformance with consumer preferences and the ability to pay. The social objective aims at a just distribution of resources among people, including future generations. The overall objective of a sustainable society is the achievement of sustainability in economic, ecological, and social systems (Herkert 1996).

SD, then, consists of maximizing the achievement of these goals across the three systems, subject to inevitable trade-offs and priority setting at any given time or place (Holmberg and Sandbrook 1992). The concept of SD maintains considerable currency in a number of circles, including engineering. In some cases engineering societies have even proclaimed sustainable development to be an ethical responsibility (Grant 1995). For example, the American Association of Engineering Societies (AAES 1994) policy states that engineering education must install in its students an early respect and ethical awareness for SD, including an understanding and appreciation of cultural and social characteristics and differences among various world communities.

The success of PP to promote SD is dependent upon achieving all three objectives of a sustainable society. However, despite proclamations that engineers have an ethical responsibility to endorse the principles of SD, questions of just distribution and other questions of equity (such as risk distribution) are often left off the table when engineers consider sustainable development policies and issues. Indeed, almost all the effort of engineers and engineering organizations on the issue of SD is focused on the need to strike a balance between economic development and environmental protection.

With respect to PP, then, the involvement of many engineers and engineering organizations in promoting SD appears to be characterized by a technocratic and/or self-serving attitude with limited concern for social ethics. In contrast, ethicists have argued from both the perspectives of environmental ethics and development ethics that sustainable development is at heart an ethical issue.

The success of PP policy to promote sustainable development is dependent upon achieving all three objectives of a sustainable society. However, despite proclamations that engineers have an ethical responsibility to endorse the principles of sustainable development, questions of just distribution and other questions of equity (such as risk distribution) are often left off the table when engineers consider sustainable development policies and issues. Indeed, almost all the effort of engineers and engineering organizations on the issue of sustainable development is focused on the need to strike a balance between economic and social development and environmental protection. It is important to note that one cannot recognize social ethics without paying due attention and respect to environmental and economic ethics.

4.7.3 Policy case: Sustainable environmental and ethical procurement policy

In 2004, the Canadian city of Calgary's council directed that a policy on sustainable environmental and ethical procurement be developed. This policy builds on the work that has been done on the City's Green Procurement Policy. The policy was passed in 2007.

The city's sustainable environmental and ethical procurement policy (SEEPP 2016) provides guidelines to govern procurement activities. This involves purchasing products and services at reasonable prices while considering key environmental and social benefits such as worker health and safety, energy efficiency, minimal packaging, or other sustainability aspects over the entire life cycle of the product or service.

Throughout the policy development and pilot implementation phase of the project, other public and private organizations alike have expressed overwhelming interest in SEEPP. These organizations include the Government of Alberta, the municipalities of Vancouver, Ottawa and Toronto, and several companies. An ethical purchasing policy aims to ensure that all purchases made by a municipality are ethically produced, with considerations of environmental and social sustainability added in the case of the SEEPP. This would include employers' respect of their employees' human rights, and for some cities may include more provisions that ensure that the workers' rights would be equal to workers' rights in the city itself.

SEEPP not only promotes awareness of environmental and ethical issues, it also encourages supply chain practices that have a positive impact on social, economic, and environmental sustainability. The City's SEEPP will be implemented in a phased approach, and will gradually be applied to all city purchases. SEEPP supports the purchase of products and services that will protect the environment and the welfare of workers while representing the best value for the corporation.

The purpose of SEEPP is to ensure that products and services purchased by the city are manufactured, produced, and provided in accordance with established international environmental standards and guidelines; environmental labeling; applicable jurisdictional legislation regarding wages, working conditions, safety, forced labor, and freedom of association, such as those embodied in the UN Declaration of Human Rights and International Labour Organization Conventions.

Calgary is committed to being a good steward of natural and economic resources. With annual expenditures of more than $2 billion in procurement alone, SEEPP demonstrates the city's commitment to creating a sustainable community by promoting worker health and safety and ensuring compliance with applicable legislation; taking a leadership role in market development for green and ethical, or otherwise sustainable products; and promoting innovation and enhancing access to green and ethical products to lower costs of sustainable purchasing over time (SEEPP 2016).

4.8 Integrating ethics and PP in engineering curriculum

> The aim of education is the knowledge not of facts but of values.
>
> **William Ralph Inge**

4.8.1 Incorporating complementary studies into engineering curriculum

Teaching of PP and ethics to engineering students is seen as part of a strategy for securing STI and the betterment of the student future. Good practices in engineering education show that the students should learn not only the technical aspects of engineering, but also the broader issues and perspectives such as the impact of engineering on society, and the roles and responsibilities of professional engineers. This training is part of what is called "complementary studies." Figure 4.8 highlights four educational questions whose answer content helps enrich future engineering curriculum.

All this gives us some insight into what PP is needed to encourage innovation. STI policies are broader than science and technology policies, but the latter must be consistent with the former to produce healthy innovation ecology. Innovation requires a predictable social structure, an open marketplace, and a business culture amenable to risk and change. STI policies when well constructed also directly address pressing basic

- Why should ethics and PP be taught to engineering students?
- What kinds of issues are to be discussed in teaching ethics and PP?
- What principles and issues are to be taught?
- How should the students relate these principles to sustainability?

Figure 4.8 Educational questions related to teaching PP and ethics.

needs in energy, transport, agriculture, social services, water and sanitation, and infrastructure.

Today, most professional engineering codes of ethics require that engineers understand and promote the principles of sustainability and/or SD and have due regard for their environmental, social, and economic obligations. Two approaches can be employed to incorporate sustainability perspectives into engineering curricula such as the center approach or the whole curricula approach (Darwish et al. 2009). The center approach requires more resources and commitment from administration. The whole curricula approach can be employed by designing new curricula which integrate more sustainable-green perspectives, cultivating sensitivity to the environmental, biodiversity, and sustainability issues in students of all engineering disciplines.

While ethics and sustainability certainly overlap, it has been pointed out that they do not coincide (El-Zein et al. 2008): "incorporating them in the same engineering course can be effective, provided that points of linkage are clearly recognized in the syllabus, a suitable combination of theory and practical applications is drawn upon and adequate teaching methods, including decision-making case problems, are used."

Not surprisingly, current education in engineering ethics is approached differently by different universities. For example, some universities offer specific courses on ethics while others opt to include an ethics component in technical courses (Darwish et al. 2009). It should also be noted that a number of universities do not explicitly or formally address ethics at all. Even where stand-alone engineering ethics courses are offered, these are not always required and thus significant portions of students pass over this elective in favor of others. Moreover, some courses suffer from either an excessively theoretical approach or from an unbalanced case-oriented approach.

The most popular tool in teaching engineering ethics is the case method (Harris et al. 2000). Ethics is best taught by looking at real-life situations through case studies. Major incidents occur and catch the public's attention, and these are recorded for future in the engineering ethics texts. Cases can be long or short, real or fictional, technical or nontechnical; they may be available in print, online, multimedia, or video formats. Most cases are self-contained, but some include documentation, such as book chapters (and sometimes entire books), journal articles, news accounts, and primary source archives (Herkert 2002).

A useful teaching method for achieving integration of engineering ethics and engineering's societal context is to broaden discussions of engineering ethics so as to include the ethical implications of PP issues relating to engineering or the development of technology. Suitable content areas include risk and product liability, SD, globalization, health care, and information technology (Herkert 2000b). Many of the cases typically used in engineering ethics courses are amenable to discussion of PP issues (e.g., the Ford Pinto case and the DC-10 case); other, lesser known cases, such as the Bjork-Shiley heart valve (Fielder 1995), also have significant PP implications.

All engineering students must be exposed to environmental ethics, with a suitable combination of theory and practical applications incorporated into existing courses. Decision-making case studies specific to engineering are a practical way of teaching environmental ethics, with lectures and weekly workshops among possible delivery vehicles (Langford 2004; Bucciarelli 2008).

Based on above, a fruitful curriculum model would simultaneously address: (1) professional and ethical responsibility and (2) the societal context of engineering. This linkage would also address the criticism of traditional engineering ethics instruction, that it focuses on microethical problems, dilemmas confronting individual engineers, but neglects the macroethical issues related to the nature and development of technology (Herkert 2002).

4.8.2 Students as partners

Engaging students and faculty effectively as partners in learning and teaching is perhaps one of the critical issues education faces today. Students as partners is a concept that interweaves through many other determinants including assessment and feedback, employability, flexible pedagogy, linking teaching and research, and retention of success. Partnership is framed as a process of student engagement to foster learning enhancement (Healey et al. 2014).

Partnership in learning and teaching can take many forms, including case studies of initiatives and active discussion groups in a transdisciplinary context. Engaging students as teachers and assessors is particularly effective. Working with professors in research projects and inquiry is another stimulating exercise. The objectives of such engagement in PP and ethics learning in particular targets several objectives, including the following:

- Design and develop student learning experience.
- Develop a sense of ethical responsibility and belonging to society and environment.

- Determine what knowledge is necessary to enable engineers to participate effectively in PP and ethics discussions.
- Develop knowledge and capabilities of students and teaching staff in the areas of sustainable PP and ethics.
- Address PP challenges engineers face within the context of sustainability.
- Create a pilot set of curricular interventions enriched by modules that discuss policies for innovation and sustainability.

The outcome of such partnership offers students opportunities to think and act independently. Instructors and students will develop different motivations and positions in determining the direction and shape of the task under consideration. It will create possibilities for discovering and learning things that cannot be realized beforehand. Such engagement activity may be conducted in phases using a mixed methods design. The first phase will consist of a series of interviews and focus groups with experts in the fields of PP and ethics. These will be used to determine the most important knowledge and skills students need to acquire in order to participate effectively in public decision-making. The second phase would be to gauge students' current level of knowledge about the subject matter and measure their level of interest. The third phase would be interaction with instructors in order to determine how PP and ethics could best be integrated into the current engineering curriculum for most effective learning (Ngambeki et al. 2011).

4.9 Sociotechnical case: Energy ethics, society, and policy

> Living in a way that reflects one's values is not just about what you do, it is also about how you do things.
>
> **Deborah Day**

4.9.1 Conflict of targets

Understanding and influencing energy systems as complex sociotechnical systems (STSs) requires a basic understanding of the nature of systems, yet they pose complex sociotechnical implications and ethical challenges that require considerations in design and implementation. In addition to advanced technology, energy-related infrastructure has an embedded societal dimension where generation and power grid are highly visible, and power demand is highly dependent on user behavior. As STSs merge to form technical clusters, the ethical issues of the new clusters are not merely the sum of the previous technologies but the emerging ones. Social implications are often overlooked or ignored as functional and nonfunctional requirements of technology design. The emergence of smart infrastructures will become the catalyst behind policy and governance issues concerning privacy, security, and equity in STSs. Examination of technological systems determines how they require and incorporate social and political institutions and manifest ethical choices, consciously or not.

Energy is vital for heating, transport, food production, and many other key services. Initially human beings focused on heating, using wood, but subsequently discovered fossil fuels in the ground and learnt how to use them for heating, to power vehicles and for electricity production. At present, globally, around 80% of the energy that we use comes from burning fossil fuel, coal, oil, and gas (Hersh 2015).

The first Industrial Revolution, which began in the second half of the eighteenth century, started in response to an energy shortage in Britain (Brinley 1985). Again today, more than two and one half centuries on humankind is facing a similar though potentially a greater problem, with a need on the one hand to generate ever greater amounts of energy and the opposing necessity to reduce the damage inflicted by industrialization, transportation and existing energy production, storage and supply technologies (Coyle and Rebow 2009). In tackling these problems, efforts are underway to recognize resources and technologies in new energy generation, including renewable forms of energy and energy efficient systems.

Renewable energy systems seem a progressive technological development from social, environmental, and ethical points of view. Many of these systems, however, are themselves not free from ethical controversies. Some controversies arise from the conceptual confusion accompanying the notion of sustainability, and some appear when sustainability is challenged with other environmental, economic, and sociopolitical values. This case is a venture into this conflict to target several objectives.

- Highlighting of the challenge of future energy systems
- Discussion of current energy policies
- Recognition of the notion of sustainability in the context of future energy systems
- Development of a sustainable ethics framework for the moral evaluation of sustainable energy systems
- Recommendations for a transdisciplinary research for integrated energy systems

4.9.2 What is energy issue?

Modern economies depend on a reliable and sufficient energy supply: energy that is secure, environmentally friendly, produced and used efficiently. All sectors of the economy—residential, commercial, and industrial—demand modern energy services. These services in turn foster economic and social development at the local level by raising productivity and enabling local income generation (IAEA 2005). However, the challenges of ensuring energy for SD are many. Today, the energy system is a cornerstone of modern life. It enables innumerable services capable of improving human, social, economic, and environmental conditions worldwide (OECD 2007).

Today, the world energy system is based on extracting concentrated forms of energy available in nature, such as fossil fuels, rivers and waterfalls, burning plants, and splitting uranium. Unfortunately, this "centralized paradigm" is not sustainable because the above concentrated forms of energy are both in short supply and critically impact the ecosystem. The practice of extracting naturally concentrated energy causes several problems including disrupting natural energy flows, depletion, and centralization. Fossil fuels, for example, signify a large share of carbon that when extracted and burned disturbs the thermal balance of the earth. When energy is extracted or generated in few centralized locations, such as oil refineries and electric power plants, energy must be brought to consumers, requiring additional energy to transport energy. "Decentralization" by using distributed generation (DG) energy sources will reduce the energy required to transport it, while also reducing the associated ecological impact; this is only a prerequisite, not an assurance, for sustainability (Wissenz 2016).

Further, the global dependence on nonrenewable energy sources is a global environmental, social, and economical problem that requires a holistic approach which integrates the expertise and perspective of diverse fields and social groups (O'Neill-Carrillo et al. 2012). The problems of unsustainable resource use, environmental degradation, and global climate change are inherently tied to energy, which, therefore, must be the key component in any solution (Besada et al. 2013). Current demographic, economic, social, and technological trends, if not counterbalanced by strong new government policies pose major challenges to the long-term sustainability of the global energy system. If governments do not implement policies beyond those already planned up to 2030, it is projected that (OECD 2007)

- Energy consumption will increase by over half.
- The energy mix will remain fairly stable and dominated by fossil fuels (80% share).

- Energy-related CO_2 emissions will increase by over half (55%); and large populations of the world will continue to lack access to electricity and modern services.

The above forecast exposes a worldwide problem with energy. However, what exactly is the energy issue? Simple, demand is or will become higher than supply. Energy supply has immense benefit, but the sheer scale of present demand does increasing harm, especially from pollutant emissions and other adverse environmental impacts. This problem has mainly been articulated to fossil fuels and mainly with regard to oil, the most important current energy source. Fossil fuels by definition are nonrenewable and are destined to run out, so economies will be forced to change as these fuels are depleted. A solution could be realized by finding ways to increase energy supply as well as to decrease energy demand. Both are, to a certain extent, technological challenges, which can be addressed by engaging additional or alternative energy sources that would help to increase the supply of energy. It is also required to develop more energy-efficient techniques to use energy that would help to decrease the overall demand (OECD 2007).

What is needed in the near future is to reduce energy demand in general and dependency on fossil fuels as quickly as possible and without major economic disruptions. Also to find new technologies that extensively reduce emissions from existing energy supply sources and develop successful economic and market procedures that reward improved environmental outcomes.

4.9.3 Energy policies

Energy policies are some of the most contentious policies any nation faces. In general, energy use needs to be looked at from two different perspectives: first, primary energy use by fuel, the forms of energy we depend on; and second, energy consumption per sector, how we use energy. Today, many policies exist that, if realized, could change several trends. Such policies include efforts to improve efficiency in energy production and use, increase reliance on nonfossil fuels, and sustain the domestic supply of energy.

Policies encouraging more efficient production and use of energy could contribute almost 80% of the avoided CO_2 emissions by 2030, with the remainder gained from fuel substitution. More efficient use of fuels, mainly by cars and trucks, accounts for almost 36% of avoided emissions; more efficient use of electricity in a wide range of applications (lighting, heating and cooling, appliances, and industrial motors) for 30%; greater efficiency in energy production for 13%; renewables and biofuels for 12%; and nuclear for the remaining 10% (OECD 2007).

Energy efficiency is crucial to sustainability. Governments should employ a range of available policy instruments, including regulations and standards, fiscal incentives, public information campaigns, labels, and public-sector leadership in procurement. The use of fossil fuels can be made more climate-friendly through implementing CCS at power plants and industrial facilities. Another critical policy response is greater investment in energy-related R&D, in part because new supply-side energy technologies need to be operating on a commercial scale by 2030 or earlier (OECD 2007).

The SD of energy systems requires policy support or at least policy attention (Jacobsson and Karltorp 2013). When choosing energy fuels and associated technologies for the production, delivery, and use of energy services, it is essential to take into account economic, social, and environmental consequences. Policymakers need methods for measuring and assessing the current and future effects of energy use on human health, society, air, soil, and water. Government policies are the key to ensuring that the energy sector advances sustainable development. There are many policy domains including environment, development, industry, transport, construction, agriculture, investment, science and technology, and education as well as energy itself, where policies guide how and how much energy is produced, converted, transported, distributed, and used (IAEA 2005).

Strong environmental, community, and policy efforts are being undertaken worldwide for a transition out of fossil fuel dependencies into renewable energy sources, technologies, and practices. Renewable energy is perceived as the alternative to fossil fuels; however, the path to an increased use of renewable energy is not clear since there are many perspectives on what is best (O'Neill-Carrillo et al. 2012).

Before considering subjective ethical motivations, developing sustainable action plans, and choosing appropriate design methods and tools, engineers must consider the objective consequences of their actions, the social and ethical problems created by their technologies, in short embracing the notion of "ethics of responsibility" (Coyle and Rebow 2009).

4.9.4 Ethics of sustainable energy

The concept of sustainability is effective for the analysis of energy issues for two main reasons. First, the concept integrates the aspect of the long-term sufficient supply of energy with the ethical aspects of the impact of energy use on the three sustainability relations. Second, the concept of sustainability makes it clear that the energy issue is embedded in a set of other related issues, and the future of energy is actually part of the larger challenge of the design of a thoroughly sustainable future. Figure 4.9 defines sustainable energy and reflects its relevance to ethics, economy and society.

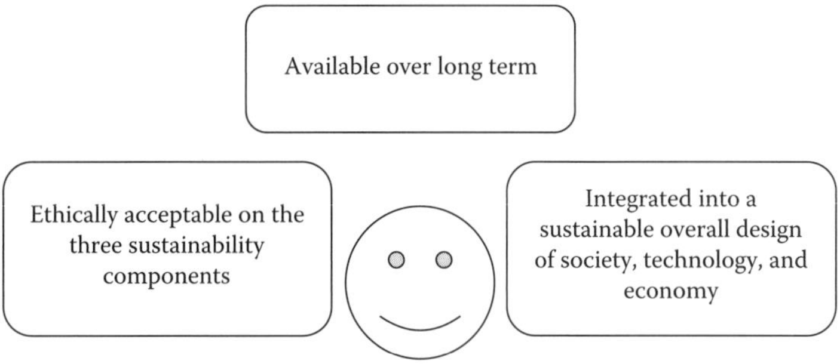

Figure 4.9 Definition of sustainable energy.

Sustainability is fundamentally about adapting to a new ethic of living on the planet and creating a more equitable and just society through a fair distribution of social goods and resources. Technology holds some potential for reducing impacts and risks. However, each energy system as a sociotechnical entity results in various impacts on other contemporaries, future generations, and the environment. Accordingly, it is needed to discuss which energy source and application produces acceptable impacts and which does not. This brings up some crucial ethical questions for discussion as shown in Figure 4.10.

The scenario of current energy supply, distribution, and consumption is related in part to individual preferences, values and activities, and the large-scale design of society, technology, and the economy. However, it is also related to the design and functioning of economic and technological systems. The association of energy with both individual behavior and societal and global systems involves crucial ethical questions and problems. An ethical approach can refer to these aspects and include them in an overall determination and analysis of the energy issue and in proposals for the design of future energy systems; a merely technological approach could not. A technological approach takes ends as a

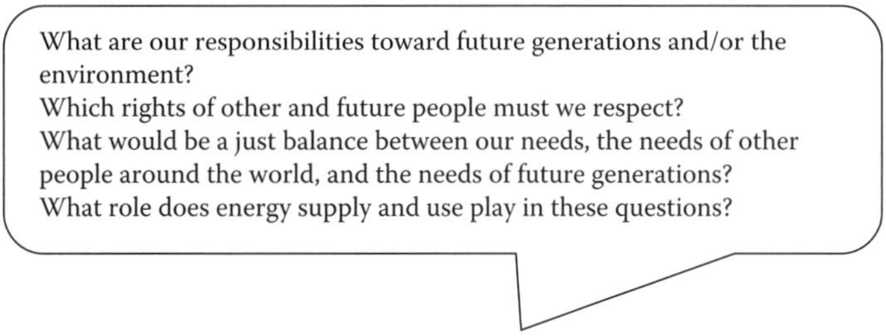

Figure 4.10 Ethical questions regarding energy.

given and seeks solutions to realize those ends. An ethical perspective considers the meaning of the ends themselves. SD is essentially about improving quality of life in a way that can be sustained, economically and environmentally, over the long term, centered on ethics and supported by the institutional structure of the country. Accordingly, SD addresses four major dimensions: social, economic, environmental, and technological. The institutional determinants are largely considered to be responses and not readily quantified as indicators. This combination is shown in Figure 4.11.

Another important dimension that should be paid full attention is energy security. At a general level, it can best be understood as robustness against (sudden) disruptions of energy supply (Grubb et al. 2006). Thinking broadly across energy systems, one can distinguish between different aspects of security that operate at varying temporal and geographic scales (Bazilian and Roques 2008). Given the interdependence of economic growth and energy consumption, access to a stable energy supply is a major political concern and a technical and economic challenge facing both developed and developing economies, since prolonged disruptions would create serious economic and basic functionality problems for most societies (Larsen and Petersen 2009). Concerns about the limited availability and distribution of resources are also a critical component of energy security in the short term. The more reliant an energy system is on a single energy source, the more susceptible the energy system is to serious disruptions.

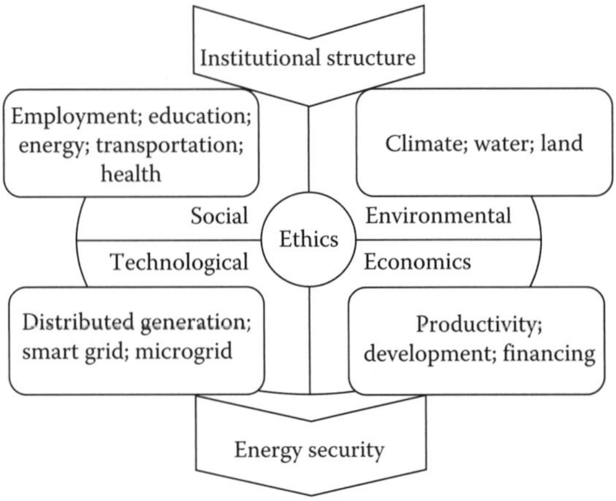

Figure 4.11 Sustainable energy framework centered on ethics and supported by the institutional structure.

4.9.5 Ethical/policy case for renewable energy

Although the contribution of renewable energy to SD has to be evaluated in a certain context, renewable energy offers the opportunity to contribute to a number of SDGs: social and economic development; energy access and security; climate change mitigation; and the reduction of environmental and health impacts. The SD concept explicitly includes ethical concerns; accordingly the inclusion of ethical concerns is crucial for an all-encompassing design of future energy systems.

A number of policies have been implemented over the past few years to promote the use of biofuels. However, biofuels are not sufficiently available and it would be difficult to produce enough of it to fully substitute the existing demand for fossil fuels. While enthusiasm over biofuels was still in full swing, problems with their large-scale production are emerging. Claims that biofuels produce significantly lower greenhouse gas (GHG) emissions when compared to fossil fuels are disputed. Some calculations show that biofuel production actually makes the GHG problem worse than by just using gasoline. Concerns were also elevated where biofuels may affect food security, deforestation, and rising food prices. This tension between biofuel production and affordable food worldwide raises a number of ethical issues.

The crucial issue for the sustainability of biofuels from crops is that they are ethically problematic against the background of the three sustainability relations. It may already be evident that biofuels from crops have an ethically problematic impact on the relationship among contemporaries. This is due to the fuel versus food problem, which has recently been discussed to some extent in Chapter 2. The production of biofuels from crops generates a trade-off in land use: use of land to produce either food or biofuels. This induces a high risk of raising food prices, potentially resulting in famine due to affordability, particularly those living in poorer countries around the world. This, in turn, could cause major new global conflicts that might exceed the current conflicts over fossil fuel distribution, ultimately resulting in negative impacts on global and national security.

Ethical issues are also present when evaluating the agricultural production of feedstocks, most of which include some form of genetic engineering. Genetic engineering is used to increase yield per acre, growing cycle, and composition (higher starch content). Biofuels are not viable on a large scale without the use of genetically engineered food and nonfood crops. Currently, genetic modification is being developed to increase the volume and turnaround time in all of the crops (Adler et al. 2007).

A second case study looked at Brazil. The country has a well-developed industry for producing ethanol from sugarcane. The main issues that came up there were centered on the environmental sustainability of this production. Deforestation is a major concern, as some rainforests have been cleared for sugarcane production. A second great concern in Brazil centers on workers' rights. There have been reports of conditions on the sugarcane mills that amount to slave labor as well as reports of unhealthy working conditions, including informal child labor in the mills (NAS 2014).

Malaysia provided another case study, which focuses mainly on palm oil diesel production. Again, environmental sustainability was a great issue there. Deforestation occurred with forest land being cleared for palm oil plantations, which led to significant losses in biodiversity. There were also a number of land grabs, both by governmental organizations and by entities in the private sector. These led to the disruption of subsistence economies in the areas in which they occurred. There are also food security concerns caused by rising prices for palm oil and for foods that use palm oil (NAS 2014).

Finally, the development of any policies on biofuels should take into account the various ethical and social issues that arise when individuals and communities feel the consequences of those policies. Policies should ensure that benefits are shared equitably, for example, through public–private partnerships. The aim should be to improve the efficiency of biofuel production at all points in the supply chain and to use less natural resources such as land, water, and fertilizers. Policies should also include respecting people's rights to food, work, and health when producing biofuels. In addition, biofuels should be environmentally sustainable, contribute to a net reduction in total GHG emissions, and adhere to fair-trade principles. The major ethical principles of biofuel production are shown in Figure 4.12.

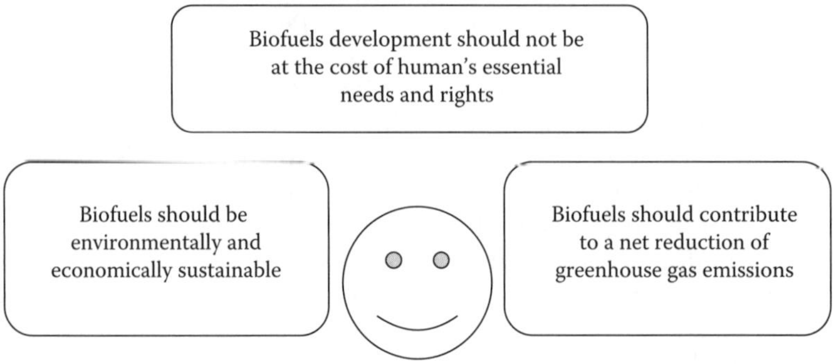

Figure 4.12 Major ethical principles of biofuel production.

4.9.6 SG: Social and ethical challenges

> Think of the Smart Grid as the internet brought to our electric system.
>
> *US Department of Energy, 2008*

The systematic integration of renewable energy resources into the power grid is driven largely by environmental regulation aimed at facilitating the use of sustainable energy resources and therefore reducing carbon emissions. Therefore, the grid is evolving from a network with relatively few and large, tightly synchronized energy resources supplying passive consumers, to a smart network driven by many distributed and central energy resources, mixed with energy storage and sensor-based user loads. The successful transformation of the passive grid to a SG to modernize the aged electric one requires choices based on solutions for major operating, economic, and PP challenges.

SG architecture is a system of systems: a large and complex network made of smaller and simpler systems distributed and interconnected. The SG comprises three fundamental structural elements: replacement of aging core physical infrastructure items including transmission lines and switching equipment with more efficient and reliable newer technologies; two-way distributed and loosely coupled supply and demand connectivity to the grid, which allows consumers to supply electricity through technologies such as photovoltaic cells, wind power; and, most importantly, highly optimized two-way information and communication technology systems architectures and networks that control the grid through process-and-rule-based programs to match power demand with supply in order to improve efficient use of energy resources. Figure 4.13 shows the typical components of the SG.

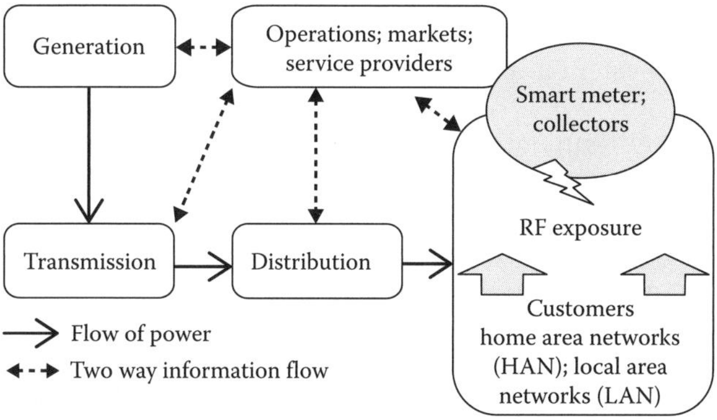

Figure 4.13 Typical components of the smart grid.

One of the major benefits of the SG is its ability to replace the use of fossil fuels with renewable energy resources. Benefits of integration of these resources will increase rapidly all along from suppliers to customers. It is an approach to control GHG emissions into the atmosphere; reduce reliance on oil and control the cost of electricity and transportation.

While these innovative features of the SG hold great potential for improved energy efficiency through better management of consumer demand, including greater utilization of renewable generation, they also pose a number of social and ethical challenges including: protecting the privacy of consumer usage information; securing the grid from attacks by foreign nations, terrorists, and malevolent hackers; and ensuring social justice in terms of both access and cost of electric power service.

As with many new technologies, the engineers engaged in developing the SG often overlook such issues or only turn to considering them once the technical standards and specifications have been settled. Failure to address these issues in a timely manner, however, may result in delays in establishing the SG and undermine its potential. Engineers and others involved in developing the SG need to examine ways to address organizational, social, and ethical dimensions that DG and more extensive efforts to influence consumer usage patterns will entail. With DG, the energy infrastructure will change bottom-up due to its large scale. Viewing this infrastructure as STS may lead to innovative insight in operational system regarding criteria like technical and social design as well as cost.

The potential privacy consequences include, among many, identity theft, tracking determining behaviors patterns, activity censorship, cyber home invasion, and performing real-time surveillance. Convergence of information and telecommunication technologies and power systems will introduce all the security issues of the Internet. Power disruption can cause a loss of infrastructure, databases and computers, and may endanger public safety and jeopardize national security. The energy management system implied by the SG may be a burden on uninformed customers.

An SG environment relies heavily on standards, mainly to ensure interoperability among systems. Standards also play a key role in SG cyber security. Security refers to the degree of protection afforded to the system from deliberate attacks. SG security involves an architecture that includes security from the beginning, consists of more than just protective devices such as firewalls, and engages processes as well as products. This includes physical assets (power, sensing, control, and communication technologies), policy, market, and management; and IT including cyber and data. Interconnection among all of the above requires secure automated information exchange and intelligent decision-making throughout the entire power grid. For the utility, security starts with defining a risk management framework that will be used to identify threats and vulnerabilities; assess the effectiveness of existing controls in mitigating the risks; recommend new controls to mitigate the risks to a level acceptable to the utility and public; periodically repeat this process to account for changes in the threat situation; and set up policies that address human behavior, which is the basis for all security risks (Habash et al. 2013).

In conclusion, the development of new energy systems, paired with improvements in both efficiency and ecological cleanliness of existing ones, will contribute to sustainable solutions. Whatever the future balance in energy development scenarios and policies, it is becoming apparent that sustainable design will be central to modern holistic engineering thinking and will be of critical importance in the delivery of education to engineers of the twenty-first century (NAE 2004).

4.9.7 Transdisciplinary research for integrated energy systems

Inter- and transdisciplinary research has been used to signify the vision of a type of scientific research that is related to society in a new way. In transdisciplinary research, the formulation of research questions, the research process, and the application of results are performed by science and society together by means of a strong set of interactions between science, engineering, politics, economics, ecology, and concerned social entities. This concept of a new type of research aims to overcome the current design of established science, which generally does not explicitly orient its research toward society and its crucial sustainability issues, but is to a substantial extent driven by internal incentives and objectives. The transdisciplinary approach in dealing with energy systems is reflected in their being defined as STSs that, in a mutual process, are shaping society while also being influenced by it.

Recognition of energy as an integrated system draws attention to the need to integrate approaches from different disciplines to tackle scientific questions about the complex STSs. Although energy research should be transdisciplinary, it should not be a type of research that becomes a mere servant of society; research that takes orders from society and works only through these to find proper solutions. Such research must undertake the task of telling society whether the simple technical solutions society wants are possible or feasible long-term solutions. It must be critical and honest toward society and should indicate whether sustainable solutions to the energy issue require substantial accompanying societal changes.

The restructuring of the modern energy systems sets a significant global challenge to foster the provision of sustainable energy. Energy research, as a new type of sustainability research, needs to challenge society. It needs to critically reflect on societal problems and societal wishes. Such integrated inter- and transdisciplinary sustainability research would be able to address the energy issue in an encompassing way. It could critically discuss sustainable solutions in the context of societal and global systems and against the background of existing forms of thought and action, and could provide revolutionary suggestions for overall sustainable solutions.

Future research efforts need to focus more directly on producing knowledge required to understand, diagnose and influence the challenges that confront societies as a result of global change. The research results should assist governments and societies to make informed decisions and continue developing PP strategies to influence human behaviors of contemporary societies.

4.9.8 Case research questions

- How can a sociotechnical system approach benefits studies of GHG emission reductions from energy systems?
- Investigate centralized and decentralized energy supply systems. What are the challenges of implementing systems of decentralized energy supply?
- How energy efficiency improvements within a large company are influenced by job design and goal setting?
- What lessons does this case hold for energy systems in your region?
- The development of SG infrastructures will become the substance behind policy and governance issues concerning privacy, security and equity in SG and other "smart" sociotechnical systems. Investigate and comment.

4.10 Knowledge acquisition

Attempting to answer the following questions involves acquisition of knowledge from this book and other books, documents, and the Internet.

- What is ethics about? Are there universal ethical principles?
- Explain the difference between values, morality, and ethics.
- Explain the difference between moral and legal responsibility.
- What are the three major ethical theories?
- What ethics can engineers learn from the past?
- Do the ends justify the means?
- How do ethics differ from law?
- What is a conflict of interest?
- Can a code of ethics be unwritten?
- What is the meaning of ethical responsibility?

- What is the direct result of ethical behavior?
- What do code and standards cover?
- What is a profession? Who is a professional?
- What is professional engineering?
- What is the purpose of professional code?
- What are the factors that influence social ethics?
- Should all engineering professors be required to have a professional engineering license? Explain.
- Why should engineers study ethics?
- What does it take to become a good engineer?
- How are competence and ethics related to engineering?
- How do ethical issues span engineering disciplines?
- What insights can engineers gain from the professional ethics of other fields, such as science, law, and medicine?
- Are engineers ethically obligated to consider sustainability in their designs? Why or why not?
- How can one practice ethics in today's business climate and in particular related to new business formation?
- Discuss how and when engineers are not the guardian of their profession?
- How much control engineers actually have over what they do?
- Can engineers contribute to PP?
- What areas of PP are most relevant to engineering?
- Why politics and PP need engineers?
- What is the role of science and engineering in guiding PP?
- What influence should engineers have in the development of technology policy?
- Where do you see the role of engineering in PP 5 years from now?
- How do science, engineering, and PP work together, and why are they important?
- The PP process has a lot in common with the engineering design process. Explain.
- What are ethical challenges within sustainability education?

4.11 *Knowledge possession*

Attempting to answer the following open-ended "not explicitly expressed" questions may require research and investigation beyond the scope of this book, mostly by engaging in conversation, class discussion, and Internet-based research.

- What is ethics in research and why is it important?
- Conduct an Internet search for landmark cases in engineering ethics. Summarize two important cases.

- What are the greatest ethical challenges engineers currently face in their workplaces?
- Are engineers ethically required to consider sustainability in their designs? Why or why not?
- How can engineers, in academics as well as in industry, work with transdisciplinary partners to promote and sustain a culture of ethics?
- Can incorporating engineering and technology case studies in non-engineering courses effectively increase technological literacy for all students?
- How can themes such as human dignity, common good, unity, integral human development, and care for the environment, be integrated with engineering practice?
- What influence should engineers have in the development of technology policy?
- How does current practice of engineering ethics education prepare students for a culture of ethics in industry?
- What successful models are in place for engineering ethics education? What are the challenges for implementing those models?
- Why and how modern societies fund science, technology, and engineering R&D?
- Some suggest embedding the concept of the "circular economy" into policies and regulations. Comment on this matter.
- Discuss around the interconnectivity among environmental policy, future economic growth, and competitiveness.
- How do science, technology, and engineering knowledge influence societal trends and public decision-making in the twenty-first century?
- What are the effective mechanisms for identifying the evidence required to inform policy-making on emerging technological issues?
- Are there cases of contemporary and historical energy transitions that reflect key ethical challenges in relation to environment and society?
- Explain by example how the development and use of technology poses ethical issues. Defend the ethical issues related to the development and use of technology.
- What are the practical approaches to incorporating PP considerations into engineering design education?
- How does current practice of engineering ethics education prepare students for a culture of ethics in industry?
- How can STI education requirements be related to the requirements of the economy?
- How can policymakers enable educational institutions to better connect with the community productive sectors?

- When judging an engineering assessment, the outcome may be projected with either: certainty; uncertainty; risky; and ignorance. Examine each of these possible outcomes in relation to engineering ethics.

4.12 *Knowledge creation*

Collaborate with peers on learning or you may work with others outside the class to narrow down the objectives of each activity. You may access class and online resources, and analyze data and information to create new ideas and balanced solutions. High-level digital tools may be used to develop multimedia presentations, simulations or animations, videos and visual displays, digital portfolios (ePortfolios), reflective practice (online publishing and blogging), or well-researched and up-to-date reports.

4.12.1 *Cases for discussion*

Case 1: Volkswagen's installation of software in 11 million Volkswagen and Audi diesel vehicles sold worldwide has led to a massive vehicle recall in the United States and an official apology from the company. The algorithm, installed in the emissions-control module, detects when the cars were undergoing emission testing. It ran the engine cleanly during tests and switched off emissions control during normal driving conditions, allowing the car to spew up to 40 times the US Environmental Protection Agency's maximum allowed level of nitrogen oxides, air pollutants that cause respiratory problems and smog (Patel 2015). Do you think that this case broke the law and is unethical?

Case 2: An electrical engineer works for a consulting company that designs small substations for wind farms. The supervisor asked engineer to replicate a set of drawings and changing the name of the client. He states that a new client has a wind farm with identical specifications to the last project the engineer worked on, so the same design will work. When the engineer asked how to bill his time, the answer was to enter the same number of hours that were billed for the initial design work. Do you agree with this process?

Case 3: Mechanical engineer X is retained by a company to design a biomedical system. Engineer X then retains the services of Engineer Y, an electrical engineer with expertise in control systems. Several months following completion of the system, Engineer X enters the biomedical engineering design into a design competition. The design wins a prize. However, the entry fails to credit Engineer Y for his part of the design.

Was it ethical for Engineer X to fail to give credit to Engineer Y for his contribution to the design?

4.12.2 *Online Ethics Center for engineering and science*

The Online Ethics Center (www.onlineethics.org) seeks submissions of high-quality ethics education resources in science and engineering for inclusion in its collection. Interdisciplinary materials are particularly welcome, as are resources that promote active learning at the undergraduate or graduate level. Materials are also needed to assist faculty new to teaching in science and engineering ethics.

Students may visit the above online resource and in particular read the "Resources" section where they will come across hundreds of case studies, essays, codes and policies, multimedia and educational activities, and subject aides. Students may work in teams to develop various tasks related to the above activities for submission to the online resource according to the class instructor's guidance. For resource submissions, see the "Submit Materials" page.

4.12.3 *Connection task on ethics for engineering design and entrepreneurship*

Groups of students (three to four, each) may be given a task to explore and analyze the design of variety of popular innovations (new products). In this regard, design is the creation and development of an economically viable product, process, or system to meet a defined need. Team members should select one product from a pool of several products and should provide a reason for the selection in a worksheet.

4.12.3.1 *Product ethical evaluation*

Each team should evaluate the assigned product designs against a decision-making criteria developed by the team. The characteristic of the criteria may depend on location, funding source, materials, technology, manufacturing, testing, users, impact, etc. Collectively, each team should decide upon a definition of "ethical" to use during evaluation process. The evaluation statement (in the worksheet) may answer the following questions (Wilson et al. 2015):

- Who does the product really benefit?
- How does it benefit users?
- Who might be harmed by the product?
- Who might be discouraged from using the product?
- Who funded the product development process?
- Who is gaining and profiting from the product?

4.12.3.2 Product redesign ethics

Later, teams apply their new point of view to redesign the assigned product (on paper) and persuasively present the outcome to the class, explaining how the team standards for ethical designs were met. To make products as ethical as possible, teams should modify the design so it better serves people and the environment, or less harm people and the environment. In quest of best design and engineering solution, engineers must follow a code of ethics that includes the protection of public health and safety. At all stages of design and product development, the best approach takes into consideration the complicated ethical issues surrounding product testing and production and their impact on community and the environment. A redesign report that takes into consideration the following issues should be developed:

- Review the engineering code of ethics.
- Decide on several principles you believe are important.
- For whom is the redesign intended to benefit?
- On whom will the redesigns be tested?
- What ideas do you have for redesigning the existing product to make it more ethical?
- What are the encountered problems?
- Reflect the above on your redesign ethical statement.

Finally, the team members should assume that they are engineers who will pitch the design to a funding agency that is looking for new projects to fund. The task is to encourage and persuade the agency to invest in the design. Details should be added to the worksheet.

This activity demonstrates to students many key considerations engineering entrepreneurs must keep in mind including engineering design, sustainability, business formation, and the involving risk and ethics challenges.

4.12.3.3 Final presentation

At the end of the assigned task, each team should prepare a 3-min presentation to be carried out in the class. The presentation should include details about the following:

- Original product design and the redesign.
- Present design as ethical as possible.
- Show how design meets the engineering ethical challenges.
- How the new design meets the requirement of entrepreneurship ethics.
- Encourage the funding agency to invest in the design.
- How do you market your design, to whom and why?

4.12.4 Debate on ethics and energy sustainability

Objective	Using a debate to help students understand argument on ethics and sustainable energy
Time	15 min for debate and 15 min for review
Level	For and against
Learning outcomes	Make an argument about a particular opinion; evaluate the arguments of peers; and understand the concept of counterarguments
Capabilities demonstrated	Developing skills on public speaking, research, teamwork, critical thinking, communication, and professional judgment
Arrangement	Students are organized into two position groups of three speaking in a specific order. Three argue for an opinion and three argue against. One or two students might each work on the opening and closing statements while the group is investigating the subject; however, the entire group should revise the statements. Each group should read an opening and closing statement for the debate.
Ideas for the topic	How to realize energy systems as complex STSs? What are the contentious ethical problems underlying energy policy? How can we analyze and cope with the potential trade-offs or ethical conflicts between the three sustainability relations? What are the ethical challenges of a large-scale transition to renewable energy systems?
Assessment	Indicate what you consider the best arguments in favor of the topic. How were they substantiated? Identify arguments that are based on poor facts, not ethical and/or sustainable or not well substantiated?

4.12.5 Energy policy on a campus demonstration project

The challenge in the United States and beyond to transition quickly from a fossil fuel-based society to one built on safe, clean renewable energy as advocated by a majority of the world's scientists is the crucible of our time (Erickson and Degan 2009).

Several groups of students are invited to be active participants in bringing renewable energy applications to their campuses, with wind, solar, geothermal, and biomass projects. Groups can be engaged in sustainability in a wide variety of ways on campus. From campus planning and behavior-change campaigns to energy conservation and waste reduction strategies, students can plan initiatives that bring about real change, both on campus and possibly in the surrounding community.

- Groups should continue to seek cost effective ways and strategies to minimize the use of fossil fuel energy sources and finite natural resources by enhancing energy conservation in existing buildings and striving toward sustainable energy practices in the renovation of current buildings and construction of new buildings.
- Continue to analyze campus activities and programs that generate waste with the goal of minimizing such waste by reducing, reusing and recycling and communicate this philosophy to the campus at large.
- Groups may consider one component as a case study. One component could be one of the University cafeterias as a key component in campus life and major contributor to the sustainability program. The strategy is to look at ensuring waste from operations is recyclable; having menus that support healthy lifestyles as well as options for special diets; and purchase of food from vendors who support local/regional farmers and organic farming methods; and exploring the possibility of on campus gardens being used by the cafeteria.
- While investigating the task, students should weigh in on energy policy initiatives at many levels, from the individual campus to the local community, even state and nation.

4.12.6 *Piece of art on engineering-policy divide*

The goal of the technical research community is to produce and disseminate the best ideas to advance technology and scientific understanding. If most good ideas come from a handful of renowned experts, that is not a problem. In contrast, the goal of the policy-making community is to create a process that is immune to domination by any single group. It is therefore better, policy-wise, to reach a suboptimal decision, or to take an excessively long time in reaching a decision, than to allow any single group to control the system for long periods of time (Peha 2011). Most policymakers are not well competent in sciences and engineering. Engineers sometimes serve as key advisors to decision-makers in both the public and private sectors and are mostly poorly taught in PP. At the core of this issue are two groups with different operational systems and different cultures. Accordingly, engineers need more than extensive technical skills; they need an enhanced understanding of the PP process and its implications for economic, social and ecological needs.

For this activity, investigate the science-policy divide, address issues that require the talents of the above both sectors, reflect the generated cause–effect, and identify approaches to bridge the gap.

4.12.7 *Workshop on future ethical engineer*

Students may work in teams to research and investigate the most critical leadership and value conflicts faced by future engineering professionals in their careers. For this task, teams are asked to develop a half-day workshop for their own class or school to understand or improve the role of ethics and values in scientific and engineering practice. Teams should identify the major ethical challenges future engineers face including personal as well as professional. Teams should prepare educational material for a session to present their approach, with time for discussion or hands-on activities. The last session may be a synthetic discussion, which might include comparative evaluations of the approaches.

4.12.8 Video contest on ethical energy

Identify an ethical issue of interest with future energy systems including renewable resources and SG. Explore and explain different views on the above matter. Propose approaches to address the issue in general and your view in particular. Ethical perspectives employed in contest range from traditional ethics (which considers whether actions are required, recommended, or forbidden) to issues of individual (microethical) and collective (macroethical) responsibility. The contest may focus throughout on responsibilities rather than outcomes. This ethical development task may include allocations of cost and benefits, risks and rewards, social and environmental justice, professional and organizational ethics.

In this task, collaborate with peers or you may work with others outside the class to narrow down the objectives of the task. You may access online resources to create a 3-min video that reflects the above subject.

References

AAES. 1994. Statement of the American Association of Engineering Societies on the role of the engineer in sustainable development. In the Role of Engineering in Sustainable Development, Washington, DC, pp. 3–6.

Adler, P., J. Stephen, D. Grasso, and W. J. Parto. 2007. Life-cycle assessment of net greenhouse-gas flux for bioenergy cropping systems. *Ecological Applications* 17(3): 675–691.

Allen, D., B. Allenby, M. Bridges, J. Crittenden, C. Davidson, C. Hendrickson, C. Matthews, C. Murphy, and D. Pijawka. 2008. Benchmarking sustainability engineering education: Final report, EPA Grant X3-83235101-0. Centre for Sustainable Engineering, Pittsburgh,PA.

Barke, R. 1986. *Science, Technology and Public Policy*. Washington, DC: CQ Press.

Bazilian, M., and F. Roques (Eds.). 2008. *Analytical Methods for Energy Diversity and Security: Portfolio Optimization in the Energy Sector: A Tribute to the Work of Dr. Shimon Awerbuch* (Elsevier Global Energy Policy and Economics Series). London: Elsevier.

Besada, H., Y. Stevens, and M. Olender. 2013. Addressing the economic cost of sustainable energy in the global south. High Level Panel on the Post 2015 Development Agenda. http://www.post2015hlp.org/wp-content/uploads/2013/06/Besada-Stevens_Addressing-the-Economic-Costs-of-Sustainable-Energy-in-the-Global-South_FINALFINAL-1.pdf.

Bobrow, D. 2006. Policy design: Ubiquitous, necessary and difficult. In Peter, G., and J. Pierre (Eds.), *Handbook of Public Policy*. Beverly Hills, CA: Sage, pp. 75–96.

Brenkert, G. G. 2009. Innovation, rule breaking and ethics of entrepreneurship. *Journal of Business Venturing* 24: 448–464.

Brinley, T. 1985. Escaping from constraints: The industrial revolution in Malthusian context. *Journal of Interdisciplinary History* XV(4): 729–753.

Bucciarelli, L. L. 2008. Ethics and engineering education. *European Journal of Engineering Education* 33(2): 141–149.

Byrne, E., C. Desha, J. Fitzpatrick, and K. Hargroves. 2010. Engineering education for sustainable development: A review of international progress. *3rd International Symposium for Engineering Education*, University College Cork, Ireland, June 30–July 2.

Capano, M. 2015. Policy design and non-design in policy-making policy formulation and the changing dynamics of public policy. Background Paper Workshop on *"Design and Non-Design in Policy-Making: When and How Policy Design Matters"* ECPR Jt. Sessions of Workshops March 31–April 3, University of Warsaw, Warsaw, Poland. https://ecpr.eu/Filestore/PaperProposal/60f6bace-4b87-40d3-8d22-c4348685f056.pdf.

Cochran, C. E., L. C. Mayer, T. R. Carr, and N. J. Cayer. 2016. *American Public Policy: An Introduction*. Ninth edition. Boston, MA: Wadsworth Cengage Learning.

Coyle, E., and M. Rebow. 2009. Sustainable design: A case study in energy systems. In Christensen, S. H., B. Delahousse, and M. Meganck (Eds.), *Engineering in Context*. Dublin Institute of Technology. Leeds, West Yorkshire, UK: Academica.

Darwish, M. M., M. Agnello, and H. Buelinckx. 2009. Why sustainable construction why now? *Fifth International Conference on Construction in the 21st Century*, Istanbul, Turkey, June 14–17.

Devon, R., and I. Van de Poel. 2004. Design ethics: The social ethics paradigm. *International Journal of Engineering Education* 20(3): 461–469.

Dieter, G. 2000. *Engineering Design: A Materials and Processing Approach*. Third edition. New York: McGraw-Hill Publishing.

Dunham, L. 2005. Entrepreneurship and ethics. University of St. Thomas, St Paul, MN. http://www.stthomas.edu/news/entrepreneurship-and-ethics/.

El-Zein, A., D. Airey, P. Bowden, and H. Clarkeburn. 2008. Sustainability and ethics as decision-making paradigms in engineering curricula. *International Journal of Sustainability in Higher Education* 9(2): 170–182.

Engineers Australia. 2010. Our code of ethics. http://www.engineersaustralia.org.au/sites/default/files/shado/About%20Us/Overview/Governance/codeofethics2010.pdf.

Erickson, C., and D. J. Degan. 2009. Generation E: Students leading for a sustainable, clean energy future. National Wildlife Federation. A Publication in Campus Ecology's Climate and Sustainability Series.

Ersdal, G., and T. Aven. 2008. Risk informed decision-making and its ethical basis. *Reliability Engineering and System Safety* 93(2): 197–205.

Fielder, J. 1995. Defects and deceptions—The Bjork-Shiley heart valve. *IEEE Technology and Society* 14(3): 17–22.

Fisscher, O., D. Frenkel, Y. Lurie, and A. Nijhof. 2005. Stretching the frontiers: Exploring the relationships between entrepreneurship and ethics. *Journal of Business Ethics* 60: 207–209.

Fleddermann, C. B. 2012. Engineering Ethics. Fourth edition. Pearson Education: Upper Saddle River, NJ.

Florman, S. C. 2002. Engineering ethics: The conversation without end. The Bridge 32(3): 14–18.

Galloway, P. D. 2005. The engineer's role in public policy. White Paper. https://ssms.jp/wp-content/uploads/PDF/issue3-vol1/SMS07-001_pd_01_Galloway.pdf

Gilder, G. 1992. *Recapturing the Spirit of Enterprise.* San Francisco, CA: ICS Press.

Gilman, S. C. 2005. Ethics codes and codes of conduct as tools for promoting an ethical and professional public service: Comparative successes and lessons, prepared for the PREM. The World Bank, Washington, DC.

Goldfarb, T., and M. S. Pritchard. 1999. Ethics in the science classroom: An instructional guide for secondary school science teachers. http://www.onlineethics.org/File.aspx?id=33732.

Grant, A. A. 1995. The ethics of sustainability: An engineering perspective. *Renewable Resources Journal* 13(1): 23–25.

Grubb, M., L. Butler, and P. Twomey. 2006. Diversity and security in UK electricity generation: The influence of low-carbon objectives. *Energy Policy* 34(18): 4050–4062.

Habash, R. W. H., V. Groza, and K. Burr. 2013. Risk management framework for the power grid cyber-physical security. *British Journal of Applied Science and Technology* 3(4): 1070–1085.

Hagenbuch, D. 2015. The 4 pillars of ethical enterprises. https://www.entrepreneur.com/article/240035.

Harris, C. E. 2008. The good engineer: Giving virtue its due in engineering ethics. *Science and Engineering Ethics* 14(2): 153–164.

Harris, C. E., M. S. Pritchard Jr., and M. J. Rabins. 1995. *Engineering Ethics.* Belmont, CA: Wadsworth Publishing Company.

Harris Jr., C., M. Pritchard, and M. Rabins. 2000. *Engineering Ethics: Concepts and Cases.* Second edition. Belmont, CA: Wadsworth.

Healey, M., A. Flint, and K. Harrington. 2014. *Engagement through Partnership: Students as Partners in Learning and Teaching in Higher Education.* Heslington, York, UK: The Higher Education Academy. https://www.heacademy.ac.uk/system/files/resources/engagement_through_partnership.pdf..

Herkert, J. R. 1996. Making connections: Engineering ethics on the World Wide Web. *Proceedings of the Frontiers on Education Conference,* Cleveland, OH, March 21–23.

Herkert, J. 2000a. *Social, Ethical and Policy Implications of Engineering.* New York: IEEE Press.

Herkert, J. R. 2000b. *Engineering Ethics and Public Policy.* Hoboken, NJ: Wiley-IEEE Press.

Herkert, J. R. 2002. Continuing and emerging issues in engineering ethics education. In *The Bridge, National Academy of Engineering,* pp. 8–13. https://www.nae.edu/File.aspx?id=7378.

Herkert, J. R. 2005. Ways of thinking about and teaching ethical problem solving: Microethics and macroethics in engineering. *Science and Engineering Ethics* 11(3): 373–385.

Hersh, M. A. 2013. Barriers to ethical behaviour and stability: Stereotyping and scapegoating as pretexts for avoiding responsibility. *Annual Review in Control* 37(2): 365–381.

Hersh, M. A. (Ed.). 2015. *Ethical Engineering for International Development and Environmental Sustainability.* London: Springer-Verlag.

Holmberg, J., and R. Sandbrook. 1992. Sustainable development: What is to be done? In Holmberg, J. (Ed.), *Making Development Sustainable*. Washington, DC: Island Press, pp. 19–38.

Hom, R. 2013. Why does public policy matter? What role can engineers play in it? https://www.quora.com/Why-does-public-policy-matter-What-role-can-engineers-play-in-it.

Hooker, J. N. 2000. Some business-related ethical issues in engineering. Tepper School of Business, Pittsburgh, PA. http://works.bepress.com/jnhooker/13.

Hyman, B. 2003. Public policy and engineering design education. *International Journal of Engineering Education* 19(1): 110–117.

IAEA. 2005. Energy indicators for sustainable development: Guidelines and methodologies. International Atomic Energy Agency, United Nations Department of Economic and Social Affairs, International Energy Agency Eurostat and European Environment Agency. http://www-pub.iaea.org/MTCD/publications/PDF/Pub1222_web.pdf.

Jacobsson, S., and K. Karltorp. 2013. Mechanisms blocking the dynamics of the European offshore wind energy innovation system—Challenges for policy intervention. *Energy Policy* 63: 1182–1195.

Keller, D. R. 2010. *Environmental Ethics: The Big Questions*. Hoboken, NJ: Wiley-Blackwell.

Kothari, R. 1990. Environment, technology and ethics. In Engel, J. R. and J. G. Engel (Eds.), *Ethics of Environment and Development*. Tucson, AZ: University of Arizona Press, pp. 27–35.

Krantzberg, G. 2009. How society gains from engineers educated in public policy. *The Journal of Policy Engagement* 1(2): 1–4.

Kropotkin, P. 1922. Ethics: Origin and development. White Paper, New York. https://theanarchistlibrary.org/library/petr-kropotkin-ethics-origin-and-development.pdf.

Ladd, J. 1980. The quest for a code of professional ethics: An intellectual and moral confusion. In Chalk, R., M. Frankel, and S. Chafer (Eds.), *AAAS Professional Ethics Project: Professional Ethics Activities in the Scientific and Engineering Societies*. Washington, DC: AAAS, pp. 154–159.

Langford, P. 2004. Engineering to shape a better world. *Engineering Sustainability* 57(ES2): 69–78.

Larsen, H., and L. S. Petersen. 2009. Riso energy report 8. The intelligent energy system infrastructure for the future. Riso-R-1695(EN), Riso National Laboratory for Sustainable Energy, Technical University of Denmark, Roskilde, Denmark.

Lawlor, R. 2013. *Engineering in Society*. Royal Academy of Engineering, Second edition.

Luegenbiehl, H. C., and M. Davis. 1992. Engineering codes of ethics: Analysis and applications. Center for the Study of Ethics in the Professions, Illinois Institute of Technology, Chicago, IL. http://ethics.iit.edu/publication/CODE--Exxon%20Module.pdf.

Luppicini, R. 2010. *Technoethics and the Evolving Knowledge Society*. Hershey, PA: Idea Group Publishing.

Mackay, M., and L. Shaxton. 2016. Understanding and applying basic public policy concepts. http://www.politicipublice.ro/uploads/understanding_public_policy.pdf.

Martin, M., and R. Schinzinger. 1996. *Ethics in Engineering*. Third edition. New York: McGraw-Hill.

Martin, M. W., and R. Schinzinger. 2005. *Ethics in Engineering*. Fourth edition. Boston, MA: McGraw-Hill.

McDonough, W., and M. Broungart. 2002. *Cradle to Cradle: Remarking the Way We Make Things*. New York: North Point Press.

McLean, G. F. 1993. Integrating ethics with design. *IEEE Technology and Society* 12(3): 18–30.

Meijknecht, T. 2014. The Engineer is professionally a person of faith: A theological-historical perspective. *Implicit Religion* 17(1): 183–195. http://www.motiv.tudelft.nl/uploads/IR%20The%20engineer%20is%20professionally%20%20a%20person%20of%20faith.pdf.

Milbrath, L. Y. 1996. *Learning to Think Environmentally (While There is Still Time)*. Albany, NY: Sunny Press.

Montpetit, E. 2003. *Misplaced Distrust: Policy Networks and the Environment in France, the United States, and Canada*. Vancouver, BC: UBC Press.

Moor, J. H. 2005. Why we need better ethics for emerging technologies. *Ethics and Information Technology* 7: 111–119.

Murray, T. 2009. An introduction to principles of ethics and morality for scientists and engineers. IDEESE: International Dimensions of Ethics Education in Science and Engineering, University of Massachusetts Amherst, Amherst, MA. http://www.umass.edu/sts/ethics.

NAE. 2004. *The Engineer of 2020: Visions of Engineering in the New Century*. National Academy of Engineering. Washington, DC: The National Academies Press.

NAS. 2014. The nexus of biofuels, climate change, and human health: Workshop summary. Roundtable on Environmental Health Sciences, Research, and Medicine, Board on Population Health and Public Health Practice. National Academy of Sciences, Washington, DC.

Ngambeki, I., D. Evangelou, S. Rao, M. Cox, and L. Behbehani. 2011. Work in progress—Developing a curriculum to prepare engineers to participate in public policy. *41st ASEE/IEEE Frontiers in Education Conference*, Rapid City, SD, October 12–15.

Nichols, S. 1997. Professional responsibility: The role of the engineer in society. *Science and Engineering Ethics* 3(3): 327–337.

NSPE. 2016. NSPE code of ethics for engineers. National Society of Professional Engineers. https://www.nspe.org/resources/ethics/code-ethics.

NSPE. 2017. History of the code of ethics for engineers. National Society of Professional Engineers. https://www.nspe.org/resources/ethics/code-ethics/history-code-ethics-engineers.

OECD. 2007. Energy and sustainable development: Canada-Energy in the context of sustainable development. Organization for Economic Cooperation and Development. http://www.un.org/esa/agenda21/natlinfo/countr/canada/energy.pdf.

O'Neill-Carrillo, E., H. R. Zamot, M. Hernández, A. A. Irizarry-Rivera, and L. O. Jiménez-Rodríguez. 2012. Beyond traditional power systems: Energy externalities, ethics and society. *IEEE International Symposium on Sustainable Systems and Technology*, Boston, MA, May 16–18.

Page, E. C. 2005. Producing public policy. In Moran, M., M. Rein, and R. E. Goodin (Eds.), *The Oxford Handbook of Public Management*. Oxford, UK: University Press.

Parkinson, S. 2010. Engineers and social responsibility. The UNESCO report, engineering: Issues, challenges and opportunities for development. http://www.sgr.org.uk/sites/sgr.org.uk/files/SGR_article_UNESCO_engr_book.pdf.

Patel, P. 2015. Engineers, ethics, and the VW scandal. IEEE Spectrum Blog. http://spectrum.ieee.org/cars-that-think/at-work/education/vw-scandal-shocking-but-not-surprising-ethicists-say.

Peha, J. M. 2011. Bridging the divide between technologists and policy-makers. https://users.ece.cmu.edu/~peha/bridging_divide.pdf.

Pritchard, M. S. 2005. Engineering ethics. http://www.blackwellreference.com/public/tocnode?id=g9781405133456_chunk_g978140513345648.

SEEPP. 2016. Sustainable environmental and ethical procurement policy. The City of Calgary. http://www.calgary.ca/CA/city-clerks/Documents/Council-policy-library/cfo008-Sustainable-Environmental-and-Ethical-Procurement-Policy.pdf.

Starcher, G. 1997. Ethics and entrepreneurship, an oxymoron: A transition to a free market economy in Eastern Europe. Paris, France: European Bahá'í Business Forum, 1997. http://bahai-library.com/starcher_ethics_entrepreneurship.

Van Gorp, A. 2005. Ethical issues in engineering design: Safety and sustainability. Simon Steven Series in the Philosophy of Technology, Delft University of Technology, Delft, Netherlands; Eindhoven University of Technology, Eindhoven, the Netherlands.

Vanderburg, W. 1995. Preventive engineering: Strategy for dealing with negative social and environmental implications of technology. *Journal of Professional Issues in Engineering Education and Practice* 121: 155–160.

Velasquez, M., C. Andre, S. J. Shanks, and M. Meyer. 1987. What is ethics? *Journal of Issues in Ethics* 1(1): 44–67.

Weil, V. 1984. The rise of engineering ethic. *Technology in Society* 6(4): 341–345.

WFEO. 2001. The WFEO model code of ethics. World Federation of Engineering Organisations, Tunis, Tunisia. https://www.wfeo.org/index.php?page=ethics.

Wilson, A. A., C. Sias, and A. Mejia. 2015. Hands-on activity: Engineering ethics. https://www.teachengineering.org/activities/view/usu_ethics_activity1.

Winner, L. 1991. Engineering ethics and political imagination. In Durbin, P. (Ed.), *Broad and Narrow Interpretations of Philosophy of Technology: Philosophy and Technology 7,* reprinted in Johnson, D. 1991. *Ethical Issues in Engineering.* Englewood Cliffs, NJ: Prentice Hall, pp. 376–385.

Wissenz, E. 2016. Decent democracy. http://www.decent-democracy.org/Energy-Problems.html.

Wujek, J., and D. Johnson. 1992. *How to Be a Good Engineer.* Washington, DC: IEEE United States Activities Board.

part two

The Creativity Landscape

chapter five

Creativity invention and innovation

Defy the Conventional

The Campaign for University of Ottawa, Canada

5.1 Objectives

- Provide an extensive historical perspective of creativity and innovation.
- Summarize innovation milestones throughout history.
- Provide knowledge about creativity, creative thinking, and creativity components in individuals.
- Introduce the notion of "Defy the Conventional" and link it to three of the greatest minds in scientific history: Newton, Darwin, and Einstein.
- Understand invention and innovation in the designed world.
- Discuss innovation, types of innovation, forms of innovation, benefits and risks of innovation, and innovation diffusion to the market.
- Differentiate and distinguish between the different types of innovation, such as radical and incremental innovation.
- Distinguish the different forms that innovation can take, such as product, process, and service innovation.
- Discuss the process of innovation and show the difference between linear and interactive models of innovation.
- Determine the relationship between invention and innovation.
- Explain the aspects of technological innovation and its determinants.
- Introduce disruptive innovation technologies and provide a brief description of the 12 related technologies.
- See the value of, and be adept at, seeing opportunities for employing engineering habits of the mind as thinking tools in every day that aid in stimulating creativity and innovation in individuals.
- Recognize that a one's ways of thinking are influenced by their profession, culture, upbringing, and context, and that a much richer realizing of a problem or system is developed by employing multiple ways of thinking.

- Define green innovation (GI) and discuss its topology and characteristics.
- Explore integrated innovation as coordinated application of scientific, technological, social, and business innovation to develop solutions to complex problems.
- Discuss strategies to enhance creativity in engineering education.
- In a case study, discuss innovation within automotive industry including Tata Nano and several self-driving cars (SDCs).
- Provide three end of chapter pedagogical knowledge strategies, namely knowledge acquisition, knowledge possession, and knowledge creation to help understand the above topics and generate new and alternative ideas and solutions.

5.2 *Historical perspective*

> Great triumphs of engineering genius: the locomotive, the truss bridge, the steel rail are rather invention than engineering proper.
>
> **Arthur Mellen Wellington**

5.2.1 *Historical approach*

Historians have had less to say about creativity than innovation. Interest has largely focused upon the end result of creativity, that is, innovation. This is in large part because of the greater interest in the economic and social consequences of innovation than its origins. In addition, creativity is not easily substantiated through historical evidence since it is not so obviously outcome based, or as easily documented, as innovation. Furthermore, there has not much been written about the reverse causality, that is, of innovation upon subsequent creativity. However, increased interest in recent years on the role of human capital in economic progress and the development of knowledge has motivated closer historical consideration of the creative origins of innovation (Ville 2011).

Most analysts of innovation emphasize the importance of a historical approach, with good reason. First, innovation is time-consuming, based on conjectures about the future, and its outcomes typically are uncertain for long periods. Therefore, analysis of any innovation requires an understanding of its history. Second, innovative capabilities are developed through complex, cumulative processes of learning. Finally, innovation processes are shaped by social contexts, as Lazonick (2002) has pointed out: the social conditions affecting innovation change over time and vary across productive activities; hence, theoretical analysis of the innovative enterprise must be integrated with historical study.

5.2.2 Early history

The controlled use of fire was a very early invention in the early Stone Age, with some of the earliest evidence dating back to somewhere between 200,000 and 600,000 years ago. Around 15,000 BC, the first animal domestication began taking place, and around 10,000 BCE, the first domestication of plants. This step was critical for the advancement of the human species. Around 4000 BCE, the ancient Egyptians were making wooden sailboats and around 1200 BCE the Phoenicians and the Greeks began to make even bigger sailing ships (Allis 2015).

The significant step in the history of innovation came with the creation of the wheel, sometime between 3300 and 3500 BC. Discovered in southern Poland, the earliest known depiction of a wheeled vehicle was on a clay pot. The next critically important innovation that contributed to the development of a strong human civilization was money. Around 3000 BC, the Sumerians were one of the first societies (if not the first) to begin using money to help the ease of commerce and exchanging of goods, replacing the barter system. The whole science of metallurgy began around 4400 BC when human civilizations began to use copper and silver, and soon thereafter it was figured out how to merge copper and tin to form bronze. Around 3000 BC an even stronger substance, iron, was found, which gave rise to a new age of human history. The next great innovation, around 200 BC, was water power—first used in the Fertile Crescent area in the Middle East. This breakthrough enabled enormous transformations in our ability as a species to harness power, and water power continued to be used into the nineteenth century, when water-powered mills were still common in England and New England (Allis 2015). Moving into the Common Era calendar, the creation of paper was most transformational. Rag paper, first used by the Chinese around the year 105, was rendered large. Invented around the sixteenth century, wood paper allowed knowledge to be spread much more easily.

5.2.3 The first Industrial Revolution

Prior to the surge of the Internet, no innovation did more for the distribution and democratization of knowledge than Johannes Gutenberg's printing press. The machine was developed around 1440 in Mainz, Germany, and improved through the use of a mold that allowed for the quick production of lead alloy-type pieces.

Moving to 1600, English scientist William Gilbert invented the term electricity, which was initiated from the Greek word for amber. Afterward, in 1752, Ben Franklin showed that lightning and the spark from amber were one and the same substance, electricity. In 1608, Hans Lippershey created a convex lens and concave eyepiece that facilitated the creation of the telescope (Allis 2013).

Most economic historians regard the developments in Britain and Northwestern Europe from around 1760 as an economic and technological watershed. Innovation during this period is best conceptualized as an economy-wide process that involved technological, organizational, and institutional change, spanning many sectors and product groups. The British Industrial Revolution (IR) was closely associated with the beginnings of a shift from a cottage system of outworkers using hand tools in cotton manufacture to the deployment of machine tools located in centralized factories (Hudson 2004).

A great deal of inventive activity during this period focused on consumer goods. According to Berg and Eger (2003), much of this consumer-goods patenting affected a vast number of small, novel products such as buckles and fasteners, cabinets and furniture, and spectacle frames. Much of the patent activity within the textiles sector, roughly one-third, involved new products (Griffiths et al. 1992). Much of the inventive activity in this key sector within the IR involved new thread types and fabrics, and focused on a consumer market.

A significant innovation of this period was the steam engine of James Watt, first introduced in 1775. Watt's innovation is commonly described as the emblematic technology of the IR.

A significant innovation of this period was the steam engine of James Watt, first introduced in 1775. Watt's innovation is commonly described as the symbolic technology of the IR. During the 1800s, external combustion granted exponential expansion in transportation, agriculture, and manufacturing, where in 1885 Karl Benz built a gasoline-engined car. The steam engine's principle of energy into motion set the platform for later innovations like internal combustion engines and jet turbines, which has powered the production of cars, trains, and aircraft during the twentieth century.

5.2.4 The second Industrial Revolution

In the late nineteenth century, industrial technologies began to change, and a range of new technologies and industries emerged. This second IR took place on the continent of Europe and in the United States. As a result, innovation was associated with both questions of spatial location and production technology. The second IR was characterized by organizational innovations that laid the groundwork for links between industry and formal science that became stronger during the course of the twentieth century. The late nineteenth century has been labeled as a second IR; major advances in new, more scientifically based industries and in different countries were driving a new expansionary phase: German chemicals, electricity, and automobiles should particularly be noted (Pierenkemper and Tilly 2004). American firms carried these advances through into the twentieth century, particularly by extending German technology into organizational and marketing innovations. Automobiles were at this point mass produced on assembly lines and sold through specialist dealers. All of this was achieved under new governance structures associated with multidivisional organizations (Chandler 2003). The technological shifts of the late nineteenth century were accompanied by changes in firm structure. Large-scale, vertically integrated enterprises emerged in Germany and the United States that incorporated specialized research and development departments or laboratories.

Bridging scientific discoveries and technological innovation typically requires considerable time. For example, in 1800, Italian Alessandro Volta made the first battery (known as a Voltaic pile). In 1803, Henry and Sealy Fourdrinier invented the papermaking machine. No significant technological applications followed Faraday's demonstration of electromagnetic induction and primitive electric generators and motors in 1831, with the exception of the telegraph. Yet this scientific discovery laid the foundations for one of the defining industries of the second IR, electrical equipment and electric power generation. In the 1830s, William Sturgeon developed the first practical electric motor. Transformers and electric lights using ac current were developed throughout the 1800s, thanks to the efforts of inventors such as Warren de la Rue, Joseph Wilson Swan, and Thomas Alva Edison, where Edison and Swan patented the first long-lasting light bulbs in 1879 and 1880 and opened the world's first power plant, liberating society from a near-total reliance on daylight. In 1888, Nikola Tesla patented the ac electric induction motor and, in opposition to Edison, became a loyal advocate of ac power. In 1905, Albert Einstein explained the photoelectric effect. When electrical power transmission was established in the last quarter of the nineteenth century, ac and dc were competing to become the standard power distribution system, a historical period known as the "war of currents."

In 1860s, James Clerk Maxwell figured out radio waves and set out basic laws of electromagnetism. Then in 1876, Alexander Graham Bell patented the telephone, though the true ownership of the invention remains controversial. In the 1880s, Pierre and Paul-Jacques Curie discovered the piezoelectric effect. In 1901, Guglielmo Marconi sent radio-frequency signals across the Atlantic Ocean from United Kingdom to Canada.

Scientific advances in physics and chemistry during the last third of the nineteenth century created considerable potential for the profitable application of scientific and technical knowledge in industry. Among important inventions in this regard, Louis Pasteur developed pasteurization, a way of preserving food by heating it to kill off bacteria. In 1895, German physicist Wilhelm Röntgen discovered x-rays. Perkin's accidental synthesis of mauveine in 1856 was the first synthetic dyestuff. Exploitation of scientific advances required the development of complex process technologies for which no scientific foundation existed.

The first in-house industrial R&D laboratories were established by German firms seeking to commercialize innovations based on the rapidly developing field of organic chemistry. Kekule's 1865 model of the molecular structure of benzene, a key component of organic chemistry and synthetic dyestuffs, provided the first scientific foundation for developing new products. Scientifically trained personnel were needed to translate Kekule's breakthrough into new products.

The rapid expansion in Germany's network of research and technical universities during the second half of the nineteenth century was critically important to the growth of industrial research, particularly in the chemical industry. German universities produced a large pool of scientifically trained researchers, many of whom sought employment in France and Germany during the 1860s. University faculty advised established firms, and university laboratories provided a site for industrial researchers to conduct scientific experiments in the early stages of the creation of in-house research laboratories. During the nineteenth century, German universities pioneered the development of the modern model of the research university, in which faculty research was central to the training of advanced degree holders. In addition, the German polytechnic institutes that had been founded during the 1830s by various German principalities were by the 1870s transformed into technical universities that played a central role in training engineers and technicians for the chemical and electrical equipment industries. According to Murmann and Landau (1998), by the 1870s Germany had nearly 30 university and technical university departments in organic chemistry, and seven major centers of organic chemistry research and teaching. Technically trained personnel moved into senior management positions within German industry, in contrast to the situation in the United Kingdom, further strengthening the links between corporate strategy and industrial research.

The institutional transformation of Germany's national innovation system was both a cause and an effect of the growth of the chemical and electrical equipment industries. Werner von Siemens of the Siemens electrical equipment firm was a founder of the German Association for Patent Protection in 1874, and the first national patent law in the new German state was passed in 1877.

5.2.5 *The following Industrial Revolution*

The development of industrial research within US manufacturing firms followed developments in the German chemical and electrical machinery industries. Many of the earliest US corporate investors in industrial R&D, such as General Electric and Alcoa, were founded on product or process innovations that drew on recent advances in physics and chemistry. The corporate R&D laboratory brought more of the process of developing and improving industrial technology into the boundaries of US manufacturing firms, reducing the importance of the independent inventor as a source of patents (Schmookler 1957).

An enormous milestone forward in the field of medicine is antibiotics that saved millions of lives by killing and preventing the growth of harmful bacteria. Scientists such as Louis Pasteur and Joseph Lister were the first to identify and attempt to combat bacteria. In 1896, the French medical student Ernest Duchesne originally discovered the antibiotic properties of penicillium; however, his research went mostly unobserved. But it was the Scottish biologist Alexander Fleming who made the first leap in antibiotics when he accidentally discovered the bacteria-inhibiting mold known as penicillin in 1928. Penicillin enabled doctors to fight bacterial infections, save lives, and heal syphilis, gangrene, and tuberculosis (Allis 2013).

The structure of the innovation process in the industrial economies was transformed after 1945. Global scientific leadership shifted decisively from Western Europe to the United States. A new set of industries, focused on several innovation areas, grew rapidly. As global trade and investment flows revived after the 1914–1945 period of war and depression, international flows of technology also expanded, and by the 1980s and 1990s enabled economies such as Japan, South Korea, and Taiwan to advance to the front rank as sources of industrial innovation. Developments in the United States illustrate these trends most vividly and highlight the development of a US National Innovation System. Just as innovation is more than science and technology, an innovation system is more than those elements directly related to the promotion of science and technology. Rather, it also includes all economic, political, and other social institutions affecting innovation (Atkinson 2014).

Moving into the second half of the twentieth century, the types and location of innovation shifted once again with its economic and industrial hegemony. From the 1950s, Japanese firms began to challenge those in Europe and North America, particularly through holistic innovation in manufacturing systems, known as lean production, new approaches to labor leadership and management, and the development of imaginative forms of interfirm transacting, especially just-in-time contracting (Fruin 1992).

In summary, Table 5.1 outlines innovation throughout the three IRs (Bruland and Mowery 2004), whereas Table 5.2 summarizes innovation milestones since the 1600s.

***Table* 5.1** Summary of innovation throughout IRs

Industrial Revolution	Outcome
First	Craft-oriented, trial-and-error process, basic woodworking and metalworking techniques
Second	German and US firms in the electrical-equipment and chemical industries, organized innovation activities in large firms interacting with a public R&D infrastructure, large enterprises of unprecedented scale
Third and the following	The role of the state as R&D funder; state actions also contributed to the spread of innovation-led development to Asia, as the military alliances and economic institutions of the post-1945 period supported the expanded international trade; Asian governments' strategies for technology transfer and industrial development were of great importance

***Table* 5.2** Selected innovation milestones

1600s	English scientist William Gilbert coined the term electricity, convex lens/Hans Lippershey, telescope/Hans Lippershey, slide rule/William Oughtred, syringe/Christopher Wren, pressure cooker/Denis Papin, pendulum clock/Christian Huygens
1700s	Steam engine/Thomas Newcomen, hot-air balloon/Bartolomeu de Gusmão, temperature scale/Gabriel Fahrenheit, lightning conductor/ Benjamin Franklin, carbonated water/Joseph Priestley, sandwich/ John Montagu, steam boat/Joffroy d'Abans, lithography/Aloys Senefelder, vaccination/Edward Jenner
1800s	First electric battery/Alessandro Volta, arc lamp/Humphry Davy, telegraph/ Samuel Soemmering
1810s	Precision Lathe/Henry Maudslay, spectroscope/Joseph von Fraunhover, Miner's lamp/ Humphry Davy
1820s	Electromagnet/William Sturgeon, waterproof clothes/Charles Macintosh, passenger railway/George Stephenson
1830s	Lawn mower/Edwin Budding, sewing machine/Barthelemy Thimonnier, electric dynamo/ Michael Faraday, propeller/Francis Pettit Smith, Morse code/Samuel Morse, mechanical calculator/ Charles Babbage
1840s	Postage stamp/Rowland Hill, antiseptic/Ignaz Semmelweis
1850s	Airship/Henri Giffard, ghyroscope/Jean-Bernard-Léon Foucault, safety lift/Elisha Otis, internal combustion engine/Jean-Joseph-Étienne Lenoir
1860s	Telephone/Johann Philipp Reis, laws of electromagnetism/Maxwell, bicycle/Pierre Michaux, plastic/Alexander Parkes, dynamite/Alfred Nobel, torpedo/Robert Whitehead, traffic lights/J. P. Knight

(*Continued*)

Table 5.2 (Continued) Selected innovation milestones

1870s	Telephone/Alexander Graham Bell, barbed wire/Joseph Glidden, four-stroke engine/Nikolaus August Otto, phonograph/Thomas Edison, light bulb/Joseph Swan and Thomas Edison
1880s	Electric current/Thomas Edison, piezoelectric effect/Pierre and Paul-Jacques Curie, metal detector/Alexander Bell, cash register/James Ritty, steam turbine/Charles A. Parsons, Coca-Cola/John Pemberton, contact lenses/F. E. Muller, photoelectric effect/ Heinrich Hertz
1890s	X-rays/Wilhelm Röntgen, basketball/James Naismith, escalator/Jesse Reno, tractor/John Froehlich, radio/G. Marconi, diesel engine/Rudolf Diesel, Oscilloscope/ Karl Braun, semiconductors for commercial purposes/Jagadish Chandra Bose, antibiotic/Ernest Duchesne, radio/ Nikola Tesla, electron/J. Thomson
1900s	Photoelectric effect/Albert Einstein, Robertson screw/Peter Lymburner Robertson, vacuum cleaner/Hubert Booth, airplane/Wilbur and Orville Wright, radar/Christian Hülsmeyer, synthetic plastic/Leo Baekeland
1910s	Electrical ignition system/Charles Franklin Kettering, neon light/ Georges Claude, electric car starter/Charles Kettering, tungsten filament/William Coolidge
1920s	Insulin/Sir Frederick Grant Banting, Q-Tips/Leo Gerstenzang and his wife, traffic light/Police Officer William Potts, Band-Aid/Earle Dickson, bulldozer/Engineer Benjamin Holt, liquid-fueled rocket/Robert Goddard, lie detector test/John A. Larson, electronic TV system/Philo Tyler Farnsworth, Penicillin/Alexander Fleming, car radio/Pail Galvin
1930s	Electron microscope/Max Knott and Ernst Ruska, FM broadcasting/ Edwin H. Armstrong, jet engine/Sir Frank Whittle, photocopier/ Chester Carlson, frequency modulation/Edwin H. Armstrong
1940s	Electronic digital computer/John Atanasoff and Clifford Berry, kidney dialysis machine/Willem Kolff, synthetic cortisone/Percy Lavon Julian, microwave oven/Percy Spencer, Velcro/George de Mestral, transistor/Bill Shankly and his team at AT&T
1950s	DNA/James Watson and Francis Crick; credit card (Diner)/Ralph Schneider; power steering/Francis W. Davis; video tape recorder/ Charles Ginsburg; bar code/Joseph Woodland and Bernard Silver; black box flight recorder/David Warren; transistor radio/Texas Instruments; solar cell/Chaplin, Fuller, and Pearson; hovercraft/ Christopher Cockerell; integrated circuit/Jack Kilby and Robert Noyce
1960s	Computer mouse/Douglas Engelbart, RAM/Robert Dennard, communication satellite/Telstar, spacewalk/Aleksei Leonov, cash dispenser/Luther Simjian, Internet/US military
1970s	Floppy disk/Alan Shugart, LCD/James Fergason, video game/Nolan Bushnell, Ethernet/Robert Metcalfe and Xerox, MRI/Raymond Damadian, artificial heart/Robert K. Jarvik, IMAX motion picture system/Canadian filmmakers/entrepreneurs, personal computer/ Steve Jobs and Bill Gates

(*Continued*)

Table 5.2 (Continued) Selected innovation milestones

1980s	Windows program/ Microsoft, Apple Macintosh /Steve Jobs, Disposable camera/Fuji, Doppler radar/Christian Andreas Doppler
1990s	The World Wide Web and Internet protocol (HTTP) and WWW language (HTML)/Tim Berners-Lee, Blackberry/Research in Motion
2000s	Segway human transporter/Dean Kamen, Braille Glove/Ryan Patterson, artificial liver/Kenneth Matsumura and Alin Foundation, Toyota's hybrid car/Toyota, iPod and iPhone/Apple, Android/Google, Adidas 1 shoes/Adidas, YouTube/Steve Chen, Facebook/Mark Zuckerberg, USB thumb drive/consortium of companies, HDTV/Digital HDTV Grand Alliance, Nintendo Wii/Nintendo Co, NASA's Ares Rockets/NASA, wireless headset/Group of companies, DNA testing kit/Sir Alec John Jeffreys, Bionic lens/Babak Parviz, the sixth sense/Pranav Mistry
2010s	iPad/Apple, Kickstarter/Perry Chen, Yancey Strickler, and Charles Adler, Google's driverless car/Google, The Stark Hand/Mark Stark, BodyGuard/David Brown, Katal Landing Pad/Aaron Coret and Stephen Slen, Liquiglide/five MIT students, Enable Talk Gloves/four Ukrainian students, Google Glass/Google, Hendo Hoverboard/Jill and Greg Henderson, Supersmart Spacegraft/Indian Space Research Organization, high beta fusion reactor/Lockhead Martin, 3D Printed Everything/3D Systems, Apple Watch/Apple, Blackphone/SGP Technologies, Coolest Cooler/Ryan Grepper, BMWi3/BMW, Microsoft Surface Pro3/Microsoft, Ringly/Christina Mercando, Super bananas/Queensland University of Technology, Copenhagen wheel/Assaf Biderman, Hemopurifier/Aethlon Medical, Blue Room/Snake River Correctional Institution in Oregon

5.3 Creativity

Genius is 1% inspiration, 99% perspiration.

Thomas Edison

5.3.1 Creativity defined

In *Webster's New World Dictionary of the American Language*, the word "creative" has three interconnected meanings: creating or being able to create, having or showing imagination and artistic or intellectual inventiveness (creative writing), and stimulating the imagination and inventive powers. Creativity is not an attribute limited to the historic greats; rather, it is something anyone can use. To a large extent, creativity is a decision. According to Taylor (1975), creativity is perceived as a hierarchy from a low to a progressively higher level as shown in Table 5.3.

Creativity is the great enabler of innovation. It is marked by the ability to bring into existence, to invent into a new form, to produce through

***Table* 5.3** Hierarchy of creativity

Expressive	The ability to develop a unique idea with no concern about its quality
Technical	The proficiency to create products with consummate skills, but with little expressive spontaneity
Inventive	The ability to develop a new use of old parts and new ways of seeing old things in an ingenious manner
Innovative	The ability to penetrate foundational principles or establish a school of thought, and formulate innovative departures
Emergent	The ability to incorporate the most abstract ideational principles or assumptions underlying a body of knowledge, as in the example of Einstein's work on general relativity

imaginative skill, and to bring into existence something new. Creativity is characterized by an extensive search for solutions, especially those that are innovative. The search can be helped by a systematic and methodical approach. Creativity is also an attitude, the ability to believe in change and novelty, a readiness to play with ideas and potentials, a flexibility of viewpoint, and the habit of appreciating the noble. One approach says that creativity, at least in the technological sense, is the ability to recognize a problem in multiple dimensions.

5.3.2 *Creative genius*

The *Oxford English Dictionary* defines genius in two ways: as having natural aptitude, ability, or capacity; quality of mind; the special endowments that fit a person for his or her peculiar work; and as having native intellectual power of an exalted type, such as is attributed to those who are esteemed greatest in any department of art, speculation, or practice; instinctive creation, original thought, invention, or discovery. The first definition comes close to what is typically meant by the term "gifted," and it implies that the gift predisposes one to high-quality thought within a specialty. The second definition focuses on the successful use of intellectual processes, primarily on creative production, which need not imply inborn talent (Paul and Elder 2010).

Genius first became the subject of scientific inquiry in the early nineteenth century, and it has continued to attract research interest to the present day. Although genius can be defined as either superlative intelligence or achieved eminence, it is further confined to creative achievement (Simonton 2012). Genius is often specialized, limited to particular intellectual domains. It is better understood in relation to talent, giftedness, aptitude, capacity, ability, and intelligence. Table 5.4 considers the following definitions (and distinctions) found in *Webster's New World Dictionary*.

Notice that talent, gift, genius, and aptitude all imply an inborn disposition to excel within some domain of thought. But intelligence, brilliance,

Table 5.4 Definitions related to the term "genius"

Genius	An inborn mental endowment, specifically of a creative or inventive kind in the arts or sciences, or one that is exceptional or phenomenal
Gifted	A special ability that is bestowed upon one, as by nature, and not acquired through effort
Talented	Native ability for a specific pursuit and connotes either that it is or can be cultivated (or left largely undeveloped) by the one possessing it
Intelligent	Learn or understand from experience or to respond successfully to a new experience
Brilliant	An unusually high degree of intelligence
Accomplished	Very skillful and conclude successfully
Proficient	Well advanced and highly competent, skilled, and adept

accomplishment, proficiency, and virtuosity need not presuppose innate tendencies. Assuming that these distinctions mirror important qualities in human development, a real possibility is suggested: a person may be highly creative, even brilliant, without having a high degree of innate talent. This possibility is borne out by empirical fact. Many highly accomplished thinkers, rightly considered geniuses, have displayed that brilliance only after investing years in perfecting potential not extraordinary to begin with (Paul and Elder 2010).

5.3.3 *Creativity process*

Creativity involves a process that starts from a problem or question that can be described in many ways (Wallas 1926). It basically contains four stages: preparation, generation, incubation, and verification (Baillie 2002). The creative process is iterative as well as recursive. It involves lots of conversations including about goals and actions, and conversations with co-creators and colleagues. The collaborators' experiences and values have an impact on the conversations.

The preparation stage includes defining, reformulating, and redefining the problem or question. The formulation of a problem is sometimes more important than its solution, which may be just a matter of mathematical or experimental skill. This stage is usually best carried out with care in order to absorb as much information as possible.

Learners, after having defined, reformulated, and redefined the problem, move toward generating solutions as many as possible. Several creative brainstorming techniques, including mind mapping, symbolic analogy, and manipulative verbs, may be used at this stage.

The incubation stage is a period of full relaxation (e.g., sleeping or showering) or relaxed attention (e.g., leisure biking), which allows one's subconscious intelligence to suggest solutions. Reportedly, people often generate a potential idea after a certain time of incubation (Tomic and Brouwers 1999). For example, many solutions to difficult challenges can often be resolved after a break away from active processing (Browne and Cruse 1988). This is an important stage because sometimes it may take days, or weeks, or even months.

The verification stage includes analyzing, clustering, and evaluating all the solutions or ideas, planning, and implementing actions (Baillie 2002). It might also be called elaboration or detail stage. This is where Edison said that it is "1% inspiration and 99% perspiration." This is in fact the 99% perspiration stage.

5.3.4 *Critical and creative thinking*

Critical and creative thought are both achievements of thought. Creativity masters a process of making or producing, criticality a process of assessing or judging. The very definition of the word "creative" implies a critical component (e.g., having or showing imagination and artistic or intellectual inventiveness). When engaged in high-quality thought, the mind must simultaneously produce and assess, both generate and judge the products it fabricates. In short, sound thinking requires both imagination and intellectual standards (Paul and Elder 2014).

Critical thinking is the art of thinking about thinking in such a way as to identify its strengths and weaknesses and recast it in improved form (where necessary). The first characteristic requires the thinker to be skilled in analytic and evaluative thinking. The second requires the thinker to be skilled in creative thinking. Thus, critical thinking has three dimensions: the analytic, the evaluative, and the creative. Though we separate these functions for purposes of theoretical clarity, we nevertheless argue that each must be involved if the other two are to be effective (Paul and Elder 2010).

On the other hand, creative thinking may be defined as the art of generating solutions to problems by the force of imagination and reasoning. It is an activity of the mind seeking to find answers to some of life's questions. Every idea is a product of thinking and every product is the manifestation of idea naked in a thinker's mind. Creative thinkers are people who see problems as opportunities to improve and do something new or something better (Okpara 2007).

5.3.5 *Creativity components*

A variety of theorists, using case studies, experiments, and a variety of research methods, have attempted to better understand the sources of creativity and innovation in individuals. While these efforts have contributed significantly to broadening the comprehension of the subject, disagreement between theorists and many hypotheses remains to be fully substantiated. It appears that the only rule is that there are no hard and fast rules concerning the sources of creativity (Adams 2005).

In general, people become more creative when they feel motivated mostly by their interest, satisfaction, and challenge of the situation and not by external pressures; the passion and interest—a person's internal desire to do something unique to showcase himself or herself; and the person's sense of challenge, or a drive to solve a problem that no one else has been able to solve.

Creative people work hard and continually to improve ideas and solutions by making gradual alterations and refinements to their works. Contrary to the mythology surrounding creativity, very few of creative excellence are produced with a single stroke of brilliance or in a frenzy of rapid activity. Much closer to the real truth are the stories of companies which had to take the invention away from the inventor in order to market it because the inventor would have kept on tweaking it and fiddling with it, always trying to make it a little better (Harris 1998).

Within every individual, creativity is a function of several components as outlined in Figure 5.1. To be creative one has to be knowledgeable: one cannot go beyond what is known without knowing it. However, knowledge can also impede creativity (Frensch and Sternberg 1989). Knowledge may be described as all the relevant information that an individual brings to bear on a problem (Adams 2005). Creativity is most likely to emerge with substantial knowledge and experience of a discipline (Hayes 2005). On the other hand, knowledge about a field can result in a closed and entrenched perspective, resulting in a person's not moving beyond the way in which he or she has seen problems in the past. Knowledge therefore can help, or it can hinder creativity (Sternberg 2006).

Problem-solving and creative thinking refer to how problems are approached and the capacity to put existing ideas together in new combinations to create solutions. The skill itself depends quite a bit on personality as well as on how a person thinks and works. Expertise and creative thinking are the entrepreneur's raw materials or natural resources.

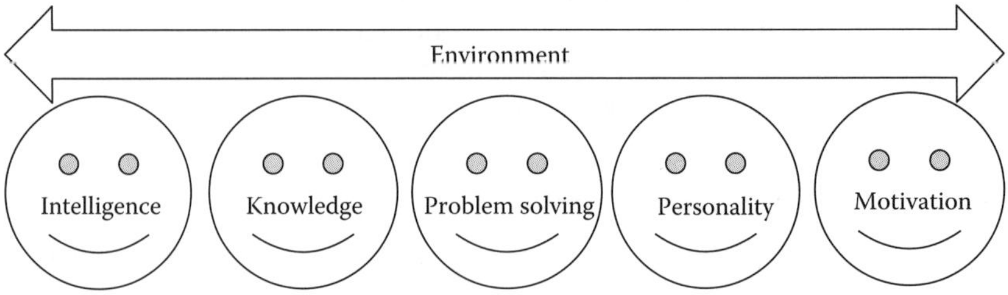

Figure 5.1 Components of creativity.

Personality and creativity are very broadly related, though specific traits tend to be associated with the exceptionally gifted (Torrance 2003). Often creative people seek opposition; that is, they decide to think in ways that countervail how others think. Note that none of the attributes of creative thinking are fixed. One can decide to overcome obstacles, take sensible risks, and so forth (Sternberg 2006).

Motivation is the drive and desire to do something, an inner passion and interest. When people are intrinsically motivated, they engage in their work for the challenge and enjoyment of it. The work itself is motivating. People will be most creative when they feel motivated primarily by the interest, satisfaction, and the challenge of the work itself—"the labor of love," "love of the work," "the enjoyment of seeing and searching for an outstanding solution," a breakthrough (Okpara 2007).

Creativity is, however, enhanced when people have some freedom, but not too much; high internal commitment to the task, but not too high; high proportion of intense rewards, but some extrinsic rewards as well; and some competition, but not winner take-all competition (Thompson 2001). Entrepreneurial activity depends on the process of innovation following creativity, not on creativity alone.

Importantly, one needs an environment that is supportive and rewarding of creative ideas. One could have all of the internal resources needed to think creatively, but without some environmental support (such as a forum for proposing those ideas), the creativity that a person has within him or her might never be displayed (Sternberg 2003a, 2006). Environmental determinants may include the context of work, task constraints, evaluation, competition, cooperation, role models, etc.

In general, however, motivation, environment, and personality are said to have a greater influence on creativity than intelligence quotient (IQ) (Amabile 1983, 1988, 1999). Three intellectual skills are particularly important (Sternberg and Beyond 1985): the synthetic skill to see problems in new ways and to escape the bounds of conventional thinking; the analytic skill to recognize which of one's ideas are worth pursuing and which are not; and the practical, contextual skill to know how to persuade others of—to sell other people on—the value of one's ideas. Research indicates a positive relationship between creativity and intelligence up to a measured IQ of 120 points (Albert and Runco 1999; Sternberg and Lubart 1999).

5.3.6 Defy the conventional

Sternberg (1997, 2003b, 2006) asserts that perhaps the one consistent attribute about successfully creative people is their explicit decision to pursue a creative path. His explanation is shown in Figure 5.2.

Let us take a closer look at the thinking of three of the greatest minds in scientific history: Newton, Darwin, and Einstein. What Newton,

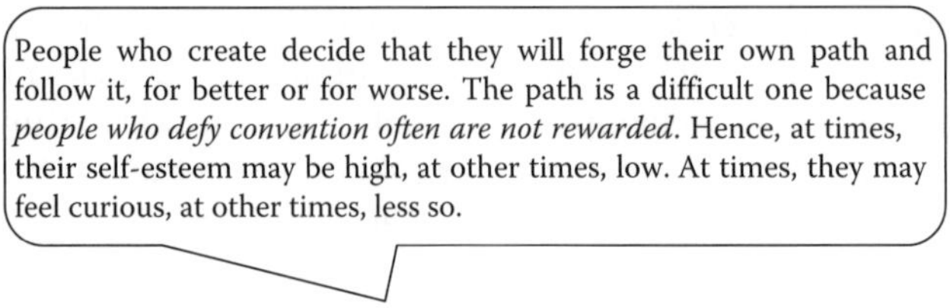

Figure 5.2 Creativity as described by Stenberg.

Darwin, and Einstein had in common was not some set of inexplicable or esoteric qualities but, rather, down-to-earth excellence in the art of questioning and an uncommon doggedness in pursuing deep answers to the questions they raised (Paul and Elder 2008).

Newton was uninterested in the set curriculum at Cambridge, and at the age of 19, he drew up a list of questions under 45 headings. His title "Questions" signaled his goal: to constantly question the nature of matter, place, time, and motion. His style was distinctly nonesoteric: to slog his way to knowledge.

In pursuing intellectual questions, Darwin relied upon perseverance and continual reflection, rather than memory and quick reflexes. "I have never been able to remember for more than a few days a single date or line of poetry." Instead, he had "the patience to reflect or ponder for any number of years over any unexplained problem. At no time am I a quick thinker or writer: whatever I have done in science has solely been by long pondering, patience, and industry" (Ockuly 2017).

Einstein did so poorly in school that when his father asked his son's headmaster what profession his son should adopt, the answer was simply, "It doesn't matter; he'll never make a success of anything."

The school system is the conventional, organized pipeline of learning, but learning without formal schooling is somewhat of a theme among known innovators and entrepreneurs. The list of dropouts is long. Michael Dell, founder of Dell computer, dropped out at 19; Steve Jobs, founder of Apple computer, dropped out at 19; Julian Assange, founder of Wikileaks, dropped out at 19; Bill Gates, founder of Microsoft, dropped out at 20; Evan Williams, cofounder of Twitter, dropped out at 20; Mark Zuckerberg, founder of Facebook, dropped out at 20; Larry Ellison, founder of Oracle, dropped out at 20; Jan Koum, founder of WhatsApp, dropped out at 21; Travis Kalanick, founder of Uber, dropped out at 21; and John Mackey, founder of Whole Foods, dropped out at 22 (Vital 2014).

By nature, creative individuals tend to defy convention. They resist thinking or doing what others are thinking or doing. Rather, they tend to go off in their own direction, seeking to propose ideas that are both novel and useful in some way. The greatest obstacle to creativity, therefore, often is not exactly criticisms from others, but rather the limits one places on one's own thinking. Such limits, however, may derive from processes of enculturation and socialization, so that it is often not clear whether restrictions on creativity are internally or externally enforced.

The investment theory of creativity (Sternberg and Beyond 1985; Sternberg and Lubart 1995; Sternberg 2003a) asserts that creative thinkers are like good investors. They buy low and sell high. Creative people generate ideas that are like undervalued stocks, and both ideas and the stocks are rejected by the public. When creative ideas are proposed, they are often viewed as bizarre, useless, and even foolish, and are rejected.

5.4 *Innovation*

Innovation = Inspiration + Perspiration + Perseverance.

Juan Roman
NASA

5.4.1 *Innovation defined*

Innovation is everywhere. In the world of goods (technology) certainly, but also in the realm of words: innovation is discussed in scientific and technical literature; in social sciences such as history, sociology, management, and economics; and in the humanities and arts. Innovation is also a central idea in popular imagination, media, and public policy (PP), and is part of everybody's vocabulary. Briefly stated, innovation has become the emblem of modern society (Godin 2008).

Innovation therefore means many things; one was simply novelty (Kallen 1930). A second meaning is social change (Stern 1927), and this meaning includes more than technological invention, that is, social invention. Innovation centers on people, culture, structure, process, and technology. It is the process through which the entrepreneur converts market opportunities into workable, profitable, and marketable ideas. Innovation results from the intersection of creativity, competence, leadership, and worldview. It is the process that combines ideas and knowledge into a new value.

Innovation and its ecology are expressed in the book *Innovation Nation* by John Kao (2007), formerly on Harvard Business School, Boston, MA. Kao defines innovation as the ability of individuals, companies, and entire nations to continuously create their desired future. Innovation depends on harvesting knowledge from a range of disciplines besides science and technology—among them design, social science, and the arts. And it is exemplified by more than just products; services, experiences, and processes. The work of entrepreneurs, scientists, and software geeks alike contributes to innovation. It is also about the middlemen who know how to realize value from ideas. Innovation flows from shifts in mind-set that can generate new business models, recognize new opportunities, and weave innovations throughout the fabric of society. It is about new ways of doing and seeing things as much as it is about the breakthrough idea.

Innovation is about risk and change, and deep forces in every society resist both of these. A striking feature of the US innovation ecology is the positive attitude toward failure, an attitude that encourages risk taking and entrepreneurship (Marburger 2011).

5.4.2 Incremental or radical

Innovations can be either incremental or radical (Dodgson et al. 2008; Christensen 1997) in degree; modifications of entities or entirely new entities; embodied in products, processes, or services; oriented toward consumers, industrial, or government use; based on various single or multiple technologies (Roberts 1998). Most innovations are incremental, meaning that they are of an evolutionary nature. Incremental innovation involves slightly upgrading a preexisting product. Examples include making a phone slightly thinner or a computer slightly faster.

Radical innovations, on the other hand, are revolutionary and imply major changes in the way a product or service works. Apple Inc. is often noted for bringing radical innovations to market (Stefik and Stefik 2006; Verganti 2007), for example, when they introduced the first generation iPod along with iTunes and accordingly kicked off the business of digital music downloads (Johnson et al. 2008).

An alternative way to distinguish radical innovation from incremental has been suggested as follows: incremental innovation involves doing of things in a new way, while radical means the doing of new things. Radical innovations are considered to be associated with high levels of uncertainty (O'Connor and Rice 2013; Enqvist 2014).

Incremental innovation is not about sweeping changes. On the contrary, firms that innovate incrementally intend to do so just a little bit at a time. The reason incremental innovation is so popular is because it has reduced risk in comparison to radical innovation. Pure radical innovation would best be described by Kim and Mauborgne (2005) as what they call the "blue ocean strategy." This strategy involves not fighting competition, but avoiding it. Rather than fighting for market share, a company steps aside and simply creates its own market.

Blue ocean strategy challenges companies to break out of the red ocean of bloody competition by creating uncontested market space that makes the competition irrelevant. Instead of dividing up existing and often shrinking demand and benchmarking competitors, blue ocean strategy is about growing demand and breaking away from the competition. An example of a blue ocean strategy is Netflix, which created uncontested marketing space by selling television (TV) shows over the Internet which no one else at the time was doing.

5.4.3 *Features and elements of innovation*

Innovation is strongly related to creativity and design. It is advanced by information gathered from insights gained by journeys into other disciplines and from active, collegial networks and open boundaries. Innovation arises from organizing circles of exchange, where information is not just acquired and accumulated but created. Definitions of innovation often suggest that it has several features as shown in Figure 5.3.

Novelty means absolute or relative to the unit or organization. Application means a presentation of an idea (Amabile 1988). Purposeful is related to intention of benefit or value, for example, the commercialization of creative ideas (West and Farr 1990). Strategic means a managed process (Gaynor 2002), and scale means identifying it as either small or large.

Innovation is the key to entrepreneurship. Entrepreneurs are the dreamers who take responsibility for creating innovation. It is the existence of innovation that differentiates the entrepreneur from others. Innovation creates competitiveness through work aimed at the transformation, renewal, and redefinition of organizations, their markets and industries, if establishment is to be deemed entrepreneurial. Figure 5.4 shows the key elements of innovation where challenge means accomplishment, attitude means turning ideas into reality, creativity means idea,

Figure 5.3 Five features of innovation.

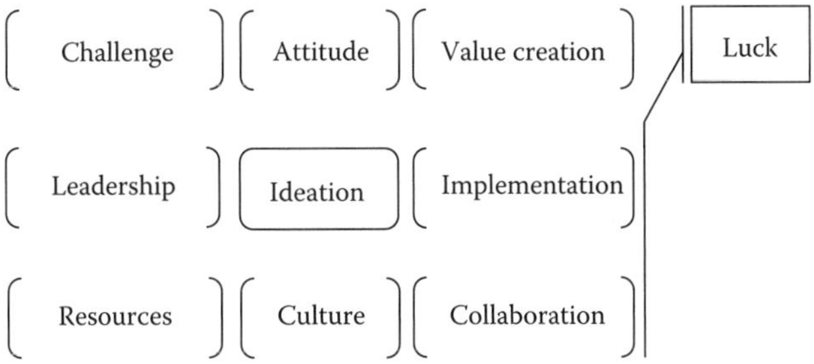

Figure 5.4 Key elements of innovation.

leadership means role model, implementation means putting ideas to use, resources should include time and people in addition to money, culture represents operation fields of innovation, and collaboration involves people who come together.

5.4.4 Forms of innovation

Innovation may take several forms, primarily including process, product, and management as shown in Figure 5.5. Innovation in processes includes changes and improvement to techniques. These contribute to increases in productivity, which lower cost and help to increase demand. In general, process innovations often have a significant impact on production.

Innovation in products or services involves opening up new markets. From a commercial viewpoint, the value of product innovations is that the innovation of a new product will encourage consumers to purchase. Examples of product innovation may include new invention, specification and quantity enhancement, and addition of new features. These lead to higher demand which increases investment and employment. Sustaining products and services are also the kinds of innovations companies often need to develop in order to remain in the market.

Innovation in management and work organization enhance human and material resources together with the capacity to anticipate techniques. Management innovation simply changes how managers do what they do. This innovation is based on new principles that challenge management convention; it should be systemic, encompassing a range of processes and methods, and be part of an ongoing program of invention, where progress mounts over time. Management processes include setting goals and laying out plans; acquiring knowledge and discipline

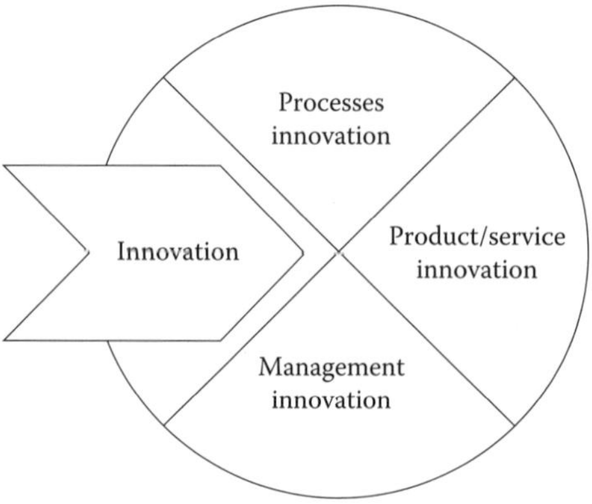

Figure 5.5 Forms of innovation.

to the chaotic process of scientific discovery; project and brand management that challenges management orthodoxy; budgeting, hiring, and promotion; employee assessment, executive development, and identifying and developing talent; and other areas that focus management principles into strategic practices.

Innovation can be characterized in a number of ways. In broad terms, it can be divided into the type of innovation that is technological (product and process) or nontechnological (corporate and marketing) (OECD 2010). Technological innovation involves the development of new technology, whereas corporate innovation encompasses innovation as a culture that permeates organizations.

Technological innovation involves three types of technological change: platform, component, and design. Platform innovation is the emergence of an entirely new technology based on scientific principles clearly different from those of the existing technologies. For example, the incandescent lamp used a new technology, electric resistance, to provide light, whereas the prior technology used combustion. Component innovation is one that uses new parts or materials within the same technological platform. For example, various types of halogen lamps such as tungsten quartz, parabolic aluminized reflector, and dichroic reflector lamp all depend on the incandescent principle but use different gases and components. Design innovation is a reconfiguration of the connections and layout of components within the original technological platform. For example, compact fluorescent lamps and circular fluorescent lamps are also design innovations based on the original principles of the fluorescent lamp that replaced the incandescent light bulb (Sood and Tellis 2005).

5.4.5 Benefits and risks of innovation

Successful innovations have many great benefits. Clark and Wheelwright (1993), for example, have summarized these as follows: improved return on investment, higher margins, expanded sales volume, increased value added, lower costs, and improved productivity. On the other hand, innovation is both risky and costly.

Risks include (Dodgson et al. 2008) market risks and uncertainty about demand, competitive risks, technological risks, organizational risks, operational risks and product delivery, financial risks and large upfront investments, and uncertain future pay-offs. These risks add up and lead to a high failure rate in new product development.

So why do firms bother to innovate? Following their discussion about the potential risks and rewards, Dodgson et al. (2008) conclude that it is more risky for a firm to choose not to innovate than to do so. Even though Kodak, for example, invented the first digital camera in 1975, it chose not to bring the technology to market for fear of cannibalization on its existing products (chose not to innovate). This decision effectively led to Kodak's downfall, when competitors such as Sony and Fujitsu eventually engineered digital cameras of their own to compete with Kodak's analog offerings (Christensen 2007).

5.4.6 Innovation process

The process of innovation goes through a number of stages, starting from laboratory inventions and ending with new products and processes appearing on the market. This process involves several stakeholders that enable the commercialization of innovation to occur. The major stages and actors involved in the innovation process are presented schematically in Figures 5.6 and 5.7 (UN 2012). While Figure 5.6 illustrates a traditional (linear) model of R&D to market, Figure 5.7 highlights an interactive or feedback approach to the corresponding processes.

The linear model (Figure 5.6) has been very influential. Academic institutions usually lobby for research funds and governments which support science use such a model. As a consequence, policies on science carried a linear conception of innovation for many decades, as well as academics studying science and technology. Still, very few people defend such an understanding of innovation anymore (Mowery 1993). The linear model, also called "technological determinism," is no longer supported by many academics but is still widely believed in general society. Research into the processes of scientific and technological discovery has shown that the linear model is not valid, disguising the role

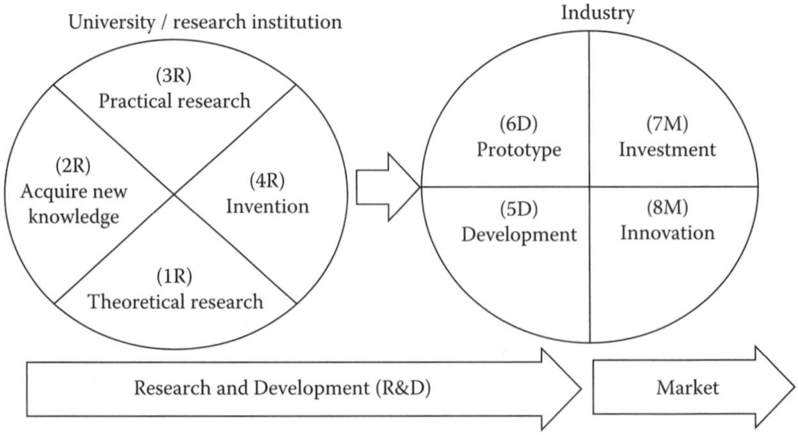

Figure 5.6 Traditional (linear) model of innovation.

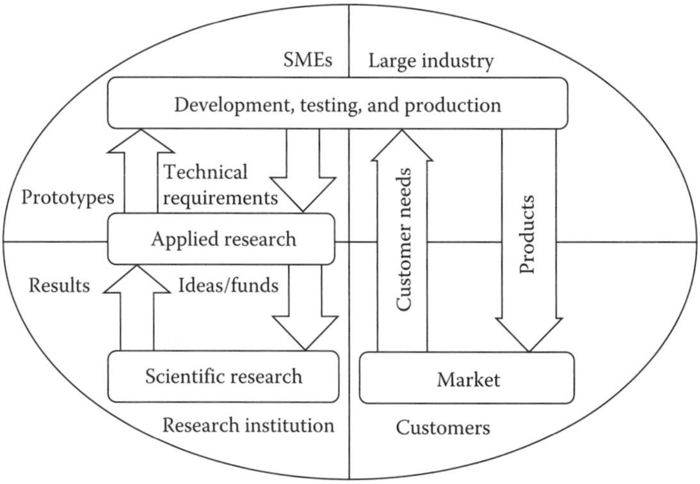

Figure 5.7 Innovative process: interaction of major actors and processes.

of human choice and values in shaping technology, as well as the social and economic interests guiding scientific inventions and technological innovations.

The nonlinear or interactive model (Figure 5.7) of innovation extends upon linear models by taking interactive terms into account. These nonlinear terms can be expected to change the causal relations between input and output. This nonlinear system can be expected to change the causal relations between input and output. The innovation process is influenced by many factors (institutional and organizational) which together can be called a system of innovation. The process is visualized as a chain, starting with the perception of a new market opportunity and/or a new invention based on novel scientific and/or technological knowledge, followed by the design for a new product, development, testing, redesign, and production, and distribution and marketing. Cooperation can take place with various combinations of internal and external actors. Under this model, technological innovation is seen to be the result of a complex interplay among various players. Technological progress is, thus, dependent on how the players interact with each other, internally and externally.

As an end process, R&D funding has been a mostly important factor in the commercialization of innovation. Today, research institutions and universities are able to manage their own intellectual property (IP) policies. They are patenting the results of their research and licensing patents to industry to commercialize the R&D results.

The evolution of innovation policies in industrialized countries has led to the emergence of a complex infrastructure of business support mechanisms. These range from the allocation of risk-free facilities, in which an entrepreneur can test a business idea, to technology transfer networks, subsidized operating premises, and venture capital funding. Most of such business support structures rely on public funding, but increasingly private for-profit and nonprofit services have also become available (UN 2012).

5.4.7 *Diffusion of innovation*

Diffusion of innovation is defined as the process by which innovations spread among users (Johnson et al. 2011). Rogers (2003) defines diffusion as "the process in which an innovation is communicated thorough certain channels over time among the members of a social system." As expressed in this definition, innovation, communication channels, time, and social system are the four key components of the diffusion of innovations. In fact, much diffusion research involves technological innovations, so Rogers (2003) usually used the words "technology" and "innovation" as synonyms. For Rogers, "a technology is a design for instrumental action that reduces the uncertainty in the cause–effect relationships involved in achieving a desired outcome." It is composed of two parts: hardware and software. While hardware is "the tool that embodies the technology in the form of a material or physical object," software is "the information base for the tool." Since software (as a technological innovation) has a low level of observability, its rate of adoption is quite slow.

As innovation typically is an expensive process, the pace of diffusion is often crucial to commercial success, and this may vary widely. A commonly used example to highlight how the pace of diffusion can vary is the TV versus the iPod. Whereas it took 37 years for the TV to sell 150 million units, the iPod reached the same number of units sold after just seven years on the market (Dodgson et al. 2008).

Rogers (2003) lists three methods to predict the rate of adoption for a forthcoming innovation. The first is to draw conclusions from past innovations which are similar in nature to the one in question. The second method is to describe the innovation to potential adopters and find out its perceived attributes, so as not to rely solely on the actual attributes. The third way is to actively investigate the acceptability of the innovation in prediffusion stages, such as test marketing or other forms of trials.

There are five important factors that decide the pace of innovation adoption on the market, on both the supply and demand sides. On the supply side, the following six product features have been identified as being important for the pace of diffusion (Rogers 2003; Sahin 2006;

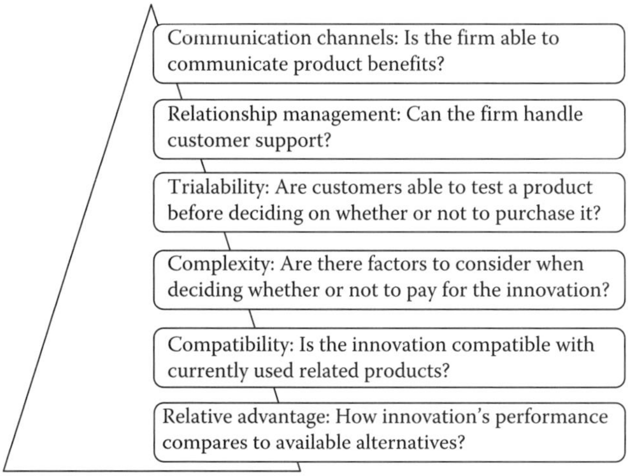

Figure 5.8 Six factors that decide the pace of innovation adoption on the market.

Dodgson et al. 2008, 66; Mohr et al. 2010, 76): relative advantage, compatibility, complexity, trialability, relationship management, and communication channels as described in Figure 5.8.

5.5 *Invention and innovation*

> To invent, you need a good imagination and a pile of junk.
>
> **Thomas A. Edison**

5.5.1 *Invention versus innovation*

An invention is a novel idea that proved workable in theory and has been transformed into reality for the first time and given a physical form. It is the discovery or development of a product or process by applying previous knowledge in new ways.

Invention comprises the process of discovery, from basic science through R&D. In general, it requires scientific skills. It may also include the repurposing of existing knowledge of technologies for new applications. The key distinguishing feature of invention is discovery and the creation of new knowledge that is made tangible and reproducible (Anadon et al. 2013). For example, Thomas Edison was an inventor. Inventions can be patented, as it provides security to the inventor, for IP rights.

On the other hand, an innovation should be distinguished from an invention or a discovery. It is the practical implementation of the new idea. An invention becomes an innovation only when it is capitalized on. An innovation has a broader definition than invention because it implies that this novel element has found, or will eventually find, a place on the market. It requires a set of technological and marketing skills. Innovation may be a development of new product, process, or service, or redesigning the existing ones to provide better solutions. In this aspect, Steve Jobs was an innovator.

5.5.2 *Path of innovation*

Invention covers all efforts aimed at creating new ideas and making them work. It often begins as prototypes in which the essential features are developed to see if they are feasible. These prototypes, or basic working models, are then improved until no other improvements can be made based on the prototype. Given that the process of innovation takes place over time, it is often not possible to be precise about the moment that an inventive idea becomes an invention. For example, Thomas Edison began to work on inventing the incandescent lamp powered by electricity in 1878. The moment at which the electric light became available on the market was the moment the invention became an innovation (Taylor 1996).

Innovation is really what drives economic growth. Joseph Schumpeter (Schumpeter 1934), who was a professor at Harvard University and was considered one of the twentieth century's major economists, said that innovation was the product of new combinations, and he proposed five combination patterns: (1) the production of a new good, (2) the introduction of a new method of production, (3) the development of a new market, (4) the acquisition of a new source of supply of raw materials, and (5) the emergence of a new organization of any industry.

Schumpeter (1934) believes that the concept of innovation, described as the use of an invention to create a new commercial product or service, is the key force in creating new demand and thus new wealth. Innovation creates new demand and entrepreneurs bring the innovations to market. This destroys existing markets and creates new ones, which will in turn be destroyed by even newer products or services. Schumpeter calls this process "creative destructions."

To sociologists, technological invention is a combination of prior art and ideas, and a complex of diverse elements: design, science, material, method, capital, skill, and management. It is a social process rather than an individual one. Certainly, without the inventor there can be no inventions, but the inventors are not the only individuals responsible for invention. Social forces such as demographic (race), geographic, and "cultural heritage" factors play a part. Second, technological invention is social in another sense: it is cumulative (or evolutionary), namely the

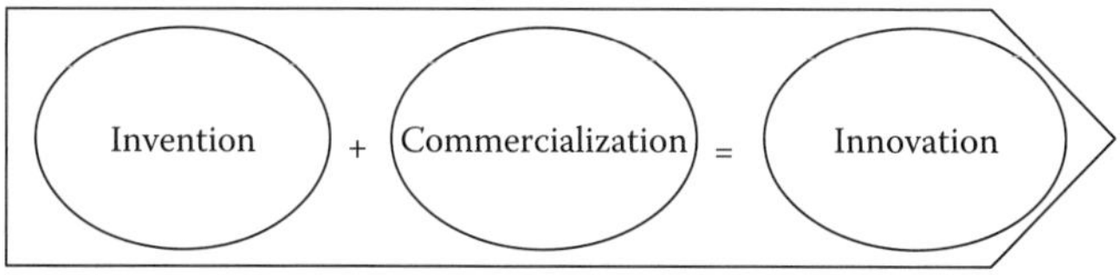

Figure 5.9 Integrating invention and commercialization to cultivate invention.

result of accumulation and accretion of minor details, modifications, perfections, and minute additions over centuries, rather than a one-step creation (Gilfillan 1935). Finally, technological invention is social in a third sense: it is increasingly systematic. It comes from organized research laboratories specifically dedicated to this end. Recalling industrialists' discourses of the time, sociologists observed a movement from the independent inventor to organized research in industrial laboratories (Hart 1931; Gilfillan 1935).

In brief, invention and innovation can be described by integrating them with commercialization as shown in Figure 5.9 (Gaynor 2014).

5.5.3 *Sources of capital*

Finding money to finance and perfecting a product, producing it, and getting it into the market concerns every inventor. Understanding the funding process requires peeling back several layers like peeling an onion. The visible outer layer consists of formal investment capital companies, including small business investment companies, the investment banking network, the stock market, and so forth. At the core of the onion lies the inventor supporting entity coming from the family income, contributing time, skills, and labor that build the "sweat equity" in the technology. Between these two extremes, the makeup of the intervening layers remains somewhat unclear (OEERE 2000).

The majority of funding for technology development projects in the phase between invention and innovation originates from angel investors, corporations, and government agencies not venture capitalists. Figure 5.10 shows the route of invention through valley of investments from which many never emerge.

If a combination of resources is put together, it will be possible to build an engineering or production prototype that works and if this is coupled to the appropriate components of a professional class plan then the product may emerge from the valley of investment on the magic wings of a licensee's technical and financial resources, or the powerful

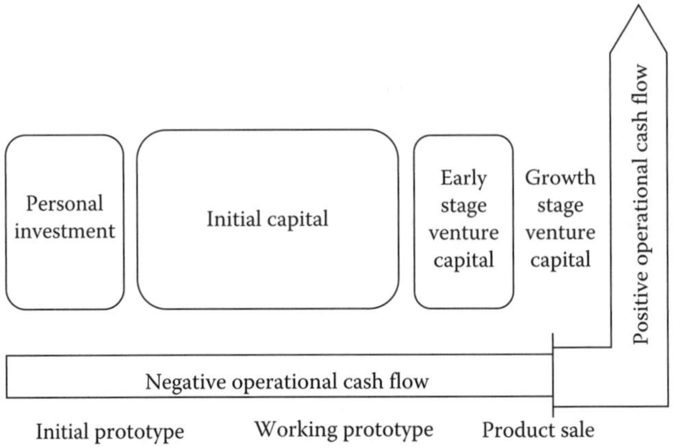

Figure 5.10 Route of invention through valley of investments.

thrust of professional investors' venture capital. All these things go hand in hand.

5.6 *Disruptive innovation*

> A disruptive innovation is a technologically simple innovation in the form of a product, service, or business model that takes root in a tier of the market that is unattractive to the established leaders in an industry.
>
> **Clayton Christensen**
> *Harvard Business School*

5.6.1 *Disruptive innovation versus disruptive technology*

> Uber, the world's largest taxi company, owns no vehicles. Facebook, the world's most popular media owner, creates no content. Alibaba, the most valuable retailer, has no inventory. Airbnb, the world's largest accommodation provider, owns no real estate. Something interesting is happening.
>
> **Visser (2016)**

The term disruptive innovation has been generally used as a replacement of disruptive technology. It is popular because market disruption is a function of not technology itself but rather of its evolving application. Sustaining innovation means innovations in technology, whereas disruptive innovations change entire markets.

The theory goes that a smaller company with fewer resources can unseat an established, successful business by targeting segments of the market that have been neglected by the incumbent, typically because it is focusing on more profitable areas. Disruption happens when the incumbent's mainstream customers start taking up the start-up's products or services in volume. Take Uber, for example, a company that is often referred to as a beacon of disruptive innovation because of its seismic impact on the taxicab industry (Hutt 2016).

Disruption theory differentiates disruptive innovations from what are called "sustaining innovations." The latter make good products better in the eyes of an incumbent's existing customers: the fifth blade in a razor, the clearer TV picture, better mobile phone reception. These improvements can be incremental advances or major breakthroughs, but they all enable firms to sell more products to their most profitable customers (Christensen et al. 2015).

5.6.2 *The technological challenge*

Historically, disruptive technological innovations have transformed societies from the horse-drawn work, to the first steam engine, to the car and airplane, to the mobile phone and the Internet. Today, the competitive pressure coming from society's need in combination with disruptive technologies means that innovation has become a driving power of economic existence. In addition, sustainability is another driver for disruption because the world needs to deal with various economic, environmental and social issues such as air and water pollution, climate change, aging population, and the need of an emerging well-to-do working class.

To set the scene, professor George Tovstiga mapped out the disruption challenge as a trichotomy: managing which implies deliberate and purposeful action; disruption which implies chaos, destruction, disorganization, and breaking up; innovation which implies unpredictability, serendipity, and novelty (Jopling 2015).

Real disruptive technologies are exciting and challenging the norm. Many retail and services have changed because of the Internet and mobile applications. Services like Uber have shaken the industry and left traditional companies and governments wondering about regulations (Gredig 2017). The parade of such new technologies and scientific breakthroughs is relentless and unfolding on many fronts. Policymakers and societies need to prepare for future technology. To do this well, they will need a clear understanding of how technology might shape the global economy and society over the coming decade. They will need to decide on how to invest in new forms of education and infrastructure, and figure out how disruptive economic change will affect comparative advantages. Governments will need to create an environment in which citizens can continue to prosper, even as emerging technologies disrupt their lives (Manyika et al. 2013).

Technologies coevolve with societies (Saviotti 2005); technological developments influence society and vice versa. The questions about who makes decisions about the development and direction of new technologies have seldom been asked and even less often answered. In academic circles during the 1960s and 1970s, questions were increasingly voiced about wanted and unwanted consequences, both foreseen and unforeseen, and the direction and steering of new technologies in science, technology, and society studies, technological forecasting, technology assessment, technology policy, and appropriate technology.

5.6.3 *The 12 potentially disruptive technologies*

The term disruptive technology was coined in 1995 by Joseph L. Bower and Clayton M. Christensen to describe the phenomena of entrenched commercial technology being replaced by new technology (Bower and Christensen 1995). It refers to a new technology having lower cost but having higher performance. On the other hand, a disruptive innovation relies on the above technologies to create a new market and value network and eventually disrupts an existing market and value network.

Important technologies can emerge in any field or from any scientific discipline, but they share four characteristics: high rate of technology change, broad potential scope of impact, large economic value that could be affected, and substantial potential for disruptive economic impact. Many technologies have the potential to eventually meet these criteria, but leaders need to focus on technologies with potential impact that is near enough at hand to be meaningfully anticipated and prepared for (Manyika et al. 2013). Today, the focus is on technologies that will have significant potential to drive economic impact and disruption by 2025. Table 5.5 outlines the 12 potentially economically disruptive technologies that may meet that vision.

The 12 technologies in Table 5.5 may not represent all potential disruptive technologies by 2025. Developing technologies may remain uneconomical on average, even as leading innovators approach breakthroughs. However, history shows that innovations in technology can cause dramatic increases in productivity and a decrease in cost, transforming industries and setting whole societies on new paths to growth. Usually, technologies that have fast declining cost curves are developed very quickly.

Table 5.5 List of the 12 innovative disruptive technologies

Technology	Details
Mobile internet	Mobile Internet refers to the provision of Internet connections using mobile phone networks
Automation of knowledge work	Intelligent software systems that can perform knowledge work tasks involving unstructured commands and subtle judgments
Internet of things	The internetworking of physical devices and objects via the Internet with the ability to function together or independently
Cloud technology	Delivery of hosted services over the Internet. It enables companies to use digital resources as a utility just like electricity rather than having to build and maintain computing infrastructures in-house
Advanced robotics	Robots with advanced sensing technologies to automate various tasks or augment humans
Autonomous vehicles	Vehicles that can navigate and operate with reduced or no human intervention
Genomics	The study of entire DNA content found in living things, including humans, plants, animals, and even viruses
Energy storage	Efficient systems that store energy for future use. Advances in energy storage technology could make hybrids, plug-in hybrids, and all-electrics vehicles cost-effective
3D printing	A process of making three-dimensional solid objects from a digital file. It enables on-demand prototyping and production, which has interesting implications for supply chains and for stocking spare parts
Advanced materials	Materials with superior properties such as toughness, hardness, durability, and elasticity. For example, graphene and nanotechnology could help create new types of applications from superefficient batteries to cancer treatment to water filtration
Oil and gas exploration and recovery	Exploration and recovery techniques that make extraction of unconventional oil and gas cleaner and economical
Renewable energy	Generation of electricity from renewable sources with reduced harmful climate impact

5.6.4 The involved risk

Disruptive innovations can hit the competitive environment in many forms, such as a revolutionary business model, a completely new technology, or a new spin on an existing product or service. These types of innovations have the potential to upset mature organizations and alter the face of entire industries (ERM 2012). This upset presents a challenge for both the new frontier of disruptive innovation and risk managers of traditional firms to protect existing markets.

Adopting disruptive technologies entails risks, and managing these risks will be critically important. Internally, organizational effectiveness and cohesion could suffer as some jobs are transformed or eliminated by technology. External risks include reputational risk and consumer resistance, as well as safety and regulatory issues. For example, new materials may have unforeseen health effects and may pose environmental risks (Manyika et al. 2013). For example, autonomous vehicles (see Section 5.11) as disruptive technology might not deliver the potential expected impact unless the safety of driverless vehicles is established, consumers accept the idea, and regulators come up with the necessary rules and standards to put these cars and trucks on the road. Business leaders need to strike a careful balance as they adopt new technologies; they must be thoughtful about risk, but they should also manage these risks without stifling potential.

Finally, disruptive innovation can be risky also because it needs embracing a very different approach to product development and marketing. Risk-taking companies should realize the potential of a disruptive technology and attempt to explore ways to integrate it into its business. A failure to realize the effects of emerging disruptive technologies may lead to losing market opportunities to competitors that have found ways to integrate these technologies into their business.

5.7 *Habits of mind*

> Imagination is more important than knowledge. For knowledge is limited to all we know and understand, while imagination embraces the entire world, and all there ever will be to know and understand.
>
> **Albert Einstein**

5.7.1 *The 16 habits of mind*

Mind-sets in general are important to understand, and cultivating habits within those mind-sets is extremely helpful. Habits of mind (HoM) is an expression used by psychologists including Resnick (1999) to describe aspects of intelligence. The term has been adopted in the United States by educationalists Costa and Kallick (2002) who suggested how the role of teachers might change if they were deliberately trying to encourage the kinds of HoM mentioned by Resnick. They came up with 16 HoM which, taken together, describe what smart people do as they go about their lives successfully dealing with whatever unexpected problems are thrown at them.

HoM means having a character toward behaving logically when challenged with problems; the answers to which are not immediately known. HoM are traits or ways of thinking that affect how a person looks at the world or reacts to a challenge. When humans experience contrasts, are puzzled by dilemmas, or come face to face with uncertainties, the most effective actions require drawing forth certain patterns of intellectual behavior. HoM are a set of 16 problem-solving, life-related skills necessary to effectively operate in society and promote strategic reasoning, depth, determination, creativity, and craftsmanship, as shown in Figure 5.11. The understanding and application of these 16 HoM serve to provide the individual with skills to work through real-life situations that equip an individual to respond using awareness, thought, and intentional strategy in order to gain a positive outcome. There are HoM of math, science, engineering, and art; however, we are discussing engineering HoM (EHoM) in Section 5.7.2.

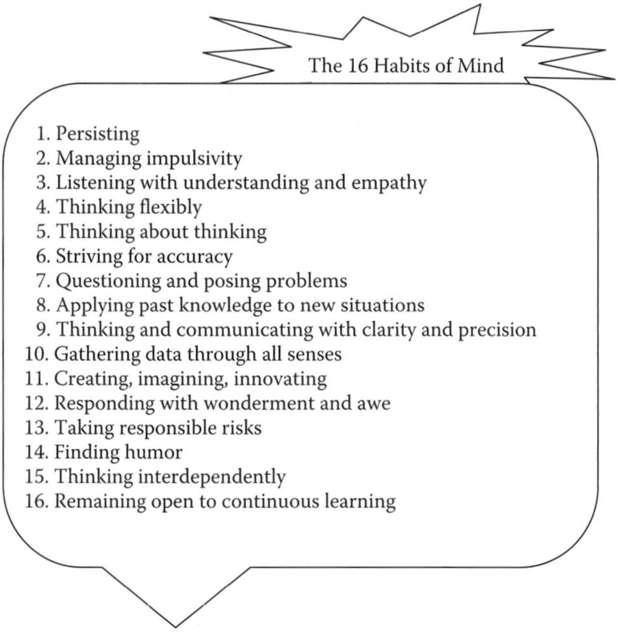

Figure 5.11 The 16 HoM.

5.7.2 Engineering HoM

A major review of engineering education within K-12 primary and secondary education has called for curriculum development to be underpinned by the promotion of six HoM (Katehi et al. 2009) to be called EHoM. These six are systems thinking, creativity, optimism, collaboration, communication, and attention to ethical considerations as shown in Figure 5.12. Australia also supports the idea that engineers need to be lifelong learners and that current educational approaches need to change (Beder 1999). EHoM provides engineers with opportunity to come up with solutions to problems or improvements to current technologies or ways of doing things. These six HoM are so encouraged, even rewarded by engineering experiences, that, over time, they become part of an engineer's everyday thinking. However, these "ways of thinking" are not exclusive to engineering. At the heart of the model is the idea that drives engineers to make things that work or work better. Engineers often engage in activities which may not involve making things. However, engineers such as software engineers, who do not usually develop physical products as such, are involved in the subelements of making such as designing and implementing.

When engineers develop EHoM, they are not just learning how to pass a test, but are learning to make meaning out of the world around them. The first EHoM, systems thinking, refers to the process of considering how each part under study relates to one other within the context of the whole. It involves the ability to recognize interconnections in the technological world (Katehi et al. 2009). On the other hand, engineers research and explain their ideas to others to gather outside input. Then they test their ideas through the creation of a prototype. Throughout the design process, professional communication and citations protocols are followed. At each phase of the project development,

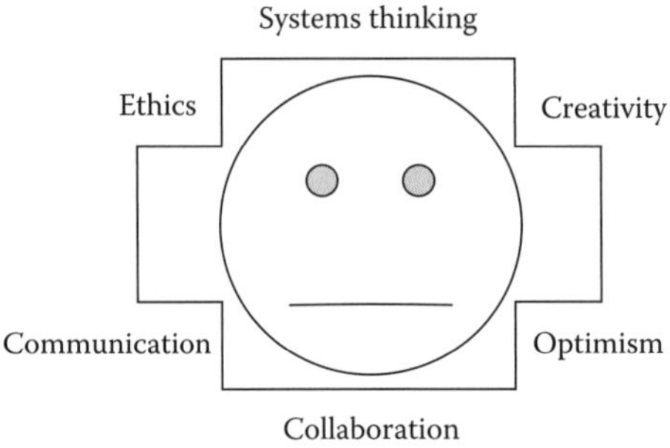

Figure 5.12 The six engineering HoM.

engineers consult, collaborate, and utilize literature reviews in order to inspire their own creativity and improve their design, a system thinking approach.

Creativity means being able to look at the big picture and identify new patterns or relationships or imagine new ways of doing things. The design process in and of itself exemplifies creativity. Finding new ways to apply knowledge and experience is essential in engineering design and is key elements of innovation.

Optimism is defined as a world view in which possibilities and opportunities can be found in every challenge and an understanding that technology can be improved (Katehi et al. 2009). Optimism means engineers should believe that things can always be improved. Just because it has not been done yet does not mean that it cannot be done. Good ideas can come from anywhere and engineering is based on the premise that everyone is capable of designing something new or different. Engineers should have a growth mind-set; believing that they can improve is an active element of being able to improve.

Collaboration in the engineering educational environment is heralded in how it leverages the perspectives, knowledge, and capabilities of team members to address a design challenge (Katehi et al. 2009). Collaboration means engineering success is built through cooperation and communication. Teamwork is essential. The best engineers are willing to work with others. They are skilled at listening to stakeholders, thinking independently, and then sharing ideas.

Communication means creating interactive approaches to document and clearly express ideas and concepts to users and learners to various audiences. Communication includes various components, such as oral, written, listening, visual, intercultural, and interdisciplinary. In engineering, communication between disciplines and beyond is very important aspect that requires attention.

Finally, in regard to ethical considerations, instead of general conversations about ethics in the outside world, ethics became personal and real for children working in the Ramps and Pathways center. For young children designing ramp structures, ethics demanded respecting each other's safety, ideas, materials, and space (Van Meeteren and Zan 2010).

5.8 Engineering innovation domain

> At its heart, engineering is about using science to find creative, practical solutions. It is a noble profession.
>
> **Queen Elizabeth II**

5.8.1 *Innovation challenges*

Engineering is a profoundly creative activity. It requires innovation and talent focused in a design process. Thomas Edison created electric light, Nikola Tesla created the ac induction motor, Alexander Graham Bell created telephony, Ove Arup created the Sydney Opera House, and Tim Berners-Lee created the World Wide Web. These great engineers all had a few characteristics in common. They were amazingly creative minded. They brought ideas to life; they were creative problem solvers.

Innovation in engineering means turning ideas into value. Most engineering projects demand creative or innovative approaches in the design of systems and services. It comes from an understanding of basic knowledge coupled with real-world experience, and it requires the right environment in which to operate and flourish. However, innovation by itself is nonetheless not sufficient to assure a successful product. If the product cannot be developed and manufactured at a reasonable cost, it will not be competitive on the market (Malmquist 2014).

What does this mean for innovation today? Innovation today is about open minds and transdisciplinary practices. Many of the problems contemporary society faces are new and complex. As a society, we have never before faced a challenge quite like capacity to produce and share so much data about the world, lives, and finances; human population that need food, water, shelter, education, employment, health care, and security. Vulnerabilities to epidemic diseases, terrorist attacks, and natural disasters require serious explorations for new techniques and approaches of protection and prevention. In each of these wide domains of human need and concern, sustainability, health, vulnerability, and delight of lifestyle, outstanding challenges anticipate engineering solutions. In such a challenging situation, engineers must constantly innovate to create solutions and invent new ways of solving problems.

All of these examples merely scratch the surface of the challenges that engineers will face in the twenty-first century. The problems described above merely illustrate the magnitude and complexity of the tasks that must be mastered to ensure the sustainability of civilization and the health of its citizens, while reducing individual and societal vulnerabilities and enhancing the joy of living in the modern world. None of these challenges will be met, however, without finding innovative ways to overcome the barriers that block their accomplishment (NAE 2008).

Although the innovation process goes well beyond engineering disciplines, it is fair to say that engineers are at the heart of most of the innovation the world is seeking. Innovation in engineering is much more than R&D. It includes an end-to-end process, such that it extracts value through implementation. The process that brings about new products and processes includes the discovery of new knowledge and inventions. The skills needed for this process are more difficult to reduce to traditional approaches and seem to require engagement with projects involving innovation and the corresponding uncertainties.

5.8.2 Engineering innovativeness

Engineering creativity is often understated and unrecognized, even by engineers themselves. Engineers depend on established science and techniques for analysis to ensure that their designs are safe and reliable. Creativity is usually associated with forward thinking, challenging, alternative seeking, and risk taking, while engineers are required to reduce risks to the public, environment, and the success of their clients and employers. However, finding innovative solutions within the boundaries of curiosity, safety, reliability, and efficiency requires considerable creativity (Lawlor 2013).

Researchers analyzing the causes of innovative behavior by engineers and entrepreneurs potentially attribute engineer innovativeness, or the level of engineer innovativeness, to several different conditions or factors. First, education and the acquisition of domain expertise are seen as crucial innovation skill factors (Andersen 2008). Second, self-efficacy, the desire and an individual's judgment of his or her ability to perform a task, strongly influences motivation and outcome expectancy during an engineering design process (Carberry 2010). Third, an individual's mind-set is a personality characteristic that influences creativity, innovation, and the willingness to take risks (Dweck 2006). Fourth, prior experience is also viewed as a key factor in innovativeness. Fifth, individuals who have created more than one new business (e.g., serial entrepreneurs) or who have worked in an industry or process for a long enough time to have developed human capital in that domain are believed to be more likely to be innovative (Pena 2002). Sixth, community influence on the production of innovations is seen as key to the generation of innovations (Rustam 2001). Finally, personality is believed to influence innovation creation (Willy and Kolvereid 2005).

Figure 5.13 shows the conceptual relationships between the major intrinsic factors and extrinsic factors of engineering innovativeness. Problem-solving processes and creative processes are similar cognitive processes and presumably require a similar set of abilities. Creativity stems from the need to solve problems. Novel ideas arise while coming up with new associations between memories, sensory input, map mental models, and memories to what is perceived. This process is much like what children do in their imaginary play games. In this way, creativity is seen as a subset, if not entirely synonymous with problem-solving (Kirton 1976).

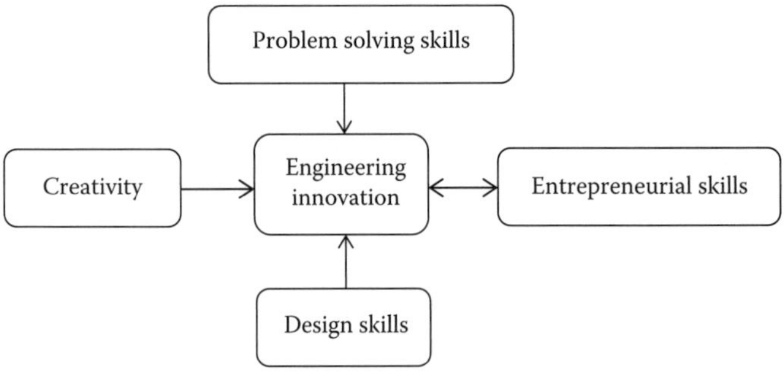

Figure 5.13 The engineering innovation domain.

Design problems challenge the engineer like no other problem because they require the engineer to exhibit skills that provide the problem structure, solution alternatives, and evaluation criteria, and to remain open to changing the proposed problem solution as new information becomes available. Creativity in solving design problems is recognized as an essential part of the engineering design process (Thompson and Lordan 1999).

The key reason for including entrepreneurial behavior within the scope of engineering innovativeness is that societal expectations for engineers are that the innovations resulting from the practice of engineering innovativeness will be implemented to benefit society. Innovative engineers therefore need to be successful entrepreneurs themselves or be able to partner with entrepreneurs to implement their new useful domain changing products, processes, or concepts (Ferguson and Ohland 2012).

5.8.3 *Engineering for integrated innovation*

Innovation is crucial for social cohesion, equality, and poverty alleviation (Dutta 2012). Innovation tends to emerge at the edges, at the boundaries between domains. Much creativity consists of a new combination of existing ideas. Where the existing ideas are present in different people, it requires some kind of interaction to produce the combination (Langrish 1985).

Integrated innovation is the coordinated application of scientific/technological, social, and business innovations to develop solutions to complex problems. This approach does not discount the singular benefits of each of these types of innovation alone, but rather highlights the powerful synergies that can be realized by aligning all three aspects to address a single challenge (Grand Challenges Canada 2010). In applying an integrated innovation approach to a complex real-world challenge, it is useful to work through three decision points: scientific/technological innovation, social innovation, and business innovation.

With respect to science and technology, serious engineering breakthroughs will be required to make innovation a reality. Science, technology, and innovation (STI) drives economic success and fuels advances that improve societal well-being. A good functioning STI ecosystem needs to include political stability and efficient institutions, reliable education infrastructure, enterprises committed to R&D, as well as a well-adjusted IP rights framework. STI is more than R&D; it generally refers to changing or creating more effective processes, products, and ideas.

Social innovation (in the context of integrated innovation) can be thought of as R&D into the ways to bring innovation to scale in specific local and regional contexts. Social innovations can include the creation and implementation of new approaches in the context of health systems, the determinants of health, ethical/social/cultural/legal frameworks, PP, leadership, human resources, and other key components of society that influence health outcomes. Goldenberg (2004) defined social innovation as the development and application of new or improved activities, initiatives, services, processes, or products designed to address social and economic challenges faced by individuals and communities. Social innovation is a worldwide phenomenon fueled by globalization and the rise of the knowledge-based economy, itself fueled by STI. At the same time, there has been an increased global awareness of complex and often intractable social problems, ranging from environmental issues to growing levels of poverty around the world and increasing socioeconomic disparities within and between countries. These problems, in turn, have highlighted the need for new and innovative approaches to address these social concerns, energizing what is becoming recognized as social innovation (Goldenberg et al. 2009). Social innovation is a novel solution to a social problem that is more effective, efficient, sustainable, and for which the value created accrues primarily to society as a whole rather than private individuals (Phills et al. 2008).

The terms "social enterprise," "social entrepreneurship," and "social finance" are often used interchangeably with "social innovation." It is clear, however, that any sophisticated understanding of how novelty transforms complex systems requires great conceptual precision (Wesley 2010). Whereas the concept of social enterprise is primary focused on organizational form and mission, social entrepreneurship is a human-centered concept that highlights the personal qualities of a person who starts a new organization (Phills et al. 2008). More definitively, social innovation is oriented toward making a change at the systemic level.

Business innovation focuses on the delivery of appropriate, high-quality goods and services where and when they are needed at an affordable price point. In practice, there will always be trade-offs between the functionality, usability, and affordability of products. Although most innovation in high-income countries focuses on the first of these three dimensions, STI can also lead to significant improvements in affordability and usability which can be as important, if not more, as drivers of global health impacts than the creation of new functionality.

5.9 Green innovation

> Great ideas rarely start great; they need to brew for a while.
>
> **Linda Gorchels**

5.9.1 Concept and topology

The concept of GI is not only about those sectors typically branded as green or clean, like renewable energy, but it is much wider, including a broad range of technical, organizational, and business innovations.

GI is a very recent as well as a very complex one. Different concepts which are mostly used interchangeably with GI are used in the literature. GI refers to innovations that are applied in products and processes that take the industry to lead to higher levels of the environmental sustainability. To these belong the concepts of eco-innovation, environmental innovation, innovation for SD, sustainable innovation, sustainable manufacturing, or clean technology (OECD 2008, 2009).

There exists no universal definition of GI, but it can generally be defined as an innovation resulting in a reduction of environmental impact, regardless of whether this impact is intended or not (OECD 2009). Kemp and Pearson (2007) define GI as the production, assimilation, or exploitation of a product, production process, service, management, or business method that is novel to the organization (developing or adopting it) and which results, throughout its life cycle, in a reduction of environmental risk, pollution, and other negative impacts of resources use (including energy use) compared to relevant alternatives.

According to Chen et al. (2006), GI is defined as hardware or software innovation that is related to green products or processes, including the innovation in technologies that are involved in energy saving, pollution prevention, waste recycling, green product design, or corporate competitive management. All these differ in some details, but at the same time all of them draw on two important strands of the eco-innovation concept: first, the innovative nature and second, the environmental compatibility.

GI incorporates technological improvements that save energy, prevent pollution, or enable waste recycling and can include green product design and corporate environmental management. This type of innovation also contributes to business sustainability because it potentially has a positive effect on a firm's financial, social, and environmental outcomes (Aguilera-Caracuel and Ortiz-de-Mandojana 2013). GI may be classified into two major items including products and processes, and organization and marketing innovation. Product innovation means product characteristics, process innovation means production methods, organizational innovation means business practice, and marketing innovation means product design and packaging. Figure 5.14 shows the typology of GIs analyzed by means of the dimensions target, mechanism, and impact which the OECD (2008) definition identified to be crucial for the classification of eco-innovations. It can be seen that the bigger the change the eco-innovation consists of, the higher the potential for environmental benefits. Further, those innovations that have only recently been acknowledged as eco-innovations-organizational and institutional ones possess a higher potential for environmental benefits than such innovations targeted at conventional levels.

According to the EIO (2012), sectors focusing on energy, transport, chemicals, bio-based products, waste management, and information and communication technology (ICT) can be put in the GI category. Their common characteristic is that they possess a high growth potential as active firms in these sectors turn to radical, more comprehensive

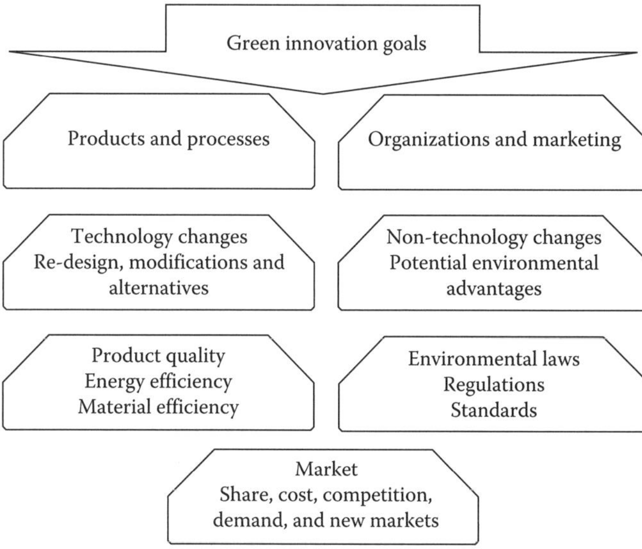

Figure 5.14 GI topology.

solutions such as innovation in products, processes, and systems that aim to increase resource productivity and can even include collaboration across sectors.

5.9.2 Green practices

Adopting green practices can be seen as a technological innovation process (Lin and Ho 2011). Innovation consists of any practice that is new to organizations, including equipment, products, processes, policies, and projects. Technological innovation pertains to products, services, and production technologies; it is related to basic activities and concerned with either product or process (Damanpour 1991).

Innovation for green growth can be characterized as frontier, adaptive, and absorptive as shown in Figure 5.15. Frontier innovations are novel solutions that have not yet been introduced to the world. They are typically adopted in the research phase of the technology development cycle. Adaptive innovations are modifications to existing technology that make them more useful in alternative conditions. They can occur across the technology development cycle. On the other hand, absorptive innovation refers to changes to an institutional environment that makes the transfer, successful implementation of, and learning from frontier and adaptive innovations easier. This applies to the final two stages of the development cycle. Examples of this type of innovation include infrastructure for knowledge and device diffusion, regulations to support IP protection, and international agreements for technology transfer (Hultman et al. 2012).

Several technological characteristics of an innovation can affect its adoption, including complexity, compatibility, relative advantage, triability, observability, ease of use, perceived usefulness, information intensity, uncertainty, and so on (Tornatzky and Klein 1982). The perceived technological characteristics of an innovation are considered as cognitive beliefs reflected in an attitude toward the innovation (Weng and Lin 2011).

Green technologies have the potential to significantly improve environmental performance relative to other technologies. To achieve this goal, green technology (GT) innovations should become integral part of

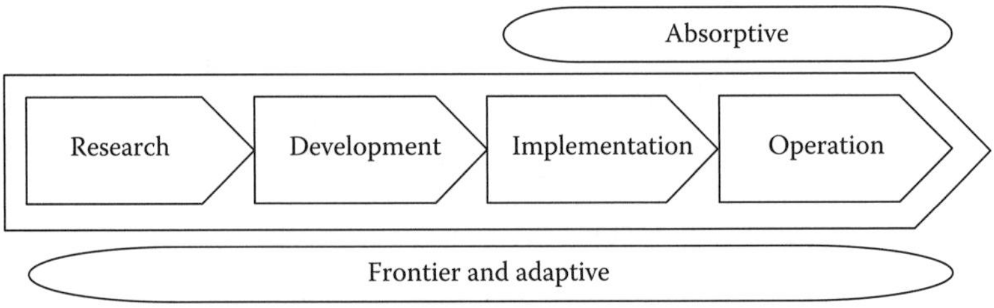

Figure 5.15 GI characteristics.

sectoral development strategies. Integration of innovation and environmental policies requires closer coordination between government agencies in charge of innovation, industry, and environmental protection agencies.

Much is expected from core and connected eco-industries to lead the way to a green economy. In order for these industries to succeed, much technological and nontechnological progress is necessary to increase the incentives for consumers and firms to eco-innovate. The incentives to eco-innovate are determined by a wide range of factors. Thus, in order to identify the main determinants of GIs one has to start at the very basics.

The contribution of innovation economics includes the role of technological development and of demand factors. What is missing is the third dimension, a regulatory framework which is crucial incentives for firms to eco-innovate. This is the contribution of environmental economics which mainly focuses on identifying optimal policy instruments to incentivize eco-innovations and ignores any market pull or technology push factors (Rennings 2000).

5.10 *Educating creativity and innovation*

> The role of the teacher is to create the conditions for invention rather than provide ready-made knowledge.
>
> **Seymour Papert**

5.10.1 *Can creativity and innovation be taught?*

Creativity and innovation are keys to a thriving economy. Why is it so difficult for us to create an environment where creativity and innovation flourish? One halt to such goal is the availability and access to education.

Numerous debates take place over the question: Can creativity and innovation be taught? Some people believe that creativity and innovation cannot be taught and is instead embedded in a person. Others are more passionate and think that they can, in fact, be taught. Creativity and innovation are cultures and almost everyone can learn and adapt to. In an academic institution, it is relatively easy to establish that culture. It is starting by engaging a fresh minded people with talents, ambitions, and motivation. It is hard to make sure that culture continues, but it is possible if the education system can provide the environment to maintain that culture, starting from blending thought processes and ending with engaging in projects. Such process is similar to critical thinking in that it is possible to train a mind in its application.

Teaching creativity and innovation cannot happen at once. It is not just about teaching design thinking for product development from customer needs to ideas to prototypes to customer validation tests as is described in Chapter 9, or teaching experimentation on established procedures and methods, or coaching teams on entrepreneurial mind-set. It is all of the above. It is about establishing a culture that involves every learner and giving the opportunity to think about innovation outcomes that are more disruptive or radical than incremental change. As an outcome, learners will understand that innovation is a holistic implementation of various elements of the culture including skills, processes, leadership, and motivation. In addition to all the above elements, innovation requires from learners to release their fear, uncertainty, and doubt while leaving the comfort zone.

Students and engineers can innovate, but they need to acquire innovation skills if they are to be involved in creating and managing innovation. These skills can be taught but only if people want to learn them and perceive value in them. The education system can accelerate the process by making their innovation activities more productive and efficient through tools and techniques, skills and HoM, and removing fear and uncertainty. Most experts agree that there are no ready-made formulas for how to innovate. But is it possible to create the suitable environment to filter ideas and execute plans, and accordingly to facilitate creativity under which innovation may thrive? Without learning, many people are generally critical of new ideas and innovative solutions. Rather than build upon promising but imperfect ideas, they just decline possibilities for innovative act.

Creative work requires applying and balancing three abilities that can all be developed (Sternberg 1985, 2003; Sternberg and Lubart 1995; Sternberg and Williams 1996). These abilities are synthetic, analytic, and practical. Synthetic comprises the skill to generate novel and exciting ideas. It is the ability to establish connections between things that other people do not recognize spontaneously. Analytic is the ability to analyze

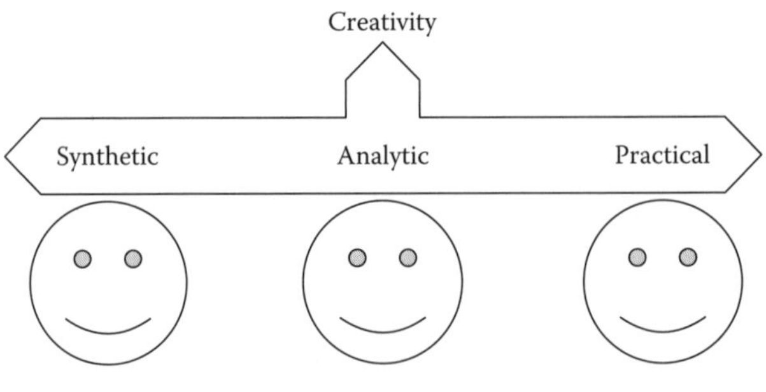

Figure 5.16 Balance among synthetic, analytic, and practical thinking.

and evaluate ideas in order to work out the implications of a creative idea. Practical is the ability to translate theory and abstract ideas into practical activities. These three abilities might not be available in one person. Accordingly, teaming people of different abilities and talents is critical in developing innovative ideas.

In the classroom environment, it is the mandate of the teacher to find a balance among synthetic, analytic, and practical thinking. A creative attitude is at least as important as creative thinking skills (Schank 1988). Figure 5.16 reflects the outcome of this balance.

5.10.2 How to develop creativity in the classroom

Based on the investment theory of creativity (Sternberg and Lubart 1995), engineering instructors can adopt a few strategies to enhance creativity. The strategies are outlined in Table 5.6.

Table 5.6 Strategies to enhance creativity in engineering education

Strategy	Details
Model creativity	Teaching for creativity requires teachers not only to support and encourage creativity, but also to role-model it and reward it when it is demonstrated. Teachers need not only to talk the talk, but also walk the walk.
Build self-efficacy	Self-efficacy is the belief in individuals' capabilities to achieve a goal. Students with a strong sense of efficacy are more likely to challenge themselves with difficult tasks and be naturally motivated. In this regard, teachers should help students believe in their own ability to be creative.
Question assumptions	Teachers generally tend to make a pedagogical mistake by emphasizing the answering and not the asking of questions. However, it is needed to teach students how to ask the right and interesting questions.
Define and redefine problems	Teachers need to promote creative performance by encouraging students to clearly feel, define, and redefine problems. Redefining a problem also means taking a problem that most people see it in one way and urging to see it in another way. It is good to encourage creative thinking by having students choose their own topics for projects, case studies, articles, or presentations; choose their own ways of solving problems; and sometimes choose again if they discover that their selection was a mistake.

(*Continued*)

Table 5.6 (Continued) Strategies to enhance creativity in engineering education

Encourage idea generation	Teachers should encourage students to generate a large number of ideas and hypothesis. The environment for generating ideas must be relatively free of criticism. Students should be praised for generating many ideas, regardless of whether some are unrelated.
Cross-fertilize ideas	Teachers should integrate transdisciplinary components into their teaching, especially in projects, case studies, and presentations to stimulate creativity by helping students to think across subjects and disciplines.
Allow time for creative thinking	Being creative takes time (Gruber and Wallace 1999). Students need to learn to allow time for incubation, reflection, and selection among alternative ideas. If they always rush, or are rushed, they will have difficulty in producing creative work.
Instruct and assess creativity	Teachers need to include some opportunities for creative thought in assignments and tests. Ask questions that require factual recall, analytic and creative thinking.
Reward creative ideas and products	Reward creative efforts; for example, a teacher may assign a project or a task and remind students that the goal is to demonstrate creativity and innovation.
Encourage sensible risks	Often, education system encourages students to play it safe. On tests safe answers are expected. When papers or reports are written, the professor's expectation is assumed. But creative people always are ones who are willing to risk something and, in the process, fail some of the time in order to succeed other times. Teachers need to encourage such risk taking.
Tolerate ambiguity	While trying creative things, it is often found that things in their early or even sometimes late stages do not work out the way they seemingly should. Yet, in order to be creative a tolerance of ambiguity is needed long enough to get ideas right.
Allow mistakes	People learn from their mistakes. However, if students become afraid to make mistakes, they will have trouble in being creative.
Identify and surmount obstacles	Because creative people "defy the conventional," they inevitably confront obstacles. The question is not whether they will confront obstacles, but whether they will have the courage to surmount them.
Teach self-responsibility	Part of teaching responsibility is to make students take responsibility to understand their creative process, criticize themselves, and take pride in their best creative work.

(*Continued*)

Table 5.6 (Continued) Strategies to enhance creativity in engineering education

Promote self-regulation	The self-regulated conceptual learning model means planning and designing, identifying priorities and allocating resources, self-monitoring, evaluating, and controlling.
Delay gratification	Part of being creative means being able to work on a project or task for a long time without immediate or interim rewards. Students must learn that rewards are not always immediate and that there are benefits to delaying gratification.
Use profiles of creative people	Teachers should encourage students to select great role models and learn from the great ones.
Encourage creative collaboration	Collaboration can prompt creativity. Teachers should encourage students to collaborate with creative people because we all learn by example. Project-based learning (PBL) is a method of exposing students to collaboration.
Imagine other viewpoints	An essential aspect of working with other people and getting the most out of collaborative creative activity is to imagine selves in other people's shoes. Broaden perspective by learning to see the world from a different point of view, and that experience enhances creative thinking and contributions (Sternberg and Beyond 1985; Sternberg 1997, 2005).
Recognize environmental fit	The very same product that is rewarded as creative in at one time or place may be scorned in another.
Find excitement	To unleash students' best creative performances, teachers must help students find what excites them. People who truly excel in a pursuit, almost always genuinely love what they do. Certainly the most creative people are intrinsically motivated in their work (Amabile 1989).
Seek stimulating environments	Teachers should help students develop the ability to choose environments that stimulate their creativity.
Play to strengths	Teachers should show students how to play to their strengths—by helping to identify the exact nature of their talents. Flexibility in assignments and a willingness to help reluctant students determine the nature of their interests and strengths are required.
Grow creatively	Once there is a major creative idea, it is easy to spend the rest of one's career following up on it.
Preach for creativity	Once teachers have mastered a few of these techniques to develop creativity and made them part of their daily teaching routine, spread the word.

5.10.3 Think outside of the box

The value that creativity and innovation offer to engineering and engineering education lies in their ability to facilitate the development of novel and effective technological solutions to problems stimulated by change. There is, however, disconnect between creativity, innovation, and engineering. Educational programs focus excessively on deep and narrow technical specifications, with little or no room in the curriculum for developing the ability to think and act creatively. Unless this disconnect is addressed through holistic changes to engineering education, we risk producing engineers who are ill-equipped to tackle the problems sparked by increasingly rapid changes in society (Cropley 2015).

Conventional education systems do not offer adequate incentives and encouragement for students to develop their creative skills. Some attributes of creative students often frustrate those teachers who do not know how to recognize them. Deliberate programs need to be introduced for students to develop their creative skills (Griffith University 2004). The implications of the concept of value innovation for education are clear: students should be encouraged to "think outside of the box" and define their own creative solutions to real-life problems, as posed by a problem or PBL approach.

Academic excellence (at least in engineering) is synonymous with skill at convergent (critical thinking) production, since engineering education (unlike engineering practice and life in general) normally involves only problems with single correct answers. On the other hand, both convergent and divergent (creative) productions are required to solve serious technological problems. The purely convergent thinker is not likely to come up with the innovative solution required when conventional approaches fail, while the purely divergent thinker will generate a great many innovative ideas but may lack both the analytical ability to carry them through to their final form and the evaluative ability to discriminate between good and bad solutions. If engineering educators cannot find enough individuals who combine these abilities, at the very least they should be turning out some who excel at one and some who excel at the other. To do this, it is necessary to provide instruction and practice in both modes of thinking. In this respect, we are failing abysmally (Felder 1988). In the educational experience we provide to our students, from the first grade through the last graduate course, never (well, hardly ever) are words breathed to the effects shown in Figure 5.17.

If the goal is to produce engineers who can solve society's most critical technological problems, then it is necessary somehow to convey these messages in order. It is necessary to provide students with opportunities to exercise and expand their natural creative abilities and to offer them environments that make these exercises effective. The potential for creative success lies among students who find their passion into it early in life. Potential implications of this viewpoint are that the educational systems should

- Some problems do not have unique solutions.
- Some problems may not have solutions at all.
- Problems in life, unlike problems in school, do not come packaged with the precise amount of information needed to solve them; some are over defined, and most are under defined.
- Problems in life, unlike problems in school, are open-ended: there is no single correct solution and any realistic answer invariably begins with, "It depends....
- The more possible solutions you think of for a problem, the more likely you are to come up with the best solution.
- Sometimes a solution that at first sounds foolish is the best solution.
- To be wrong is not necessarily to fail.

Figure 5.17 Problems of life and problems of school (From Felder, R.M., *Chemical Engineering Education* 22(3), 120–125, 1988.)

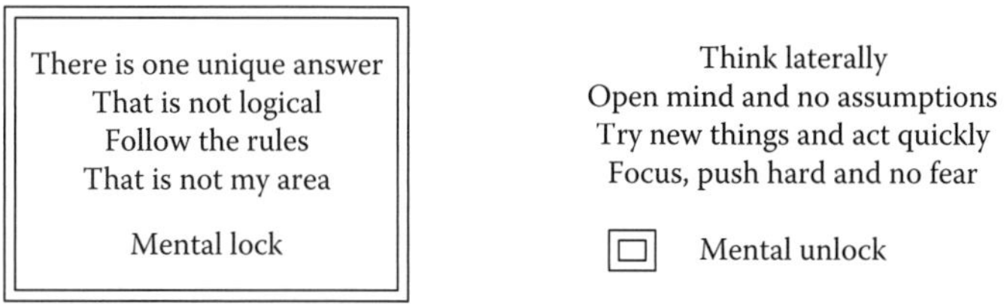

Figure 5.18 Outside-the-box thinking.

provide greater emphasis on helping students recognize areas of interest, areas where they can achieve a state of flow which leads to growth of skill and confidence, the states under which creativity develops.

For students to be creative, they need to be able to view things from different perspectives; they must have flexibility and a tolerance of ambiguity. Figure 5.18 highlights the difference between thinking within the box and thinking out of the box.

5.10.4 T-Shaped innovation forward strategy

People, knowledge, and innovation drive the prosperity of nations. Any strategy to move forward should be based on these three pillars. Talented researchers should be inspired, developed, attracted, and retained to meet the demands of the modern global economy. World-leading education and research through legacy investments should be supported. The innovation pillar encourages greater partnerships among businesses, universities, and colleges to drive innovation and encourages the adoption of new processes and technologies that help businesses prepare to compete and win in the global marketplace (NESTA 2011).

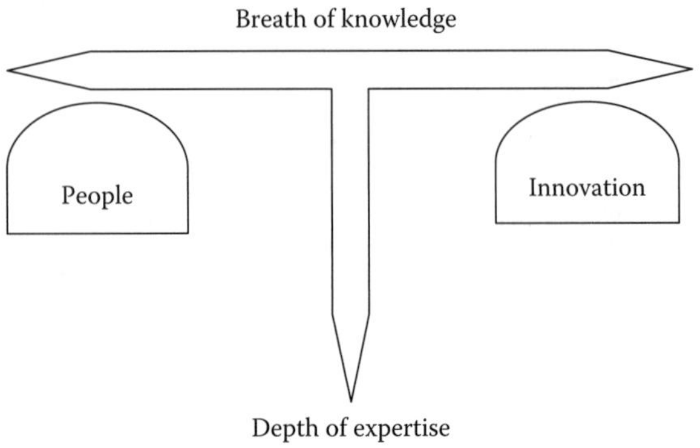

Figure 5.19 The T-shaped innovation forward strategy.

The educational system should help students strike the right balance of depth and breadth of knowledge (the T-shaped mind, Figure 5.19) (Adams 2005). It is increasingly recognized that the linear model of innovation, which gave prominence to R&D as the key phase in innovation (a "technology push" strategy) is failing the needs of technological learning. Increasingly, firms and universities are deepening their relationships, and this is leading to a hybrid community that engages scientists, researchers, and entrepreneurs in an environment more akin to a network than a hierarchical or compartmentalized structure (UN 2010).

- What is creativity?
- What contributions does creativity make to engineering and society?
- What are the stages in the development of an engineering creative solution
- What factors affect the role of creativity in the engineering process?
- What role does creativity play in innovation?
- How do engineers solve problems?
- How is creativity measured?
- How are creative ideas generated?
- How is creativity fostered in people?
- How is creativity managed?
- When and why do engineers use creativity to solve problems?
- When and why do different thinking styles play a role in creative problem solving?
- Why is creativity valuable in products?
- When and why are different tools used to support engineering creativity?
- When and why are different factors active in fostering/inhibiting creativity?

Figure 5.20 Aspects of creativity and innovation that should be addressed by an engineering learning module.

Also needed are educational programs that cover the whole spectrum of innovation, from supporting innovation culture, capacity building to knowledge creation. Greater emphasis on knowledge creation is likely to generate further innovations in education and technology. It is particularly important in an engineering course that creativity be the focus on functioning knowledge. Functioning knowledge in the context of engineering creativity is demonstrated by the ability to develop novel and effective solutions to practical, realistic technological problems. The necessary declarative knowledge for engineering creativity should address several questions as stated in Figure 5.20 (Cropley 2015).

5.11 *Disruptive innovation case: Powering future cars*

> Transportation is the center of the world! It is the glue of our daily lives. When it goes well, we don't see it. When it goes wrong, it negatively colors our day, makes us feel angry and impotent, curtails our possibilities.
>
> **Robin Chase**
> *Transportation Entrepreneur*

5.11.1 *Revolutionary or disruptive innovation?*

The growth of the modern city is based on mobility. The evolution from the medieval city in which all movements were on foot to today's sprawling agglomerations has only been possible with, first, the railway and, later, the automobile.

The first self-propelled vehicles were powered by steam engines. In France, Nicolas Joseph Cugnot (1725–1804) built the first automobile in 1769. In between 1832 and 1839, Scotsman Robert Anderson is credited with inventing the first electric car that achieved a speed of 6 km/h. In 1842, both Thomas Davenport and Robert Anderson invented electric cars. English inventor Thomas Parker built the first electric car in London in 1884, using his own designed high-capacity rechargeable batteries (see Figure 5.21). The first stationary gasoline engine developed by Carl Benz (1844–1929) was a one-cylinder two-stroke unit which ran for the first time on New Year's Eve 1879. The beginning of automotive goes to 1886 with the first patented vehicle in Germany. The first car was produced by German engineers Gottlieb Daimler (1834–1900) and Wilhelm Maybach (1846–1929), who founded Daimler Motoren Gesellschaft. The first car was sold in 1892.

The above cars were revolutionary but not a disruptive innovation, because they were expensive luxury items that did not disrupt the

Figure 5.21 Electric car built by Thomas Parker (Photo from 1895, Wikipedia.)

market for horse-drawn vehicles. The market for transportation essentially remained intact until the debut of the lower-priced Ford Model T in 1908 (Christensen 2003). The mass-produced automobile was a disruptive innovation, because it changed the transportation market, whereas the first 30 years of automobiles did not.

5.11.2 *Debate on futuristic transportation*

To continue the evolution and build alternatives requires both technological innovation and price incentives to induce changes in mobility and location behavior (Wegener 2010). However, more also the negative sides of evolution may become apparent. As transportation grows to perfection, it might destroy not only the very preconditions for its success but also the comfort it promised. In addition, implications on other modes and sectors should be considered. At the heart of the debate over futuristic transportation reside several important questions.

Within the transport sector, the roads account for the highest volume of CO_2 emissions. Hereby, some countries classify CO_2 emissions related to road transport into freight and passenger transport. In general (with the exception of China), passenger transport is responsible for about 60%–70% of the road sector CO_2 emissions (OECD/ITF 2010). This high share of the road passenger transport on global CO_2 emissions shows the high potential the passenger vehicle sector possesses to lower global environmental impact and simultaneously puts pressure on the sector to do so. The International Energy Agency (IEA) has set the ambitious target of cutting global energy-related CO_2 emissions to half by 2050 compared to the 2005 level, whereby the transport sector is expected to contribute 23% to the required reductions (IEA 2010, 2012). An important mean to reach this target is the development and widespread introduction of electric vehicles (EVs). The recently published IEA report estimates that 75% of all vehicles sales by 2050 will need to be EVs to reach the ambitious environmental goals (IEA 2013).

Thus far, the ongoing development of such energy-efficient cars has been achieved due to technological progress in processes and products. It is technology that has typically been at the center of innovation toward SD in the transport sector. But increasingly eco-innovations in the transport sector become more integrated such that they also include formal and informal institutional arrangements (OECD 2009). These institutional arrangements can be diverse as they include whole alternative business models (e.g., bike or car sharing), new policies, and green lifestyles. Such changes are influenced by a range of complex factors including measures which are difficult to quantify such as institutions, environmental policies, cultural behavior, and individual environmental awareness. They are also certainly dependent on the income level but also on the degree of inequality in an economy.

There exist different types of EVs, including battery EVs (BEVs), hybrid EVs (HEVs), and fuel cell EVs (FCEVs). In its definition, the IEA distinguishes the three types of EVs and terms these categories together as advanced vehicles. The BEVs are plug-in, battery vehicles, which are typically referred to as EVs. HEVs have an internal combustion engine as well as a motor with battery pack, while the FCEVs use a fuel cell system to convert hydrogen into electricity (IEA 2011). Over the past few years, efforts to accelerate the introduction and the adaption of EVs have increased. To these efforts belongs the Electric Vehicles Initiative (EVI), a multigovernment policy forum introduced in 2010 under the Clean Energy Ministerial. The EVI includes BEVs, PHEVS, and FCEVs in its definition for EVs.

5.11.3 Nano Tata: Thinking outside the patent box

How could Tata Motors make a car so inexpensively? It started by looking at everything from scratch, applying what some analysts have described as "Gandhian engineering" principles—deep frugality with a willingness to challenge conventional wisdom. A lot of features that Western consumers take for granted—air conditioning, power brakes, radios, etc.—are missing from the entry-level model (Hagel and Brown 2008).

5.11.3.1 Creativity in innovation

The Tata Nano has been one of the revolutionary products of our age. It has broken the price barrier and created a new market while giving life to the aspirations of a large segment of people in India. This considerable feat was achieved by striking an innovative balance between cost and features. This balance gave rise to a number of excellent attributes such as economy, space, style, and safety.

More fundamentally, the engineers worked to do more with less. The car is smaller in overall dimensions than the popular Indian car Maruti, but offers about 20% more seating capacity as a result of design choices like putting the wheels at the extreme edges of the car. The Nano is also much lighter than comparable models as a result of efforts to reduce the amount of steel in the car (including the use of an aluminum engine) and the use of lightweight steel where possible. The car currently meets all Indian emission, pollution, and safety standards, though it only attains a maximum speed of about 65 mph. The fuel efficiency is attractive—50 miles to the gallon.

5.11.3.2 Innovative modular design

Tata Motors has filed for 34 patents associated with the design of the Nano, which contrasts with the roughly 280 patents awarded to General Motors (GM) every year. Admittedly that figure tallies all of GM's research efforts, but if innovation is measured only in terms of patents, no wonder the Nano is not of much interest to Western executives. Measuring progress solely by patent creation misses a key dimension of innovation: some of the most valuable innovations take existing, patented components and remix them in ways that more effectively serve the needs of large numbers of customers (UKessays 2017).

But even this broader perspective fails to capture other significant dimensions of innovation. In fact, Tata Motors itself did not draw a lot of attention to what is perhaps the most innovative aspect of the Nano: its modular design. The Nano is constructed of components that can be built and shipped separately to be assembled in a variety of locations. In effect, the Nano is being sold in kits that are distributed, assembled, and serviced by local entrepreneurs (Hagel and Brown 2008).

5.11.3.3 Innovation in nanotechnology

By using a new pretreatment process based on nanotechnology at its paint shop, Tata Motors is reaping green benefits and cost savings. Not only did the new process cut down energy use and water consumption, it also reduced the generation of highly toxic effluent sludge, a severe health hazard. The positive environmental effect of the new pretreatment process is so high that it has won the "Tata Innovista Promising Innovation" award.

The nanotechnology process has several advantages over the conventional method: it uses less energy, water, and chemicals; it reduces water and air pollution drastically; and, most significantly, it generates no toxic sludge, saving the company the need to invest in a new landfill. The waste products from the new process are useful iron hydroxides, which are a raw-material input for the pigment industry.

5.11.3.4 *Nano's innovative engine*

The Nano development team was divided into a number of engineering excellence centers, and each was given cost targets within which to work. With the engine as the aggregate, the boundaries included the exhaust system, the cooling system, the intake system, and the fuel system. These had to achieve acceptable performance criteria, be economical on fuel consumption and comply with tough emission regulations.

The development of a one-cylinder engine, as is used in an auto rickshaw, would have helped the team adhere to the budget. Yet they consciously chose to develop a two-cylinder engine. The team began working on the concept design of the engine. The first design was for a 538 cc engine that could deliver 16 hp. The plan at that time was not to exceed 20 hp, thereby fitting in the budget. "With some modifications, we could have had 18 hp," adds Mr. Jain. "We made an engine that cost us approximately 40 per cent more than our target. But we believed there was scope for improvement. So we went back to work." Subsequently, the team made another design, with a 554 cc engine that could deliver 26 hp. A third upgraded version consisted of a 586 cc engine which could deliver 31 hp. The improvement enhanced the confidence of the team members. They had managed to double the power while maintaining cost. It was now time to make a full-fledged design.

5.11.4 *SDCs: Disruptive innovation*

> They won't have a steering wheel, accelerator pedal, or brake pedal because they don't need them. Our software and sensors do all the work. The vehicles will be very basic—we want to learn from them and adapt them as quickly as possible—but they will take you where you want to go at the push of a button. And that's an important step toward improving road safety and transforming mobility for millions of people.
>
> **Chris Urmson**
> *Director of Google's Self-Driving Car Project*

There is much assumption concerning the effects of autonomous vehicles. Consumers will soon be able to purchase reasonable self-driving vehicles that can critically reduce traffic and parking costs, accidents and pollution emissions, and chauffeur nondrivers around their communities, reducing roadway costs, eliminating the need for regular public transit services.

Car crashes kill 1.3 million people every year. Safe autonomous cars will save many lives, and make transportation affordable and more convenient. But neither automakers nor technology companies can realize this vision alone (Ng and Lin 2016).

The SDC is a hot topic. Take, for instance, the Google self-driving project or Tesla's recent announcement of its newest software update, with which their vehicles would be able to drive on their own. It can surely be predicted that the popularity of the SDC will only increase in the future (Schrijver and Fraeyenhoven 2016). SDCs will improve our lifestyles and make the world smaller. They will prevent tens of thousands of fatalities every year. There will also be ugly public debates, efforts by incumbent businesses to create legislative barriers and a lot of confusion. But the technology is coming, whether we are ready or not (Wadhwa 2016).

5.11.4.1 Incentives

The political, environmental, and social impacts of autonomous vehicles are wide ranging; there is much incentive behind enabling autonomous vehicles that will move the industry toward their use. However, there are several concerns about autonomous vehicles that may impede adoption of the technology (Hudda et al. 2012). One of the main advantages of SDCs is improved safety. Since over 90% of traffic accidents are caused by human error, such as fatigue and distracted driving (Hadi 2014), SDCs are expected to be radically safer than traditional cars. They are also expected to increase fuel efficiency and reduce travel times. Complexity in SDCs is expected to be very low, probably lower than traditional cars. As time goes on and the market sees and experiences more and more semiautonomous cars, technology readiness will rise further (Enqvist 2014).

5.11.4.2 Challenges

Several barriers have been preventing fully autonomous cars from hitting the road: high technological component prices, varying degrees of consumer trust in the technology, and relatively nonexistent regulations. Technology has been improving as new market entrants find innovative ways to expand on existing fully autonomous car technology. As a result, the price of the components required for fully autonomous cars has been dropping. Consumer trust in fully autonomous vehicle technology has increased in the past two years. California became the first US state to propose regulations. California's regulations stipulate that a fully autonomous car must have a driver behind the wheel at all times, discouraging Google's and Uber's idea of a driverless taxi system (BI Intelligence 2016).

5.11.4.3 Landscape

The existing landscape for autonomous vehicles consists of two players. The first includes traditional players, companies, and industries already in the automotive business that is introducing autonomous features as a natural evolution of their products. The second includes disruptive players, companies, and industries that currently have no existing business model or revenue stream attached to the automotive industry. They typically favor pursuing innovation that moves directly to fully autonomous vehicles (Hudda et al. 2013).

5.11.4.4 Disruptive Google

The most deeply involved player in the autonomous automobile market outside of the automobile industry is Google (Hudda et al. 2012). Google is investing in building SDC ahead of a regulatory framework, driving regulators to strategically balance their priorities around promoting innovation and ensuring public safety (Los Angeles Times 2014). Google's position in the development of SDCs so far is unique, following a different model than those given by companies such as Volvo and Tesla. Volvo is developing its autonomous feature set by partnering with tech vendors, such as Nvidia, with its graphic processing unit-based deep-learning self-driving system, and is preparing to test the system with volunteers in its cars in various countries.

5.11.4.5 Traditional players

Technology that paves the way to SDCs is available from several manufacturers such as Mercedes-Benz, BMW, Cadillac, Ford, Tesla, Volvo, and others in more mainstream vehicles. Tesla is developing its own technology and uses autonomous systems from tech developers such as Mobileye. Tesla is conducting long-term testing with customers who purchase its sensor-equipped cars and more than 1 billion miles of autonomous driving data to date. Volvo still envisions a steering wheel in the vehicle, but wants drivers being able to reclaim time when they hand over control to the car in dull driving conditions.

5.11.4.6 Enabling technologies

The car industry relies on a number of distance measurement and satellite positioning technologies such as sensing technologies, software-based networks, virtual machines, machine intelligence, GPS, GPS-enabled devices as well as other advanced disruptive technologies. Significant investments are ongoing to develop solutions to address current limitations of these technologies. Moreover, existing systems of roads have been built with human drivers in mind. With only modest changes, existing roadways can support safe computer and human-driven cars.

5.11.4.7 Motivators

Over the coming years, the SDCs could disrupt the entire automotive ecosystem. The industry will undergo major change as semiautonomous driving emerges, followed by an eventual shift to full self-driving. Auto makers that have always seen themselves as product suppliers will take on a new identity as providers of mobility services. This will open the door to lucrative new digital revenue streams, especially as they begin to explore opportunities in other digital areas such as entertainment, commerce, and monitoring a driver's health and fatigue level (Viereckl et al. 2015). The CEO of Tesla Motors, Elon Musk, envisions that 90% of all driving could be automated by 2017, but adds that truly full automation is not feasible due to the complexity involved in preparing automated responses for every possible situation (Waters and Foy 2013).

There are many strong socioeconomic motivators for the adoption of innovative SDCs. Human safety, infrastructure efficiency, quality of life, and a ready customer base are just a few of the key factors that will help make SDCs a reality. Technology is converging rapidly, both incrementally from existing vendors and from new entrants. A car equipped with existing systems can take in more information quickly and reliably, and then process it to implement a correct decision about a complex situation. Yet to be solved are the complex issues associated with the legal and liability infrastructure. Gradual introduction of these features combined with strong economic motivators is sure to overcome such obstacles. The future will surely include autonomous vehicles—the only question is how quickly (Araujo et al. 2012; Hudda et al. 2012).

5.11.5 Case research questions

- Do you think SDCs are just the beginning?
- What makes an innovation disruptive?
- How disruptive are SDCs?
- Who are the stakeholders in the transportation system?
- What are trends and future challenges in transport policies?
- What does futuristic transportation mean for cities?
- Research sustainable transportation technologies from a systems point of view to realize benefits on energy security, job creation, and emission reductions.
- How will higher transport costs affect mobility and location patterns?
- Will there be a rebirth of public transport?
- Will there be a social gap between those who can sustain their mobility and those who must give up their cars?

- Would SDCs make a big difference to our lives? Is it feasible to expect a market for them to emerge?
- Do you think sustainable cars are feasible? Why or why not?
- What are the potential benefits of SDCs?
- What are the challenges to the growth of market for SDCs?
- What are the enabling technologies for the growth of SDCs?
- Research and identify possible solutions to launch a self-driving vehicle project.
- Highlight challenges of educating governments, users, and industry on sustainability and trends of relevance to transportation.

5.12 *Knowledge acquisition*

Attempting to answer the following questions involves acquisition of knowledge from this book and other books, documents, and the Internet.

- How do you define innovation? What are main types of innovation?
- Why is innovation so important?
- Compare invention to innovation.
- What is the payoff to innovation?
- How do innovative technologies evolve?
- Should firms make or buy innovations?
- Who is a creative genius and who is a creative thinker?
- How do you perceive the notion of "Defy the Conventional"?
- What do you think "Defy the Conventional" refers to?
- How would seeing an ad with "Defy the Conventional" change your impression of a university?
- What are the components of creativity?
- How do you attract creative and highly driven engineers to your company?
- Is engineering creativity different from creativity in other fields?
- What is creative about being an engineer? What sorts of activities do engineers do that are creative?
- What drives takeoff of innovations?
- How do you find opportunities for innovation?
- How do you know if an idea is good?
- Do you think innovation drives the wealth of nations?
- How can you get other people behind an innovative idea?
- What is disruptive innovation?
- What is a "habit of the mind" or a "way of thinking"?
- Do engineers have a unique "way of thinking"?
- Can engineering HoM be applied to nonengineering settings?

- What aspects of a recent engineering project excite you because of its innovative and efficient design?
- How do engineers employ ST as a way of thinking?
- How do you encourage innovation in an organization?
- How should you form and manage innovation teams?
- What is the role of innovation in emerging markets?
- How can you quickly turn good ideas into good businesses?
- Why is creativity important to engineering and engineering education?

5.13 Knowledge possession

- Describe, analyze, and evaluate the impacts that five major inventions and innovations have had on humans.
- What do you think it means to "Defy the Conventional" in the framework of an academic approach founded on innovation?
- What are the top disruptive technologies that will change the world in the coming few years?
- Identify examples of how technology impacts human life. Describe, analyze, and evaluate the influences that technological innovations have had on humans. Explain how economic, political, and cultural issues are affected by the development and usage of technology.
- What innovation is required for a green economy?
- What are the most recent promising technological innovations and how can they be effectively implemented in cities and communities?
- What approaches of teaching encourage students to be creative? What contexts for learning enable students to be creative? How could creativity instructions be integrated into scientific and engineering teaching?
- What factors impede students' creativity in engineering education?
- What are the knowledge, skills, and characteristics that enable engineers to transform creative ideas into innovations that advance society?
- What could engineering education do better to improve creativity and innovation culture? Investigate pedagogical techniques that have proven successful in promoting the innovation culture.
- What are the essentials of effective teaching and learning at a time of disruption and innovation for academic institutions?

5.14 Knowledge creation

Collaborate with peers on learning or you may work with others outside the class to narrow down the objectives of each activity. You may access class and online resources, and analyze data and information to create new ideas and balanced solutions. High-level digital tools may be used to develop multimedia presentations, simulations or animations, videos and visual displays, digital portfolios (ePortfolios), reflective practice (online publishing and blogging), or well-researched and up-to-date reports.

5.14.1 *Reflection practice on engineering innovation*

In terms of real innovation, how could you use Thomas Edison, Nikola Tesla, Henry Ford, or Steve Jobs, to stimulate classroom discussions about engineering innovation? For this task, form four teams, each three to five students to prepare a presentation about the innovation personality traits of each of the most accomplished engineering inventors. What is the greatest legacy of each?

5.14.2 *Engineering communication on innovative views on smart cities*

Communication skills in engineering programs can be developed as part of an effort to improve students' awareness, problem-solving skills, creativity, interpersonal and group skills, assessment skills, and self-directed learning skills. To improve the targeted process skills, the instructors may conduct process skills workshops, facilitate in-class exercises, and give take-home assignments. The focus of this communication task is mastery of the fundamental elements of effective communication: reading the communicative situation, understanding the audience, creating a well-crafted message, and projecting confidence and competence through an appropriate communication style.

For this topic of smart cities, there is a short overview, followed by in-class activities and take-home assignments. By the end of the task, students are expected to demonstrate an ability to identify the key elements of effective oral and written communication, write clear and accurate summaries or proposals, and make an effective oral presentation that addresses the audience wants and needs appropriately.

The following communication skill task on smart cities provides an introduction and cluster of questions which cover the major themes of smart cities innovation including technological, social, and business innovations.

5.14.2.1 *Integrated innovation of smart cities*

Over 50% of the world's 7.2 billion people live in cities, and in a few short decades, the world's population is expected to exceed 10 billion. This rapid pace of change in our world presents challenges to our ability to adapt. Increasing population size and density have an impact on housing, food supply, transportation, education, and health. Increasing diversity, in turn, has an impact on cultural norms, notions of community, and ideas of citizenship (University of Calgary 2015).

A smart sustainable city is an innovative city that uses ICTs and other means to improve quality of life, efficiency of urban operation and services, and competitiveness, while ensuring that it meets the needs of present and future generations with respect to economic, social, environmental as well as cultural aspects.

Smart cities are viewed as the futuristic cities, but considering today's rate of innovation it is highly likely that smart city will become reality within few years. Smart cities are a miniature for initiatives that integrate typical infrastructure with technology to ease traffic, congestion, pollution, and energy consumption. However, a truly smart city starts with a purpose and an economic vision defined by planners and citizens, not by technology. An attempt to apply social innovation concepts and approaches to the smart city trend is now in the plan. It is realized that much of the focus of the smart city movement to date, in which the city authorities and other organizations deploy sensors, networks, and data to improve the efficiency and effectiveness of urban systems, is only part of the initiative.

Smart cities are sociotechnical systems. Technologies must be understood broadly, including not only ICTs, transportation technologies, energy systems, etc., but also all tools, devices, and material structures that affect human activities and capacities (University of Calgary 2015).

5.14.2.2 General innovation questions

- State five competing definitions of a smart city.
- Name ten cities that are adopting the notion of smart cities?
- What are some innovative ideas in making a city smart?
- How to design cities that function efficiently?
- How to understand the evolving innovation ecosystems of smart cities?

5.14.2.3 Theme 1: Technological innovation

- What are the building blocks and interconnection technologies that will make smart cities a reality?
- How to ensure that the adoption of smart technologies is clearly beneficial, without unintended negative consequences?

- What are some specific disruptive technologies and applications for smart cities?
- How can water, storm water, and wastewater infrastructure be designed to optimize water supply, demand management, and treatment?
- How can smart power grids maximize the collection, storage, and distribution of low-carbon electricity?
- How can smart power grids and building control systems be designed to provide more efficient heating, cooling, lighting, and appliance use?

5.14.2.4 Theme 2: Social innovation

- What are the forms of social innovation for accelerating transition to smart cities?
- How can societal problems in cities be addressed through social innovation platforms and which examples can be presented?
- How to use social media and big data resources to improve the quality and effectiveness of the built environment?
- How do you recognize the societal vision on smart cities?
- How could social networks act as laboratory of social innovation?
- What are the barriers of social innovation?
- What are the scaling dynamics, drivers, and barriers for urban social innovation?
- How may the future Internet enable social innovation platforms that enhance participation, well-being, and sustainability in smart cities?
- What are the impacts of smart city design and ICTs on elderly and less privileged populations and how can we ensure that they have equal access to enhanced connectivity and mobility?
- How do we create cities that provide high-quality education, training opportunities, cultural competence, and exposure to the arts for all citizens?

5.14.2.5 Theme 3: Business innovation

- What are possible partnerships, collaboration frameworks, and business models for platforms of change stimulating integrated innovation and the development toward smarter cities?
- What could be the role of living labs or other innovation models or ecosystems to create, guide, and manage such platforms?

5.14.3 Debate on innovation in futuristic transportation

Objective	Introducing an open-ended debate in the classroom to help students understand argument on the concepts of futuristic transportation
Time	15 min for debate and 15 min for review
Level	For and against
Learning outcomes	Make an argument about a particular opinion, evaluate the arguments of peers, and understand the concept of counterarguments
Capabilities demonstrated	Developing skills on public speaking, research, teamwork, critical thinking, communication, and professional judgment
Arrangement	Students are organized into two position groups of three speaking in a specific order. Three argue for an opinion and three argue against. One or two students might each work on the opening and closing statements while the group is investigating the subject; however, the entire group should revise the statements. Each group should read an opening and closing statement for the debate.
Ideas for the topic	What are the challenges and opportunities in the transportation system? Are we opting for transportation alternatives; where people can choose to walk, bike, drive, or take public transportation; or is it about developing places that cater to a single transportation option to drive? What kind of transportation and mobility systems should we have for the future?
Assessment	Indicate what you consider the best arguments in favor of the topic. How were they substantiated? Identify arguments that are based on poor facts, not ethical and/or sustainable or not well substantiated.

5.14.4 Pitch communication on how to enhance interest of youth in engineering

This activity may be held where students present their term-project ideas via a poster gallery walk. Students may form teams of two around the project ideas they like the best. The term-project teams should work together for four weeks on their chosen project. Teams should work separately to design learning activities that enhance the interest of high school students in engineering. The team should adopt one EHoM only while building the solution model. These innovative models may take different forms such as course project, after-school program, Saturday school, summer school or camp, or any other new idea.

The project poster should consist of two main sections: the model and the written summary of the design. To create their models, the students should employ many of the EHoM that are identified in this chapter. In addition, students should conduct literature reviews to deepen their understanding of the relevant theories regarding their system of interest, and to help them to identify the key factors that must be considered in order to capture their system's behavior in a model.

The model of EHoM may provide a framework for developing a better understanding of engineering among the young people. The model may examine the importance of science, technology, engineering, art, and math (STEAM) approach in high schools. In fact, the idea of STEAM is a mind-set, one that expands with ideas and opportunity. The nuts and bolts are not enough; what is need is broader and deeper thinkers.

5.14.5 Innovation pitch competition on biomechatronics devices

Biomechatronics is an interdisciplinary field that applies mechatronics to biological systems and microsystems to innovative ways of solving emerging engineering problems and to develop biomedical and rehabilitative products for the medical industry. The recognition of mechatronic devices has been growing in recent years and the market of biomechatronics is expanding as well. Heart pacemakers and cochlear implants are examples of simple biomechatronic devices while more advanced examples are orthotics and prosthetics, autonomous robotic systems, and implant devices.

For this task, select, research, and think of GT and mechatronics engineering to help your team invent or innovate a product that will assist or extend human life. Develop an early concept idea and a creative proposal (up to three pages) that is useful and comprehensive. The innovation proposal might be based on the following outlines: problem, feasible technology solution, innovation, expected impact, management and financing, funding and grant resources.

5.14.6 Consulting study on designing an innovative class of the future

The objective of this study is to design an innovative solution for an ideal engineering class of the future, a concept that responds to emerging trends around the recent Finnish educational notion of scrapping traditional "teaching by subject" in favor of "teaching by topic." The design should offer solutions to merging transdisciplinary learning functions as well as to digital and analog supporting technologies. This concept may focus on a specific theme or a whole system as well and should be integrated with the concept of sustainability.

The project may be done in teams of three to four students. Students from different disciplines levels may be represented across all teams. Each team will be responsible for further defining the direction of the project assignment, based on their research into class environment and future learning needs. Teams also need to define clear responsibilities for each team member, so that everyone is clear on the contribution of individual students.

5.14.7 *Piece of art on indicators of future STI and SD policies*

The increasing diffusion of new digital technologies and the increased globalization of markets have contributed to changing the nature of STI and the interface between the three areas. The complementary, interdependent nature of these different concepts means that they must be considered in their mutual performance. Understanding the nature of these concepts will help to address the need for new indicators for STI. The purpose of this task is to examine current efforts to measure STI indicators and understand how these indicators identify new paths that could be explored. Students may develop a digital piece of art that briefly outlines the mostly needed STI indicators that would aid in creating and assessing innovation policies in a global comparable approach and promote a culture for achieving SDGs.

5.14.8 *Class poster competition on smart vehicles*

During the past few years, smart vehicles have received attentions and developments from both research and industry community. In such vehicles, smart systems are able to sense and realize surrounding environment based on several types of sensors.

For this task, several class teams may be formed to solve specific real-life interdisciplinary problem in smart vehicles. Each team is assigned a task. The teams use problem-solving strategies to generate ideas, choose the best solution, complete comprehensive patent and marketability searches, and design prototypes. Each team should present its design in a poster format. Examples for tasks that may be suggested to the teams include the following:

- Sensors calibration methods
- Radar and sonar sensors for driving assistance system
- 3D laser-based technology to obtain a 360° view of the area around the vehicle
- Sensor fusion system for detecting obstacles
- Sensor-based vehicle localization in GPS-denied environment
- Sensor-based scene analysis and understanding

- Smart bumpers to minimize collision effects
- Advanced collision-warning system
- Radar-based system for controlling traffic lights
- Alternatives to speed bumps

References

Adams, K. 2005. The sources of innovation and creativity. National Center on Education and the Economy, Washington, DC.

Aguilera-Caracuel, J., and N. Ortiz-de-Mandojana. 2013. Green innovation and financial performance: An institutional approach. http://journals.sagepub.com/doi/abs/10.1177/1086026613507931.

Albert, R. S., and M. A. Runco. 1999. A history of research on creativity. In Sternberg, R. J. (Ed.), *Handbook of Creativity*. Cambridge: Cambridge University Press, pp. 16–31.

Allis, R. 2013. The starter guide: Creating a better world through entrepreneurship. http://startupguide.com/world/the-40-greatest-innovations-of-all-time/.

Allis, R. 2015. The 40 greatest innovations of all time. http://startupguide.com/tag/history-of-innovation/.

Amabile, T. M. 1983. *The Social Psychology of Creativity*. New York: Springer-Verlag.

Amabile, T. M. 1988. From individual creativity to organizational innovation. In Gronhaug, K., and G. Kaufmann (Eds.), *Innovation: A Cross-Disciplinary Perspective*. London: Norwegian University Press, pp. 139–166.

Amabile, T. M. 1989. The motivation to be creative. In Isaksen, S. G. (Ed.), *Frontiers of Creativity Research: Beyond the Basics*. Buffalo, NY: Bearly, pp. 223–254.

Amabile, T. M. 1999. How to kill creativity: Keep doing what you're doing. Or, if you want to spark innovation, rethink how you motivate, reward and assign work to people. *Harvard Business Review* 76(5): 76–87.

Anadon, L. D., G. Chan, E. Kempster, L. Vinsel, J. Stephens, K. Araujo, and P. Guardabassi. 2013. Background paper for the project on innovation and access to technologies for sustainable development. Harvard Kennedy School of Business. https://www.hks.harvard.edu/content/download/68197/1245598/version/1/file/2013-03.pdf.

Andersen, O. J. 2008. A bottom-up perspective on innovations: Mobilizing knowledge and social capital through innovative processes of bricolage. *Administration and Society* 40: 57–78.

Araujo, L., K. Mason, and M. Spring. 2012. SDCs: A case study in making new markets. BIG Innovation Centre, Lancaster University, Lancaster.

Atkinson, R. D. 2014. Understanding the U.S. national innovation system. The Information Technology and Innovation Foundation. http://www2.itif.org/2014-understanding-us-innovation-system.pdf.

Baillie, C. 2002. Enhancing creativity in engineering students. *Engineering Science and Education Journal* 11: 185–192.

Beder, S. 1999. Beyond technicalities: Expanding engineering thinking. *Journal of Professional Issues in Engineering Education and Practice* 125(1): 12–18.

Berg, M., and E. Eger. 2003. *Luxury in the Eighteenth Century*. New York: Palgrave.

BI Intelligence. 2016. Google's SDC project close to becoming its own business. Business Insider. http://www.businessinsider.com/googles-self-driving-car-project-close-to-becoming-its-own-business-2016-4.

Bower, J. L., and C. M. Christensen. 1995. Disruptive technologies: Catching the wave. *Harvard Business Review* 1(January–February): 43–53.

Browne, B. A., and D. F. Cruse. 1988. The incubation effect: Illusion or illumination? *Human Performance* 1: 177–185.

Bruland, K., and D. Mowery. 2004. Innovation through time. http://www.globelicsacademy.net/pdf/DavidMowery_2.pdf.

Carberry, A. R. 2010. Measuring engineering design self-efficacy. *Journal of Engineering Education* 99(1): 71–79.

Chandler, A. D. 2003. *Strategy and Structure: Chapters in the History of the Industrial Enterprise*. Garden City, NY: Martino Fine Books.

Chen, Y.-S., S.-B. Lai, and C.-T. Wen. 2006. The influence of green innovation performance on corporate advantage in Taiwan. *Journal of Business Ethics* 67(4): 331–339.

Christensen, C. M. 1997. *The Innovator's Dilemma: How New Technologies Cause Great Firms to Fail*. Boston, MA: Harvard Business School Press.

Christensen, C. M. 2003. *The Innovator's Solution: Creating and Sustaining Successful Growth*. Boston, MA: Harvard Business Press.

Christensen, C. M., M. E. Raynor, and R. McDonald. 2015. What is disruptive innovation? https://hbr.org/2015/12/what-is-disruptive-innovation.

Clark, K., and S. Wheelwright. 1993. *Managing New Product and Process Development: Text and Cases*. New York: The Free Press.

Costa, A., and B. Kallick. 2002. Discovering and exploring habits of mind. Association for Supervision and Curriculum Development, Alexandria, VI.

Cropley, D. H. 2015. Promoting creativity and innovation in engineering education. *Psychology of Aesthetics Creativity and the Arts* 9(2): 161–171.

Damanpour, F. 1991. Organizational innovation: A meta-analysis of effects of determinants and moderators. *Academic Management Journal* 34(3): 555–590.

De Schrijver, S., and O. V. Fraeyenhoven. 2016. SDCs in Belgium: A clash between innovation and liability? Information technology, who is who legal? http://astrealaw.be/backend/admin/sites/default/files/wwl_tmt_2016_-_article.pdf.

Dodgson, M., D. Gann, and A. Salter. 2008. *The Management of Technological Innovation: Strategy and Practice*. New York: Oxford University Press.

Dutta, S. 2012. The global innovation index 2012: Stronger innovation linkages for global growth. INSEAD and World Intellectual Property Organization.

Dweck, C. S. 2006. *Mindset: The New Psychology of Success*. New York: Ballentine Books.

EIO. 2012. The Eco-innovation gap: An economic opportunity for business. Eco-innovation observatory. Funded by the European Commission, DG Environment, Brussels, Belgium.

Enqvist, H. 2014. SDCs: Diffusion of radical innovations and technology acceptance: A new framework for measuring technology acceptance for SDCs. Master's Thesis, Lunds University, Lund, Sweden.

ERM. 2012. Managing the risk of disruptive innovation. Enterprise Risk Management Initiative. https://erm.ncsu.edu/library/article/managing-the-risk-of-disruptive-innovation.

Felder, R. M. 1988. Creativity in engineering education. *Chemical Engineering Education* 22(3): 120–125.

Ferguson, D. M., and M. W. Ohland. 2012. What is engineering innovativeness? *International Journal of Engineering Education* 28(2): 253–262.

Frensch, P. A., and R. J. Sternberg. 1989. Expertise and intelligent thinking: When is it worse to know better? In Sternberg, R. J. (Ed.), *Advances in the Psychology of Human Intelligence*. Hillsdale, NJ: Erlbaum.

Fruin, W. M. 1992. *The Japanese Enterprise System: Competitive Strategies and Cooperative Structures*. Oxford: Clarendon Press.

Gaynor, G. H. 2002. *Innovation by Design: What It Takes to Keep Your Company on the Cutting Edge*. New York: American Management Association (AMACOM).

Gaynor, G. H. 2014. Innovation: Transition from bottom-up to top-down innovation. *IEEE Engineering Management Review* 42(1): 5–7.

Gilfillan, S. C. 1935. *The Sociology of Invention*. Cambridge, MA: MIT Press.

Godin, B. 2008. Innovation: The history of a category. Project on the Intellectual History of Innovation Working Paper No. 1. http://www.csiic.ca/PDF/IntellectualNo1.pdf.

Goldenberg, M. 2004. *Social Innovation in Canada: How the Non-profit Sector Serves Canadians and How It Can Serve Them Better*. Ottawa, ON: Canadian Policy Research Networks. http://www.cprn.org/documents/51684_EN.pdf.

Goldenberg, M., W. Kamoji, L. Orton, and M. Williamson. 2009. Social innovation in Canada: An update. Research Report, Canadian Policy Research Networks, Inc. (CPRN). https://pdfs.semanticscholar.org/802b/91efebaa5d3da4462321879a8af28cd97272.pdf.

Grand Challenges Canada. 2010. Integrated innovation, Canada. http://www.grandchallenges.ca/wp-content/uploads/integratedinnovation_EN.pdf.

Gredig, P. 2017. Uber disruptive technologies challenge old business models. https://www.fcc-fac.ca/en/ag-knowledge/technology-and-innovation/uber-disruptive-technologies-challenge-old-business-models.html.

Griffith University. 2004. Response to Queensland's future–Building on the smartstate.https://www.griffith.edu.au/__data/assets/pdf_file/0011/38486/smartstate_review.pdf.

Griffiths, T., P. A. Hunt, and P. K. O'Brien. 1992. Inventing activity in the British textile industry, 1700–1800. *Journal of Economic History* 52: 881–906.

Gruber, H. E., and D. B. Wallace. 1999. The case study method and evolving systems approach for understanding unique creative people at work. In Sternberg, R. J. (Ed.), *Handbook of Creativity*. New York: Cambridge University Press.

Hadi, S. 2014. Autoliv active safety: Vision systems. http://www.autoliv.com/Investors/Presentations/2014-05-06%20Handelsbanken%20Active%20-%20Safety, %20Salah%20Hadi.pdf.

Hagel, J., and J. S. Brown. 2008. Learning from Tata's Nano: The innovations of the $2,500 car carry important lessons for Western executives. Bloomberg. http://www.bloomberg.com/news/articles/2008-02-27/learning-from-tatas-nanobusinessweek-business-news-stock-market-and-financial-advice.

Harris, R. 1998. Introduction to creative thinking. http://www.virtualsalt.com.

Hart, H. 1931. *The Technique of Social Progress*. New York: Henry Holt and Co.

Hayes, M. 2005. Creativity in consulting engineering: How civil engineers talk about design? MSc Thesis, Queensland University of Technology, Creative Industries Research and Applications Centre, Brisbane, Australia.

Hudda, R., C. Kelly, G. Long, J. Luo, A. Pandit, D. Phillips, L. Sheet, and I. Sidhu. 2012. Self-driving cars. Fi Fung Institute for Engineering Leadership. College of Engineering, University of California, Berkeley, CA.

Hudson, P. 2004. Industrial organization and structure. In Floud, R., and P. Johnson (Eds.), *The Cambridge Economic History of Britain*, Vol. 1, Industrialisation, 1700–1860. Cambridge, MA: Cambridge University Press, pp. 28–56.

Hultman, N., K. Sierra, and A. Shapiro. 2012. Innovation and technology for green growth. The 2012 Brookings Blum Roundtable Policy Briefs. https://www.brookings.edu/wp-content/uploads/2016/06/10-green-growth-hultman-sierra.pdf.

Hutt, R. 2016. What is disruptive innovation? https://www.weforum.org/agenda/2016/06/what-is-disruptive-innovation/.

IEA. 2010. Energy technology perspectives 2010: Scenarios and strategies to 2050. Paris, France: OECD Publishing.

IEA. 2011. Electric vehicles: Are they a passing fad, or here to stay? http://www.iea.org/newsroomandevents/news/2011/may/name, 19824, en.html.

IEA. 2012. Energy technology perspectives: Pathways to a clean energy system. Paris, France: OECD Publishing.

IEA. 2013. Global EV outlook: Understanding the electric vehicle landscape to 2020. Paris, France: OECD/International Energy Agency.

Johnson, M. W., C. M. Christensen, and H. Kagermann. 2008. Reinventing your business model. *Harvard Business Review* 86(12): 50–59.

Johnson, G., R. Whittington, and K. Scholes. 2011. *Exploring Strategy: Text and Cases*. Ninth edition. Essex: Pearson Education.

Jopling, K. 2015. Disruptive innovation in the creative industries: Killer questions from key industry players. White Paper, Henley Business School, University of Reading, Reading. https://www.henley.ac.uk/files/pdf/mba/White_paper_pdf.

Kallen, H. M. 1930. Innovation. In Seligman, E. R. A., and A. S. Johnson (Eds.), *Encyclopedia of the Social Sciences*. New York: Macmillan. pp. 58–61.

Kao, J. 2007. *Innovation Nation: How America Is Losing Its Innovation Edge, Why It Matters, and What We Can Do to Get It Back*. New York: Free Press.

Katehi, L., G. Pearson, and M. Feder. 2009. Engineering in K12 education: Understanding the status and improving the prospects, Committee on K-12 Engineering Education, National Academy of Engineering and NRC. Washington, DC: National Academies Press.

Kemp, R., and P. Pearson. 2007. *Final Report of the MEI Project Measuring Eco-innovation*. Maastricht, the Netherlands: UM-MERIT.

Kim, W. C., and R. Mauborgne. 2005. *Blue Ocean Strategy*. Boston, MA: Harvard Press.

Kirton, M. 1976. *Adaption-Innovation Inventory* (1998 Edition), L. L. Brown and R. Brown, Eds. Highlands, Gravel Path, Berkhamsted, Hertfordshire: Occupational Research Centre.

Langrish, J. 1985. Innovation management: The role of creativity. Institute of Advanced Studies, Manchester Polytechnic, Manchester.

Lawlor, R. (Ed.). 2013. Engineering in society. Royal Academy of Engineering, Second edition.

Lazonick, W. 2002. Innovative enterprise and historical transformation. *Enterprise and Society* 3: 3–47.

Lin, C. Y., and Y. H. Ho. 2011. Determinants of green practice adoption for logistics companies in China. *Journal of Business Ethics* 98(1): 67–83.

Los Angeles Times. 2014. California isn't ready for driverless cars. http://www.latimes.com/opinion/editorials/la-ed-driverless-cars-20141228-story.html.

Malmquist, D. 2014. A tool for holistic optimization of mechatronic design concepts. Doctoral Thesis. KTH Royal Institute of Technology, School of Industrial Engineering and Management, Department of Machine Design, Stockholm, Sweden.

Manyika, J., M. Chui, J. Bughin, R. Dobbs, P. Bisson, and A. Marr. 2013. Disruptive technologies: Advances that will transform life, business, and the global economy. New York: McKinsey Global Institute.

Mohr, J., S. S. Sengupta, and S. Slater. 2010. *Marketing of High-Technology Products and Innovations*. Third edition. Upper Saddle River, NJ: Prentice Hall.

Mowery, D. C. 1993. Economic theory and government technology policy. *Policy Sciences* 16: 27–43.

Murmann, J. P., and R. Landau. 1998. On the making of competitive advantage: The development of the chemical industries of Britain and Germany Since 1850. In Arora, A., R. Landau, and N. Rosenberg (Eds.), *Chemicals and Long-Term Economic Growth*. New York: John Wiley & Sons.

NAE. 2008. Grand challenges for engineers. National Academy of Engineering. http://www.engineeringchallenges.org/File.aspx?id=11574&v=ba24e2ed.

NESTA. 2011. Creating innovation in small and medium-sized enterprises: Evaluating the short-term effects of the creative credits pilot. Working Paper: National Endowment for Science, Technology and the Arts, 2011. http:// www.nesta.org.uk/publications/assets/features/creating_innovation_in_smes.

Ng, A., and Y. Lin. 2016. SDCs won't work until we change our roads-and attitudes. Transportation. https://www.wired.com/2016/03/self-driving-cars-wont-work-change-roads-attitudes/.

Ockuly, M. D. 2017. Joy of quotes. http://www.joyofquotes.com/inspirational-quotes-by-author.html.

O'Connor, G. C., and M. P. Rice. 2013. A comprehensive model of uncertainty associated with radical innovation. *Journal of Product Innovation Management* 30(1): 2–18.

OECD. 2008. Sustainable manufacturing and eco-innovation: First steps in building a common analytical framework. DSTI/IND (2008)16/REV1, Paris, France.

OECD. 2009. Eco-innovation in industry: Enabling green growth. Paris, France: OECD Publishing.

OECD. 2010. The OECD innovation strategy: Getting a head start on tomorrow. Paris, France: OECD Publishing.

OEERE. 2000. From invention to innovation. Office of Energy Efficiency and Renewable Energy, US Department of Energy. http://www.nrel.gov/docs/fy00osti/26620.pdf.

Okpara, F. O. 2007. The value of creativity and innovation in entrepreneurship. *Journal of Asia Entrepreneurship and Sustainability* III(2): 1–14.

Paul, R., and L. Elder. 2010. The miniature guide to critical thinking concepts and tools. Dillon Beach: Foundation for Critical Thinking Press. https://www.criticalthinking.org/files/Concepts_Tools.pdf.

Paul, R., and L. Elder. 2014. Critical thinking: Concepts and tools. Dillon Beach: Foundation for Critical Thinking Press.

Pena, I. 2002. Intellectual capital and business start-up success. *Journal of Intellectual Capital* 3(2): 180–198.

Phills, J. A., K. Deiglmeier, and D. T. Miller. 2008. Rediscovering social innovation. *Stanford Social Innovation Review* 6(4): 34–43.

Pierenkemper, T., and R. Tilly. 2004. *The German Economy during the Nineteenth Century*. New York: Berghahn Books.

Rennings, K. 2000. Redefining innovation-eco-innovation research and the contribution from ecological economics. *Ecological Economics* 32: 319–332.

Resnick, L. 1999. Making America smarter. *Education Week Century Series* 18(40): 38–40.

Roberts, E. B. 1998. Managing invention and innovation. *Research Technology Management* 31(1): 11–29.

Rogers, E. M. 2003. *Diffusions of Innovation*. Fourth edition. New York: The Free Press.

Rustam, L. 2001. Technology business incubators to help build an innovation-based economy. *Journal of Change Management* 3(2): 167–176.

Sahin, I. 2006. Detailed review of Rogers's diffusion of innovation theory and educational technology related studies based on Rogers' theory. *The Turkish Online Journal of Educational Technology* 5(2): 14–23.

Saviotti, P. P. 2005. On the co-evolution of technologies and institutions. In Weber, K., and J. Hemmelskamp (Eds.), *Towards Environmental Innovation Systems*. Berlin, Germany: Springer, pp. 9–32.

Schank, R. C. 1988. *The Creative Attitude: Learning to Ask and Answer the Right Questions*. New York: Macmillan.

Schmookler, J. 1957. Inventors past and present. *Review of Economics and Statistics* 57: 321–333.

Schumpeter, J. A. 1934. *The Theory of Economic Development*. Cambridge, MA: Harvard University Press.

Simonton, D. K. 2012. Creative genius as a personality phenomenon: Definitions, methods, findings, and issues. *Social and Personality Psychology Compass* 6(9): 691–706.

Sood, A., and G. J. Tellis. 2005. Understanding the seeds of growth: Technological evolution and product innovation. PhD Thesis, Marshall School of Business, University of Southern California, Los Angeles, CA.

Stefik, M., and B. Stefik. 2006. *Breakthrough: Stories and Strategies of Radical Innovation*. Boston, MA: MIT Press.

Stern, B. J. 1927. *Social Factors in Medical Progress*. New York: Columbia University Press.

Sternberg, R. J. 1997. *Successful Intelligence*. New York: Plume Books.

Sternberg, R. J. (Ed.). 2003a. Psychologists defying the crowd: Stories of those who battled the establishment and won, American Psychological Association, Washington, DC.

Sternberg, R. J. 2003b. Creative thinking in the classroom. *Scandinavian Journal of Educational Research* 47(3): 325–338.

Sternberg, R. J. 2005. WICS: A model of positive educational leadership comprising wisdom, intelligence, and creativity synthesized. *Educational Psychology Review* 17(3): 191–262.

Sternberg, R. J. 2006. The nature of creativity. *Creativity Research Journal* 18(1): 87–98.

Sternberg, R. J., and I. Q. Beyond. 1985. *A Triarchic Theory of Human Intelligence*. New York: Cambridge University Press.

Sternberg, R. J., and T. I. Lubart. 1995. *Defying the Crowd: Cultivating Creativity in a Culture of Conformity*. New York: Free Press.

Sternberg, R. J., and T. I. Lubart. 1999. The concept of creativity: Prospects and paradigms. In Sternberg, R. J. (Ed.), *Handbook of Creativity*. Cambridge: Cambridge University Press, pp. 3–15.

Taylor, I. A. 1975. An emerging view of creative actions. In Taylor, I. A., and J. W. Getzels (Eds.), *Perspectives in Creativity*. Chicago, IL: Aldine.

Taylor, E. 1996. *Innovation Design Environment and Strategy. Block 1: An Introduction to Innovation*. Milton Keynes, UK: The Open University.

Thompson, J. L. 2001. *Strategic Management*. London: Thomson Learning.

Thompson, G., and M. Lordan. 1999. A review of creativity principles applied to engineering design. Proceedings of the Institution of Mechanical Engineers, Part E. *Journal of Process Mechanical Engineering* 213(1): 17–31.

Tomic, W., and A. Brouwers. 1999. Idea generating among secondary school teachers. *Creativity and Innovation Management* 8: 262–268.

Tornatzky, L. G., and K. J. Klein. 1982. Innovation characteristics and innovation adoption-implementation: A meta-analysis of findings. *IEEE Transactions on Engineering Management* 29(1): 28–45.

Torrance, E. P. 2003. Reflection on emerging insights on the educational psychology of creativity. In Houtz, J. C. (Ed.), *The Educational Psychology of Creativity*. Cresskill, NJ: Hampton Press, pp. 273–287.

UKessays. 2017. Impact of Tata Nano on automobile industry. https://www.ukessays.com/essays/marketing/impact-of-tata-nano-on-automobile-industry-marketing-essay.php.

UN. 2010. Entrepreneurship education, innovation and capacity-building in developing countries. Trade and Development Board Investment, Enterprise and Development Commission, Geneva, Switzerland. http://unctad.org/en/docs/ciimem1d9_en.pdf.

UN. 2012. Fostering innovative entrepreneurship: Challenges and policy options. United Nations Economic Commission for Europe (UNECE). Geneva, Switzerland: United Nations.

University of Calgary. 2015. Human dynamics in a changing world: Smart and secure cities, societies, and cultures. pp. 1–30. http://www.ucalgary.ca/publications/files/publications/human-dynamics-research-strategy.pdf.

Verganti, R. 2007. Design, meanings and radical innovation: A meta-model and a research agenda. *Journal of Product Innovation Management* 25(5): 436–456.

Viereckl, R., D. Ahlemann, A. Koster, and S. Jursch. 2015. Connected car study 2015: Racing ahead with autonomous cars and digital innovation. Strategy & tracks the growth of connected car technologies and its impact on pricing, sales, and innovation in the European automotive industry. http://www.strategyand.pwc.com/reports/connected-car-2016-study.

Ville, S. 2011. Historical approaches to creativity and innovation. In Mann, L., and J. Chan (Eds.), *Creativity and Innovation in Business and Beyond: Social Science Perspectives and Policy Implications*. New York: Routledge, pp. 64–81.

Visser, S. 2016. 5 disruptive technologies that are challenging the traditional banking model. http://www.ibmbigdatahub.com/blog/5-disruptive-technologies-are-challenging-traditional-banking-model.

Vital, A. 2014. Entrepreneurs who dropped out. Funders and Founders. fundersandfounders.com/entrepreneurs-who-dropped-out/.

Wadhwa, V. 2016. I feared my SDC–Until I let go of the wheel. *The Washington Post.* https://www.washingtonpost.com/news/innovations/wp/2016/07/07/after-the-fatal-tesla-crash-i-still-feel-safe-in-my-self-driving-car/.

Wallas, G. 1926. *The Art of Thought.* New York: Harcourt and Brace.

Waters, R., and H. Foy. 2013. Financial times. http://www.ft.com/intl/cms/s/0/70d26288-1faf-11e3-8861-00144feab7de.html#axzz2y8h8BtD0.

Wegener, M. 2010. The future of mobility in cities: Challenges for urban modelling. *Transport Policy* 29(September): 275–282.

Weng, M.-H., and C.-Y. Lin. 2011. Determinants of green innovation adoption for small and medium-size enterprises (SMES). *African Journal of Business Management* 5(22): 9154–9163. DOI: 10.5897/AJBM11.273. http://www.academicjournals.org/AJBM.

Wesley, F. 2010. Making a difference: Strategies for scaling social innovation for greater impact. *The Innovation Journal: The Public Sector Innovation Journal* 15(2): 2–19, Article 2.

West, A. M., and J. L. Farr. 1990. Innovation at work. In *Innovation and Creativity at Work: Psychological and Organizational Strategies.* New York: John Wiley & Sons, pp. 3–13.

Willy, B. A., and L. Kolvereid. 2005. Organizational strategy, individual personality and innovation behavior. *Journal of Enterprising Culture* 13(1): 7–19.

chapter six

Engineering leadership

> To me, a leader is someone who holds her- or himself accountable for finding potential in people and processes. And so what I think is really important is sustainability.
>
> **Brene Brown**
> *University of Houston*

6.1 Objectives

- Provide a historical perspective to the concept of leadership.
- Understand leadership and its relationship to creativity and vision.
- Discuss several leadership theories.
- Describe leadership theories of motivation and management.
- Develop an understanding of what emotional intelligence (EQ) is and its value to life and leadership development.
- Discuss the concept of positive psychology, positive leadership, and their impact on innovation.
- Introduce the concept of authentic leadership and its strong relevance to management.
- Introduce the six common styles of leadership.
- Explain the three levels of leadership model as a practical tool for developing leaders' leadership presence, know-how, and skill.
- Know about innovation leadership and its two components.
- Know why leaders should practice systems thinking (ST).
- Raise the potential of engineers as major contributors to society by integrating the teaching of leading and leadership skills.
- Discuss the qualitiy of leadership for meeting sustainability development goals.
- Know about the role of leadership in engineering education.
- Enhance leadership culture that promotes increased engagement of students in curricular and extracurricular activities.
- Present a case of greening buildings through leadership by design. Realize the role of design as a tool for leadership. Understand what integrated design process (IDP) is and how it is different from

traditional approaches. Realize its benefits, and why it is critical for achieving sustainable design.
- Provide three end of chapter pedagogical knowledge strategies, namely knowledge acquisition, knowledge possession, and knowledge creation to help understand the above topics and generate new and alternative ideas and solutions.

6.2 Historical perspective

> A leader is one who knows the way, goes the way and shows the way.
>
> **John C. Maxwell**

6.2.1 Ancient leadership

Leadership is a topic that originated long back in history when people started understanding the importance of leaders' role in various facets of life such as politics, governmental issues, foreign policy, and war. Social and political scholars have recognized the importance of leadership throughout human history (Chemers 1997).

According to Bass (1990a), "Leadership is one of the world's oldest preoccupations." This proves how the study of leadership is one of the keystones of civilization, a building block off of which humanity has built itself.

Leadership, and its study, has roots in the beginning of civilization. Egyptian rulers, Greek heroes, and biblical patriarchs all have one thing in common, that is leadership. Ancient leadership approach comprises the writings of early philosophers and thinkers who put together their thoughts on leaders, leadership, and necessity of leadership development. *Encyclopedia of Leadership* (Goethals et al. 2004) lists Confucius and Sun Tzu, Aristotle, Plato, Cleopatra, Niccolo Machiavelli, Pareto, Thomas Hobbes, Mary Parker, Bertrand Russell, and several other philosophers and thinkers who have contributed their thoughts to development of leadership theoretical base. These efforts and other philosophical approaches constitute a rich and ongoing normative approach to understanding leadership and seek to provide ethical and constructive views of good leadership. Many of the modern theories of leadership also borrow some ideas from classical thoughts on leadership. Though these theories mostly discuss leadership in very general terms at government, regime, and military levels, modern theories of leadership try to implement these ideas in modern business and organizational leadership.

The Republic by Plato is arguably the first attempt to shed light on the theory of politics and leadership. It was written over 2000 years ago.

The Republic is not a play or a novel; it may be seen as a kind of debate, a fitting description for most of the dialogues. It is Plato's intent in this dialogue to establish, philosophically, the ideal state, a state that would stand as a model for all existing and emerging societies. According to Williamson (2008), Plato's theory of the good life challenges contemporary consumer culture and the definition of the good life as desire satisfaction; his critique of democracy raises difficult questions about the ways democracies train and choose leaders; and his account of the ideal regime illustrates the importance and difficulty of taming endemic conflicts between private interests and the public good. At the same time, Plato offers an account of leadership as benefiting the entire community that remains morally attractive.

Nicomachean Ethics and *Politics* are two of Aristotle (384–322 BC: Greek philosopher, logician, and scientist who studied in Plato's Academy in Athens)'s books, which shed some light on the politics and art of leadership among the early most writing on the subject. Currently, ethics and politics are two distinct study areas. Aristotle, however, unified both areas as part of the practical sciences that are concerned with good action.

The other famous writings come from Sun Tzu *The Art of War,* Niccolo Machiavelli *The Prince,* Vilfredo Pareto *The Treatise on General Sociology,* and so on. These are only some examples of ancient approaches to leadership. Many modern scholars of leadership have written about the wisdom of these ancient approaches offer for a deeper understanding of leadership. Several ideas offered by these approaches still hold. However, increased complexity of business world due to industrialization of early twentieth century rejuvenated the interest in scholarship of leadership.

In recent years, leadership has evolved into an independent field of study that organizes the ideas of historical theorists into proper theories. The following sections are dedicated to the theories that were presented after the dawn of twentieth century.

6.2.2 *The rise of modern leadership*

The Industrial Revolution created a paradigm shift to a new theory of leadership in which "common" people gained power by virtue of their skills (Clawson 1999). New technology, however, was accompanied and reinforced by mechanization of human thought and action, thus creating hierarchical bureaucracies. One major contributor to this era of management and leadership theory was Max Weber, a German sociologist who observed the parallels between the mechanization of industry and the proliferation of bureaucratic forms of organization (Morgan 1997). He noted that the bureaucratic form routinized the process of administration in the same manner that the machine routinized production.

Weber's concerns about bureaucracy, however, did not affect theorists who set the stage for what is now known as "classical management theory" and "scientific management." Classical theorists focused on the design of the total organization while scientific managers focused on the systematic management of individual jobs. In contrast to Weber, classical theorists such as Henri Fayol and F. W. Mooney, loyal advocates of bureaucratization devoted their energies to identifying methods through which this kind of organizational structure could be achieved (Bass 1990a; Morgan 1997). Collectively, these theorists set the basis for many modern management techniques, such as management by objectives.

Scientific management, an approach heralded by Frederick Taylor, was technological in nature (Hersey et al. 1996). Taylor fused the perspective of an engineer into management with a strong emphasis on control, ruthless efficiency, quantification, predictability, and deskilled jobs. He initiated time and motion studies to analyze work tasks to improve worker productivity in an attempt to achieve the highest level of efficiency possible. Consequently, he has been accused of viewing people as instruments or machines to be manipulated by their leaders. The function of the leader under scientific management theory was to establish and enforce performance criteria to meet organizational goals; therefore, the focus of a leader was on the needs of the organization and not on the individual worker.

Although the classical and scientific approaches were different, the goals were similar; organizations are rational systems and must operate in the most efficient manner possible to achieve the highest level of productivity (Morgan 1997). Both theories relied on the machine metaphor with a heavy emphasis on mechanization of jobs, which undermined the human aspect of the organization and failed to recognize organizations as complex organisms (Stone and Patterson 2005).

6.3 *Understanding leadership*

> Logic will get you from A to B. Imagination will take you everywhere.
>
> **Albert Einstein**

Leadership, although largely talked about, has been depicted as one of the least realized concepts across all cultures and nations. Many researchers have emphasized the dominance of this misunderstanding, stating that the existence of several unsound assumptions concerning leadership often interferes with the general conception of what leadership is all about.

There are numerous definitions and theories of leadership; however, there are enough similarities in the definitions to conclude that leadership is an effort of influence and the power to induce compliance (Wren 1995). Our work, work environment, the motivation to work, leaders, leadership, leadership style, and a myriad of other work-related variables have been studied for almost two centuries (Stone and Patterson 2005).

Leadership is one of social science's highly examined phenomena. Nonetheless, leadership is often easy to recognize in practice but it is difficult to describe accurately. Even in this absence of precise description, the literature on leadership offers several theoretical perspectives and a broad definition of leadership that are required before introducing the phenomenon to engineering domain.

Leadership is a quality that all of us can relate to, but it is difficult to describe in a broad context in a way that is applicable to all professionals at all levels. It means different things to different people, for example, getting others to follow, the use of authority in decision-making, a personal characteristic, an ability to achieve effective performance, and a relationship through which one person influences the actions of others.

Leadership comes from the passion of the change agents and identifying who those people are becomes very important. A change agent is any individual or group that initiates and/or facilitates change (Duncan 1978). The definition of leadership is strongly associated with creativity. Good leaders are able to create an environment that will encourage all the members of their team to develop their skills and imagination, so that they can contribute to the common project and vision of the group. Good leaders sets the bar high for their people, because they want to reach the goals and make the best of their teams. Only a demanding leader will achieve good results. In addition to carefulness, the leader must know how to listen, in order to know the needs of the coworkers, and then provide the necessary time and resources for them to do their job properly, and therefore meet what is demanded of them. The leader must be at the forefront to lead and guide his or her team throughout the whole process until the goal is reached. The leader is the one who is responsible for taking the risks that others are not able to take. The leader knows how to motivate better than anyone else. Through motivation, the leader can feed the energy and potential of coworkers, in order to achieve the objectives. Most leadership is by example, and the effect of example is long-tern and seeks endless improvement.

Leadership makes or breaks establishments. Therefore, every aspect of leadership should be examined thoroughly—styles, types, reasons, and rewards—so that effective leadership can be appropriately placed and applied.

Leadership is a skill comprised of many traits and qualities. Some of these qualities include vision, mission, values, commitment, motivation, and consensus building. An important purpose of leadership in our modern age is to provide vision, direction, and motivation for a team of individuals to accomplish a mission that otherwise could not be accomplished by a single individual. Leading means having a vision and sharing it with others. It requires providing the resources and infrastructure for today and the future. It is about creating change. It involves thinking about the imaginable and the unimaginable. As noted in Gaynor (1993), "The whole concept of leadership means creating change as contrasted to maintaining the status quo. It implies thinking of the future, influencing, persuading, changing minds, doing what those above and below may consider unacceptable, sticking your neck out, taking calculated risks, risking yourself as a person in championing a controversial point of view or approach, and having the confidence and ability to speak out and support unpopular but necessary issues."

Today, the field of leadership focuses not only on the leader, but also on followers, peers, supervisors, work setting/context, and culture, including a much broader array of individuals representing the entire spectrum of diversity, public, private, and not-for-profit organizations. Leadership is no longer simply described as an individual characteristic or difference, but rather is depicted in various models as dyadic, shared, relational, strategic, global, and a complex social dynamic (Yukl 2006; Avolio 2007).

There is an ever-growing variety of theories to realize the concept and practice of leadership. A brief overview of the well-known theories is provided in the following section. The goal is to provide an overview that keeps the subject simple, abstaining from any evaluation of the theories.

6.4 Leadership theories

> Leadership is not about knowing all the answers.
>
> **Bruce Rhodes**

6.4.1 Great man theory

> Great leaders are rare, exceptional people, born to lead.

The scientific study of leadership began at the turn of the twentieth century with the "great man" perspective, which saw history as being shaped by exceptional individuals (Bass 1990). The term "great man" was used because, at the time, leadership was understood of primarily as a male quality, especially in terms of military leadership. Developed in the mid-1800s to early 1900s by early Darwinists studying inheritable traits, great man theory became one of the first and most widely held ideas about leadership. This theory stated that leadership is directly dependent on an individual who is rather than what he knows or what he can do, thus basing leadership on inherited factors.

The mythos behind some of the world's most famous leaders such as Julius Caesar, Alexander the Great, Abraham Lincoln, Mahatma Gandhi, Winston Churchill, and Martin Luther King Jr. helped impact the notion that great leaders are born and not made. In many examples, it seems as if the right man for the job seems to emerge almost magically to take control of a situation and lead a group of people into safety or success (Cherry 2016).

Early research on leadership in the beginning of twentieth century examined the leaders who had achieved a level of greatness, and later on, this approach became famous as great man theory. The underlying idea behind this approach was that some individuals are born with certain characteristics and qualities which make them leaders eventually. Research focused on individual characteristics that universally differentiated leaders from nonleaders. Bass (1997) argued that leaders during the early twentieth century were considered to be superior individuals different from the others around them because of skills, capabilities, inherited money, and social standing. The aim was to prepare a master list of traits which would eventually result in an ideal leader.

6.4.2 *Trait theory*

> Leaders are born, not made. Effective leaders have specific sets of innate traits.

The great man approach was followed by a new set of approaches called trait theory. Similar in some ways to great man theory, trait theory assumes that people inherit certain qualities and traits that make them better suited to leadership (Cherry 2016). A leader is thought to have a natural affinity for leadership, with more superior qualities than others that set them apart from their followers. The basic premise behind trait theory was that effective leaders are born, not made—thus the name sometimes applied to early versions of this idea, the great man theory.

The trait theory of leadership is generally considered the first modern theory of leadership. It became popular during the second half of the twentieth century and, despite academic criticism, it has continued to be popular (Cherry 2017). It began with a focus on the traits of effective leaders. Many leadership studies based on this theoretical framework were conducted between 1930s and 1950s. The trait leadership theory believes that people are either born or made with certain qualities that will make them excel in leadership roles. That is, certain qualities such as intelligence, sense of responsibility, creativity, and other values put anyone in the shoes of a good leader (Hattangadi 2015).

The trait theory assumes that people are either born or not born with the qualities that prompt them to success in leadership roles. It is certainly inherited qualities, such as personality and cognitive ability, that inspire effective leadership. The fundamental notion of trait theory was that leaders have certain characteristics that are utilized across time to enhance organizational performance and leader prestige. The idea was that traits affected behaviors and behaviors affected effectiveness. Traits are the distinguishing personal characteristics of a leader, such as physical characteristics, aspects of personality, and aptitudes.

Trait approaches dominated the initial decades of scientific leadership research. Later, they were disdained for their inability to offer clear distinctions between leaders and nonleaders and for their failure to account for situational variance in leadership behavior. Recently, driven by greater conceptual, methodological, and statistical sophistication, such approaches have again risen to prominence. However, their contributions are likely to remain limited unless leadership researchers who adopt this perspective address several fundamental issues (Zaccaro 2007).

According to trait leadership theory, effective leaders have in common a model of personal characteristics that help their ability to mobilize others toward a shared vision. Stogdill (1974) reviewed leadership traits and identified qualities that included age, physique, and appearance; intelligence (verbal, perceptual, and reasoning capabilities); knowledge management (high productive knowledge sharing); responsibility (the art of motivating and encouraging people to engage); integrity (honesty and trustworthiness); emotional control (manage emotions within oneself and relationship); sociability (inclination to seek out pleasant social

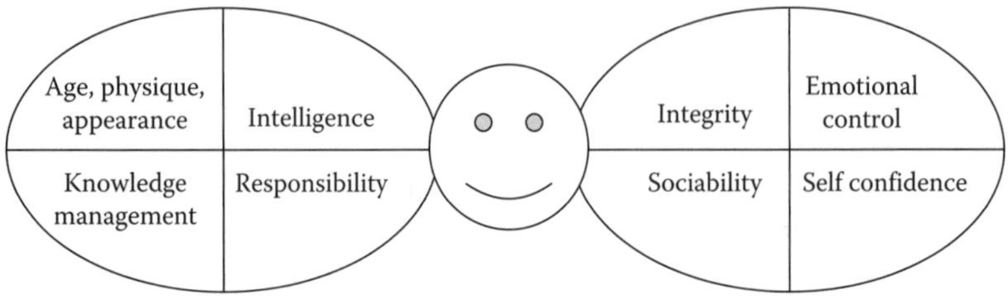

Figure 6.1 Major leadership traits and characteristics.

relationship); and self-confidence (competencies and skills). Figure 6.1 shows major leadership traits and characteristics.

6.4.3 Behavioral theory

> Leaders are made, not born. Effective leaders use specific sets of behaviours or styles. This style focuses on the behavior of the leader and what leaders do and how they act.

The trait movement gave way to the behavioral theory of leadership in the 1950s. It became clear that success of the style of leader behavior performed was depending on the situation. The behavioral theory of leadership relies upon the idea that great leaders are made, not born. It considers the recognizable actions and reactions of leaders and followers in a given situation. Accordingly, people can learn to become leaders through learning and reflection.

Behavioral theory promotes the value of leadership styles with an emphasis on concern for people and collaboration. It promotes participative decision-making and team development by supporting individual needs and aligning individual and group objectives. Behavior theory consists of skills and styles that differ from traits since behaviors can be developed. The theory focuses on the study of specific behaviors of a leader. A typical view was that leadership involved in two kinds of behaviors: those that were mission oriented and that led to productivity, and those that were person oriented and that were sensitive to people's feelings (Sternberg 2005).

There are two key strengths of behavior theory. First, it stands for the idea that leadership traits can be learned through training and experience. This is unlike the great man theory by claiming that leaders can be made, and are not really born with unique traits. Second, this new idea that leaders can be made has inspired more recent leadership theory research that is based on developing the leaders to generate desired outcomes. However, there is still a lack of knowledge on how behavior theory can be used in various cultural situations. One behavior that may works in certain cultural situation may not work in another situation.

6.4.4 Situational theory

> The most effective leadership style is dependent on situational variables.

The situational leadership model is arguably the most recognized in the history of the behavioral sciences. Developed by Dr. Paul Hersey in the late 1960s, the model is a powerful, yet flexible tool that enables leaders of all kinds—managers, salespeople, peer leaders, teachers, or parents—to more effectively influence others (Hersey and Blanchard 1969; Hersey 1985). Situational leadership is a theory of leadership that is part of a group of theories known as contingency theories of leadership which hold that a leader's effectiveness is related to the leader's traits or behaviors in relation to differing situational factors. The situational leadership suggests that there is no "one size fits all" approach to leadership. Depending on the situation, varying levels of leadership and management are necessary.

The situational view is generally viewed today as oversimplified. First, whereas it rightfully acknowledges the importance of situations, it fails to acknowledge the importance of individual differences. Some leaders, in a given situation, fail, and hence are replaced, often by people who succeed better. For example, Steve Jobs took over the ailing Apple Company from Gil Amelio, as Lou Gerstner took over the diminishing IBM Corporation from John Akers. The successors saved their companies from further decline, showing that individuals matter, not just situations. Second, the situational approach fails to recognize the interaction between persons and situations. A given situation may work for one person and not for another (Sternberg 2005). Depending on situation, leaders must utilize the most suitable leadership style. Four basic situational leadership styles are given in Figure 6.2. Coaching is a type of situational leadership style that comprises a vast deal of involvement by providing guidance in a working process. This style is effective when coworkers are more responsible and experienced. The directing leadership style involves providing direct instructions while providing tasks over a challenging situation and applying specific knowledge to handle the situation. The delegating style places more of the responsibility on the shoulders of the coworkers by transferring decision making power to one or more coworkers. In the supporting style, leaders play more kind of a motivational role by reducing coworkers stress and frustration.

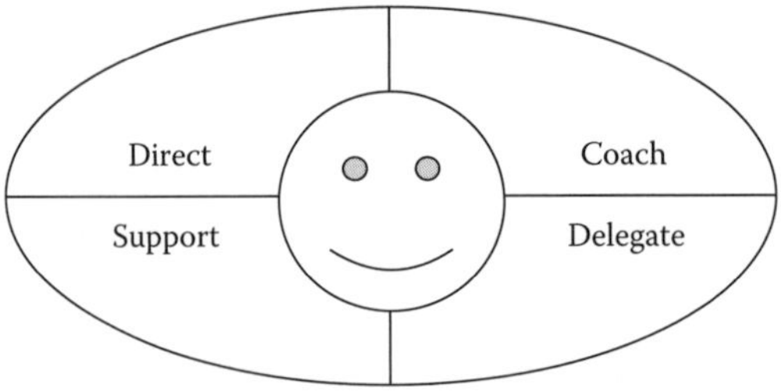

Figure 6.2 The four basic situational leadership styles.

6.4.5 Charismatic theory

> Leaders have a divinely inspired gift that inspires followers via special relationships.

The credit of introducing the word "charisma" goes to German sociologist Max Weber (Weber 1958). Charisma is a Greek word meaning "divinely inspired gift," which shows an unusual quality to charismatic individuals by which they can influence others, do miracles, or foresee the future. According to Max Weber, charisma is a quality of an individual by virtue of which he or she is treated as endowed with supernatural, superhuman, or exceptional powers or qualities. Charismatic leaders enjoy position and personal power. They are able to express a convincing vision and are able to stimulate convincing emotions in followers.

The main issue that can spoil leaders is arrogance and a lack of concern to followers. They may prone to narcissism which may lead them to self-serving ways. They, at times, become forgetful of the organizational needs and start pursuing their personal goals. The overwhelming power and importance lead them to ignore the followers.

Depending on the nature of the organization, leaders can be labeled as affective, instrumental, or both. Like leadership in general, charismatic leadership has a wide range of definitions in different cultural and temporal settings. Charismatic leaders are essentially very skilled communicators, individuals who are both verbally expressive and also able to communicate to followers in a deep, emotional way (Epley 2015).

6.4.6 Transactional theory

> Leaders who focus on ways to maintain the status quo and manage the daily operations.

Transactional leadership is a theory considered to be value-free. This theory was first described by Max Weber in 1947 and then by Bernard Bass in 1981. The underlying theory of this leadership method was that leaders exchange rewards for employees' compliance, a concept based in bureaucratic authority and a leader's legitimacy within an organization (Tracey and Hinkin 1994). This style is most often used by the managers. Examples of this reward exchange included the leader's ability to fulfill promises of recognition, pay increases, and advancements for employees who perform well (Bass 1990b).

Transactional leadership focuses on ways to maintain the status quo and manage the day-to-day operations of a business. It does not focus on identifying the organization's goals and how employees can work toward and increase their productivity in alignment with these goals, thus increasing organizational profitability (Avolio et al. 1991).

Transactional leaders approach followers with a goal of exchanging one thing for another (Burns 1978). The concept of transactional leadership is narrow in that it does not take the entire situation, employee, or future of the organization in mind when offering rewards. Transactional leadership focuses on control, not adaptation (Tracey and Hinkin 1994).

6.4.7 *Transformational theory*

> Leaders use charisma, intellectual stimulation, individualised consideration and inspirational motivation to inspire colleagues to reach organisational goals.

Transformational approaches to leadership originate in the work of Burns (1978). Burns suggested that there are essentially two ways of performing leadership functions. One is where there is an implicit or explicit contractual relationship between the leader and his or her followers. This type of leadership, which has come to be called transactional leadership, is characterized by followers agreeing to do certain stipulated things in exchange for the leader (usually a boss) doing other things. A second and more powerful kind of transformational leadership tries to gain converts to ideas.

This theory focuses on the interactions that occur between leaders and followers. It is based on the notion that a leader's job is to create structures that make it clear what is expected of coworkers and also the consequences (e.g., rewards and punishments) for meeting or not meeting the expectations. This theory is often likened to the concept of management and continues to be an extremely common component of many leadership models and organizational structures. It states that the process is by which a person interacts with others and is able to create a firm relationship that results in a trust, that will later result in an increase of motivation, both intrinsic and extrinsic, in both leaders and coworkers. The essence of transformational theories is that leaders transform their coworkers through their inspirational nature and charismatic personalities. Rules and regulations are adaptable, driven by group standards. These attributes provide a sense of belonging for the coworkers as they can easily recognize with the leader and goal.

With transformational leadership, the leader's focus is directed toward the organization, but leader behavior builds follower commitment toward the organizational objectives through empowering followers to accomplish those objectives. While transactional leaders focus on exchange relations with followers, transformational leaders inspire followers to higher levels of performance for the sake of the organization (Burns 1978).

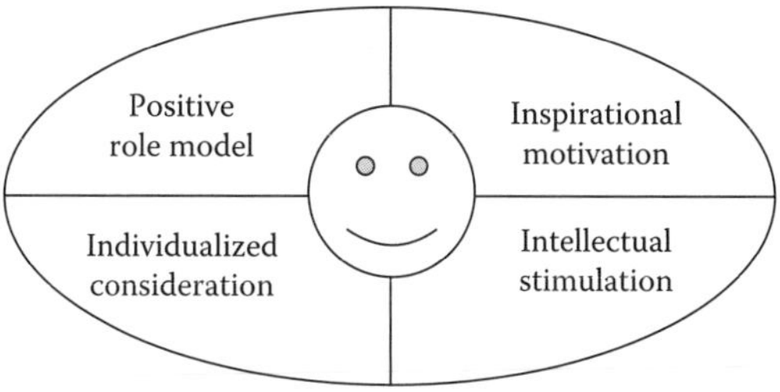

Figure 6.3 The four basic elements that make transformational leaders.

The very definition of transformational leadership states the building of commitment to the organizational objectives. The primary focus is on the organization, with follower development and empowerment secondary to accomplishing the organizational objectives. The result, nonetheless, is enhanced follower performance (Stone and Patterson 2005). In general, transformational leadership theory is proactive dominant theory that works to change the organizational culture by implementing new ideas.

Transformational leaders in general focus on followers, motivating them to high levels of performance, and in the process, help followers develop their own leadership potential. Four basic elements make transformational leaders (see Figure 6.3). The first element is to be a "positive role model." The second is "inspirational motivation," where transformational leaders have the ability to inspire and motivate followers. The next two elements are about the leader–follower relationship. "Individualized consideration" is when transformational leaders demonstrate genuine concern for the needs and feelings of followers. The final element is called "intellectual stimulation," where leaders challenge followers to be innovative and creative. A common misunderstanding is that transformational leaders are "soft," but the truth is that they constantly challenge followers to higher levels of performance (Riggio 2014b).

6.5 *Leadership theories of motivation and management*

> One secret of leadership is that the mind of a leader never turns off. Leaders even when they are sightseers or spectators, are active; not passive observers.
>
> **James Humes**

6.5.1 Motivation and management

Motivation comes from the word "motive" which implies needs, desires, or drives within the individuals. It is driven by aspiration and ambition. It involves the biological, emotional, and social forces that activate behavior. Motivation is the driving force which helps to achieve goals. It may also be defined as the degree to which determined effort is directed toward a goal.

In management, motivation depicts ways in which managers stimulate productivity in their coworkers. The initial phase of motivation refers to the extent of effort being applied to complete the task. The second characteristic relates to the enthusiasm of the individual to stay with a task until it is finished. The third is whether the effort is directed toward the organization's goals or related to the individual's selfishness.

Several theories have been intended in relation to the psychological interaction between an organization management and the coworkers. In this section, we will only consider three theories that deal with the motivational factors that influence coworkers' behavior, a critical approach in an engineering organization.

6.5.2 Theory X and Theory Y

Theory X and Theory Y are theories of human motivation and management. Douglas McGregor of Massachusetts Institute of Technology (MIT), an American social psychologist, proposed his famous XY theories in his 1960 book "The Human Side of Enterprise." Theory X and Theory Y are still referred as two contrasting theories in the field of management and motivation, and while more recent studies have questioned the rigidity of the model, McGregor's XY theories remain a valid basic principle from which to develop positive management style and techniques. They are central to organizational and cultural development. The XY theory is a constructive and simple reminder of the natural rules for managing people, which under the pressure of day-to-day business are all too easily forgotten. McGregor's ideas propose two important approaches to managing people.

Theory X and Theory Y created by McGregor has been a valid basic principle from which to develop positive management style and techniques. McGregor's ideas suggest that there are two fundamental approaches to managing people. Several managers influenced by Theory X, and generally get poor results. On the other hand, liberal managers use Theory Y, which produces better performance and results, and allows people to grow and develop (Haji Mohamed and Mohamad Nor 2013).

Theory X assumes that people are naturally lazy, try to avoid work as much as possible, attempt no responsibility, and prefer to be supervised. It is more suitable for an organization in which the employees do not like their work situation and will avoid work whenever possible. In such case, the employees have to be forced, controlled, or reminded in order for the organization to meets its objectives. In brief, Theory X is an authoritarian style, where the emphasis is on productivity, on the concept of a fair day's work. The problem with Theory X view is that treating people that way may lead them to act that way, creating a self-fulfilling prophecy.

Progressive managers use Theory Y, which generates better performance and allows people to grow and develop. Theory Y is appropriate for an organization in which the employees like their jobs and they tend to be self-directed. It is indeed a democratic form of leadership, where people will apply self-control and self-direction in the quest of organizational goals, without external control or the threat of punishment. Theory Y assumes that people will actively participate in the realization of organizational objectives. It is management's main task in such a system to boost that commitment.

McGregor's ideas significantly relate to modern understanding of the psychological contract, which provides many ways to appreciate the unhelpful nature of Theory X leadership, and the useful constructive beneficial nature of Theory Y leadership. Theory X assumes that individuals are base, work-shy, and constantly in need of a good prod. It always has a ready-made excuse for failure—the innate limitations of all human resources. Theory Y, however, assumes that individuals go to work of their own accord, because work is the only way in which they have a chance of satisfying their high-level need for achievement and self-respect. Figure 6.4 shows Theory X and Theory Y leadership scenarios (Hersey et al. 1996).

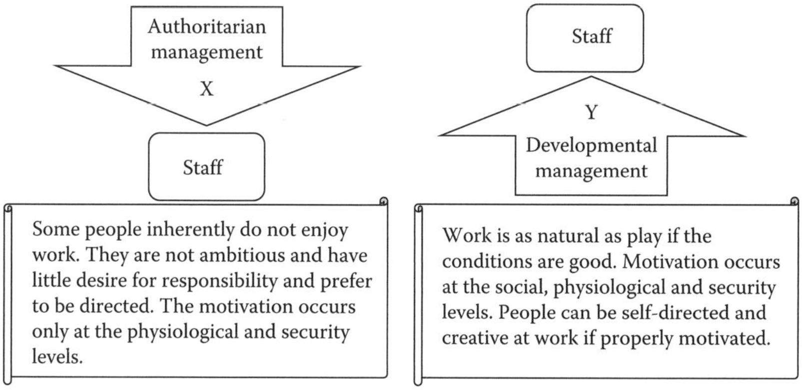

Figure 6.4 Theory X and Theory Y leadership scenarios.

6.5.3 Theory Z approach to management

Theory Z was formed by William Ouchi of UCLA, Los Angeles, in his book "Theory Z: How American Management Can Meet the Japanese Challenge (Ouchi 1981)." Theory Z is often referred to as the Japanese management style, which is basically what it is. Professor Ouchi spent years researching and examining Japanese and American companies using the Theory Z management styles. By the 1980s, Japan was known for the highest productivity in the world, while America's productivity had fallen significantly.

It is interesting that Ouchi chose to name his model "Theory Z," which apart from anything else tends to give the impression that it is a McGregor idea. For Ouchi, Theory Z focused on increasing employee loyalty to the company by providing a job for life with a strong focus on the well-being of the employee, both on and off the job. Theory Z essentially advocates a combination of all that is best about Theory Y and modern Japanese management, which places a large amount of freedom and trust with workers, and assumes that workers have a strong loyalty and interest in teamworking and the organization.

According to Ouchi, Theory Z, management tends to promote stable employment, high productivity, and high employee morale and satisfaction. Theory Z also places more reliance on the attitude and responsibilities of the workers, whereas McGregor's XY theory is mainly focused on the management and motivation from the manager's and organization's perspective. One of the most important pieces of this theory is that management must have a high degree of confidence in its workers in order for this type of participative management to work. There is no doubt that Ouchi's Theory Z model offers excellent ideas, although it is lacking the simple elegance of McGregor's model, where thousands of organizations and managers around the world have still yet to embrace. For this reason, Theory Z may for some be like trying to manage the kitchen at the Ritz before mastering the ability to cook a decent fried breakfast (Ouchi 1981).

6.6 Emotional intelligence

All learning has an emotional base.

Plato

6.6.1 EQ defined

Emotions are important segments of information that explain human behaviours. The talent to express, understand, and manage emotions is critical. In general, emotions can be constructive or destructive. Examples of constructive emotions include accomplishment, belonging, resolution, resolve, empathy, trust, positivity, and, sometimes, a bit of healthy fear. On the other hand, examples of destructive emotions include anger, apathy, disrespect, despair, envy, fear, negativity, and doubt.

The term EQ was first defined by Peter Salovey and John Meyer in 1990. Their work has since been significantly expanded by Daniel Goleman, an American psychologist (in his role as a science reporter at The New York Times) who helped to popularize EQ (Goleman 1995, 1998). Those were times when the notion of IQ as the standard of excellence in life was unquestioned; a debate raged over whether it was set in our genes or due to experience. Goleman identified that IQ is actually less important for success in life and work than EQ, a set of skills that are not directly related to academic ability. Goleman has asserted that EQ abilities were about four times more important than IQ in determining professional success and prestige, even for those with a scientific background (Golemen 1998).

6.6.2 EQ domains and competencies

EQ is the ability to understand and manage oneself emotions, and those of others. EQ differs from what the people think of intellectual ability, in that EQ is learned, not acquired. EQ is the ability to understand and manage emotions in positive ways to ease stress, communicate efficiently, understand others, overcome challenges, and resolve conflict.

The good values are essential in preparing one to be happy and succeed in their career later. Therefore, EQ needs to be developed and nurtured to create high-quality and competitive graduates. People with a high degree of EQ know what they are feeling, what their emotions mean, and how these emotions can affect other people. For leaders, having EQ is essential for success. After all, who is more likely to succeed is a leader who shouts at his team when under stress, or a leader who stays in control and calmly assesses the situation. According to Goleman (1998), effective leaders are alike in one crucial way: they all have a high degree of EQ. He identified five competencies of EQ along two major domains including self-mastery skills and social mastery skills as shown in Table 6.1.

6.6.3 EQ for engineers

It should be noted that EQ is not the opposite of IQ (Riemer 2003). Many elements of EQ might be more familiar to engineers as "soft skills" and "people skills," but those skills alone minimize the value of EQ which

Table 6.1 EQ competencies and domains

Self-mastery	
Self-awareness	Ability to identify own emotions and their impact. Being emotionally aware is just the first step to emotional management.
Self-regulation	Ability to control emotions and behavior. Exert greater self-control: like a traffic signal, stop (red)/think (yellow)/act (green).
Self-motivation	Stay positively self-motivated with a desire to do things by an interest in learning. It is also self-improvement versus a pursuit of wealth and status.
Social mastery	
This is feasible when one has achieved self-mastery	
Empathy	Ability to understand and share others experiences and emotions. It is about compassion and caring. Empathy is a matter of projecting outside of oneself to appreciate what other people are going through.
Social skills	Ability to recognize and understand the emotions of others and connect with them.

provides us with the characteristics we need to be more successful human beings. EQ skills can be applied across different work environments including engineering. Such skills are mostly applicable in this era of globalization, which is taking place at an ever-increasing speed and provides an environment in which the modern engineer must engage.

It has been stated that in industry, IQ gets you hired, but EQ gets you promoted. For example, a manager at AT&T Bell Labs was asked to rank his top performing engineers. High IQ was not the deciding factor, but instead how the person performed regarding answering e-mails, how good they were at collaborating and networking with colleagues (rather than lone wolf), and their popularity with others (rather than socially awkward) in order to achieve the cooperation required to attain the goals (Gibbs 1995).

This example highlights the benefits of high EQ regarding communication skills, time management, teamwork, leadership skills, customer services, operational results, moral and retention, earning, and business acumen. Such important skills flow on from EQ, like the skillful recognition of others' emotional reactions and empathy to come across as genuine and warm, which will achieve greater cooperation from others, rather than being perceived as oblivious and boorish (Salovey and Meyer 1998).

EQ's benefits broaden beyond the workplace. It helps at home, saves marriage, and provides the ability to talk to teenager. It helps us be successful, well-rounded people in all aspects of our lives (Platt 2015).

Given the principles stated above, it becomes quite apparent that encouraging EQ abilities should be a component of engineering education. This becomes especially relevant given that the skills that employers value include a willingness to learn, flexibility, communication skills, teamwork, and other forms of working with others. Because such skills fall into the category of EQ, universities need to be aware of industry demands on graduates (Riemer 2003).

6.7 *Positive psychology and leadership*

> A dream you dream alone is only a dream. A dream you dream together is reality.
>
> **John Lennon**

6.7.1 *Positive psychology*

The word psychology is made up of two Greek roots, psyche, meaning mind, and logos, meaning word (Kalat 2013), and is translated as the study of the mind (Kalat 2013). The American Psychological Association (APA) defines the field of psychology as follows: "Psychology is the study of the mind and behavior. The discipline embraces all aspects of the human experience … in every conceivable setting … the understanding of behavior is the enterprise of psychologists" (APA 2013). Psychological knowledge is essential to scientific and technological innovation. Technology requires the use of human operators, and understanding human capacities and limits is essential for implementing technological advances.

After World War II, the field of psychology was driven to adjust behavioral and psychological issues. In the 1950s, Abraham Maslow identified this shortfall in psychology science. This was reintroduced by the American psychologist Martin Seligman who advocated that another branch of psychology should be created, one that focuses on the building of positive qualities and study of what is going right in individuals who are happy, positive, and satisfied.

The subfield of positive psychology, a relatively new branch within the field of psychology, focuses on helping people and organizations create well-being and meaning in their lives in order to understand and enhance their human experiences (Adams 2012). Positive psychology focuses on thriving individuals, especially on their strengths and virtues, on their subjective experiences, and on living a good life.

6.7.2 Positive leadership

When positive psychology is applied in the workplace, the outcome is positive organizational behavior (POB). POB leads to increased optimism, positivity, resiliency, and efficacy. Positive emotions have been found to predict happiness, well-being, improved outcomes, satisfaction, and success at work. Positive leaders drive positive emotions within themselves and their followers, and may lead to more work satisfaction, better engagement, increased performance, and better atmospheres.

Managers who infuse positive psychology into their leadership style can create a positive psychological work culture within the employee base and allow employees to begin to innately practice antecedents of innovation such as creativity and engagement that leads to innovation itself. One way for leaders to infuse their leadership style with positive psychology is to cultivate the use of their innate character strengths. Leaders can use positive psychology to create and support a more positive organizational culture that in turn creates positive psychological states within the employees of an organization. Within this new psychological paradigm, employees are not merely products of cultural events and experiences; they become invested in or take psychological ownership of their work (Avey et al. 2009), working from a more innovative mind-set.

Positive leadership can be defined as the application of character strengths to leadership and management situations in order to create a virtuous organization where employees can flourish. Virtues are values that have been identified by various philosophical systems. Positive psychology's framework encompasses six virtues, which include wisdom and knowledge, courage, humanity, justice, temperance, and transcendence (Peterson and Seligman 2004). Character strengths are behavioral ways in which the virtues are expressed.

Conventional wisdom supports the idea that leaders whose leadership style entails positive attributes would elicit positive outcomes (Avey et al. 2011). While interest in both positivity and leadership has grown over the last decade (Avey et al. 2011), it has only been recently that researchers have given direct attention to the interplay of positivity and the leader–follower dynamic (Carmeli et al. 2009). Several theories have been put forward, but no specific definition of positive leadership exists (Youssef and Luthans 2012). Attributes that are usually assigned to positive leadership include a charismatic nature, positive directive, and a participatory style when engaging subordinates (Linley et al. 2010). These attributes can be found in various leadership theories. Examples of these include the authentic leadership theory, the transformational leadership theory, the charismatic leadership theory, the altruistic leadership theory (Donaldson and Ko 2010), and the spiritual leadership theory (Youssef and Luthans 2012).

6.7.3 Authentic leadership

Don't find fault, find a remedy.

Henry Ford

Webster defines authenticity as "real or genuine; not copied or false; true and accurate." It comes from the Greek word for author. Authenticity as a construct dates back to at least the ancient Greek philosophy, as captured by their timeless reproach to "be true to oneself" (Harter et al. 2002). The Greek philosophy focused on the development of core, or cardinal, virtues. These virtues are prudence (fair-mindedness, wisdom, seeing all possible courses of action), temperance (being emotionally balanced and in control), justice (being fair in dealings with others), and fortitude (courage to do the right thing) (Riggio 2014a).

Authentic leadership brings together the concept of authenticity with positive psychology; it focuses on whether leadership is genuine. Luthans and Avolio (2003) defined authentic leadership as "a process that draws from both positive psychological capacities and a highly developed organizational context, which results in both greater self-awareness and self-regulated positive behaviors on the part of leaders and associates, fostering positive self-development."

A theory of authentic leadership has been emerging over the last several years from the intersection of the leadership, ethics, and POB and scholarship literature works (Luthans 2002; Cameron et al. 2003; Luthans and Avolio 2003; Avolio et al. 2004). As conceptualized within the emerging field of positive psychology (Seligman 2002), authenticity can be defined as "owning one's personal experiences, be they thoughts, emotions, needs, preferences, or beliefs, processes captured by the injunction to know oneself" and behaving in accordance with the true self (Harter et al. 2002). The definition and the theory of authentic leadership were developed to define four components as shown in Table 6.2.

Authentic leaders have insight, sometimes referred to this as vision, but that usually has exclusive reference to the future. They demonstrate initiative. They go first. They do not sit on the sidelines. They do not ask others to do what they are unwilling to do themselves. Authentic leaders are people of extraordinary integrity who lead with purpose, meaning, and values with strong people relationships. Authentic leadership involves insight, initiative, impact, influence, and integrity. The most important skill a leader can master is the ability to listen in a way that surfaces the underlying concerns of another and finds the intersections between seemingly opposing points of view. Authentic leaders instill work values that are personal convictions about the outcomes that are expected from work and the type of behavior of followers (George and Jones 2008).

Table 6.2 Four components of an authentic leadership

Self-awareness (Know thyself)	To know your strengths, limitations, and values. Self-awareness is needed in order to develop the other components of authentic leadership.
Relational transparency (be genuine)	Being honest and straightforward in dealing with others. An authentic leader does not play games or have a hidden agenda.
Balanced processing (be fair-minded)	Soliciting opposing viewpoints and considering all options before choosing a course of action.
Internalized moral perspective (do the right thing)	Having an ethical and fairness core and knowing the right thing to do.

There are many prominent theories about authentic leadership, including one from a well-known book written by former Medtronic CEO and Chairman Bill George, called "Authentic Leadership: Rediscovering the Secrets to Creating Lasting Value," (George 2003), which was published in 2003. This book boosted the authentic leadership management style into prevalent acceptance. It urged the new generation of twenty-first century leaders to lead with their hearts as well as their minds, with passion and an ethical code. George emphasizes that anyone can become an authentic leader through hard work and developing their leadership qualities. He proposes that to begin developing authentic leadership style, it is necessary to follow five qualities including understanding purpose, practicing values, leading with heart, establishing connected relationships, and demonstrating self-discipline.

The essence of authentic leadership is EQ. People with high IQs and low EQs can hardly be called authentic leaders. In contrast to IQ, which basically does not change in one's adult life, EQ can be developed. The first and most important step on this journey is gaining self-awareness (George 2016).

6.7.4 *Innovation leadership*

Innovative leadership is the ability to think differently and motivate others to create new and better ideas to move toward positive results. It has two approaches. First is an innovative approach to leadership to bring new thinking and different actions to how to lead, manage, and go about work. Second is about leadership for innovation where leaders must learn how to create an organizational climate where others apply innovative thinking to solve problems and develop new products and services. It is about growing a culture of innovation, not just hiring a few creative outliers (Horth and Buchner 2014).

Innovation is identical to creativity which involves a great deal of risk. An innovation leader has high-risk tolerance, and has the ability to consider all possibilities to make well-calculated risks that often pay off. The innovative leader ignores the conventional order and tries new ones. Such a leader rallies the entrepreneurial energy, champions innovation, and, if required, push past the drives of negativity. Innovative leaders may not always be successful each time they try something, but they set up a direction and they often cause large changes.

Innovation leadership in organizations is not about micromanagement, but it is about the big picture and works well with creative thinking that adds to the vision and makes it greater rather than focusing far too much on details. Today, innovation leaders are generally required in technology-based organizations that evolve rapidly. Therefore, such leaders must have the required skills and knowledge to communicate effectively with their teams, deliver a cohesive vision, and realize the inherent risks and benefits of creativity and innovation.

Innovation needs certain antecedents for the desired change or innovation to take place. These antecedents include creativity (Eisenbeiß and Boerner 2010), an engaged workforce (Bhatnagar 2012), a positive work culture (Shipton et al. 2006), and positive emotions (Fredrickson and Cohn 2008). Management that allows leadership to be creative and deploy various leadership styles plays a large role in creating an environment of innovation within their organizations (Hsiao and Chang 2011). This is true no matter what model of leadership is relevant to the organization: hierarchy and responsibility or impact and influence.

The application of character strengths in the workplace correlates with engagement, an antecedent of innovation and positive experiences (Harzer and Ruch 2013), which are associated with the creation of an environment where innovative work behavior can flourish. Positive experiences within a workplace can arguably generate a positive culture, which is an antecedent for innovative work behavior. Figure 6.5 shows the impact of positive psychology on creating innovation environment.

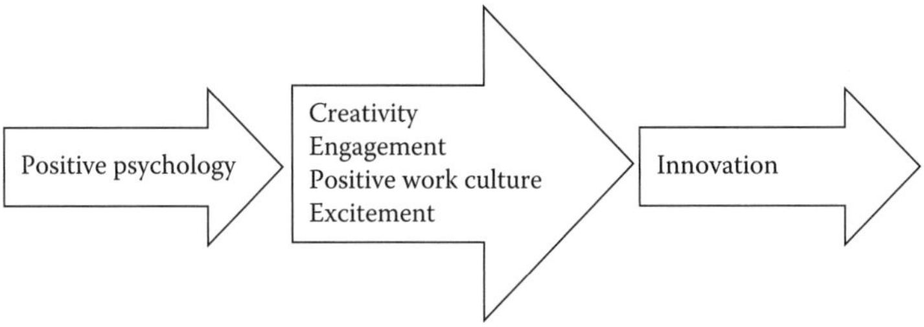

Figure 6.5 Impact of positive psychology on creating innovation environment.

Many of today's leadership problems are critical and pressing; they demand quick and decisive action, but they are complex. Because the organization, team, or individual does not know how to act, there is a need to slow down, reflect, and approach the situation in an unconventional way using innovative thinking (intuitive), which is a crucial addition to traditional business thinking (logical). It allows us to bring new ideas and energy to leader and to solve challenges. It also paves the way to bring more innovation into organizations (Horth and Buchner 2014).

6.8 Leadership styles

> At one time leadership meant muscle; but today it means getting along with people.
>
> **Indira Gandhi**

Leadership style is the manner and approach of providing direction, implementing plans, and motivating people. As seen by the employees, it includes the total pattern of explicit and implicit actions performed by their leader (Newstrom and Davis 1993). On the other hand, it is the result of philosophy, personality, and experience of the leader.

The first major study of leadership styles was performed in 1939 by Kurt Lewin who led a group of researchers to identify different styles of leadership (Lewin et al. 1939). This early study has remained quite influential as it established the three major leadership styles: authoritarian or autocratic, participative or democratic, and delegative or laissez-faire (free rein). A good leader uses all three styles, depending on what forces are involved between the followers, the leader, and the situation.

Leaders approach their employees in different ways. Positive leaders use rewards, such as education, new experiences, and independence, to motivate employees, while negative employers emphasize penalties (Newstrom and Davis 1993). On the other hand, negative leaders act dominant and superior with people. They believe the only way to accomplish is through penalties. They believe their authority is increased by alarming everyone into higher levels of productivity.

Different situations call for different leadership styles. Most leaders do not strictly use one or another, but are somewhere on a continuum ranging from extremely positive to extremely negative. People who continuously work out of the negative are bosses, while those who primarily work out of the positive are considered great leaders. Figure 6.6 reflects the impact of knowledge and skills in leadership styles.

Figure 6.6 Leadership styles as a function of knowledge and skills.

Today, there are many distinct leadership styles that have evolved over the last years, each embodying a different set of traits and skills. Different styles are needed for different situations and each leader needed to know when to exhibit a particular approach. The following are the six leadership styles that psychologist and author Daniel Goleman uncovered among the managers he studied, as well as a brief analysis of the effects of each style on the corporate climate. Table 6.3 shows the six common styles of leadership (Benincasa 2012).

6.9 *The three levels of leadership model*

> I alone cannot change the world, but I can cast a stone across the water to create many ripples.
>
> **Saint Teresa**

The three levels of leadership is a modern leadership model formulated by James Scouller (2011). The model was designed as a practical tool for developing a person's leadership presence, with an aim to summarize what leaders have to do, not only to bring leadership to their organization, but also to develop themselves technically and psychologically as leaders. The three levels referred to in the model's name are public, private, and personal leadership as shown in Figure 6.7. It is sometimes known as the 3P model of leadership (the three Ps standing for public, private, and personal leadership).

Table 6.3 The six common styles of leadership

Leader style	Description
Pacesetting *Do as I do, now*	This style works well when the team is already motivated and skilled, and when the leader needs quick results.
Authoritative *Come with me*	This style works well when the team needs a new vision. It inspires an entrepreneurial spirit and vibrant enthusiasm for a mission. It is not the best fit when the leader is working with a team of experts who know more than him or her.
Affiliative *People come first*	This style works best in times of stress when the team needs to rebuild trust. It should not be used exclusively, because a sole reliance on praise and nurturing can foster mediocre performance and a lack of direction.
Coaching *Try this*	This style works well when the leader wants to build lasting personal strengths that make them more successful overall. It is least effective when teammates are defiant and unwilling to change or learn, or if the leader lacks proficiency.
Conceiving *Do what I tell you*	This style is most effective in times of crisis or during an actual emergency. However, it should be avoided in almost every other case because it can alienate people and suppress flexibility and inventiveness.
Democratic *What do you think*	This style is most effective when the leader needs the team to buy into or have ownership of a decision, plan, or goal. It is not the best choice in an emergency situation.

The "three levels of leadership" model combines the strengths of older leadership theories. These include the traits, behavioral, situational, and functional models. It addresses their limitations. Also, it offers a foundation for leaders who want to apply the philosophies of servant leadership. Hence, it is for those who are committed to "authentic leadership" (Smith 2016).

The first two levels, public and private leadership, are "outer" or "behavioral" levels. Scouller (2011) distinguished between the behaviors involved in influencing two or more people simultaneously (what he called "public leadership") from the behavior needed to select and influence individuals one to one (which he called "private leadership"). The third level, personal leadership, is an "inner" level and concerns a person's leadership presence, know-how, skills, beliefs, emotions, and unconscious habits. He listed 34 distinct "public leadership" behaviors and further 14 "private leadership" behaviors.

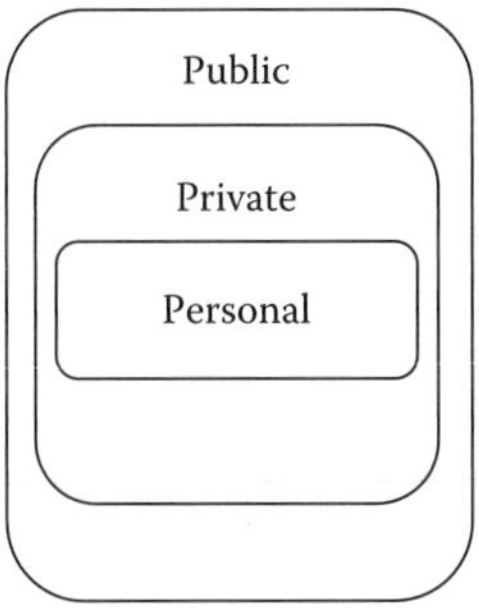

Figure 6.7 The three levels of leadership.

At the heart of the model is the leader's self-awareness, his progress toward self-mastery and technical competence, and his sense of connection with those around him. It is the inner core, the source, of a leader's outer leadership effectiveness. The idea is that if leaders want to be effective they must work on all three levels in parallel. The two outer levels, public and private leadership are what the leader must do behaviorally with individuals or groups to address the "four dimensions of leadership." These dimensions are as follows:

1. A shared, motivating group purpose or vision
2. Action, progress, and results
3. Collective unity or team spirit
4. Individual selection and motivation

The inner level, personal leadership refers to what leaders should do to grow their leadership presence, know-how, and skills. It addresses the leader's technical, psychological, and moral development and its impact on his or her leadership presence, skills, and behavior. It has three aspects:

1. Developing technical know-how and skill
2. Cultivating the right attitude toward other people
3. Working on psychological self-mastery

Scouller (2011) argued that self-mastery (a psychological process) is the key to growing one's leadership presence, building trusting relationships with followers, and enabling behavioral flexibility as circumstances change, while staying connected to one's core values. To support leaders' development, he introduced a new model of the human psyche and outlined the principles and techniques of self-mastery. The assumption in this model is that personal leadership is the most powerful of the three levels. He likened its effect to dropping a pebble in a pond and seeing the ripples spreading out from the center, hence the four arrows pointing outward in the diagram. According to Scouller (2011), "The pebble represents inner, personal leadership and the ripples the two outer levels. Helpful inner change and growth will affect outer leadership positively. Negative inner change will cause the opposite."

The importance and development of leadership presence is a central feature of the model. Scouller (2011) suggested that it takes more than the right know-how, skills, and behaviors to lead well; that it also demands "presence." Presence is not the same as charisma; it is something deeper, more authentic, more fundamental, and more powerful and does not depend on social status. It comes from personal power and command over one's thoughts, feelings, and actions; high, real self-esteem; and the drive to be more, to learn, to grow.

6.10 Leadership and ST

> We can't solve problems by using the same kind of thinking we used when we created them.
>
> **Albert Einstein**

6.10.1 ST defined

Systems, like the human body, have parts, and the parts affect the performance of the whole. All of the parts are interdependent (Reed 2006). The order in which parts are arranged affects the performance of a system.

ST is a discipline for seeing wholes. It is a framework for seeing interrelationships rather than things, for seeing patterns of change rather than taking static snapshots (Ambler 2013). ST is a perspective, a language, and a set of tools (Monat and Gannon 2015). It is a thinking set that may help leaders identify and remedy complex problems by understanding not only the full extent of the problem but also the reason(s) of the problem.

The concept had been developed by Professor Jay W. Forrester at MIT Sloan School of Management. Many theories associated with systems theory come from its mathematical offshoots, but general ST applications and advancements can be seen in disciplines ranging from medicine and engineering to psychology, political science, and art (Haines 1998). Thus, the ST approach even from its historical origin point is complex of science with possibility to understand reality from more than one point. It is a method of critical thinking by which relationships between the system's parts are analyzed in order to realize a situation for better decision-making.

The concept of ST was popularized by Peter Senge in his book *The Fifth Discipline* (Senge 1990) where he describes ST as: "A discipline for

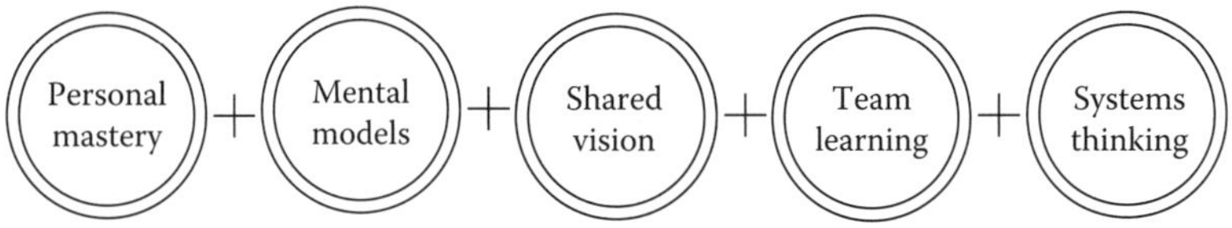

Figure 6.8 Five components of learning transformation.

seeing wholes. It is a framework for seeing interrelationships rather than things, for seeing patterns of change rather than static snapshots." ST is a management discipline that concerns an understanding of a system by assessing the linkages and interactions between the components that encompass the wholeness of that defined system. It is useful for seeing and understanding wholes, interrelations, and change patterns. ST in the context of this book assesses how the world's ecosystems work together with organizational, political, and societal systems as well.

Senge stated that there are five components that, when put together in a company, transform it into a learning organization. Each provides a vital dimension in building organizations that can truly learn and continually enhance their capacity to realize their highest aspirations. The five components are shown in Figure 6.8. Personal mastery is the discipline of continually clarifying and deepening our personal vision, of focusing our energies, of developing patience, and of seeing reality objectively. One key to change success is in surfacing deep-seated mental models including beliefs, values, and mind-sets. Building shared vision occurs when there is a genuine vision; people excel and learn, not because they are told to, but because they want to. Team learning starts with dialogue, the capacity of team members to suspend assumptions and enter into a genuine process of thinking and talking together. ST is the final component of the learning organization concept. Senge believes this discipline to be the key to hold all the five concepts together as a coherent whole (Crooks 2007).

6.10.2 *Managing complexity*

ST provides a great deal of power and value. It can be used to solve complex problems that are not solvable using conventional reductionist thinking, because it focuses on the relationships among system components, as well as on the components themselves; those relationships often dominate system performance (Monat and Gannon 2015). Complex systems usually interact with their environments and are, therefore, open systems. A project is usually an open, complex, and social system made up of many subsystems including administrative and management functions, teams, and individuals, and it operates within the larger system that comprises the performing organization. Figure 6.9 shows phases of ST which is comprised of perspective, set of tools, and action taking.

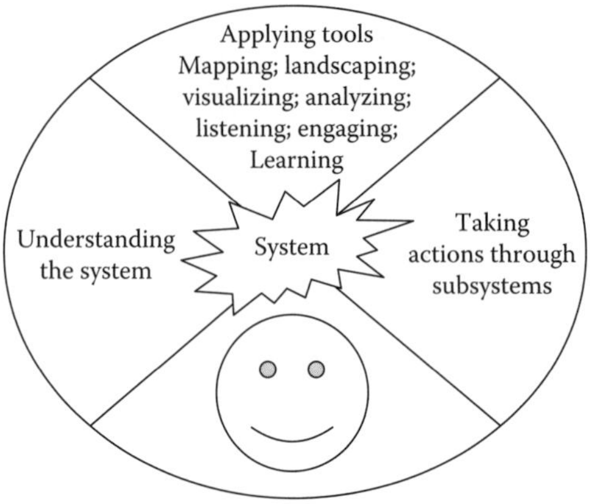

Figure 6.9 Phases of ST.

ST provides an approach for managing complexity. It is an approach to help decision-makers recognize and understand the cause-and-effect relationships among all types of subsystems by applying various tools. ST is an innovative means of the leader's activity; it has an impact on quality of leadership performance, which in turn enhances results of an organization. It develops individual and collective thinking skills and advances decision-making by focusing attention on the causes of performance problems and the systems changes that will generate better outcomes.

ST is the opposite of linear thinking, and it focuses on the relationships among system components, as opposed to the components themselves (Monat and Gannon 2015). It recognizes that the parts of any whole system may only exist or be understood in relation to the whole. This way of thinking about a system helps to move away from focusing on a simple and linear cause and effect relation to one that is multidimensional and emphasizes on the parts of the system to the whole system. Linear way of thinking may be useful when there is a cause that can be identified. When considering a whole system, the term "complex" is frequently used to describe them. The term is used loosely to mean that dealing with them is not straightforward. In the language of a relatively new science of complexity theory, the term "complex systems" acquires a very specific meaning from which a whole series of consequences follows (Page 2011).

6.10.3 Feedback loop

Successful leaders need to be systems thinkers who are open to developing a variety of new skills. They must gain consent across broad and unfamiliar constituencies. They must deal with the complexity of unintended

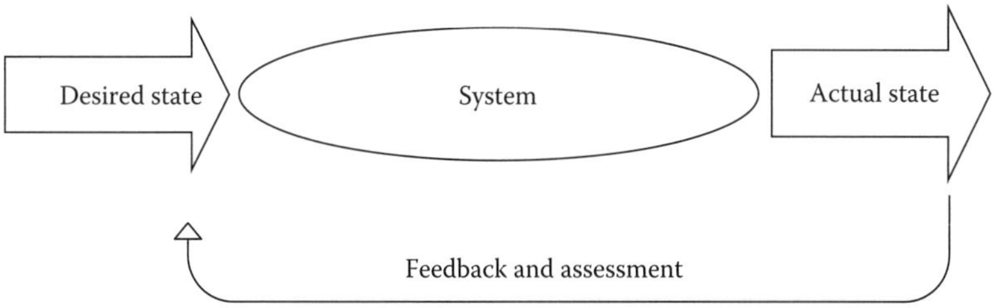

Figure 6.10 ST feedback loop.

consequences. They must follow a systems approach that requires multiple perspectives.

The practice of ST starts with the idea of feedback loop as shown in Figure 6.10. ST acknowledges that systems are dynamic. Feedback loops take the system output into consideration, which enables the system to adjust its performance in order to meet a desired output state. Leaders should work on the system not in the system. Leaders should be aware of what is going on inside the system but also outside. It is necessary to use research and feedback loops to collect and constantly renew relevant information so that learning and adaptation are possible. They must ask the right questions and acquire the new rules of engagement in order to establish the future leadership.

6.11 Imbedding leadership in engineering profession

> Producing change is about 80% leadership … and 20% management… In most change efforts, those percentages are reversed. We continue to produce great managers; we need to develop great leaders.
>
> **Kotter (2006)**

6.11.1 Leadership in engineering practice

In preparation for leading opportunity, engineers must understand the principles of leadership and be able to practice them in growing proportions as their careers advance. In addition to the necessity for strong leadership, ability is the need to also acquire a working framework upon which high ethical standards and a strong meaning of professionalism can be developed.

Contemporary engineering practice and leadership around the world is ordered as a profession, and as such, requires a knowledge base and expertise that extends beyond technical competencies (Beder 1999; Lemaitre et al. 2006). Leadership first comes from the human heart and second from the mind. It is a human experience and process, both emotional and intellectual (see Figure 6.11). A profession as a specific form of collective organization of humans is inseparable from two aspects, an articulation of a vision and an appropriate set of practices and relations.

Some comments from Robert Lutz (Tobia 1999), then former president and vice chair of Chrysler Corporation, and now retired General Motors vice chair, noted in an IEEE-USA's Today's Engineer article, "Robert Lutz Gives Engineers The Nod": "Engineers need to be, like anybody else in business, proactive and somewhat outgoing. And they need to reach outside technical areas. Mainly, engineers need to be good communicators, because there is no point in achieving an engineering breakthrough; having a new idea; or coming up with a new material, if you cannot get your colleagues excited about it."

In an engineering context, leadership integrates a number of capabilities which are critical in order to function at a professional level. These capabilities include the ability to assess risk and take initiative, the willingness to make decisions in the face of uncertainty, a sense of urgency and the will to deliver on time in the face of limitations or difficulties, resourcefulness and flexibility, trust and loyalty in a team setting, and the ability to relate to others. Leadership skills are also important to allow

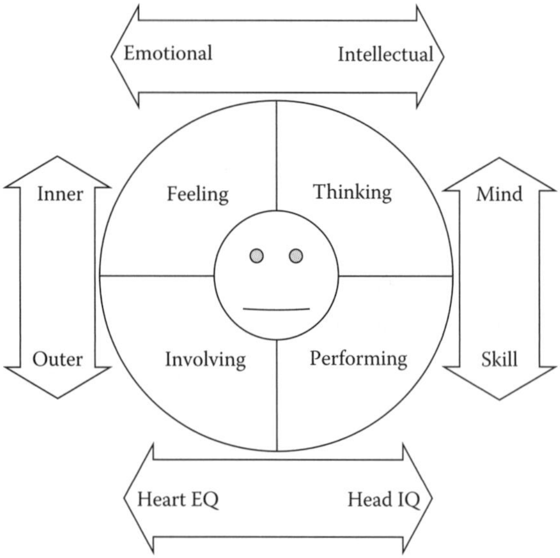

Figure 6.11 Leadership as a human process.

engineers later in their careers to help develop and communicate vision for the future and to help shape Public Policy (PP). These leadership competences are essential for the professional practice of engineering and for the protection of public health, safety, and welfare (Butcher 2013).

Engineering leadership consists of abilities and morals that transform technical people from individual contributors into leaders to deliver a complex interdisciplinary product. Leadership is a process and there is a two-way relationship between the leader and the team. Leaders today should become fluent in ST because, in various ways, ST is leadership. They inspire and influence teams to accomplish things that they otherwise would not have done on their own.

6.11.2 Leadership and management

Many engineering managers came to management through the technical ranks. Although they may have had enough of engineering training and mentoring, they frequently learn management skills the hard way, through trial and error (Rothman 1999). Effective managers are capable of employing a blend of supervisory and technical skills to the direction and wrap-up of complex projects. This always demands that they are able to work well within teams and have the strength of character to lead a team of engineers who may come from a range of various disciplines.

Some would define management as an art, while others would define it as a science. Whether management is an art or a science is not what is most important. Management is a process that is used to accomplish organizational goals. That is, a process to achieve what an organization wants to achieve. Managers are the people to whom this management task is assigned, and it is generally thought that they achieve the desired goals through the key functions of planning and budgeting, organizing and staffing, and problem-solving and controlling. Leaders on the other hand set a direction, align people, motivate, and inspire (Kotter 2001). Other researchers consider that a leader has soul, the passion, and the creativity, whereas a manager has the mind, the rational, and the persistence. A leader is flexible, innovative, inspiring, courageous, and independent, and at the same time, a manager is consulting, analytical, deliberate, authoritative, and stabilizing (Capowski 1994).

Sometimes, it is necessary to distinguish between management and leadership behaviors. Both should be carried out in parallel. They are not the same, but they are necessarily linked and complementary. Any attempt to separate the two is expected to cause more problems than it solves. Still, considerable differences exist. The manager's job is to plan, organize, and coordinate within the organization, while the leader's job is to inspire and motivate from top. Figure 6.12 shows the difference between a manager and a leader.

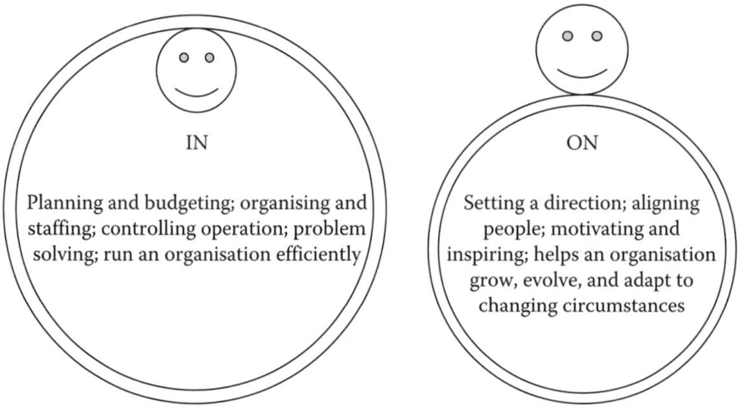

Figure 6.12 Elements of both management and leadership.

6.12 *Leadership for sustainable development*

> The pessimist sees difficulty in every opportunity. The optimist sees the opportunity in every difficulty.
>
> **Winston Churchill**

Leadership for sustainability is a relatively new idea that represents "a radically expanded understanding of leadership that includes an enlarged base of everyday leaders in all walks of life who take up power and engage in actions with others to make a sustainable difference in organizations and communities" (Ferdig 2007).

Leadership skills are critical in driving processes of change. Leadership is often assumed, intuitively, to be an important driver of Sustainable development (SD). Leadership is recognized as important for motivating a change in human behavior toward more sustainable practice. Engaging political leaders is considered imperative for the success of global and regional SD (Walker et al. 2009). At more localized scales, interactions between contemporary and traditional leadership structures are recognized as important (Johannes 2002). As environmental problems escalate, the impetus for understanding where and how effective leadership can be found and fostered has increased (Evans et al. 2015).

Sustainability leaders view and carry out their work through an economic, social, and environmental lens with an appreciation for the finite nature of our planet and the need for social justice. They make proactive decisions, are innovative in how they put together knowledge and skill sets, and they influence decision-making, often at multiple scales from local to global (Timmer 2007).

Sustainability requires leaders to organize the knowledge and expertise within an organization in new ways. ST, where the leader considers the relationships between segments of the business and how to leverage those relationships, requires a different strategy than "business as usual." Thinking systemically means the business leader is looking for cross-functional, cross-departmental collaboration, and ways to work with the value chain, life cycle costs, technologies, suppliers, customer issues, and employee contributions simultaneously (Hughes and Hosfeld 2005).

As discussed in Chapter 2, there are three connecting aspects of sustainability: environmental, social, and economic. Elkington (1997) who coined the triple bottom line (TBL), the people, the planet, and profits, indicated that society depends on the economy and the economy depends on the global ecosystem, whose health represents the ultimate bottom line. Driven by sustainability, TBL provides a framework for measuring the performance of the business and the success of the organization using three lines: economic, social, and environmental (Goel 2010).

Hughes and Hosfeld (2005) led a project on leadership of sustainability. They selected potential interviewees from businesses involved in and/or recognized for environmental or sustainability projects. The research illustrates that the process of adopting sustainability into business operations is very much a young field. The notion of corporate social responsibility or sustainability has been around by that name for only about 20 years and sustainability as a business model is fairly new. The five-stage SD pattern proposed by Hughes and Hosfeld (2005) to help business leaders by making visible the inevitable stages of adopting sustainability is shown in Figure 6.13. Regardless of the path taken, the key message is that leaders must possess the values of and passion for sustainability, or the effort will not succeed. They begin with a single idea

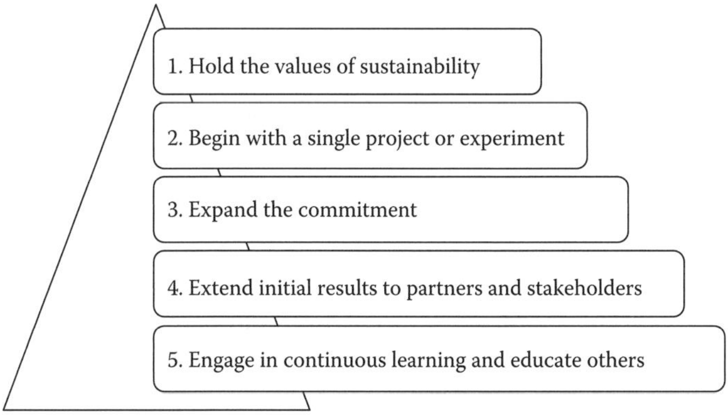

Figure 6.13 The five-stage sustainability development pattern.

or experiment. The key message is to start small and build gradually. In this stage, personal integrity causes the leader to look at the whole system of their business and find other ways to incorporate sustainability. Regardless of where they begin, these business leaders find ways to go beyond the initial incentives, innovations, or regulations to implement more practices over time.

To broaden the initiative, leaders should engage stakeholders, identify and address obstacles, and often get external help from peers or consultants. This stage demonstrates a movement from experimentation to readiness for adopting ST within the business. The key message at this stage is to stay the course and include other innovative thinkers and doers. Leaders must stay current with the knowledge and share what they know with others. The key message is that leaders must make time and create ways to educate themselves and others about sustainability.

6.13 Engineering leadership education

> A leader is one who knows the way, goes the way and shows the way.
>
> **John Maxwell**

6.13.1 Grooming graduates as leaders

Traditional engineering education does not prepare engineers to lead people and organizations. This is particularly true in their early career, since many are ill-prepared to lead projects or organizations. To meet the industry's expectations, engineering schools across the world started aligning their mission statements to educate engineering leaders. According to Stanford University, "engineers and technology professionals with strong analytical, managerial, and business skills will be the innovative leaders of tomorrow" to counter technological changes (Stanford 2010). Employers are looking for employees who are adaptable and flexible. They need those who have initiative and acquire leadership skills. These are essential skills in leadership and management, but today they are becoming increasingly important across disciplines.

Bernard Gordon, the pioneer of engineering leadership programs at the MIT, advocates for the urgent need of engineer leaders, leaders who can contribute to society as technological innovators. Engineers invent and bring to market new technologies that enable advances in virtually every field including health care, manufacturing, infrastructure, transportation, communications, and energy production (MIT 2010).

It has been observed that employers while recognizing technical knowledge as of prime importance also consider leadership as instrumental in realizing organization's goals. A report on graduate employment at the MIT highlighted that in the US, employers, while deciding between two equally qualified candidates, wanted to select the one who had held a leadership position (Hastings et al. 2010). In this regard, one of the aims of the engineering education should be a focus on "bringing the employee visions and values into line with those of the organization" (Bullen and Goh 2010). Engineering leaders need to articulate a vision for the organization they are working for. They need to think like leaders since goals need to be achieved by the concerted action of many individuals who are aligned and rallied by such effective leadership.

6.13.2 Can leadership be learned?

From the above, it is clear that leadership is an integral part of engineering profession. Now the question for consideration here is: can leadership be taught? The answer is simple. Yes, leadership, like creativity, is not a scientific concept; however, leadership, like other skills, can be taught. The literature is positive on the essential mechanisms, styles, and dynamics. Leadership like other terms—education, creativity, and discovery—refers both to certain tasks or activities, on one hand, and to certain outcomes, on the other hand. Educational materials and programs are currently thriving. But the challenging question is: can leadership be learned? The answer to that question is not clear; leadership is not learned easily or acceptably. Much propositional knowledge about leadership can be taught. However, much procedural knowledge (know-how) can also be taught by a variety of instructional means: lectures, case study, and practice. Many critical skills are, therefore, communicable that may or may not result in the realization of leadership by those who grasp them. What is difficult to teach is leadership as such any more than creativity or discovery as such can be directly taught as procedures (Howard 1992).

Teaching leadership has special challenges. Some would argue that it cannot be taught—that it must be learned through experience—and they are not wholly wrong. It cannot be taught by lecture alone; it requires a number of different strategies to engage students in a number of ways: intellectual, social, psychological, and emotional, and with a number of formats such as experiential workshops, design laboratories, team projects, field excursions, mentoring, coaching, guided reflection, service learning, discussion tutorials, and visioning exercises (UoT 2010).

Much of leadership education is devoted to teaching style and technique. Much of what is taught is, in fact, not leadership at all but management. It is entirely possible to learn and even to put into practice what is taught and still fail at being a good leader. The essential components of leadership have remained more or less constant: intelligence, insight, instinct, vision, communication, discipline, courage, and constancy. All can be studied and studied again. The ability to ace leadership principles and practices does not, however, mean that leadership has been learned. Because what is being taught does not necessarily help leadership candidates learn the essentials. Knowing is one thing; doing is quite another (Sarner 2007).

Leadership requires vision and developing a vision requires the ability to feel, see, think, listen, speak, and know. Usually engineers lead in unique ways that reflect engineering knowledge and thinking. An emerging model that encompasses three paradigms of engineering leadership is shown in Figure 6.14. Technical mastery is characterized by insightful, detail-oriented problem-solving, and communication. Engineers with a technical mastery orientation often play informal mentoring roles as the go-to specialist for their expertise. Collaborative optimization is characterized by strong team skills that balance high-quality work with efficiency and engagement. Engineers with a collaborative optimization orientation are excellent at building bridges across organizational units and leveraging team members' strengths. Organizational innovation means the implementation of a new method in the undertaking's business practices.

Ramsden (1998) elaborates that teaching leadership refers, for example, to bringing new ideas or creating excitement about teaching. Research leadership can be evidenced, for example, by inspiring respect as a researcher, or leading by example. Strategic vision and networking are demonstrated through furthering interests across the university. Collaborative and motivational leadership is demonstrated among others by honesty and integrity and openness.

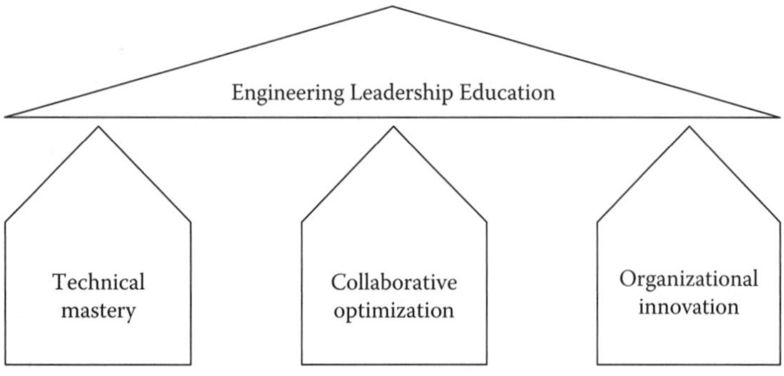

Figure 6.14 Engineering leadership education.

When learning in an interprofessional environment, learners build cooperative, collaborative, and social learning skills. They can develop interprofessional knowledge of team and leadership competencies (Berland 2010; Posnick-Goodwin 2011). As Widhalm (2011) explains, "we need to re-learn identity formation as a process of radical interdependence." Because nothing can exist outside of its relationships with others, to truly understand our existence and our purpose, we need to understand our interconnectedness and interdependency. Strengthening learners' sense of community, and deepening their sense of connectedness, will facilitate the development of collaboration (Burns et al. 2015).

Incorporating elements of leadership learning in studies, rather than as a separate study unit or module, will link learning and work attitudes, including motivation, creativity, and interpersonal skills, with the tasks at hand, such as project work. Learning leadership skills seems to be in line with experiential learning and a constructivist approach to studies, as leadership by nature implies an experiential approach.

6.13.3 Academic leadership

> Who are the people who become positive educational leaders?

Leadership in an academic setting is a concept that requires largely incompatible attributes. Therefore, it is a property of the institution and not a property of an individual or a unit. The basic form of academic leadership is "intellectual leadership." This is the development of leading ideas and the formation of new academic directions. Intellectual leaders provide "exemplary leadership," which is leadership through the power of personal example marked with insight, rigor, dedication, openness, and generosity. Exemplary leadership can be exercised across research, teaching, external engagement, and service, but generally, given the nature of academia, research attainment is a prerequisite. According to Ramsden (1998), effective academic leadership in higher education is a function of several factors or characteristics. These include leadership in teaching and learning; leadership in research, strategic vision, and networking; collaborative and motivational leadership; fair and efficient management; development and recognition of performance; and interpersonal skills.

Resource leadership is another critical type of leadership in a university setting. The momentum of innovation, at least in science and engineering, is powered by money to provide research facilities, studentships, effective groups that can undertake challenging problems and infrastructure on which education depends. Recognizing opportunities and building cases, generally through collaboration, secures the funding and then project manages the whole to a successful ending. These all call for an academic leadership of a high order and of a type that does not always relate to a slim model of leadership but is nonetheless vital.

Hersey and Blanchard (1988) proposed four leadership styles that include telling, selling, participating, and delegating appropriate to the ability and willingness of followers to perform the assigned tasks. Also effective academic leaders require leadership competencies to perform the necessary leadership roles in a university, especially when operating in a global context. A systematic leadership development program needs to be developed to ensure academic leadership effectiveness in universities (Shahmandi et al. 2011).

Sternberg (2005) proposed a model, WICS, standing for wisdom, intelligence, and creativity synthesized (see Figure 6.15). He argues that educational leaders exhibit a synthesis of the three attributes of wisdom, intelligence, and creativity. To a large extent, he argues that the development and display of these attributes is a decision over which one has substantial control, not merely some kind of innate set of predispositions.

Wisdom may be the most important attribute to seek in educational leaders. People can be intelligent or creative but not wise. Wisdom is about balancing various self-interests (intrapersonal) with the interests of others (interpersonal) and about other aspects of the context in which one lives (extrapersonal), such as one's city or country or the world. The WICS theory views intelligence, creativity, and wisdom as different, but as involving fundamental similarities. The basis for "intelligence" narrowly defined, as it is measured by successful intelligence, is the analytical aspect

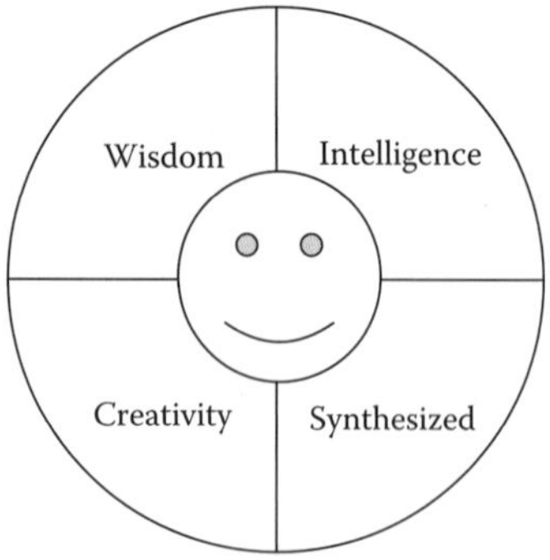

Figure 6.15 WICS model for educational leaders.

of successful intelligence. The basis for creativity is the creative aspect of successful intelligence. Successful intelligence lies at the basis of conventional intelligence, creativity, and wisdom.

The success of various organizations, in particular, educational institutions, depends on effective and efficient leaders. Effective academic leaders such the deans and head of departments in universities should benefit from various leadership styles. Such institutions need to embrace the different ways that leadership can establish itself and have a vision of how they can match. That vision, essentially combined, builds institutional leadership.

6.14 Green building case: Leadership by design

> There is little in the architecture of a city that is more beautifully designed than a tree.
>
> **Jaime Lerner**

6.14.1 Sustainable green building design

Governments have a unique and important role to play in the efforts on climate change and energy efficiency. The operation of governments is extensive and presents a unique leadership example. Also such role presents an opportunity to find new ways to use less energy, reduce waste, and lower greenhouse gas (GHG) emissions.

In 1993 the US Green Building Council (USGBC) was established by Rick Fedrizzi, David Gottfried, and Mike Italiano to promote sustainability in the building and construction industry (USGBC 2014). Once the USGBC was formed, discussions with over 60 firms and nonprofit organizations surrounding a sustainable construction industry began and ultimately led to the development of a new green building rating system. In 2000, the Leadership in Energy and Environmental Design (LEED) program was created by the USGBC to precipitate change in the construction industry by providing guidance on how to implement green building practices, within the framework of a green rating system (Vanry 2015). In 2004, "the first LEED rating system adapted for Canada-wide use was launched, LEED Canada for New Construction and Major Renovations version 1.0" (www.cagbc.org), abbreviated as LEED NC v1. The Canadian rating system, which was an adaptation of the US rating system, was changed to reflect Canada's climate, regulation, and building codes.

The LEED system is not a code or standard. It is rather a voluntary method by which building owners may demonstrate their commitment to

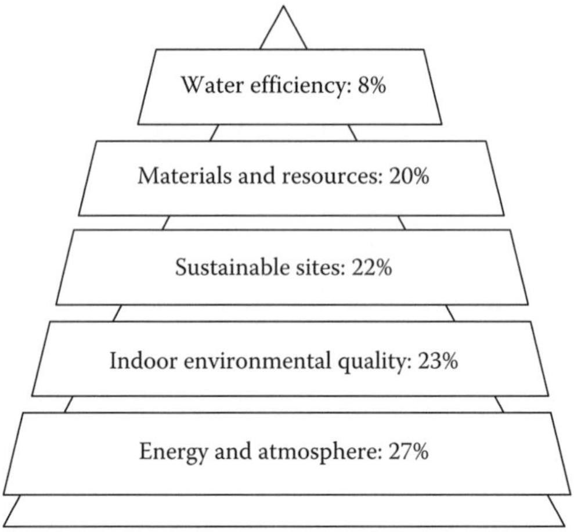

Figure 6.16 Categories of LEED.

energy-efficient and environmentally friendly building design, construction, operations, and maintenance practices that are better than minimum code requirements. LEED is an ecology-oriented building certification program that is revolutionizing the manner we think about how buildings and communities are planned, built, preserved, and functioned. Accordingly, leaders worldwide have made LEED the most widely used third-party verification for green buildings. LEED concentrates its efforts on improving performance across key areas of environmental and human health, energy efficiency, indoor environmental quality, materials selection, sustainable site development, and water savings. Figure 6.16 shows categories of LEED.

6.14.2 Why green building is important?

The transition to green buildings is an important component of the wider transition to a low-carbon economy. Buildings are responsible for a significant share of energy-related carbon emissions, accounting for 8.1 gigatons (Gt) of the current total of 29.0 Gt. To tackle climate change and reduce GHG emissions say to the 14 Gt level for the year 2050 in the International Energy Agency's Blue Map scenario requires a radical greening of buildings globally. It requires green construction techniques to be used for new buildings. Because many existing buildings will remain in use, it also requires retrofitting existing buildings with energy-efficient and renewable energy technologies (IEA 2005).

The construction of green buildings has the potential to deliver many other benefits, beyond that of reduction of GHG emission. In many cases, green buildings improve the comfort of the occupants of buildings, for example, by supplying hot water which would otherwise not be available, by reducing the cost of maintaining a comfortable temperature, and by improving sanitation arrangements. It provides opportunities for enterprise, not only for construction firms but also for businesses offering the technologies, materials, and services required. Investment in green buildings offers considerable scope for generating employment opportunities, a key PP concern in many countries.

Although training in green building skills has increased over recent years, employers still face difficulties in finding qualified people to undertake certain jobs. In the case of green buildings, the main reason for labor shortages is that skill requirements change as green building technologies and practices are introduced or changed, so that previously satisfactory skills sets are no longer adequate. Labor shortages may also come about because there are not enough people interested in working in an area, or because there are deficiencies in training which make it difficult for those who are interested to attain the necessary skills. In most countries, there are enough people interested in working in the building sector. Skills gaps therefore are mainly a consequence of the rapid greening of building activity and of advances in techniques and technologies which change skill requirements faster than education and training systems can respond. There is a strong argument that skills-led strategies which are based on developing skills and capabilities ahead of current practice can make a very strong contribution to the policy objective of driving green building forward (ILO 2011).

6.14.3 Design as a leadership tool

Design and understanding of design can make people better leaders. That is a belief held by Sheila Danko, the J. Thomas Clark Professor of Entrepreneurship and Personal Enterprise in the Department of Design and Environmental Analysis at Cornell University. "Good design, like good leadership, is transformative. Both empower people to reach their own potential and improve the world around them," she says (Stone 2005). The essential goal of design is to see change happen and move beyond the known. And change, lasting and meaningful change, is the yardstick by which leadership is gauge (Collar 2012).

Design for design's sake does not work. For it to truly change an organization, it has to be baked into every decision the product team makes. For this, embedded designers are key for the way they stay connected with the other design efforts to bring a layer of cross-product convergence that gives the entire product suite a solid identity and feel. This matter is all about leadership. What leaders do to influence the entire organization through a solid practice of providing sound user research, communicating effectively, being open and transparent, and making it safe to push the boundaries (Spool 2015).

Given the rapidity of change, there is a requirement for adaptability to change. There is a need for adequate environmental awareness. Green building also calls for interdisciplinary skills, including the ability to work effectively with people from other disciplines as well as individually having skill sets which cross traditional occupational boundaries. Finally, teamworking, coordination, and leadership skills are important core skills in all areas of green buildings.

6.14.4 An institutional model of leadership in sustainable design

ACCE: Algonquin College

Cost	\$79 million
Size	190,000 ft^2
Sustainability certification	LEED Platinum
Project completion	2011
Source	Algonquin 2012

6.14.4.1 Algonquin Centre for Construction Excellence

Opened in September 2011, the LEED Platinum accredited Algonquin Centre for Construction Excellence (ACCE), Ottawa, Canada brings the next generation of carpenters, plumbers, civil engineering technologists, interior designers, and many other trades and professions under a single, green roof in this one-of-a-kind living laboratory. The ACCE is a model of SD, says Algonquin College President Kent MacDonald. The building functions as a living laboratory which supports the learning of students in a way unlike any other postsecondary institution in Canada.

A five-storey biowall made up of living plants filters the air, providing oxygen to the atrium space and all five connected floors. The ACCE building's design also includes features like the green roof, solar panels, a storm water recovery system, and many other green innovations incorporated into the building's design. With its open demonstration spaces and many design and sustainability features, ACCE is a natural hub for the entire trades and design industry as well as its associations. Bringing together students, professors, and researchers with local builders and tradespeople will create synergies, increase industry awareness, and produce highly skilled graduates who are familiar with the collaboration requirements of today's construction and design sectors (Algonquin 2012).

6.14.4.2 Integrated design process

With limited knowledge about designing to LEED standards, the team sought external expertise on sustainability and green buildings. Algonquin implemented the IDP, a holistic systems approach to designing high-performance buildings that are cost-effective and have an exceptional return on investment. The big-picture goal of IDP was incorporating sustainability into the project.

IDP is a term that is not exclusively associated with high-performance building design; in principle, it is a flexible approach that can be applied to almost any type of design or decision-making process (Busby Perkins+Will and Stantec Consulting 2007). It is an interdisciplinary team approach, which facilitates thinking of the building as a system, and considers competing and complimentary aspects of the local site including code and bylaw requirements; climate; building form and space planning; envelope; energy efficiency; renewable energy; mechanical, electrical, and other systems; landscaping and user preferences in the design of a building or community. It relies upon every member of the project team sharing a vision of sustainability, and working collaboratively to implement sustainability goals (CMHC 2012).

Generally, IDP is an iterative process, not a linear or silo-based approach; a flexible method, not a formula; different each time, not predetermined; and an iterative process with ongoing learning and emergent features, not a preordained sequence of events. The IDP is as much a mind-set as it is a process. Having the right mind-set without the process is unlikely to lead to success, and following the process without the right mind-set is almost certain to fail. The importance of mind-set is evident in a set of principles which underpin a successful IDP (Busby Perkins+Will and Stantec Consulting 2007).

The IDP has impacts on the design team that differentiate it from a conventional design process in several respects. The client takes a more active role than usual; the architect becomes a team leader rather than the sole form-giver; and the structural, mechanical, and electrical engineers take on active roles at early design stages. The team always includes an energy specialist and, in some cases, an independent design facilitator (Larsson 2009).

IDP begins by engaging all stakeholders to create a shared understanding of the project that is articulated in a document that lists the project vision, guiding principles and objectives, as well as a definition of sustainability tied to the project. All the stakeholders are brought together in the same room at the same time throughout the project to collaborate on innovative solutions to the design challenges that focus on achieving the shared understanding. Traditional practices that focused on budgets, scope, and schedule are transformed to an integrated perspective that factors in operating costs, life cycle costs, and occupant experience influenced by air quality, lighting, and other features.

The planning team of the ACCE invested in a change management exercise that involved an initial training workshop session in IDP methodology that included all the stakeholders. Participants who had a range of needs collaborated to ensure the design accommodated all people. Best practices for access were incorporated throughout the ramp.

6.14.4.3 ACCE for sustainability education

More than offering a handful of green-focused courses or programs of study, Algonquin College is embedding sustainability in all its Ontario College Credential programs by leveraging existing processes for program renewal or development. The College was already seeing a demand for new programs with a specific focus in sustainability and the environment, relating to professional development in green building design and performance, and technical assessments for environmental and water resources. While these programs met the need for students with an interest in sustainability, the College needed to determine how to best ensure that a foundational understanding of sustainability would be a part of the curriculum to reach a broader base of students.

6.14.5 Case research questions

- What is leadership sustainability?
- What is leadership for sustainability?
- What is LEED certification? What are the different LEED rating systems? How are the rating systems structured?
- What are the phases that make up the integrated design process?

6.15 Knowledge acquisition

Attempting to answer the following questions involves acquisition of knowledge from this book and other books, documents, and the Internet.

- What is leadership? Is leadership a technical model? Or is leadership a behavior model? Or is leadership more a matter of style, or philosophy?
- What is a leader's best asset?
- What qualities make an effective leader?
- What is the relationship between leaders and followers?

- What advice would you give someone going into a leadership position for the first time?
- What are you doing to ensure you continue to grow and develop as a leader?
- What are leadership skills?
- How do you lead others in your field?
- What are qualities a person possesses that make him or her an effective leader?
- Who are innovative leaders and what are their qualities?
- What is EQ, compared with regular intelligence? What does it measure?
- Why should leaders practice ST?
- What is ST? Why it is needed?
- How should leaders practice ST?
- How do you handle resistance to ideas and policies that you propose to others?
- Does every manager need to be a leader?
- What should leaders keep in mind about success and failure?

6.16 *Knowledge possession*

Attempting to answer the following open-ended "not explicitly expressed" questions may require research and investigation beyond the scope of this book, mostly by engaging in conversation, class discussion, and Internet-based research.

- In which way do you see that new technologies will affect leadership and leaders?
- Identify an economic, environmental, or social problem, which currently exists that impacts one as an effective leader?
- In aspect of EQ self-control, how to slow down your response through active listening?
- Does a leader need to be motivated? How can leaders maintain themselves to stay motivated?
- Can someone be a good leader, but not a good manager? Which is better for a company?
- How an academic leader can become more effective?

6.17 *Knowledge creation*

In the following tasks, collaborate with peers on learning or you may work with others outside the class to narrow down the objectives of each task. You may access class and online resources and analyze data and information to create new ideas and balanced solutions. High-level digital tools may be used to develop multimedia presentations, simulations or animations, videos and visual displays, electronic portfolios, feasibility studies or consulting reports, apps or other computer program, website or recording, debates or innovation pitches, original piece of art, or well-researched and up-to-date reports that reflect the objectives of each task.

6.17.1 *Reflection practice on developing an innovation mind-set*

You may often sense the habits and attitudes of people in a learning community like a school or university by the way they interact with each other. Where and how do decisions get made? What does the learning space look like? How does the space support students and faculty? Who lead and manage the learning environment? How does that influence the flow of knowledge about topics, subjects, projects, goals, and deadlines? What fields of disciplines suit more as a source of knowledge and motivation? Based on all above, try to create an image of innovation mind-set or piece of art that is required by every player in the learning environment. You can use metaphors and symbols, as long as those are clearly communicated or explained in supplementary materials.

6.17.2 *Leadership portfolio*

The leadership portfolio is a communication task of where you have come from, where you are currently, and where you are headed on your leadership route. It expresses your story and documents your development in the engineering leadership path. The portfolio is your opportunity to pose yourself as a leader with knowledge, insight, and experience.

For this task, prepare your leadership portfolio that summarizes your leadership style including key qualities you possess, your strengths, and areas for improvement. Specify examples of your leadership experience in class, projects, sports, community, etc., and how you have exhibited specific leadership qualities in each. Finally, what you see as your unique advantage over others? Develop the portfolio using a high-level digital tool of your preference.

6.17.3 *Write–talk communication on leadership in energy efficiency*

Engineers know that communication, both oral and written, is an essential part of their jobs. The need to communicate with customers, managers, technicians, and other engineers is something they become aware of early in their careers. Therefore, engineering educators must ask themselves whether they are preparing students in this area which is as important to their careers as is their technical training.

Energy efficiency is an enormous field with numerous different stakeholders, such as utility companies, nongovernmental organizations, and the general public. Worldwide, initiatives in energy policy are on the rise. These include areas as building codes, appliance standards, combined heat and power (CHP), state-owned facilities, tax incentives, transportation, and utility efficiency programs.

6.17.3.1 Building codes

Buildings are considerable global energy consumption loads and their energy efficiency is important for future sustainability. Energy efficiency in buildings is crucial for SD, climate and resource protection, and a low-risk worldwide energy system. Approximately 40% of global final energy demand and one-third of the energy-related emissions are related to buildings (IEA 2005). Up to 90% of energy can be saved through energy efficiency in new buildings and in retrofits, and various co-benefits achieved at the same time.

Energy codes set minimum-efficiency standards for new buildings, typically covering heating, ventilation, air-conditioning, and cooling; lighting efficiency; safety and security; and other appliances and subsystems. Building codes, which by addressing design, affect long-term energy demands. They are an essential part of government efforts to transform the long-term market for energy efficiency.

- What are the components of the recommended policy package for energy efficiency in new buildings?
- What is the most accepted as the leading and best enforced energy building code in your country or state?
- Investigate a success story on building codes that illustrates the link between codes and voluntary programs.
- How will existing buildings be made more efficient and how will new buildings fit in with old?
- Do you think energy harvesting like buildings that are installed with renewable energy generators can replace energy conservation? Energy conservation means that the building is installed with intelligent energy-saving devices, or applied with certain policies to control the energy consumption and minimize waste.

6.17.3.2 Combined heat and power

Unused thermal energy could be used to generate electricity. CHP technologies put otherwise-wasted heat from power generation to productive use in power plants, manufacturing plants, and commercial buildings.

- What are the leading examples of policies that encourage CHP in your country or state?
- What methods can best be used to measure impacts of CHP energy policies?

6.17.3.3 *Tax incentives*

Several countries and/or states offer income tax credits or deductions, sales tax exemptions, and other tax related incentives for energy-efficient products and practices. Incentives, both financial and nonfinancial, include programs such as tax credits and expedited permitting for efficient products and systems.

- What are the leading tax incentive examples in your country or state?
- What are the goals of energy-efficiency incentives?
- What are the types of incentive for end-users on energy efficiency?
- What are the types of incentive for utilities on energy efficiency?

6.17.3.4 *Transportation*

The transportation sector is critical to energy policy because it accounts for over two-thirds of oil consumption, 30% of total energy use, and one-third of GHG emissions. While law governs fuel economy, countries and states can reduce transportation energy use and emissions through a wide range of policies, from encouraging efficient vehicle purchases to reducing transport demand through growth policy.

- Categorize transportation policies.
- Provide leading examples of energy policies on transportation in your country or state.
- Discuss strategies for reducing transportation's dependence on oil in the next years.
- What are the alternative energy transportation options?
- Is public investment in refueling transportation infrastructure necessary and/or appropriate for enhancing policies?

6.17.4 *Piece of art on understanding feedback*

The practice of ST starts with the idea of "feedback" that illustrates how actions can reinforce or balance each other. "Feedback" means a reciprocal flow of influence. In ST every influence is both cause and effect. As a result, for each situation a systems diagram with feedback loops may be drawn. Reinforcing and balancing feedback and delays are the crucial elements of ST (Akay 2015). In this task, create a piece of art that reflects the above on engineering leadership.

6.17.5 Debate on engineering by design practice and design leadership

Objective	Introducing an open-ended debate in the classroom to help students understand argument on the concepts of engineering practical and leadership design skills.
Time	15 min for debate and 15 min for review.
Format	For and against.
Learning outcomes	Make an argument about a particular opinion, evaluate the arguments of peers, and understand the concept of counterarguments.
Capabilities demonstrated	Developing skills on public speaking, research, teamwork, critical thinking, communication, and professional judgment.
Arrangement	Students are organized into two position groups of three speaking in a specific order. Three argue for an opinion and three argue against. One or two students might work on the opening and closing statements while the group is investigating the subject; however, the entire group should revise the statements. Each group should read an opening and closing statement for the debate.
Ideas for the topic	Engineers like to be taught practical, tangible skills which can be immediately applied. This impacts the way that curriculum is developed and facilitated. Whereas with students studying social work for example, the emphasis or motivation for students may be to develop strong relationships with clients. For a professional faculty such as engineering, high value is placed on career success and problem-solving which may also come at the expense of relationship building (Simpson et al. 2012). What counts as leadership and what counts as engineering? Who are leaders and who are engineers? Should faculties teach technical design skills only or should teach in addition personal development, management, and leadership design skills? Are high successful companies looking for engineers with high practical design skills or for engineers with high design leadership skills?
Assessment	Indicate what you consider the best arguments in favor of the topic. How were they substantiated? Identify arguments that are based on poor facts, not ethical and/or sustainable or not well substantiated.

6.17.6 Poster on leadership as highly EQ

Communication skills are no longer just nice to have. They are must-haves. The industry is going toward more collaboration-integrated project delivery (Walpole 2016). Many elements of EQ might be more familiar to us as "soft skills" and "people skills." But those terms minimize EQ's value. While talking about those terms, it does not sound as important, but when while thinking of it as an intelligence trait, it looks as a set of competencies that can bring to the job and to life (Platt 2015). Investigate this topic and reflect the outcome as well as answer to the following questions in a poster format.

- Do you think that engineers tend not to concern themselves with emotions, for being more interested in technical ideas, problems, and solutions?
- Do you think improving one's EQ does come instantly or easily?
- What drives and motivates you?

6.17.7 Piece of art on critical evaluation of management and leadership

Leadership and management are two distinct but complementary systems. While managers promote stability, leaders press for change. Only organizations that can embrace both sides of that contradiction can thrive in turbulent times (Kotter 1996). Make a list of what you believe are leadership tasks and what are management tasks? Then critically evaluate the above quotation. Reflect the outcome in a piece of art.

6.17.8 Video contest on leadership in public libraries

Libraries are a center for knowledge. Both public and academic libraries have a broad range of sources for information needs. However, such libraries are currently facing drastic changes due to technological advances (e.g., smartphones and e-readers), and the changing information-seeking behavior of library users. These libraries are also facing additional changes brought on by the continued economic downturn, which has forced many of them to undergo budget cuts that have resulted in the reduction of facilities, staff, hours, and resources. Yet public library use has increased as more people are coming to the library to take advantage of the services and resources offered. Today, public libraries function in a climate where budget cuts and the realignment of services are a reality. They have to find a balance between providing core services and offering new ones that meet the information needs of their communities (Jusic 2013).

For this contest, develop a 3 min video that answers the following question: what public libraries should be doing in a changing environment. The content should reflect on required leadership that creates and promotes visionary scenarios for developing spaces to support innovation, experiential learning and entrepreneurial opportunities, start-ups, clubs, and labs as well as a way to generate revenues. You may propose a mission statement for an innovative entrepreneurial model of revenue-generating activities as a minor role for libraries. The business income must be earned in a way that advances the purposes for which the library was established. Specific policies and procedures for evaluating income-generating projects should be explored.

References

Adams, N. 2012. Skinner's walden two: An anticipation of positive psychology? *Review of General Psychology* 16(1): 1–9.

Akay, E. 2015. Why and how should leaders practice systems thinking? https://www.linkedin.com/pulse/why-how-should-leaders-practice-systems-thinking-esin-akay.

Algonquin. 2012. Algonquin Centre for Construction Excellence receives LEED platinum certification. http://myalgonquin2.algonquincollege.com/articles/announcements/algonquin-centre-for-construction-excellence-receives-leed-platinum-certification/.

Ambler, G. 2013. Systems thinking as a leadership practice. http://www.georgeambler.com/systems-thinking-as-a-leadership-practice/.

APA. 2013. How does the APA define psychology? http://www.apa.org/support/about/apa/psychology.aspx#answer.

Avey, J. B., B. J. Avolio, C. D. Crossley, and F. Luthans. 2009. Psychological ownership: Theoretical extensions, measurement and relation to work outcomes. *Journal of Organizational Behavior* 30(2): 173–191.

Avey, J. B., B. J. Avolio, and F. Luthans. 2011. Experimentally analyzing the impact of leader positivity on follower positivity and performance. *The Leadership Quarterly* 22(2): 282–294.

Avolio, B. J. 2007. Promoting more integrative strategies for leadership theory-building. *American Psychology* 62: 25–33.

Avolio, B. J., W. L. Gardner, F. O. Walumbwa, F. Luthans, and D. R. May. 2004. Unlocking the mask: A look at the process by which authentic leaders impact follower attitudes and behaviors. *Leadership Quarterly* 15: 801–823.

Avolio, B. J., D. A. Walderman, and F. J. Yanimarina. 1991. Leading in the 1990s: The four is of transformational leadership. *Journal of European Industrial Training* 15: 9–16.

Bass, B. M. 1990a. Concepts of leadership: The beginnings. In Wren, J. T. (Ed.), *The Leaders Companion: Insights on Leadership through the Ages*. New York: The Free Press, pp. 49–52.

Bass, B. M. 1990b. *Bass and Stogdill's Handbook of Leadership: Theory, Research, and Managerial Applications*. New York: Free Press.

Bass, B. M. 1997. Does the transactional-transformational leadership paradigm transcend organizational and national boundaries? *American Psychologist* 52: 130–139.

Beder, S. 1999. Beyond technicalities: Expanding engineering thinking. *Journal of Professional Issues in Engineering Education and Practice* 125(1): 12–18.

Benincasa, R. 2012. 6 leadership styles, and when you should use them. https://www.fastcompany.com/1838481/6-leadership-styles-and-when-you-should-use-them.

Berland, M. 2010. What America cares about: Compassion counts more than ever? http://www.parade.com/news/what-america-cares-about/featured/100307-compassion-counts-more-than-ever.htm.

Bhatnagar, J. 2012. Management of innovation: Role of psychological empowerment, work engagement and turnover intention in the Indian context. *The International Journal of Human Resource Management* 23(5): 928–951.

Bullen, F., and S. Goh. 2010. Engineering management education post GFC and Copenhagen: Are we on the right track? *40th ASEE/IEEE Frontiers in Education Conference*, Washington, DC, October 27–30.

Burns, J. M. 1978. *Leadership*. New York: Harper and Row Publishers.

Burns, H., H. D. Vaught, and C. Bauman. 2015. Leadership for sustainability: Theoretical foundations and pedagogical practices that foster change. *International Journal of Leadership Studies* 9(1): 131–143.

Busby Perkins+Will and Stantec Consulting. 2007. Roadmap for the integrated design process. http://www.greenspacencr.org/events/IDProadmap.pdf.

Butcher, D. 2013. 5 must-have soft skills for engineers' career success. http://news.thomasnet.com/imt/2013/03/18/5-must-have-soft-skills-for-engineers-career-success.

Cameron, K. S., J. Dutton, and R. Quinn (Eds.). 2003. *Positive Organizational Scholarship*. San Francisco, CA: Berrett-Koehler.

Capowski, G. 1994. Anatomy of a leader: Where are the leaders of tomorrow? *Management Review* 83(3): 10–18.

Carmeli, A., B. Ben-Hador, D. A. Waldman., and D. E. Rupp. 2009. How leaders cultivate social capital and nurture employee vigor: Implications for job performance. *Journal of Applied Psychology* 94(6): 1553–1561.

Chemers, M. M. 1997. *An Integrated Theory of Leadership*. Hove, UK: Psychology Press.

Cherry, K. 2016. The great man theory of leadership. https://www.verywell.com/the-great-man-theory-of-leadership-2795311.

Cherry, K. 2017. What is the trait theory of leadership? https://www.verywell.com/what-is-the-trait-theory-of-leadership-2795322.

Clawson, J. G. 1999. *Level Three Leadership: Getting below the Surface*. Upper Saddle River, NJ: Prentice-Hall.

CMHC. 2012. Integrated design process. EQuilibriumTM Housing InSight. Canada Mortgage and Housing Corporation, Ottawa, ON. https://www.cmhc-schl.gc.ca/odpub/pdf/67612.pdf.

Collar, S. 2012. Drawing: A leadership tool? http://www.stantec.com/blog/2012/10/leadership-tool.html#.WGmLSfkrLcs.

Crooks, D. 2007. Systems thinking in learning organizations, Society of Petroleum engineers. http://www.spe.org/twa/print/archives/2007/2007v3n3/twa2007_v3n3_Soft_Skills.pdf.

Donaldson, S. I., and I. Ko. 2010. Positive organizational psychology, behavior, and scholarship: A review of the emerging literature and evidence base. *The Journal of Positive Psychology* 5(3): 177–191.

Duncan, W. J. 1978. *Essentials of Management*. Second edition. New York: Dryden.

Eisenbeiß, S. A., and S. Boerner. 2010. Transformational leadership and R&D innovation: Taking a curvilinear approach. *Creativity and Innovation Management* 19(4): 364–372.

Elkington, J. 1997. *Cannibals with Forks: Triple Bottom Line of 21st Century Business*. Stoney Creek, CT: New Society Publishers.

Epley, J. L. 2015. Weber's theory of charismatic leadership: The case of Muslim leaders in contemporary Indonesian politics. *International Journal of Humanities and Social Science* 5(7): 7–17.

Evans, L. S., C. C. Hicks, P. J. Cohen, P. Case, M. Prideaux, and D. J. Mills. 2015. Understanding leadership in the environmental sciences. *Ecology and Society* 20(1): 50. http://dx.doi.org/10.5751/ES-07268-200150.

Ferdig, M. A. 2007. Sustainability leadership: Co-creating a sustainable future. *Journal of Change Management* 7(1): 25–35.

Fredrickson, B. L., and M. A. Cohn. 2008. Positive emotions. In Lewis, M., J. M. Haviland-Joens, and L. F. Barrett (Eds.), *Handbook of Emotions*. Third edition. New York: Guilford Press, pp. 120–134.

Gaynor, G. H. 1993. *Exploiting Cycle Time in Technology Management*. New York: McGraw-Hill.

George, B. 2003. *Authentic Leadership: Rediscovering the Secrets to Creating Lasting Value*. San Francisco, CA: Jossey-Bass.

George, B. 2016. HBS: The truth about authentic leadership. http://www.bill-george.org/page/hbs-the-truth-about-authentic-leaders.

George, J. M., and G. R. Jones. 2008. *Understanding and Managing Organisational Behavior*. Upper Saddle River, NJ: Prentice Hall.

Gibbs, N. 1995. The EQ factor. *Time* 146: 60–68.

Goel, P. 2010. Triple bottom line reporting: An analytical approach for corporate sustainability. *Journal of Finance, Accounting, and Management* 1(1): 27–42.

Goethals, G. R., G. J. Sorenson, and J. M. Burns. 2004. *Encyclopedia of Leadership*. Thousand Oaks, CA: SAGE Publications.

Goleman, D. 1995. *Emotional Intelligence: Why It Can Matter More Than IQ?* New York: Bantam Books.

Goleman, D. 1998. *Working with Emotional Intelligence*. London: Bloomsbury Publishing.

Haines, S. G. 1998. *Systems Thinking And Learning*. Amherst, MA: HRD Press.

Haji Mohamed, R. K. M., and C. S. Mohamad Nor. 2013. The relationship between McGregor's X-Y theory management style and fulfillment of psychological contract: A literature review. *International Journal of Academic Research in Business and Social Sciences* 3(5): 715–720.

Harter, J. K., F. L. Schmidt, and T. L. Hayes. 2002. Business-unit-level relationship between employee satisfaction, employee engagement, and business outcomes: A meta-analysis. *Journal of Applied Psychology* 87: 268–279.

Harzer, C., and W. Ruch. 2013. The application of signature character strengths and positive experiences at work. *Journal of Happiness Studies* 14(3): 965–983.

Hastings, D., S. Lerman, and M. Parker. 2010. The demand for MIT graduates. http://web.mit.edu/fnl/volume /223/hastings.html.

Hattangadi, V. 2015. The lasting leadership theories. http://drvidyahattangadi.com/the-lasting-leadership-theories/.

Hersey, P. 1985. *The Situational Leader*. New York: Warner Books.

Hersey, P., and K. H. Blanchard. 1969. Life cycle theory of leadership. *Training and Development Journal* 23(5): 26–34.

Hersey, P., and K. H. Blanchard. 1988. Management of organizational behavior: Utilizing human resources. Fifth edition. Englewood Cliffs, NJ: Prentice-Hall.

Hersey, P., K. H. Blanchard, and D. E. Johnson. 1996. *Management of Organizational Behavior: Utilizing Human Resources*. Upper Saddle River, NJ: Prentice Hall.

Horth, D., and D. Buchner. 2014. Innovation leadership: How to use innovation to lead effectively, work collaboratively, and drive results. Centre for Creative Leadership. White Paper. http://www.ccl.org/wp-content/uploads/2015/04/InnovationLeadership.pdf.

Howard, V. A. 1992. Can leadership be taught? *Paideusis* 6(1): 3–15.

Hsiao, H.-C., and Chang, J.-C. 2011. The role of organizational learning in transformational leadership and organizational innovation. *Asia Pacific Education Review* 12(4): 621–631.

Hughes, P., and K. Hosfeld. 2005. The leadership of sustainability: A study of characteristics and experiences of leaders bringing the "triple-bottom line" to business. Center for Ethical Leadership, Seattle, WA.

IEA. 2005. Evaluating energy efficiency policy measures and DSM programmes. Volume I, Evaluation Guidebook. International Energy Agency, Paris, France. https://www.bigee.net/s/1nawwk.

ILO. 2011. Research brief—Greening of the building sector is held back by skill shortages. Skills-led strategies can drive green building forward. International Labour Office, European Union, Geneva. http://www.ilo.org/wcmsp5/groups/public/---ed_emp/---ifp_skills/documents/publication/wcms_168353.pdf.

Johannes, R. E. 2002. The renaissance of community-based marine resource management in Oceania. *Annual Review of Ecology and Systematics* 33: 317–340. http://dx.doi.org/10.1146/annurev.ecolsys.33.010802.150524.

Jusic, J. 2013: Entrepreneurial leadership in public libraries. http://publiclibrariesonline.org/2013/08/entrepreneurial-leadership-in-public-libraries/.

Kalat, J. 2013. *Introduction to Psychology*. Tenth edition. Belmont, CA: Wadsworth Cengage Learning.

Kotter, J. P. 1996. *Leading Change*. Boston, MA: Harvard Business School Press.

Kotter, J. P. 2001. What leaders really do? *Harvard Business Review* 79(11): 85–96.

Kotter, J. P. 2006. Leadership vs. management: What's the difference? *Journal for Quality and Participation* 29(2): 13–17.

Larsson, N. 2009.The integrated design process; history and analysis. International Initiative for a Sustainable Built Environment (iiSBE). http://www.iisbe.org/system/files/private/IDP%20development%20-%20Larsson.pdf.

Lemaitre, D., R. Le Prat, E. De Graaff, and L. Bot. 2006. Editorial: Focusing on competence. *European Journal of Engineering Education* 31(1): 45–53.

Lewin, K., R. Lippit, and R. K. White. 1939. Patterns of aggressive behavior in experimentally created social climates. *Journal of Social Psychology* 10: 271–301.

Linley, P. A., S. Harrington, and N. Garcea. 2010. *Oxford Handbook of Positive Psychology and Work*. New York: Oxford University Press.

Luthans, F. 2002. The need for and meaning of positive organizational behavior. *Journal of Organizational Behavior* 23: 695–706.

Luthans, F., and B. J. Avolio. 2003. Authentic leadership development. In Cameron, K. S., J. E. Dutton, and R. E. Quinn (Eds.), *Positive Organizational Scholarship*. San Francisco, CA: Berrett-Koehler, pp. 241–258.

MIT. 2010. MIT Engineering leadership program. http://engineering.mit.edu/education/specialprograms/gordon.php.

Monat, J. P., and T. F. Gannon. 2015. What is systems thinking? A review of selected literature plus recommendations. *American Journal of Systems Science* 4(1): 11–26.

Morgan, G. 1997. *Images of Organizations*. Thousand Oaks, CA: Sage Publications.

Newstrom, J. W., and K. Davis. 1993. *Organizational Behavior: Human Behavior at Work*. New York: McGraw-Hill.

Ouchi, W. 1981. *Theory Z: How American Management Can Meet the Japanese Challenge*? Reading, MA: Addison Wesley.

Page, S. E. 2011. *Diversity and Complexity*. Princeton, NJ: Princeton University Press.

Peterson, C., and M. E. P. Seligman. 2004. *Character Strengths and Virtues: A Handbook and Classification*. New York: Oxford University Press.
Platt, J. R. 2015. Why emotional intelligence is key to your success? http://theinstitute.ieee.org/career-and-education/career-guidance/why-emotional-intelligence-is-key-to-your-success.
Posnick-Goodwin, S. 2011. The most technologically advanced yet. California Teachers Association, Burlingame, CA. http://www.cta.org/Professional-Development/Publications/Educator-Feb-10/Most-technologically-advanced-generation.aspx.
Ramsden, P. 1998. Managing the effective university. *Higher Education Research and Development* 17(3): 347–337.
Reed, G. E. 2006. Leadership and systems thinking. Defence AT&L, May–June, 10–13. http://www.au.af.mil/au/awc/awcgate/dau/ree_mj06.pdf.
Riemer, M. J. 2003. Integrating emotional intelligence into engineering education. *World Transactions on Engineering and Technology Education* 2(2): 189–194.
Riggio, R. E. 2014a. What is authentic leadership? Do you have it? https://www.psychologytoday.com/blog/cutting-edge-leadership/201401/what-is-authentic-leadership-do-you-have-it.
Riggio, R. E. 2014b. The 4 elements of transformational leaders. https://www.psychologytoday.com/blog/cutting-edge-leadership/201411/the-4-elements-transformational-leaders.
Rothman, J. 1999. Successful engineering management: 7 lessons learned. http://www.jrothman.com/articles/1999/01/successful-engineering-management-7-lessons-learned/.
Salovey, P., and J. M. Meyer. 1998. Emotional intelligence. In Jenkins, J. M., K. Oatley, and N. L. Stein (Eds.), *Human Emotions: A Reader*. Malden, MA: Blackwell.
Sarner, M. 2007. Can leadership be learned? https://www.fastcompany.com/73352/can-leadership-be-learned.
Scouller, J. 2011. The three levels of leadership: How to develop your leadership presence, knowhow and skill. Cirencester, UK: Management Books 2000.
Seligman, M. E. P. 2002. *Authentic Happiness: Using the New Positive Psychology to Realize Your Potential for Lasting Fulfillment*. New York: Free Press.
Senge, P. M. 1990. The Fifth discipline: The art and practice of the learning organization, currency. Crown Business, Revised edition. 0385517254.
Shahmandi, E., A. D. Silong, I. A. Ismail, B. B. Abu Samah, and J. Othma. 2011. Competencies, roles and effective academic leadership in world class university. *International Journal of Business Administration* 2(1): 44–53.
Shipton, H. J., M. A. West, C. L. Parkes, J. F. Dawson, and M. G. Patterson. 2006. When promoting positive feelings pays: Aggregate job satisfaction, work design features, and innovation in manufacturing organizations. *European Journal of Work and Organizational Psychology* 15(4): 404–430.
Simpson, A. E., G. J. Evans, and D. Reeve. 2012. A summer leadership development program for chemical engineering students. *Journal of Leadership Education* 11(1): 222–232.
Smith, W. 2016. Leadership theories—The three levels of leadership model. http://coachingwolf.com/2016/08/12/leadership-theories-three-levels/.
Spool, J. M. 2015. The new design leader emphasizes leadership. https://articles.uie.com/new_design_leader/.
Stanford. 2010. Stanford engineering management and leadership. Stanford Centre for Professional Development.

Sternberg, R. J. 2005. WICS: A model of positive educational leadership comprising wisdom, intelligence, and creativity synthesized. *Educational Psychology Review* 17(3): 191–262.

Stogdill, R. M. 1974. Handbook of leadership: A survey of theory and research. New York: Free Press.

Stone, C. 2005. Design as a tool for leadership and social change. *Human Ecology* 32(3): 2–5.

Stone, G., and K. Patterson. 2005. The history of leadership focus. School of Leadership Studies, Regent University, Virginia Beach, VA. https://www.regent.edu/acad/global/publications/sl_proceedings/2005/stone_history.pdf.

Timmer, D. 2007. Becoming a sustainability leader IISD's role in shaping the next generation of sustainable development leadership. https://www.iisd.org/pdf/2007/sustainability_leader.pdf.

Tobia, P. M. 1999. Robert Lutz gives engineers the nod. *IEEE Todays Engineer* 2(1): 6–11.

Tracey, J. B., and T. R. Hinkin. 1994. Transformational leaders in the hospitality industry. *Cornell Hotel and Restaurant Administration Quarterly* 35(2): 18–24.

UoT. 2010. Proposal to establish an Institute for Leadership Education in Engineering in the Faculty of Applied Science and Engineering at the University of Toronto as an Extra-departmental Unit C. Report No. 3249. http://www.engineering.utoronto.ca/wp-content/blogs.dir/28/files/2015/02/Rpt3249_ILead-Recommendation.pdf.

USGBC. 2014. United States green building council. http://www.usgbc.org/about/history.

Vanry, J. 2015. A case study assessment of the energy consumption of LEED certified academic buildings in Ontario: Is LEED certification necessarily better? MSc Thesis, Waterloo University, ON, Canada.

Walker, B., S. Barrett, S. Polasky, V. Galaz, C. Folke, G. Engström, F. Ackerman, K. Arrow, S. Carpenter, K. Chopra, G. Daily, P. Ehrlich, T. Hughes, N. Kautsky, S. Levin, K. G. Mäler, J. Shogren, J. Vincent, T. Xepapadeas, and A. de Zeeuw. 2009. Looming global-scale failures and missing institutions. *Science* 325(5946): 1345–1346.

Walpole, B. 2016. Leadership is communication, rising engineers told. http://news.asce.org/leadership-is-communication-rising-engineers-told/.

Weber, M. 1958. The three types of legitimate rule. *Berkeley Publications in Society and Institutions* 4(1): 1–11.

Widhalm, B. 2011. Educators as architects of living systems: Designing vibrant learning experiences beyond sustainability and systems thinking. *Journal of Sustainability Education* 2: 1–17.

Williamson, T. 2008. The good society and the good soul: Plato's republic on leadership. *The Leadership Quarterly* 19(4): 397–408.

Wren, J. T. 1995. *The Leaders Companion: Insights on Leadership through the Ages*. New York: The Free Press.

Youssef, C. M., and F. Luthans. 2012. Positive global leadership. *Journal of World Business* 47(4): 539–547.

Yukl, G. A. 2006. *Leadership in Organizations*. Upper Saddle River, NJ: Pearson/Prentice Hall.

Zaccaro, S. J. 2007. Trait-based perspectives of leadership. *American Psychologist* 62(1): 6–16.

chapter seven

Engineering entrepreneurship

To win big, you sometimes have to take big risks.

Bill Gates

7.1 Objectives

- Give a historical perspective about entrepreneurship.
- Define entrepreneurship and discuss its relation to innovation.
- Explore the domains of an entrepreneurial ecosystem.
- Show the pathway between apprenticeships and entrepreneurship.
- Define skills needed for entrepreneurship and what makes someone an entrepreneur.
- Know about the interactive dimension of the entrepreneurial space.
- Explain the complete cycle of entrepreneurial process.
- Discuss the process of entrepreneurship and its phases of development.
- Explain the notion of entrepreneurial brain and the four-quadrant model of the human brain.
- Highlight the value of apprenticeships as a pathway to entrepreneurship.
- Emphasize the importance of intrapreneurial innovation as a potent process for corporate renewal and success.
- Present five inspirational examples including from the history context to show what entrepreneurial/intrapreneurial innovation entails and what they can deliver to the community and organization. These include Thomas Edison, Nikola Tesla, Henry Ford, Harvey Firestone, and Steve Jobs.
- Learn how to develop an entrepreneurial business plan.
- Introduce the concept of Timmons model of entrepreneurship.
- Understand the nature of technology entrepreneurship and whether it is for engineers.
- Discuss the notion of sustainability entrepreneurship and its related concepts of sustainopreneur and ecopreneurs.
- Discuss leadership education and know about Kern Entrepreneurship Education Network (KEEN) pyramid mind-set concept.

- Explain the basic entrepreneurial marketing (EM) concepts, goals, and methods, as well as operational functions from a technology perspective.
- Describe the skills engineers need to acquire to become entrepreneurs.
- Illustrate the means and importance of embedding innovation and entrepreneurship into formal education and curriculum development to develop engaging entrepreneurial ecosystem.
- Present an entrepreneurship case that provides a glimpse into the Ottawa-based company "Med-Eng Systems" and the engineer entrepreneur behind the company's flagship products, primarily bomb suits and protective gear.
- Provide three end of chapter pedagogical knowledge strategies, namely knowledge acquisition, knowledge possession, and knowledge creation to help understand the above topics and generate new and alternative ideas and solutions.

7.2 Historical perspective

> The best time to plant a tree was 20 years ago. The second best time is now.
>
> **Chinese proverb**

The origins of enterprise are often associated with the Industrial Revolution. However, this chapter presents some evidence of entrepreneurial activities from a much earlier date, the medieval period. By concentrating on individuals rather than organizations, it may be possible to push back the study of entrepreneurship beyond the Industrial Revolution and early-modern trade to an era that perceived the beginnings of the modern state.

The ancient and medieval worlds seem not to have developed a concept of entrepreneurship that could plausibly be seen as similar to the modern notion. Philosophers gave only limited attention to economic matters and, in so far as agriculture, industry and trade were discussed, much thinking would have been a subbranch of politics or ethics. In the Aristotelian tradition, economic thought was highly normative. Trade was a suspect activity liable to undermine the good order of society. Even if the reality was more complicated, early social thought concerned static societies built upon caste or social position where justice was the outcome of each group faithfully performing its allotted function. In the hierarchy of social esteem, the noble warrior took pride of place, agriculture was respected and compatible with the inculcation of certain virtues, industry in support of military power was too useful to neglect, but commerce was the province of less respected if not completely despised social groups. It was, however, the agricultural and industrial revolutions of the eighteenth and nineteenth centuries that finally produced the modern multifaceted image of the entrepreneur (Ricketts 2008).

7.2.1 Early period

This early type of entrepreneurship continued for millennia. Hunter-gatherer tribes would trade goods from different parts of their respective regions to provide an overall benefit for their tribe. Between the agricultural revolution and 2000 BC, cities started to appear around the world. Early areas of civilization were concentrated around rivers, particularly the Nile, the Tigris and Euphrates, the Indus, and the Yellow and Yangtze. By 3000 BC, cities in Sumeria (Iraq) contained tens of thousands of people. The city of Uruk, found on the banks of the Euphrates, was home to 50,000 people in the same amount of space that would have previously supported just one tribe of hunter-gatherers (Pushprofile 2016).

An early example of the earliest definition of an entrepreneur as a go-between is Marco Polo, who attempted to establish trade routes to the Far East. As a go-between, Marco Polo would sign a contract with a money person (forerunner of today's venture capitalist) to sell his goods. A common contract during this time provided a loan to the merchant adventurer at a 22.5% rate, including insurance. While the capitalist was a passive risk bearer, the merchant adventurer took the active role in trading, bearing all the physical and emotional risks. When the merchant adventurer successfully sold the goods and completed the trip, the profits were divided with the capitalist taking most of them (up to 75%), while the merchant adventurer settled for the remaining 25% (Hisrich 2005).

7.2.2 The Middle Ages

The term entrepreneurship may be traced back to as early as the Middle Ages, when the entrepreneur was just a person who performed tasks, such as acting or managing projects. In such large production projects, this individual did not take any risk, but simply managed the project using the resources provided. However, it was during the sixteenth century when business was used as a common term, and the entrepreneur came into focus as a person who is responsible for undertaking a business venture. A typical entrepreneur in the Middle Ages was a person in charge of large architectural works, such as castles and defences, public buildings, convents, and cathedrals.

7.2.3 Seventeenth century

Many people see the past 200 years of entrepreneurship as being fuelled by "machines and markets." The reemergent connection of risk with entrepreneurship developed in the seventeenth century, with an entrepreneur being a person who entered into a contractual arrangement with the government to perform a service or to supply stipulated products. Since the contract price was fixed, any resulting profits or losses were the entrepreneur's. One entrepreneur in this period was John Law, a Frenchman, who was allowed to establish a royal bank. The bank eventually evolved into an exclusive franchise to form a trading company in the New World, the Mississippi Company. Unfortunately, this monopoly on French trade led to Law's downfall when he attempted to push the company's stock price higher than the value of its assets, leading to the collapse of the company (Herbert and Link 1982).

In 1607 the Virginia Company sent three ships across the Atlantic and unloaded 109 passengers at what became Jamestown, Virginia. They were embarked on a new business enterprise that they hoped would be profitable, American plantations. The Virginia Company was a joint-stock company, a relatively new invention that allowed people to invest in enterprises without running the risk of losing everything if the business did not succeed. By limiting liability, corporations greatly increased the number of people who could dare to become entrepreneurs by pooling their resources while avoiding the possibility of ruin. Thus the corporation was one of the great inventions of the Renaissance, along with printing, bookkeeping, and the full-rigged ship (Gordon 2014).

Richard Cantillon, a noted economist and author in the 1700s, understood law's mistake. Cantillon developed one of the early theories of the entrepreneur and is regarded by some as the founder of the term. He viewed the entrepreneur as a risk-taker, observing that merchants, farmers, craftsmen, and other sole proprietors buy at a certain price and sell at an uncertain price, therefore operating at a risk (Herbert and Link 1982).

During the seventeenth century, market forces such as inflation had become more pronounced with the influx of Spanish gold from the New World and the Price Revolution. Individual agency or the "entrepreneur" sat between poles, in a century in which entire populations began to be defined by the mechanistic processes of the first Industrial Revolution and the ideals of the French Revolution (Bennet 2014).

7.2.4 Eighteenth century

The Industrial Revolution marked another profound shift in the history of entrepreneurship. Starting in the eighteenth century, entrepreneurship moved from small-scale production in small towns to large-scale production in big cities. During that period, the person with capital was differentiated from the one who needed capital. In other words, the entrepreneur was distinguished from the capital provider (the present-day venture capitalist). One reason for this differentiation was the industrialization occurring throughout the world (Nandy and Kumar 2014). Many of the inventions developed during this time were reactions to the changing world, as was the case with the inventions of Eli Whitney (1765–1825) and Thomas Edison (1847–1931). Both Whitney and Edison were developing new technologies and were unable to finance their inventions themselves. Whitney is known as the inventor of the cotton gin which is a mechanical device that removes the seeds from cotton, a process which, until the time of its invention, had been very labor-intensive. He financed his project with expropriated British crown property. On the other hand, Edison raised capital from private sources to develop his experimentation in the fields of electricity and chemistry. Both Whitney and Edison were capital users (entrepreneurs), not providers (venture capitalists). A venture capitalist is a professional money manager who makes risk investments from a pool of equity capital to obtain a high rate of return on the investments (Hisrich 2005).

The entrepreneur was first detailed by Irish-French economist Richard Cantillon during the eighteenth century in his essay Essai sur la Nature du Commerce en Général, in which he described the entrepreneur as someone who bought at one price and sold at another uncertain price, in doing so managing risk. Cantillon's essay and lone surviving work took economic theory away from the domain of the philosopher and religious thinker, offering a counter point to the economics of the merchant trader. His work introduced individual agency during the beginning of the eighteenth century (Bennet 2014).

7.2.5 Nineteenth and twentieth centuries

The Industrial Revolution transformed nineteenth century commerce and industry while producing significant progresses in transportation, industrial production, communications, and energy. The resulting effect was a profound alteration of society from both an economic and cultural standpoint. This fertile environment created extraordinary business opportunities for entrepreneurs with vision and fortitude. This was encouraged by a system of government that promoted risk-taking and minimal regulatory interference (Busse 2011).

In the late nineteenth century, classical economics evolved into neoclassical economics with its emphasis on mathematical and scientific precision and its preoccupation with resource allocation and pricing decisions. During this period and the early twentieth century, entrepreneurs were frequently not distinguished from managers and were viewed mostly from an economic perspective. In the twentieth century, the cult of the entrepreneur initially receded. The large-scale organizations established in the nineteenth century and the corporations developing in the newer electrical, chemical, communications and motor industries began to look more managerial and professional than heroically entrepreneurial. The entrepreneurs having blazed their pioneering trail, it began to be seriously considered that professional scientists, technicians and managers would be able to maintain momentum.

Allowing incorporation as a matter of law, rather than requiring an act of the executive or of the legislature, began in the United States as early as 1811, when New York State passed a general incorporation law for certain businesses, including anchor. Soon enlarged in scope, the ability to incorporate simply by filling out the right forms freed the process from politics, and the number of corporations exploded. There had been only seven companies incorporated in British North America, but the state of Pennsylvania alone incorporated more than 2000 between 1800 and 1860 (Gordon 2014).

In the middle of the twentieth century, the notion of an entrepreneur as an innovator was established: the function of the entrepreneur is to reform or revolutionize the pattern of production by exploiting an invention or, more generally, an untried technological method of producing a new commodity or producing an old one in a new way, opening a new source of supply of materials or a new outlet for products, by organizing a new industry (Schumpeter 1978).

The notion of innovation and novelty is an essential part of entrepreneurship in the above definition. Certainly, innovation, the performance of introducing something new, is one of the most challenging tasks for the entrepreneur. It takes not only the ability to create and conceptualize but also the ability to realize all the influences at work in the environment. The newness can consist of anything from a new product to a new distribution system to a method for developing a new organizational structure (Hisrich 2005).

During this period, many people developed and built successful businesses. For example, Thomas Edison made a fortune through the electric light bulb and many other inventions. Also, Edward Harriman (1848–1909), who reorganized the Ontario and Southern railroad through the Northern Pacific Trust, and John Pierpont Morgan (1837–1913), who developed his large banking house by reorganizing and financing the nation's industries, are examples of successful entrepreneurs.

7.2.6 Post–World War II entrepreneurship

World War II clearly changed the political, technological, and economic environment. After the war, entrepreneurship began to change for a few different reasons. First, the economy was increasingly becoming more global every decade. Better means of shipping and communication made it easy for entrepreneurs to sell products and services to a global audience. During the 1940s and 1950s, the economy in the United States was inspired by Joseph Schumpeter's concept of entrepreneurship as an agent of disruptive economic change (Jones and Wadhwanim 2007). The war-related products such as computers, radars, and jet engines that emerged from the war were commercialized through the military and then converted into civilian products; accordingly few firms were created. Entrepreneurial activities in terms of firm formation declined or stagnated between 1950 and 1965 and remained at a low level until 1980 (Carlsson et al. 2013).

7.2.7 The 1980s and the 1990s

The year 1980 represents something of a turning point for entrepreneurship activity. A number of institutional reforms in the United States including strengthening of intellectual property rights mark a transition to a new technological regime in which new business formation plays an increasing role in converting new knowledge into economic growth (Carlsson et al. 2013). Entrepreneurial activity began to pick up as the dynamism of the economy increased. It became evident that large firms were not always superior in promoting technological development and economic growth. The "twin oil crises" in the 1970s triggered a reappraisal of the role of small firms. Many large companies were hit by severe economic difficulties. Large companies were increasingly seen as inflexible and slow to adjust to new market conditions (Carlsson 1989). The increased interest in smaller firms can be attributed to a fundamental change in the world economy, related to the intensification of global competition, the resulting increase in the degree of uncertainty, and greater market fragmentation; and changes in the characteristics of technological progress giving large firms less of an advantage (Carlsson 1992).

7.2.8 Modern entrepreneurship

Today, entrepreneurs are the essence of economies all over the world. The global economy, combined with modern infrastructure, has introduced a new range of competition to the domain of entrepreneurship. No longer are you competing with entrepreneurs in your school, town, village, or city: you are competing with entrepreneurs all over the world. Many of these entrepreneurs can access inexpensive means of production. They may have better access to raw resources of cheap materials and labor. This has made modern entrepreneurship more challenging and probably more rewarding than ever before but harsh challenges nonetheless.

From the above brief historical review it is obvious that entrepreneurship has evolved over time. The slightly varied notions that still exist reveal this history. The small-scale trader and dealer, the self-employed craftsman, the innovator and improver as well as the founder of entirely new technologies and industries are all counted as entrepreneurs. Today, interest in developing entrepreneurial ecosystems is growing up faster, and thus new approaches are emerging to keep up with demand and also keep up with the evolving nature of entrepreneurship education.

7.3 The entrepreneurship landscape

> Nothing encourages entrepreneurial activity more than the freedom to take risk. A second great spur to entrepreneurship is the freedom to fail.
>
> **Gordon (2014)**

7.3.1 Entrepreneurship defined

The word "entrepreneur" may be referred to one who undertakes, manages, and assumes the risk of a new enterprise. It comes from the French, where it literally means "undertaker." The word was loanword into English in the mid-nineteenth century, perhaps the golden age of the entrepreneur when the number of new economic niches was exploding and the hand of government was at its lightest in history. The activity of entrepreneurship, of course, is much older, going back to ancient times. As for America, the nation was founded, quite literally, by entrepreneurs (Gordon 2014).

Entrepreneurship has been examined by a variety of scholars and organizations in recent decades. It is not a concept that has a tightly agreed definition. In modern common usage an "entrepreneur is a person who undertakes an enterprise, especially a commercial one, often at personal financial risk" (Ricketts 2008). According to the Canadian Organization for Economic Co-operation and Development, entrepreneurship is defined as the phenomenon associated with enterprising human action in pursuit of the generation of value, through the creation or expansion of economic activity, by identifying and exploiting new products, processes or markets. It is a dynamic concept, and can be measured through various indicators, such as enterprise start-up and survival rates, enterprise duration, and high-growth firm rates.

Joseph Schumpeter introduced the modern definition of "entrepreneurship" in 1934. According to Schumpeter, the carrying out of new combinations we call "enterprise," and the individuals whose function it is to carry them out we call "entrepreneurs." Schumpeter tied entrepreneurship to the creation of five basic "new combinations" namely: introduction

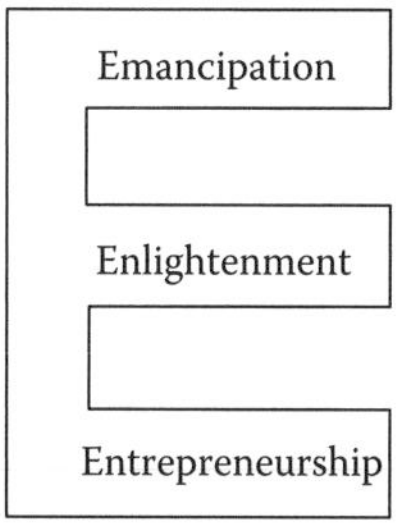

Figure 7.1 The three "E's."

of a new product, introduction of a new method of production, opening of a new market, the conquest of a new source of supply, and carrying out of a new organization of industry (Schumpeter 1934).

Stevenson and Gumpert (1985) define entrepreneurship as the process of creating value by putting together a unique package of resources to exploit an opportunity. Entrepreneurship is the ability to create and build something from practically nothing. It is initiating, doing, achieving, risk-taking, and building an enterprise.

Entrepreneurship is the process of creative destruction that is essential to the sustainability of economic development (Kalantarian et al. 2012). It is the dynamic process of creating incremental wealth. The wealth is created by individuals who assume the major risks in terms of equity, time, and/or career commitment or provide value for some product or service. The product or service may or may not be new or unique, but value must somehow be infused by the entrepreneur by receiving and locating the necessary skills and resources (Odeigah 2012).

Finally, Fleischmann (2006) introduces an interesting model called the "Three Es," which nicely position and connect entrepreneurship as shown in Figure 7.1. Emancipation carries the meaning of self-sufficiency or freedom, probably from poverty or from dependence where most people prefer dependence over thinking and acting for themselves, which is why dependence has become almost second nature. Enlightenment historically is a movement of the 18th century that emphasized the idea that science and logic provide people more knowledge and understanding than belief and religion. In the context of entrepreneurship, it may be understood as the state of having knowledge or understanding and basing on experimentation, observation, and evidence. Enlightenment may stress the confidence in human reason and promise of economic self-realization.

7.3.2 Innovation and entrepreneurship

Entrepreneurship is a field that involves creativity, integrated learning, and openness to experience. Therefore, the concept of innovation and newness is an integral part of entrepreneurship in this aspect. Innovation, the act of introducing something new, is one of the most difficult tasks for the entrepreneur. It takes not only the ability to create and conceptualize but also the ability to understand all the forces at work in the environment. The newness can consist of anything from a new product to a new distribution system to a method for developing a new organizational structure. Entrepreneurship spans opportunity recognition, resource acquisition, and innovation as shown in Figure 7.2.

Entrepreneurship is seen as a critical link between new knowledge and economic growth as it facilitates the transfer of knowledge. These factors distinguish entrepreneurship from more simple forms of management and ordinary business activities. According to Cogliser and Brigham (2004), entrepreneurship and innovation have a two-way relationship. Entrepreneurship comes into play in innovation in the place where a person comes across something but may not have the capability of translating the same into a proposition that is commercial. In the opinion of Currie et al. (2008), innovation relates to entrepreneurship as it is its particular instrument, being an act that leads to the provision of resources with fresh ability for wealth creation. An entrepreneur is an individual, with the willingness and the capability of transforming inventions into innovations. While invention refers to generating new concepts, innovation makes the concept alive, and entrepreneurs take risks in the process of making the concepts alive all of which determine business success. Today small entrepreneurial firms are responsible for half of all innovation and credited with 95% of all radical innovation.

Entrepreneurs rely upon innovation to create new markets and to differentiate themselves in highly competitive markets (Schumpeter 1947; Amabile 1993; Shane 2003). Innovation is the cornerstone of successful entrepreneurship within dynamic emerging markets and requires both expert level domain knowledge and the ability to acquire and apply new knowledge to solve problems (Shane 2000; Jemmell 2012).

Peter Drucker (1985) in his book *Innovation and Entrepreneurship* has outlined an approach to entrepreneurship as the practice driven of innovation: entrepreneurship is neither a science nor an art. It is a practice. It is

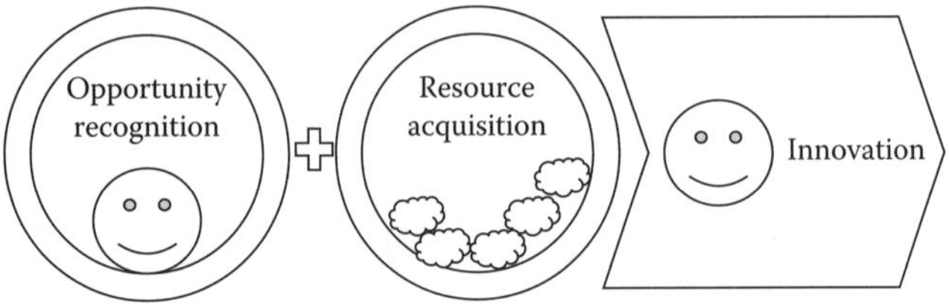

Figure 7.2 The three aspects of the entrepreneurial process.

a knowledge base but as in all practices—medicine, for instance, or engineering—knowledge in entrepreneurship is a means to an end. Indeed what constitutes knowledge in a practice is largely defined by the end, that is, by the practice. Innovation is the specific tool of entrepreneurship, the means by which change is exploited as an opportunity for a different business or a different service. It is capable of being presented as a discipline, capable of being learned, capable of being practiced.

7.3.3 *Entrepreneurial activity*

The sole entrepreneurial activity is recognizing and exploiting an opportunity which is the creative human action in search of the generation of value, through the creation or growth of economic activity, by identifying and developing new products, processes, or markets. There are several types of entrepreneurial activity, from corporate venturing to social change enterprises. Value created by entrepreneurs may be either taken by the entrepreneur or traded or shared with employees, stakeholders, and society.

There are three basic ideas that explain the appearance of entrepreneurial activity. The first focuses on the individual, in other words, entrepreneurial action is conceived as a human attribute, such as the willingness to face uncertainty (Kihlstrom and Laffont 1979), accepting risks, the need for achievement, which differentiates entrepreneurs from the rest of society. The second fundamental idea emphasizes economic, environmental factors that motivate and enable entrepreneurial activity, such as the dimension of markets, the dynamic of technological changes (Tushman and Anderson 1968), and the structure of the market normative and demographic (Acs and Audretsch 1990) or merely the industrial dynamic. The third factor is linked to the functioning of institutions, culture, and societal values. These approaches are not exclusive (Eckhardt and Shane 2003), given that entrepreneurial activity is also a human activity and does not spontaneously occur solely due to the economic environment or technological, normative, or demographic changes.

Entrepreneurship goes beyond starting a new business. It implies the whole process whereby individuals become aware of the opportunities that exist to develop ideas, and carry out initiative with responsibility. In a broader sense, entrepreneurship helps young individuals develop new skills and experiences that may be applied to many other challenges in life. It is therefore a key priority area with the potential to stimulate job and wealth creation in an innovative and independent way. At the end, entrepreneurial activity does not impact economic situation only but there exist considerable social and cultural impacts.

The idea of entrepreneurship is about looking for, finding, and exploiting opportunities. This is encompassed in the entrepreneurial spirit for being proactive, innovative, creative, accretive, and enthusiastic.

This entrepreneurial philosophy establishes itself in different ways. For example, in terms of attitudes (failure is learning, embrace innovation, and change), behavior (pursue opportunity, persevere, manage risk), professionally (being ambition in a number of different ways and seek change over career life cycle), or personally (family and community involvement). Therefore, entrepreneurship in this context is not simply about the creation of entrepreneurs, but rather the development of the ability to manage risk and change, to see and seize opportunity, and to be the drivers of creativity and innovation.

7.3.4 Entrepreneurial ecosystem

Entrepreneurial ecosystem means the combined structure and general nature of entrepreneurship. It is just like any natural ecosystem; it is balanced only when all of its components are in coherence. Increasing numbers of firms emerge and develop not only because of talented and visionary entrepreneurs but because these firms are situated in an environment or ecosystem which encourages and sustains them, making the task of entrepreneurs simpler and stronger.

Stories about entrepreneurship often focus on heroic individuals. However, entrepreneurship is not an activity undertaken by lonely heroes: it takes an ecosystem to enable productive entrepreneurship. An entrepreneurial ecosystem is "a set of interdependent actors and factors coordinated in such a way that they enable productive entrepreneurship within a particular territory" (Stam and Spigel 2016). Enormous differences in national and regional entrepreneurial ecosystems are reflected in enormous differences in the nature and occurrence of entrepreneurship between countries and regions.

There are now a number of models of entrepreneurial ecosystems, but all models have to be a blend of "top-down" and "bottom-up" approaches based on appropriate structure and general conditions. One model is the process by which individuals create opportunities for innovation. This innovation will eventually lead to new value in society (aggregate value creation), which is the ultimate outcome of an entrepreneurial ecosystem, while entrepreneurial activity is an intermediary output of the system (see Figure 7.3). This entrepreneurial activity has many manifestations, such as innovative start-ups, high-growth start-ups, sometimes referred to as "scale-ups," and entrepreneurial employees (Stam 2014).

Two distinct layers of the entrepreneurial ecosystem are structure conditions and general conditions (Stam 2015). Both are summarized in Figure 7.3. The structure conditions include the social (informal and formal institutions) and physical conditions enabling or constraining human interaction. In addition, access to a more or less exogenous demand for new goods and services is also of great interest. This access to buyers of goods

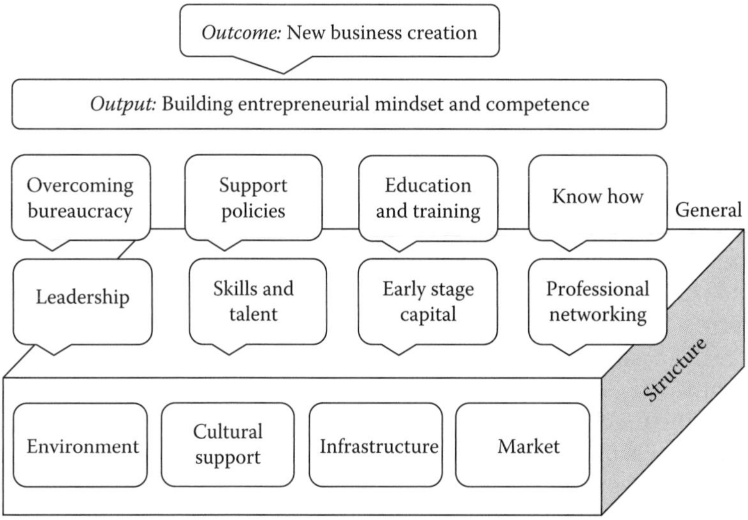

Figure 7.3 Domains of the entrepreneurial ecosystem.

and services, however, is likely to be more related to the relative position of the ecosystem than to the internal conditions of the ecosystem. However, the role of demand might also be endogenous, with the role of venturesome consumers (Bhidé 2009) or even user innovators (Von Hippel 2005).

The policy implications are twofold. First, efforts to stimulate high-growth entrepreneurship cannot be restricted to top-down efforts which simply focus on structure conditions. Bottom-up efforts, involving other tiers of government as well as nongovernment actors, are also required. Second, it needs a distinctive set of policies from those that are targeted at business start-up in general. Merely focusing policy efforts on increasing the number of new businesses has little effect because extremely few firms achieve significant growth (Mason and Brown 2014).

Successful entrepreneurs are created through nurture not just by nature. This happens, through the development of their own core human and leadership capacity. To achieve this, they need real support and inspiring mentoring throughout this journey. Their impact goes beyond job creation and economic growth. They will also be the driving force behind social development and act as role models for many future generations to emulate (Bury 2016).

Other pillars of entrepreneurial ecosystem may include a risk-taking culture and ambition to change through new technology and products; the generosity of accomplished entrepreneurs and venture capitalists who give back their time, money, and advice; close connection between learning institutions and industry; and government support of cutting-edge research.

7.3.5 Government support policies

As discussed in Chapter 4, policies provide guidance for the implementation of government programs. They guide the thinking of government in the execution of programs and define its objective. Among the most successful strategies for encouraging entrepreneurship and small business, are changes in tax and regulatory policies, access to capital, and the legal protection of property rights. Tax mechanisms including tax rate reductions, tax credits for investment and education, and tax deductions for businesses are all proven approaches for encouraging business growth.

In the case of government support policies, it is assumed that since government is in the lead for entrepreneurial development, it should provide the much needed resources within its capability. Such resources include provision of environment conducive to business that will highly promote entrepreneurship. Government policy in this context is any course of action which aims at regulating and improving the conditions of small and medium-sized enterprises (SMEs) in terms of supportive, implementation, and funding policies by the government. Based on this definition, government policy as it relates to entrepreneurial practice is targeted at encouraging entrepreneurship by making a favorable environment for the entrepreneurs. It does this through enactment of guidelines that will regulate entrepreneurial activity generally for the reason that entrepreneurship is the bedrock of a nation's path to industrialization. Furthermore, government needs to enact policies that would be user-friendly to the entrepreneurs (Obaji and Olugu 2014).

Innovation and growth largely depend on entrepreneurship, which in turn may require financing in the form of seed venture capital. Availability of financing services is one of the significant factors for sustaining the newly formed firms as well as an essential element in entrepreneurship. Therefore, the most important activity a government may undertake is to assist potential entrepreneurs with finding money for start-ups.

The following programs are among the suite of broader government policy choices: direct government investment programs, such as the United States' Small Business Innovation Research Program, Australia's Innovation Investment Funds, Canada' NSERC Idea to Innovation, and Canada' Labour-Sponsored Venture Capital Corporations. While these programs may involve public and industry subsidization of venture capital, the examples have generated records indicating effectiveness in fostering innovation and economic development.

Small business may succeed where there is respect for individual property rights and a legal system to protect those rights. Without property rights, there is little motivation to create or invest. For entrepreneurship to flourish, the law needs to protect intellectual property. If innovations are not legally protected through patents, copyrights, and trademarks, entrepreneurs are unlikely to engage in the risks necessary to invent new products or new methods. Figure 7.4 shows the components of government support policies to promote entrepreneurship.

7.4 The entrepreneurs

> A person who never made a mistake, never tried anything new.
>
> **Albert Einstein**

Entrepreneurship refers to an individual's ability to turn ideas into action. It includes creativity, innovation, and risk-taking, as well as the ability to plan and manage projects in order to achieve objectives.

7.4.1 Who is an entrepreneur?

> An entrepreneur is someone who jumps off a cliff and builds a plane on the way down.
>
> **Reid Hoffman**

An entrepreneur is entrepreneurial, as differentiated from managerial, who is someone who organizes, manages, assumes the risks of a business or enterprise, and is prepared to sacrifice time, effort, and money to turn a good idea into a marketable product. An entrepreneur is the person who is able to actualize innate potentials and develop a character that is not dependent but independent.

Entrepreneurs identify needs or opportunities and then undertake the business venture themselves. The opportunity may encompass pioneering a truly innovative product; devising a new business model; creating a

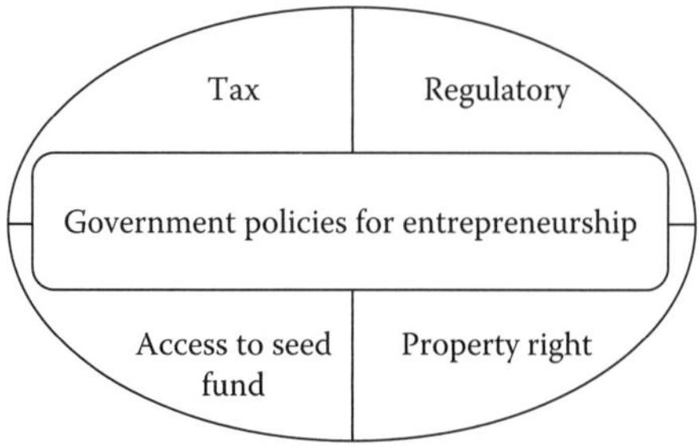

Figure 7.4 Government policies as a lead entity to promote entrepreneurship.

Table 7.1 The 12 Overlapping Definitions of the Entrepreneur

The entrepreneur is a	Person who assumes the risk associated with uncertainty
	Supplier of financial capital
	Innovator
	Decision-maker
	Leader
	Manager or superintendent
	Organizer or coordinator of economic resources
	Proprietor of an enterprise
	Employer of factors of production
	Contractor
	Arbitrageur
	Person who allocates resources to alternative uses

better or cheaper version of an existing product; and targeting an existing product to new sets of customers (Eisenmann 2013).

The term entrepreneur is often used as a substitute for business owner, starter, self-employed, or sole trader (Lans 2009). The perception of an entrepreneur as an innovator is based on the paradigm which puts the entrepreneur as a person involved in the identification of opportunities and employs the innovation tool for developing successfully new business (Meyer 2003). In reviewing this terminological history, Herbert and Link (1982) came up with a list of 12 overlapping definitions as shown in Table 7.1.

The entrepreneurs seek out new opportunities; look for the chance to profit from change and disruption in the way business is done; track opportunities with discipline and persistence; pursue only the very best opportunities and avoid draining themselves and their organizations by chasing after every option; and concentrate on implementation, specifically, adaptive execution. The commonly shared characteristics of successful entrepreneurs are the behavioral styles of leadership. By combining the above thoughts it may be generalized that entrepreneurs are risk bearers, coordinators and organizers, gap fillers, leaders, innovators, and creative followers.

7.4.2 What makes someone an entrepreneur?

Entrepreneurs are success oriented, enjoy taking responsibility for decisions, and dislike routine work. Creative entrepreneurs possess high levels of energy and great degrees of persistence which, combined with a willingness to take moderate and calculated risk. This enables them to transform what began as a very simple ill-defined idea into a concrete project.

Although an entrepreneur is generally defined as an individual, a group or an organization may also be entrepreneurial. Just as an individual can add other disciplines to technical base, groups can do the same. When individuals of different skills come together and collaborate to pursue a common goal, the team can be entrepreneurial (Crawford 2012).

Successful entrepreneurs may come from various ages, income levels, gender, and race. They may differ in education and experience. Figure 7.5 shows that most successful entrepreneurs share certain personal attributes, including creativity, dedication, determination, flexibility, leadership, passion, self-confidence, and smarts. This consists of common sense joined with knowledge or experience in a related business or endeavor.

7.4.3 The entrepreneur domain

Entrepreneurs have been described as people who have the ability to see and evaluate business opportunities, gather the necessary resources to take advantage of them and initiate appropriate action to ensure success (Meredith et al. 1991). The defining characteristic for an entrepreneur is the ability to act on opportunities. Other main characteristics are drive, passion, resourcefulness, risk-taking, and importantly the belief that one can be successful. The characteristics of an entrepreneurial mind-set can be learned, including the ability to explore opportunities, learn from failures, and solve problems, as well as acquiring technical, business, interpersonal, and communication skills.

Figure 7.6 shows the three main domains of the entrepreneur. The process incorporates the "how" of doing business. The purpose is rooted in the desire to solve problems. Achieving this purpose is an extractable part of the business model. On the human side, intension and excitement are crucial elements of success.

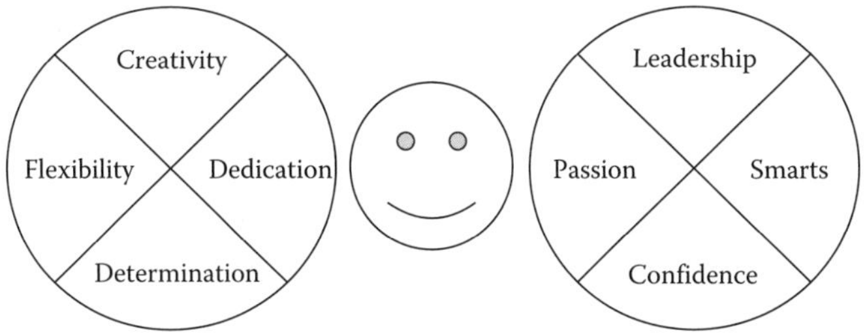

Figure 7.5 Personal attributes that makes someone an entrepreneur.

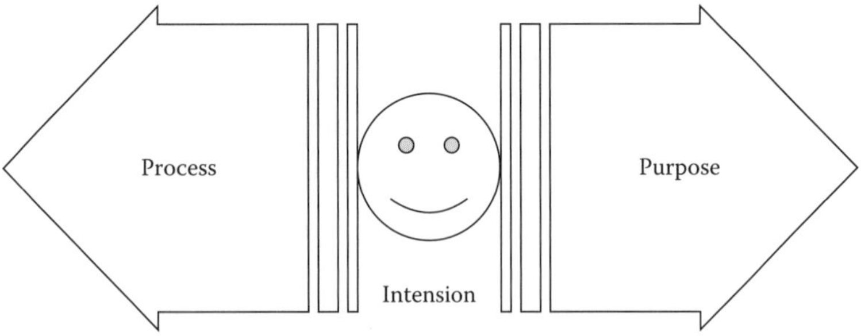

Figure 7.6 The three main domains of the entrepreneur.

7.4.4 Entrepreneurial brain and traits

Success is not what you have, but who you are.

Bo Bennett

Whole-brain thinking is a must for envisioning the future while designing a strategic plan. Ned Hermann (Lumsdaine and Lumsdaine 1995) divides the brain into four quadrants: two on the left (quadrants A and B) and two on the right (quadrants C and D), as shown in Figure 7.7. The left half works more with logic, words, structures, and analysis. The right half works more with emotions, pictures, whole entities, relationship

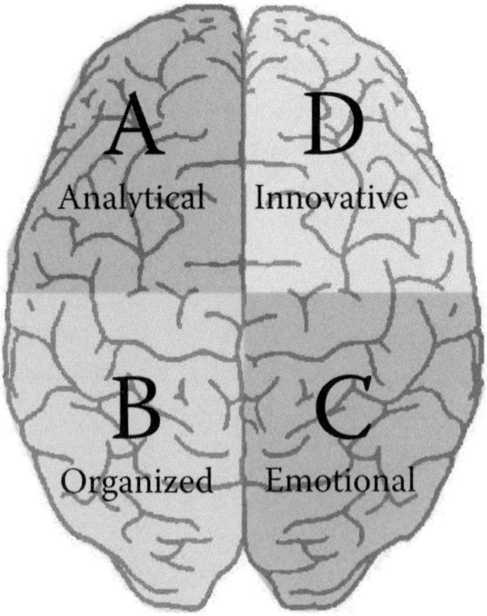

Figure 7.7 The four-quadrant model of the human brain.

among parts, and synthesis. The left half is sequential and time-bound (masculine) and the right half is holistic and time-less (feminine). In Asian philosophy, these two aspects form the yin-yang (feminine-masculine) combination. Most of us are trained to quadrant "A" thinkers, who think in terms of numbers and words. Quadrant "B" thinkers are task oriented and result driven. Quadrant "C" and "D" thinkers think in terms of systems or images, relying heavily on a holistic model of a situation. A number of activities can be designed to move a quadrant "A" thinker (knowledge worker) to a quadrant "D" designer (entrepreneur).

Professions requiring quadrant "A" dominance are those of engineers, computer scientists, analysts, hankers, lawyers, and physicians, practicing external activities. Quadrant "B" dominant professions, with procedural activities, are those of administrators, tactical planners, bureaucrats, and bookkeepers. Teachers, nurses, social workers, and musicians are interactively involved with people and are dominated by quadrant "C" thinking. Entrepreneurs, explorers, playwrights, research and development (R&D) personnel, detectives, and artists are dominated by quadrant "D" thinking (internal creativity).

It is recommended that the persons desiring the enhancement quadrant "D" of thinking should practice the following activities:

- Looking at the big picture and the context
- Participating actively, simulating, and asking "what-if?"
- Respecting multiplicity and aesthetics
- Brainstorming for and playing with wild ideas
- Exploring unobvious facts and figures
- Thinking about present and future trends
- Synthesizing to come up with innovations

Leadership in a knowledge organization deals with direction (production capability), while management deals with speed (production). Leaders derive their strength from the top line: vision, mission, values, effectiveness, and moral principles. They are dominated by quadrant D thinking. They develop this thinking by their own effort and in their own style, after going through stages of quadrants A, B, and C, usually in the same order.

Psychologist David McClelland (1987) characterized high achievers/entrepreneurs as possessing several traits as outlined in Figure 7.8. Other characteristics of entrepreneurs include high degree of commitment; willingness to accept risk, work hard, and take action; and flexibility. Another important characteristic that entrepreneurs have to acquire is effectual reasoning. This term is defined by Sarasvathy (2001) as the word "effectual" is the inverse of "causal." Effectual reasoning, however, does not begin with a specific goal. Instead, it begins with a given set of means and

- Desire for responsibility
- Preference for moderate risk
- Confidence in their ability to succeed
- Desire for immediate feedback
- High level of energy
- Future orientation
- Skill in organization
- Value of achievement over money

Figure 7.8 Entrepreneurial traits.

allows goals to emerge contingently over time from the varied imagination and diverse aspirations of the founder (entrepreneur) and the people they interact with. While both causal and effectual reasoning call for domain-specific skills and training, effectual reasoning demands something more imagination, spontaneity, risk-taking, and salesmanship.

7.5 Apprenticeships as a pathway to entrepreneurship

> Choose a job that you like, and you will never have to work a day in your life.
>
> **Confucius**

Apprenticeship is an ancient system of on-the-job training that goes back to the guilds of the Middle Ages. Usually, apprentices had to sign a contract known as an "indenture," requiring them to serve a master for 7 years in exchange for learning the trade. Along the history, the apprenticeship system has changed, but a few trades still use it.

7.5.1 Apprenticeship as a model of learning

Apprenticeship is a combination of on-the-job training and school-based education. It provides a supportive framework for the development of occupational expertise and the broader attributes required to work and continue learning in different occupational contexts (Fuller and Unwin 2010). The metaphor of the apprentice journey is universally understood, making it possible to discuss the concept of apprenticeship across the world. The concept transcends occupational boundaries and hierarchies; hence artists, journalists, surgeons, chefs, and carpenters will often refer to the way they served an "apprenticeship." Some would agree with Collins et al. (1989) that "Apprenticeship" is the way we learn most naturally.

In many countries, including the United Kingdom, apprenticeships are also an instrument of state policy, forming part of national systems of education and training. They are generally seen as programs for young people making the transition to the labor market. Prior to the creation of engineering schools, engineering was taught in an apprenticeship style. However, from the onset of formal education, engineering curricula have been based largely on science and mathematical knowledge.

Unfortunately, undergraduate engineering education is based in universities, rather than in the engineering workplace, as opposed to "trade apprenticeships," such as those for plumbers or electricians, who receive a large part of their training in their workplace (Lindsay et al. 2008). The activity in which knowledge is developed and deployed is an integral part of what is learned; in other words, learning is situated (Brown et al. 1989). This means that engineering students undergo a transition from engineering study to professional engineering while completing their degree and enter the workplace (Lindsay et al. 2008). This transition is all too often very difficult as many graduates are incapable of formulating creative solutions to problems they have never encountered before and, therefore, do not have the ability to solve real-world problems (Aparicio and Ruiz-Teran 2007).

Apprenticeships are ideal for people who want to learn how a business is run and eventually start their own. Apprenticeships do not provide specific boundaries or clear instructions and guidance to form an entrepreneur. However, the apprenticeship will allow young people to gain work experience while building their entrepreneurial skills and key character traits such as resilience, determination and self-management, as well as project management, customer service, and networking. Attracting and training young people for the world of work requires teachers to have coaching competences to foster entrepreneurship and business awareness among students.

7.5.2 Entrepreneurial apprenticeships

Lalande (2016a) describes an apprenticeship model of entrepreneurship education at the University of Ottawa (uOttawa) Entrepreneurship Hub. The model is based on the famous quote from management guru Peter Drucker (1985): "Entrepreneurship is neither a science nor an art. It is a practice." Another deep influence in adopting the apprenticeship model of "learning-by-doing" was Lalande's involvement with the global maker movement and more specifically maker education. He was also intrigued by the constructionist learning paradigm pioneered by Seymour Papert, one of the founding faculty members of the MIT Media Lab. The core tenet of his constructionist theory of learning is that people build knowledge most effectively when they are actively engaged in constructing things in the world: "I imagined that Papert's model could be applied to entrepreneurship education by way of apprenticeships. Practice-based entrepreneurship, I thought, would be the next best thing to actually starting up your own venture."

According to Justin McLeod, one of Lalande's students at the uOttawa Entrepreneurial Apprenticeships, apprenticeships allow students to learn every day. Instead of being in a classroom learning theory around entrepreneurship, they are doing hands-on learning and creating their own blueprint, they are defining themselves (McLeod 2016). In 2015, Lalande and McLeod started a "proto-venture" called StudIoT, a team of engineers and computer scientists that focus on Internet-of-things (IoT) hardware R&D. It is the true definition of an entrepreneurial apprenticeship co-op. According to Mcleod, apprenticeships and PBL are a match made in heaven. An electrician does not learn by simply reading a book, he is in the field, wiring up a house. He is learning while doing. Here at StudIoT, everything is project driven. With a good mix of client-driven projects and internal exploratory projects, there is never a dull moment. We learn while we do.

7.6 Intrapreneurship

Come to work each day willing to be fired.

(Pinchot 2016)

7.6.1 Term defined

The term intrapreneurship (entrepreneurship within an existing business structure) (also termed "corporate entrepreneurship" or "corporate venturing") is derived from a combination of "intra" or internal, and "entrepreneurship." Intrapreneurship is defined as a collection of formal and informal activities within an organization leading to the implementation of innovative ideas and behaviors (Toftoy and Chatterjee 2005). It is often associated with larger companies that have taken notice of the rise in entrepreneurial activity. Intrapreneurship practices have been developed in response to the modern world's rapidly changing marketplace (Gale 2007). Large companies are built around structure and process and this is not the best environment to spark creativity and entrepreneurship. However, there are an increasing number of larger companies who are now talking about "intrapreneurship," referring to the entrepreneurial efforts made within their companies aimed at exploiting new markets and products (Baker 2013).

The term "intrapreneurship" has become part of the business lexicon for the past 30 years. *TIME* and *Newsweek* articles about intrapreneurship were both published in 1985. But 3 years earlier, Howard Edward Haller's completed formal academic case study and Master's thesis documented the terms "intrapreneurship" and "corporate entrepreneurship." In June 1982, Haller successfully defended his Master's thesis which was an intrapreneurship case study. He wrote about the Super Mini Computer firm, PR1ME Computer Inc. (1977–1980). Haller's Master's thesis research was published by the University in 1982 (Cited in Wikipedia.org's History of Intrapreneurship). Three years later the term "intrapreneuring" was popularized by management consultant Gifford Pinchot III in his book "*Intrapreneuring*" which was published in 1985 (Haller 2016).

7.6.2 Who is an intrapreneur?

The word "intrapreneur" is iconic for many millennials. They use it as shorthand for the freedom to pursue their own ideas and the chance to make a meaningful difference early in their careers. This is what millennials are demanding. As millennials spread the word, companies are implementing intrapreneuring to recruit and retain the best and the brightest (Pinchot 2016). The main difference between entrepreneurs and intrapreneurs is that the outcome of success default to the organization rather than to the intrapreneur.

An intrapreneur is defined as a person within an existing company who takes a responsibility for turning an idea into a product or service through risk-taking and innovation. Intrapreneurs are usually highly self-motivated, proactive, and action-oriented people who are willing to taking the initiative, even within the system of an organization, in pursuit of an innovative product or service. They have the ability to "think outside the box," and are risk-takers and importantly leaders. They have all traits that are also possessed by successful entrepreneurs.

Intrapreneurs are employees who work within a business in an entrepreneurial capacity, creating innovative new products and processes for the organization (Gale 2007). Intrapreneurs are also internal trendsetters who are also referred to as "innovation heroes," "catalysts," "innovation champions," and by other such flattering titles. Compared to an entrepreneur, the intrapreneur will have more resources available and does not need to begin from scratch, which means taking less risk.

At the level of an individual intrapreneur, the trigger for innovation could arise from the aspiration to challenge oneself beyond the obvious.

Intrapreneurs seeking to reinvent a company in order to increase efficiency may do so by removing "unproductive layers" of bureaucratic hierarchy, harnessing the power of technology, proper delegation of authority and power or find other ways to improve efficiency and effectiveness (Seshadri and Tripathy 2006).

As Harvard Business Review contributors Vijay Govindarajan and Jatin Desai assert, "Intrapreneurs can transform an organization more quickly and effectively than others because they are self-motivated freethinkers, masters at navigating around bureaucratic and political inertia."

In summary, an intrapreneur is someone who operates like an entrepreneur but has the backing of an organization in terms of resources.

7.6.3 The secret weapon of success

Intrapreneurship has been called the secret weapon for success with an objective of cost reduction and improved customer focus. It can serve as the growth engine for the organization and helps to expand or deepen its core competencies. Stopford and Baden-Fuller (1990) describe corporate entrepreneurship as "rejuvenation" within an existing organization. It acts as an effective solution toward the various growing complexities within companies. It is when employees have an entrepreneurial mind-set and spirit internally, healthy sense of competition, and tries to foster creativity within corporate environment. Usually, existing businesses have the financial resources, business skills, and frequently the marketing and distribution systems to commercialize innovation successfully. Through intrapreneurship they may bridge the gap between science, technology, and the marketplace.

Intrapreneurship has been implemented in high-tech firms such as 3M, Anaconda-Ericsson, Apple Computer, AT&T, Corona Data Systems, Data General, DuPont, GE, Genentech, Lockheed, Prime Computer, Rubbermaid, Sony, Texas Instruments, Toyota, and many other successful firms. While Steve Jobs promoted the notion "intrapreneurship" in a 1985 Newsweek article, this topic was still growing as the entrepreneurial mind-set was increasingly required within companies as much as a disrupt way of doing business.

One way that large companies can encourage and support entrepreneurial activity is through corporate venturing. Corporate venturing is where a company provides the necessary funding to a new venture in return for a portion of the equity. The new venture would almost certainly be in an area in which the corporate venturing company had some interest and would often be seen as a way of getting into a new business area as an alternative to acquiring an existing business, without all the problems that often beset acquisitions (Baker 2013).

7.6.4 Intrapreneurship innovation pathway

The primary difference between the two types of innovators is their context; just as innovation in start-ups requires entrepreneurs, innovation in big companies requires intrapreneurs. At the root of every successful corporate innovation, there are one or more passionate intrapreneurs. Intrapreneurs have the persistence, courage, and cunning to get through the inevitable corporate immune system resistance and turn opportunities into profitable realities (Pinchot 2016).

Many scholars have highlighted the importance of pervasive innovation across the organization (as opposed to centralized innovation by specifically created groups/teams) as one of the important strategies for long-term marketplace success, especially in large organizations. However, most large organizations experience a severe gap between intent and reality in this regard (Pinchot 1985; Hamel 2002).

There is a strong relationship between innovation and employees taking on psychological ownership of the company's growth, thereby manifesting entrepreneurial behavior. Since this is done within the framework of a large organization rather than as an autonomous entrepreneur, it is more appropriate to look at these innovators as corporate entrepreneurs or intrapreneurs. Intrapreneurism enables employees of an organization to unleash their passion that often results in generating new avenues for business growth or alternately provides radically different ways of doing existing business (Seshadri and Tripathy 2006).

Most companies would require that the intrapreneur should seek permission before attempting to create a future development; however, most intrapreneurs are more inclined to act first and then ask for permission later if they succeed, rather than asking for permission before acting. Intrapreneurs are also typically the intraorganizational revolutionary challenging the status quo and fighting to change the system from within. This ordinarily creates a certain amount of organizational friction. In such case, a healthy environment of mutual respect is required in order to ensure that such friction can be positively handled. Figure 7.9 shows the steps toward a successful intrapreneurship within an organization.

7.6.5 Innovative climate for intrapreneurship

The most important aspect in establishing an "intrapreneur-friendly" organization is making sure that employees are placed in an innovative working environment. An effective intrapreneuring climate is a cultural shift. What works to support intrapreneurs is a proper climate of achieving innovation, not a process only (Gale 2007). In addition to

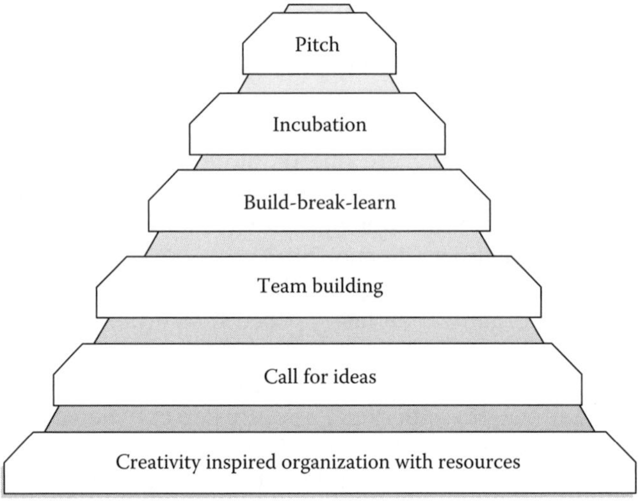

Figure 7.9 Intrapreneurship pathway within an organization.

resources and funding, the organization culture should allow failure, trial and error and accept new ideas. Active support through accepting new ideas and mentoring from top management is crucial. It is also necessary to recognize that the style of the encouraged intrapreneurialism needs to be compatible with business operations and the organization's culture.

A comprehensive model of intrapreneurship environment where, apart from the culture of innovation in the organization that the top management is responsible for creating, there are other major components including the role played by autonomous corporate entrepreneurs, innovation as a capability whereby people from different disciplines in the organization are trained for allowing error and failure, and, finally, a process which ensures that new ideas are encouraged, rewarded, and progressively ramped up from creativity to experimentation, test and assessment, scale-up, and, finally, realization. Figure 7.10 summarizes the components of an ideal environment for intrapreneurship within an organization.

7.7 From engineers to entrepreneurs

> Innovation is the specific instrument of entrepreneurship. The act that endows resources with a new capacity to create wealth.
>
> **Drucker (1985)**

Figure 7.10 An ideal environment for intrapreneurship.

7.7.1 *Innovation-driven thinking*

The expertise of the engineering profession is critical to transform innovative ideas into reality. Engineers drive technology and are therefore at the foreground of innovation. Based on above, engineering entrepreneur is one who organizes, manages, and assumes the risk of an engineering business or enterprise. This may require engineering research, analysis, development, design, and assessment that require proper education.

Engineers are trained to think logically and to follow a methodology to uncover useful solutions. This is how engineering brings value to consumers. This is the basis of a successful business. In the same way, an engineer is always willing to search for a solution to a problem, even one in the business world. When a person is not afraid to learn and gain more knowledge, growth will occur. Not every entrepreneur has a background in engineering, but it certainly can be a positive influence. An engineer will possess the necessary characteristics that make a new business successful (Turner 2012).

Entrepreneurs need to be flexible and patient. Sometimes highly valued strengths of engineers may actually become weaknesses: intelligence and precision. In general, engineers are highly intelligent and can become discouraged when others with whom they interact are not as knowledgeable about their subject. Entrepreneurs sometimes need to make prompt decisions and cannot wait for precise solutions; however, engineers value precise solutions that sometimes may not be worth the effort, time, or money necessary to complete. Engineers also tend to devalue soft skills or visionary thinking. They may also resist change once on a given path. They want to finish something, even if it adds no more value (Crawford 2012).

Essential aspects to being entrepreneurial are vision and opportunity. Being entrepreneurial allows an engineer to be more strategic in a project or in an organization. Being involved at multiple, cross-disciplinary levels can give an engineer a broader perspective on the end result of the project, sometimes resulting in engineering insights and decisions that improve the final product (Crawford 2012). An engineering entrepreneur should be able to deal with uncertainty and ambiguity; be flexible; understand and be able to speak the language of other professionals and other entrepreneurs, not necessarily just that of his or her own engineering or business domain; possess effective functional thinking and vertical (in-depth) thinking, as well as a team player's attitude; be able to be both a convincing "speaker" and an attentive "listener"; be able to understand, explain, and persuade, and possess courage to take on reasonable risks and responsibilities as a leader.

In addition, an engineering entrepreneur should be a people-oriented person; have a creative and an inquiring mind; be conscious; and be knowledgeable in languages and cultures. Communication skills, both oral and written, are important, and computer skills are vital. Such skills have become part of the modern culture, and are no longer only an element of education. Figure 7.11 shows how engineers become entrepreneurs.

Additionally, engineers also need to understand a new concept called "thrivability," which is simply defined, as the "ability to thrive," but is an extremely complex phenomenon that cannot be grasped easily. Thrivability emerges from each person holding the persistent intention to be generative: that is to create more value than we consume. When practiced over time, this builds a world of ever-increasing possibilities, which is an important aspect of being an innovative society (Russell 2013).

7.7.2 KEEN pyramid of mind-set

The KEEN is a collaboration of US universities that strive to instill an entrepreneurial mind-set in undergraduate engineering and technology

Figure 7.11 How engineers become entrepreneurs?

students. This definition of engineering was coined by Robert Kern who stressed the fact that engineering pedagogy must sustain engineering education's strict technical depth, while adding the breadth of all that is encompassed in the entrepreneurial mind-set. The distinct desire of the KEEN initiative is to change the engineering educational process such that the resulting graduates have both the competencies currently expected of graduates and an entrepreneurial mind-set that not just fosters innovation, but also results in contributing to the manufacturing of goods and services that are competitive in a global marketplace (Petersen et al. 2012).

In terms of the words "entrepreneurial," "intrapreneurial," and "engineer," a pyramidal image (tetrahedron) is proposed which is separated into three horizontal sections (Figure 7.12). The bottom section of the KEEN pyramid represents engineers, the majority of engineering graduates, while the somewhat smaller middle section represents those graduates who become intrapreneurs, and the smallest, uppermost section represents those who become entrepreneurs.

The base section of the pyramid represents most engineers who graduate from college, those who are, in the traditional sense, just engineers. Engineers are intellectual, tool-carrying, technical problem solvers. They excel in problem analysis and design synthesis. Good engineers are able to artistically express mathematics and science through their problem solutions. They are skilled in knowing what to do and how to do it once the problem has been described to them. They tend to have little to no interest in interacting with external-to-the-company, end-use customers. They also tend to be happy working for large or medium-sized companies where long-term security is more likely (compared to small startups). They tend to be motivated by having an intellectually challenging job for which they are adequately compensated; and, by nature, they tend not to be risk-takers.

The middle section of the pyramid represents a smaller number of engineers who become intrapreneurs, those who desire to be engaged in more of the creative process of new product development. They will be motivated to change the rules of competitive engagement through product redesign. They may also redefine the boundaries of competition by

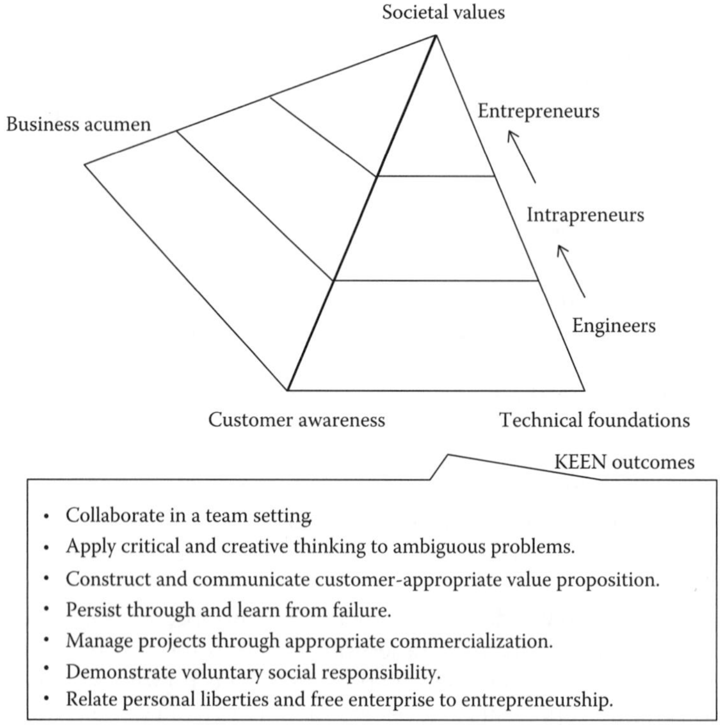

Figure 7.12 The KEEN pyramid depicting the attributes of an entrepreneurial engineer.

leveraging the core competencies of their employer into new market areas in order to gain market share for their employers. They will have a strong desire to speak directly to end-use customers of the company.

At the top of the pyramid will be a far smaller number of engineering graduates, those who seek to be market innovators. They will be motivated (even more than intrapreneurs) to redefine the rules of competitive engagement, redefine the boundaries of competition, or create entirely new markets through the application of disruptive technologies. They will understand the concepts of risk management and competing for the future, and the importance of business development. Engineers in all three of these sections have four defining attributes: working insights into technical fundamentals, customer awareness, business acumen, and societal needs.

Entrepreneurial engineers need to be able to convey their new-product story in business terms. They need to be able to negotiate organizational management obstacles by effectively collaborating in a team setting. They need to be able to recognize opportunities that have a technical solution. They need to construct and effectively communicate a customer-appropriate value proposition. They need to apply critical and creative thinking to solving ambiguous problems. Entrepreneurial engineers need to be able to see the value of their work as it affects society, preserves freedom and liberty, and maintain a standard of living which far too many of us simply take for granted (Kriewall and Mekemson 2010).

7.7.3 Technology entrepreneurs

Technologies are rules and ideas that direct the way goods and services are produced (Kemeny 2010). It results in technological innovations when new rules and ideas find practical use through being applied and/or commercialized by entrepreneurs. The transdisciplinary nature of technology entrepreneurship demands more focus and understanding among disciplines.

Technology entrepreneurship is a special vehicle of entrepreneurship that facilitates prosperity of an individual or a firm based on exploring technological advances. It is a combining of business and engineering venture that has its roots based on some kind of technology to take ideas to markets. Technology entrepreneurship is an investment in a project that builds and engages individuals and assets that are related to technological knowledge for the purpose of creating value for a growing firm.

Technology entrepreneurship can also be defined as the setting up of new enterprises by individuals or corporations to exploit technological innovation (Bailetti 2012). It may also be described as the commercialization of emerging technological discoveries or innovation. The interdependence between scientific and technological change, as well as the selection and development of new products, assets, and their attributes, differentiates technological entrepreneurship from other entrepreneurship types. Technical entrepreneurship is different from normal entrepreneurship in the scope of venture because many individuals with the tech skills to develop and create new technology might not have the skills to run a business. Technology entrepreneurship is where individuals with ideas for improvement in technology can come together with individuals who have skills to make the venture successful. New ideas can originate from inspirations such as improving an existing product, making a process easier and faster, or just an arbitrary thought. Technology entrepreneurship applies equally well to newly formed or established firms as well as small or large firms. Established and large firms can engage in technology entrepreneurship just as well as start-ups do (Bailetti 2012).

Technology entrepreneurs still involve traditional business but require a style of business leadership that is based on the process of recognizing high-potential, technology-intensive entrepreneurial opportunities, mobilizing resources such as talent and cash, and managing rapid growth using principled, real-time decision-making abilities (Dorf and Byers 2007). Entrepreneurs, whether in the technology domain or not, are successful when they really fill a need with a good product or service. A start-up company should think about what its target market needs (desirability) before they consider the technology (feasibility) that they might use and the business model (viability) that they might develop. A technology entrepreneur generally tries to solve a market problem by employing new technologies. Whether that means developing a better mechatronic tool, a better switching device, or a better biomedical system, an opportunity is always identified and then attempted to be filled. Many people misperceive this process by identifying known technology product or service in the market and try to imitate them.

The market is moving toward building more technology-focused business ventures, and this creates a wide spectrum of opportunities for potential entrepreneurs who are trying to build an enterprise. With the various technological advancements in the twenty-first century, it is becoming increasingly easier for people to set up a business from the comfort of their home.

7.7.4 Sustainability entrepreneurs

The recognition of entrepreneurship as a solution to, rather than a cause of, environmental degradation and social inequality has moved the field toward the identification of a new type of entrepreneurial activity, namely sustainable entrepreneurship. Sustainable entrepreneurship focuses on the preservation of nature, life support, and community in the pursuit of perceived opportunities to bring into existence future products and processes with economic as well as noneconomic gains to individuals, the economy, and society (Schumpeter 1934; Shepherd and Patzelt 2011).

Sustainable entrepreneurs destroy existing conventional production methods, products, market structures and consumption patterns, and replace them with superior environmental and social products and services. Therefore, sustainable entrepreneurs create the market dynamics of environmental and societal progress. This approach takes a different perspective from the traditional focus of entrepreneurship by emphasizing additional goals of promoting sustainable living and environmental improvement. This type of entrepreneurship involves searching for opportunities for new products or services or new technologies or production processes that help to solve social or environmental challenges (Brazdauskas 2015).

There is currently a growing recognition of the equal importance of issues such as the links between sustainability and innovation; the role of SMEs; the importance of sustainability in strategic business development; the emergent significance of green consumer demands on firms; sustainable practices in particular industries; and how firms can utilize the opportunities that market-based environmental policies provide (Schaper 2010). Moreover, sustainability in engineering projects is an extremely important phenomenon, since a sustainable business is an enterprise that has a positive net impact on the global and local environment as well as the social and economic spheres. It can be thought of as a business that strives to meet the triple bottom line (TBL). Often, sustainable businesses have progressive environmental and human rights policies (Diesendorf 2000). Engineers who are involved in entrepreneurship need to understand that there is no alternative to sustainability development (SD) (Nidumolu et al. 2009). This is a revolutionary understanding that needs to be accepted throughout the industry.

7.7.4.1 Sustainopreneurs

Sustainopreneurship is a business model that aims to solve problems related to SD agenda through creative organizing. Therefore, sustainopreneurs create the market dynamics of environmental and societal progress. This approach takes a different perspective from the traditional focus of entrepreneurship by underlining further goals of advocating sustainable living.

Although our understanding of sustainopreneurship has evolved through two separate streams, social as well as environmental entrepreneurship, sustainable entrepreneurship can be considered as a unique perspective that combines economic, social, and environmental value creation, with an overall concern for the well-being of future generations (Hockerts and Wüstenhagen 2010). The adoption of environmentally responsible business practices can conceivably open up an additional range of opportunities for entrepreneurs. The move to a sustainable business framework provides numerous niches which enterprising individuals and firms can successfully identify and service. These include the development of new products and services, improving the efficiency of existing firms, new methods of marketing, reconfiguring existing business models, and practices and so forth (Schaper 2010).

7.7.4.2 Ecopreneurs

The term ecopreneur is emerging as a new class of entrepreneurs with considerable consideration toward green products and services. Ecopreneurs are usually motivated by their green values, earning a living, passion toward sustainability, and realizing a need for green business in the market.

Ecopreneurship is different from sustainability entrepreneurship which integrates the three strands of the TBL (economic, social, and environmental). Tillery and Young (2009) argue that sustainability entrepreneurship goes further than "environmental" or "social" entrepreneurship as it encompasses a more comprehensive range of the TBL. Ecopreneurs play an important role in framing SD. They frame SD as an emotional problem, relating it to subjects that are dear to them, to profitability, and to the premise that it can be done.

Clearly there are some characteristics shared by all ecopreneurial activity. First, it is entrepreneurial in some way, shape, or form. All green entrepreneurs undertake business ventures which involve a measure of risk, whose outcomes are never predictable and for which the possibility of failure is always present. And, like other entrepreneurs, they must also identify a feasible business opportunity, research it, harness resources to turn the idea into reality, develop and execute a plan for business development, and oversee its growth. A second feature common to all ecopreneurs is that their commercial activities have an overall positive effect on the natural environment and the move toward a more sustainable future. A third factor that appears to be common to many environmental entrepreneurs is their intentionality. Their personal belief system and their set of values and aspirations usually see protection of the natural environment, and a desire to move on to a more sustainable future pathway, as important goals in themselves (Schaper 2010).

7.8 Inspirational role models

> Ideas are easy. Implementation is hard.
>
> **Guy Kawasaki**

History is full of wild spirits who were able to see the potential in ideas and then implement the ideas in ways that no one ever had before. We may explore history from the late nineteenth century to the late twentieth century to consider five inspirational entrepreneurs: four from the late nineteenth century who played a pivotal role in changing our ability to live and travel and Steve Jobs (with the help of Steve Wozniak) from the late twentieth century who revolutionized the way we use computers, listen to music, and more (Sellers 2014).

Interest in entrepreneurship extends beyond education. Along decades, technology entrepreneurs have become heroes, and the entrepreneurial process has been embraced as a key element of future success and global leadership.

A central commitment to engineering entrepreneurship was not limited to the following five pioneers in particular. Those great entrepreneurs not only put innovative ideas in motion but commit to them mentally as well. Their success was built on combination of innovations; for example, any new product was delivered with a new message.

7.8.1 The fathers of modernity

7.8.1.1 Thomas Edison

> Invention is 95 percent perspiration and 5 percent inspiration.
> I have not failed. I've just found 10,000 ways that won't work.

Figure 7.13 The fathers of modernity and members of the Millionaire's Club: Henry Ford, Thomas Alva Edison, and Harvey Samuel Firestone. (Courtesy of Wikimedia.)

> Our greatest weakness lies in giving up. The most certain way to succeed is always to try just one more time.
>
> **Thomas Edison (1847–1931)**
> *inventor, founder of Edison General Electric and the Edison Illuminating Company, among others*

Edison was born in the town of Milan, Ohio, on February 11, 1847. At age 12, Edison was selling newspapers on the Grand Trunk Railway. By age 15, he was publishing the Grand Trunk Herald in a freight car which also served as a laboratory for electrical and mechanical experimentation. He saved the life of the rail corporation employee's son and in return as a reward he was taught telegraphy. He then went on to develop several products such as a stock ticker and vote recorder that were not accepted as practical in use. Edison earned $40,000 for his work on the telegraph and opened a research laboratory in Menlo, New Jersey, where he and a team of scientists developed the carbon telephone transmitter to greatly enhance the work of Alexander Graham Bell (LTBN 2007).

In 1877, Edison announced the invention of the phonograph recorded on a tinfoil cylinder. In 1879, the incandescent light bulb was introduced (LTBN 2007). He later introduced the world's first electric lighting system in the commercial blocks of lower Manhattan. Edison's system, a coal-burning electric generating station, was limited to supplying electricity to about a one-mile radius at the time. His use of dc, however, was replaced by the work of others including George Westinghouse and Nikola Tesla using ac.

Edison is recognized by being the inventor of the industrial research laboratory (at Menlo Park in 1876), and most of the use-driven R&D that translated basic research into innovative products came from this model and similar industrial laboratories over the past century.

7.8.1.2 Henry Ford

> If I had asked people what they wanted, they would have said faster horses.
>
> Whether you think you can or you think you can't, you're right.
>
> I will build a motor car for the great multitude. It will be so low in price that no man will be unable to own one.
>
> **Henry Ford (1863–1947)**
> *machinist, inventor, entrepreneur, industrialist, and business tycoon*

Henry Ford was born in a farm near Dearborn suburb of Detroit, Michigan. At age 16, he came to Detroit to be hired as a student engineer and in a brief period of time, he went back to his home in Dearborn working only part time for Westinghouse Engine Company and working some time in a machine shop that he established on the family's land. In 1891 he started as an engineer for Edison Illuminating Company and was later promoted to Chief Engineer. In his on-call time he began to research and test internal combustion engines. In 1893, he finished the construction of his first experimental car that moved in four bicycle tires. After 3 years, the entrepreneur Henry Ford successfully demonstrated his second prototype car. In 1901, he established his first business "Detroystuyu Car Company." The company was held for 2 years, where he, as a chief engineer built about 20 cars. In 1903, he started Ford Motor Company.

Henry Ford is regarded as one of the greatest innovators and industrialists of all time. He is often referred to as the quintessential American inventor/entrepreneur (Anastakis 2008). As the founder of the Ford Motor Company and the mastermind behind assembly line technology, Ford has been ranked as the number one greatest entrepreneur by both the Nation's Business survey in 1971 and the Business History Review in 2003 (McCormick and Folsom 2003).

At the time, horseless carriages were expensive and available only to wealthy people. Yet in just four decades, Ford's innovative vision of mass production would not only produce the first reliable, affordable automobile for the masses, but would also spark a modern Industrial Revolution.

While most other automakers were building luxury-laden automobiles for the wealthy, Ford had a different vision. His dream was to create an automobile that everyone could afford (Caruso 2017). Ford shipped his first car, the Model A, in 1903. It would not be until 1908 that the infamous Model T, Ford's masterpiece, would be released. In the 5 years that spanned between the A and the T, the Ford plant engineered, and sold, more than 20 different models before finally getting it right.

The Model T made this dream a reality. Simpler, more reliable, and cheaper to build than the Model A, the Model T—nicknamed the "Tin Lizzie"—went on sale in 1908 and was so successful within just a few months that Ford had to announce that the company could not accept any more orders; the factory was already swamped. Ford had succeeded in making an automobile for the masses, but only to create a new challenge, how to build up production to satisfy demand (Entrepreneur 2017).

Contrary to common belief, Ford was neither the pioneer of the automobile nor the first person to ever create an automobile. He was, however, the person who took the steps to make automobiles an available and affordable mode of transportation. His entrepreneurial efforts in the automobile industry and his innovation with assembly lines granted him millions in profit and international acclaim even to this day (McCarthy 2002). Henry Ford succeeded because he understood the nature of innovation and entrepreneurship. Neither innovation nor entrepreneurship is about a faster horse, Ford's conclusion about what his customers might want if left to their own imaginations (Pech 2016).

7.8.1.3 Harvey Firestone

The growth and development of people is the highest calling of leadership.

Harvey Firestone (1868–1938)
inventor, entrepreneur, industrialist, and business tycoon

Harvey Firestone was an inventor and innovator, as well as a smart businessman. He was born on December 20, 1868, in Columbiana, Ohio. After working for an Ohio buggy company, Firestone started his own business selling rubber tires for carriages.

Firestone's sales talents earned him accountabilities. In 1892, he was in charge of the Michigan district, however, the buggy company went bankrupt in 1896, and Firestone determined that the future was in wheels rather than buggies. With a friend's help, Firestone established a rubber wheels company in Chicago in 1896, which he sold in 1899 and profited $40,000. Gaining this amount and a patent for attaching rubber tires to wheels, Firestone transferred to Akron, Ohio, then the hub of rubber tire production. With his own money and patent, he established the Firestone Tire and Rubber Company, holding 50% of the ownership.

At the beginning, entrepreneur Firestone relied on other manufacturers to produce his tires, but the company did not perform well. In 1903 the company began manufacturing its own tires and started improving its performance. Firestone decided to meet the needs of the new automobile industry, and he started producing a pneumatic tire for autos. Firestone initiated the manufacture of pneumatic tires for the Ford Model T car, and a significant sale of tires to Ford in 1906 boosted Firestone to the best of the American tire industry. The company was innovative in design and manufacturing, initiating several new tires and treads. Ford and Firestone established a strong personal and business relationship that survived for many years.

Firestone promoted the use of motor-driven trucks, the building of the American highway system, and the elimination of railroad grade crossings. In 1923 he introduced the balloon tire, which shortly became the standard for motor vehicles. Another of the innovations Firestone brought to the tire and rubber industry was that of the "one-stop master service store," which he designed to provide tires, gasoline, oil, batteries, and brake service through a single outlet (Gale Encyclopedia 2000).

7.8.1.4 Nikola Tesla

> I do not think there is any thrill that can go through the human heart like that felt by the inventor as he sees some creation of the brain unfolding to success... such emotions make a man forget food, sleep, friends, love, everything.
>
> **Nikola Tesla (1856–1943)**
> *inventor, electrical engineer, mechanical engineer, and futurist*

Nikola Tesla (Figure 7.14) was born on July 10, 1856, in the Smiljan village of Austrian Empire (currently Croatia). In 1881, Tesla started working for a telegraph company called the Budapest Telephone Exchange where he was promoted to the position of chief electrician. During his time there, Tesla made many positive changes to the Central Station equipment. This was the period when the idea for a rotating magnetic field flashed through his

Figure 7.14 Nikola Tesla. (Courtesy of Wikimedia.)

mind. Tesla got a job at the Continental Edison Company based in France, where he had to design electrical equipment. He excelled at this job and made many improvements. Two year later, he was sent to New York City where on the recommendation of his former employer, he was hired by Thomas Edison. Soon Tesla became an important part of the company solving several of their problems. He resigned after having differences with Edison and in 1886 formed his own electrical company by the name of Tesla Electric Light and Manufacture. He established another company called Tesla Electric Company in 1887. Tesla is known for pioneering some of the most significant inventions in history including the Tesla Coil and the alternating current electrical system of generators, transformers, and motors which are commonly used nowadays (Badal 2015).

The rivalry between Nikola Tesla and Thomas Edison during the rapid advance of technology in the late nineteenth and early twentieth centuries is widely known. How that rivalry started is less well known, though, and it may teach today's technology innovators and entrepreneurs something remarkably simple and important. Tesla challenged Edison's claim that current could only flow in one direction (dc). Tesla claimed that energy was cyclic and could change direction (ac), which would increase voltage levels across greater distances than Edison had pioneered (Schwartz 2017).

Tesla left Edison looking for investors to back him. Seeing an opportunity, George Westinghouse (an American industrialist, inventor, corporate entrepreneur, and a rival of Edison in his own right) bought Tesla's 40 US patents for the polyphase alternating current system of generators, motors, and transformers. In 1888, Tesla went to work for Westinghouse in order to develop the ac system. At this time, electricity was still new and feared by the public due to fires and electric shocks.

In 1893, Westinghouse outbid Edison in lighting up the Columbian Exposition in Chicago, which allowed Westinghouse and Tesla to show the public the marvels and advantages of electric light and appliances via ac. This demonstration of ac convinced J.P. Morgan, an American investor who had originally financed Edison, to back Westinghouse and Tesla in their design and construction of the first hydroelectric power plant in Niagara Falls in 1895 (Schwartz 2017).

7.8.2 The intrapreneur Steve Jobs (and Steve Wozniak)

> Innovation distinguishes between a leader and a follower. Your time is limited, so don't waste it living someone else's life.
>
> **Steve Jobs (1955–2011)**
> *innovative entrepreneur and intrapreneur, pioneer in the computer and communications business*

Engineers typically grow by partnering with entrepreneurs, someone who pushes them constantly beyond their comfort zone. Or the entrepreneurial part of the engineer takes over (like Steve Jobs or Bill Gates) and the ability to recognize opportunity overtakes the often more conservative, perfectionist side.

This entrepreneurial legend was born on February 22, 1955, in San Francisco, California. He was adopted by Paul and Clara Jobs who were a middle class couple. He lived most of his childhood in Mountain View, Santa Clara; the place which is currently known as the Silicon Valley. The growing years of Steve's life consisted of gizmos and other electric systems being worked by the neighboring engineers. He joined Reeds College but soon dropped out after realizing that he was more interested in fruitarian diet and philosophy.

Although incredibly smart, Jobs lacked right guidance. During his high school years he befriended Steve Wozniak who was a computer engineer not knowing that this friendship would alter his life course. Steve took up a job at Atari as a video game designer for a short while. At the age of 21 years, he started Apple Computers along with Steve Wozniak. Their computers were user-friendly, smaller, cheaper, and easily available for the common people. The first model, the Apple I made sales of $774,000. This figure increased by 700% to $139,000 million (Fe 2013).

Steve Jobs was the iconic CEO of Apple Computers Inc. (one of the most successful startups of all time) and the company's public face for more than a decade.

In early 1980s, Steve Jobs and his handpicked group of 20 Apple Computer engineers separated themselves from the other Apple employees to innovatively and intrapreneurially create the Apple Macintosh Computer (Mac. Under Steve Jobs' personal leadership the MAC group operated totally independently and without interference from anyone at Apple (Haller 2016).

This separate Apple intrapreneurship venture would ultimately compete with Apple's mainstay products. This competition was part of what led Apple's CEO, John Scully, and venture capitalist Arthur Rock to become displeased with Jobs' leadership style and his intrapreneurial independence. Scully and Rock led the Apple board of directors to fire Steve Jobs (which John Scully later admitted was mistake on his part). Several years later Steve Jobs returned to save Apple as its Chairman until his death in 2012.

Steve Jobs as the CEO of Apple was particularly adept at changing market expectations and demand by means of "i-products" in rapid and steady succession throughout the latter half of his career. Underpinned by his expertise in computer technology, the iPod was not so much a musical device as a handheld computer that stored music digitally and reproduced it; the iPhone was not so much a telephone as another handheld computer that could relay spoken communications. And that same miniaturized digital technology also came to encompass a camera and musical recording capability, with access to commercial broadcasts, as well as numerous links to special functions conveniently referred to as "apps." All this in one well-designed appliance, with the promise of more digital feats to follow.

Jobs was not the first person to have an idea to create a user-friendly computer, and he was not the first person to come up with an idea about music players or smartphones, but he was the first person to implement them. Jobs spotted potential in ideas and then implemented them in ways that no one had ever dreamed of before (Pech 2015).

To be a successful intrapreneur takes much more than just creativity or an idea. A successful intrapreneur has to be willing to take real risks at sharing and pushing a unique idea. An intrapreneur has to be willing to go into work focused on a mission and be willing to be fired at any moment in defense of the intrapreneurial objective. Steve Jobs clearly demonstrated that virtue.

7.9 The entrepreneurship process

> The more risk you're able to tolerate, then, generally, the bigger innovation opportunity you can create.
>
> **Stephen Hoover**

The process of developing a new venture is expressed in the entrepreneurial process, which entails more than just problem-solving in a typical management position (Stevenson et al. 1985). An entrepreneur must find, evaluate, and develop an opportunity by overcoming the forces that resist the creation of something new. The process is opportunity driven, is driven by a lead entrepreneur and an entrepreneurial team, and has four distinct phases as shown in Figure 7.15. The process is integrated and holistic; although these phases proceed progressively, no one stage is dealt with in isolation or is totally completed before work on other phases occurs. For example, to successfully identify and evaluate an opportunity (phase 1), an entrepreneur must have in mind the type of business desired (phase 4).

7.9.1 Identification and evaluation of the opportunity

Opportunity identification and evaluation is a very demanding task. Most good business opportunities do not suddenly emerge, but rather result from an entrepreneur's readiness to possibilities or, in some cases, the establishment of mechanisms that identify potential opportunities. Opportunities do not appear quickly but they are results of the entrepreneur's devotion

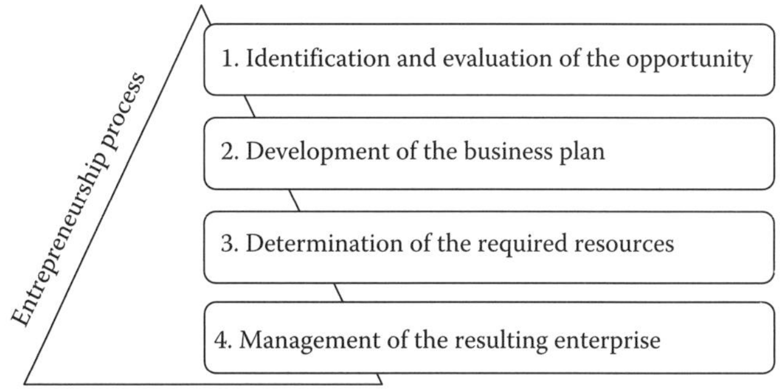

Figure 7.15 The four distinct phases of entrepreneurship process.

to these opportunities. Often luck might help. The decision-making on in which area a business shall be carried on usually requires careful consideration and collaboration.

In general, inspiration for opportunities comes from monitoring of market atmosphere and its factors such as demand for products and export prospects. The idea may also arise from a discovery of vital commodity's sources, new products and technologies. Ideas and inspiration may be also taken from real life situations. There are many sources for new ideas and venture opportunities for individuals. Ideas for business may come from own skills, hobbies and interests; flourishing ideas from others; separation from a company; inefficiency in the market; and ability to correct that inefficiency. Many sources of ideas may come also from existing businesses, such as franchises. Perhaps the most likely source of ideas for new business comes from listening to customers.

Assessment of product ideas is the initial stage of evaluation. It marks the potential value of a product into the marketplace. Business opportunity evaluation means that the resources will be allocated on further enhancement and development of the opportunity. This evaluation of the opportunity is the most critical element of the entrepreneurial process, as it allows the entrepreneur to assess whether a specific product or service makes the profit needed compared to the needed resources. The evaluation process involves looking at the duration of the opportunity, its real and perceived value, its risks and profits, its suitability to the personal skills of the entrepreneur, and its differential gain in its competitive environment.

The next step is to conduct what is called "customer analysis" for market identification and to realize the existence of marketplace. This may be accomplished by conducting surveys or simply by trying to answer series of questions:

- Who will buy the product?
- What and when does the customer aim to buy?

- Does a market exist for the innovation?
- What specific customer needs does the product satisfy?
- List specifically the people or companies that can be considered likely customers?
- Why will they buy? What product characteristics encourage these customers to buy?
- What price is the customer willing to pay?,

There is a second question to ask. Who else is in the market and who is supplying the same market? That is what is called "competitor analysis." A competitive analysis allows us to assess competitor's strengths and weaknesses in the marketplace and to implement effective strategies to improve competitive advantage. From a practical viewpoint, an entrepreneur needs to be able to live in the competitors' strategic shoes.

Also, there is a need to conduct a broader "industry analysis" to understand the attraction of the industry going to enter.

- Is the industry growing or shrinking?
- What power do the suppliers have in this industry?
- How many buyers are there? Are there alternate products?
- Are there any obstacles to entry? If so, what are they? Are there any regulations that would be subject to?

Finally, the opportunity must fit the personal desire, skills, and goals of the entrepreneur. Sometimes, people are hesitant to start new businesses, because they think they do not possess the characteristics of what would make them successful entrepreneurs. It is particularly important that the entrepreneur be able to accept involved risk and put forth the necessary time and effort required to make the venture succeed. The development of a business plan allows entrepreneurs to recognize risks in order to avoid unforeseen mistakes; to estimate sales and expenses and generate revenue forecasts in order to understand the return on investment; and to realize customers and competition better through a detailed marketing plan. Figure 7.16 reflects the components of identification and evaluation of the opportunity process.

7.9.2 Development of business plan

Planning is a key to any business throughout its existence. Determining whether an idea is a credible and feasible business option requires a well thought-out and orderly business plan. The business plan (typically about 30 pages long) is a written statement intended to crystallize business objectives, inform readers about the business, and provide a guidebook for managing the company. Importantly, preparing a business plan requires

Figure 7.16 Identification and evaluation components of the opportunity process.

entrepreneurs to split the excitement and emotion from the new idea and verbalize the concept, mission, and feasibility of the new business.

A good business plan must be developed in order to utilize the defined opportunity. This is a very time-consuming phase of the entrepreneurial process. A good plan should set the course of a business over its lifespan. It plays a key role in allocating resources throughout the business. A business plan is regarded as a filter for cleaning and screening of ideas with absence of potential for building a successful entrepreneurship. Moreover, a business plan helps the entrepreneur to formulate and remain committed to the long-term goals of the product or service offering. Figure 7.17 describes what a business plan needs to include. The marketing objectives entail how many new customers to gain and the anticipated size of customer base at the end of the period; better understand the characteristics and preferences of customers; identify opportunities to increase sales and grow business; monitor the level of competition in the market; and so on. Operational information includes details such as where the business is based, who suppliers are, and the premises and equipment needed. Financial information, including profit and loss forecasts, cash flow forecasts, sales forecasts, and audited accounts, should be listed. And finally, a summary of the business objectives, including targets and dates should be included.

The business plan should evolve in much the same way as technology. At first, it should be simple and brief, then more detailed and complex as the marketplace evolve. Despite its inevitably greater complexity, the plan must remain framed in plain, simple, declarative sentences that tell what you want to achieve, and how you plan to achieve it. Above all, the plan

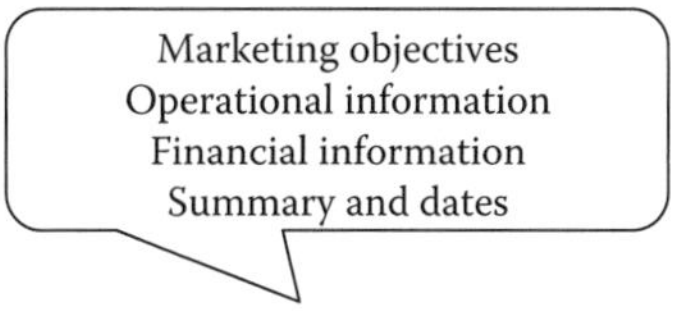

Figure 7.17 Elements of a business plan.

must always reflect you and your objectives (OEERE 2000). The kind of plan and its level of complexity depend on several factors, including, but not limited to, the following:

- Stage of technical development
- Commercialization strategy
- Growth strategy (bootstrap, slow and steady, or high growth)
- Amount of capital needed for development
- Sources of capital to approach (family, informal investors, bankers, institutional equity investors)

The plan may begin as a simple description of your project, not just the technology, the whole project, including information on management, commercialization strategy, resources required for development, and so forth. Any basic plan will contain the following components, and will prove an invaluable tool for making decisions about the commercialization strategy (OEERE 2000):

- Cover page
- Table of contents
- Executive summary
- Detailed discussions of the project, product, and market

The cover page should include the contact information: the name of the business, address, phone numbers, principals, and date of plan. A very brief (one or two sentence) synopsis on the company purpose, or any other appropriate information about company and plan may be added.

The table of contents should be one page. It is an essential part of a good business plan. It should be specific enough to let readers get to the sections they want to find very quickly. Subheadings may be included as far not to go beyond that single page.

The executive summary should be brief; one page is enough, certainly not more than two pages. However, the executive summary needs to give a genuine overview and tell the reader what to expect in the remainder of the plan.

The discussion part should be written in simple technical language where the story is told, remembering that nontechnical people like potential investors and prospective licensees need to understand the plan. The description should be reduced to the simplest terms that will convey a full understanding of the technology, including the following:

- What it is?
- What it does?

- What potential applications it has?
- What tasks remain to make it market ready?

The size and nature of market should be demonstrated to convince the reader that the project is a good bet. This is how the validity and business potential of product definition is documented.

To produce and/or sell the invention, a business plan is must. An effective, polished commercialization plan can serve as a strong foundation; however, a business plan demands a significant step upward in sophistication of information and presentation. Thus, if intention is to venture the invention, some sections should be added to commercialization plan, this includes the following:

- Marketing strategy
- Operations plan
- Management plan
- Financial information and risk analysis

7.9.3 Determination of the required resources

Leadership, as discussed in Chapter 6, requires organizing resources toward a goal while simultaneously preserving and encouraging a strategic vision. To successfully execute a business plan, in order to translate the business concept into an action, entrepreneurs have to surround themselves with the right mix of resources, which includes people, capital, and partners. The entrepreneur must determine the resources needed for addressing the opportunity. This process starts with an appraisal of the entrepreneur's present resources. Any resources that are critical need to be differentiated from those that are just helpful. Care must be taken not to underestimate the amount and variety of resources needed. The entrepreneur should also assess the downside risks associated with insufficient or inappropriate resources. The next step in the entrepreneurial process is acquiring the needed resources in a timely manner while giving up as little control as possible. An entrepreneur should strive to maintain as large an ownership position as possible, particularly in the start-up stage. As the business develops, more funds will probably be needed to finance the growth of the venture (Edward 2012). Figure 7.18 outlines the required resources.

Human resources include talents, one of the most important resources, outstanding team work of team members and physical labor. Opportunity resources include intellectual property like patents, trade secrets, trademarks, confidentiality agreements, exclusive customer relationships, technological know-how, knowledge capital, and relationship capital which is

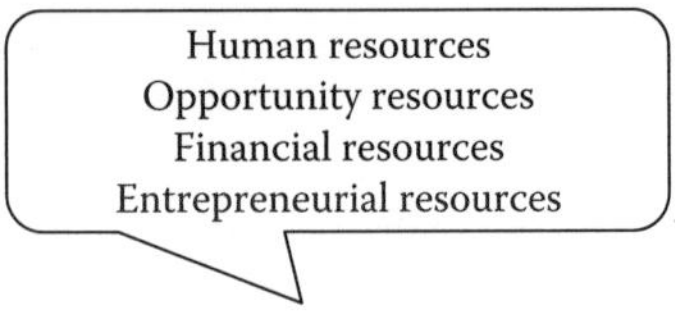

Figure 7.18 Required resources for the entrepreneurship process.

especially important for strategic partnerships and outsourcing. Financial resources include cash and cash equivalents as well as access to funding and financial backing to pursue the opportunity. Entrepreneurship resources include the collective domain of expertise and combined intrinsic motivation of a venture team.

7.9.4 Management of the resulting venture and entrepreneurial risk

After resources are acquired, the entrepreneur should use them to implement the business plan. The management part of any planning remains the most important factor in the success of a new business. Understanding how to operate a business is essential to long-term planning strategies and success. The operational problems of the growing business must also be explored. This involves employing a management structure and approach, as well as establishing the key factors for success. A control scheme must be established, so that any problem can be quickly identified and solved. The most important task an entrepreneur should take into consideration is to distinguish tasks that can be performed from those that should not be assigned. The next step involves determining what additional management skills are needed and then engaging those skills.

In their book, *Entrepreneurship*, Hisrich and Peters (2002) say that managing a new venture differs from managing an existing operation along five key management issues: strategic orientation; commitment to opportunity; commitment of resources; control of resources; and management structure. The entrepreneurs born with these management skills come from a rare breed of people with intelligence, great heart, and creative skills. They are visionary and self-confident, good communicators with unlimited energy, and have a strong passion for what they do. The heart of entrepreneurial management is continually juggling the vital management issues outlined in Figure 7.19 (Price 2011).

There are many risks an entrepreneur and an investor in an entrepreneurial venture are faced with. Do they have the means not just to start the company, but also develop the company? A main source of risk is technology risk. To the extent that company employs technology, there are

- What is this venture about? (mission and values statement)
- Where should it go? (goals and objectives)
- How will it get there? (growth strategy)
- What does it need to get there? (people and resources)
- What structure is best? (organizational capabilities)
- How much money does it need and when? (financing strategy)
- How will it recognize the final destination? (vision of success)

Figure 7.19 Entrepreneurial management issues.

obviously issues of its leading edge, intellectual property, and the product risk. If a product is not developed yet, can it be produced? Will it function? Another risk is associated with the industry and availability of supply. Moreover, there are financial risks including raising the initial money to carry out the task and possibility to raise the follow-up money. All the above matters are under the notion of investor risk. The market size and the length of the opportunity window are the initial bases for determining the risks and rewards. The risks reflect the market, competition, technology, and amount of capital engaged.

The most common mistake with entrepreneurs, and especially entrepreneurs with an engineering and technology background, are too motivated by the technical features of the product, rather than on the need that they are attempting to fulfill. Customers do not buy engineering and technology. Customers buy products that they need, in order to satisfy some issues that they wish to add value to. It is not engineering and technology; it is their services that really matter.

Entrepreneurship, risk, and uncertainty are long-time bedfellows, and they push the entrepreneur to the limit. Peter Bernstein (1998), in *Against the Gods, the Remarkable Story of Risk,* describes that the modern concepts of risk dates back more than 800 years with the early principles of gambling. According to Bernstein, "The revolutionary idea that defines the boundary between modern times and the past is the mastery of risk."

7.9.5 Timmons model of entrepreneurship

According to the Timmons model of entrepreneurship (Timmons 1978), the three critical factors of a successful venture are opportunities, teams, and resources. The successful entrepreneur is one that can balance these critical factors. The process starts with opportunity, not strategy, resources, or planning. Opportunity recognition results from creativity, which is shared by the entrepreneur and the entrepreneurial team. Creativity results from impact between academic learning and real-world practice. Value creation results from integration of opportunity and effective use of resources. Blend of people, opportunity, and resources coming together at a particular time determine the chance for success.

At the center of the framework is a business plan, in which the three basic components are integrated into a complete strategic plan for business. Creation and recognition of opportunities are at the heart of the process. In understanding of opportunity first is to focus on market readiness, the consumer trends and behaviors that seek new products or services. The model holds that a sound business opportunity would readily receive financing, and identification of the opportunity first makes the business plan successful.

The Timmons model discounts that the popular notion is to reduce the risk of starting a venture and encourages bootstrapping or starting with the bare minimal requirements as a way to attain competitive advantages. Once the entrepreneur identifies an opportunity, he/she works to start a business by putting together the team and gathering the required resources. The nature of the opportunity determines the size and shape of the team.

The Timmons model gives the team a special attention and considers a good team as crucial for success. Among all resources, only a good team can unlock a higher potential with any opportunity and manage the pressures related to growth. Figure 7.20 shows the Timmons model of the entrepreneurship process.

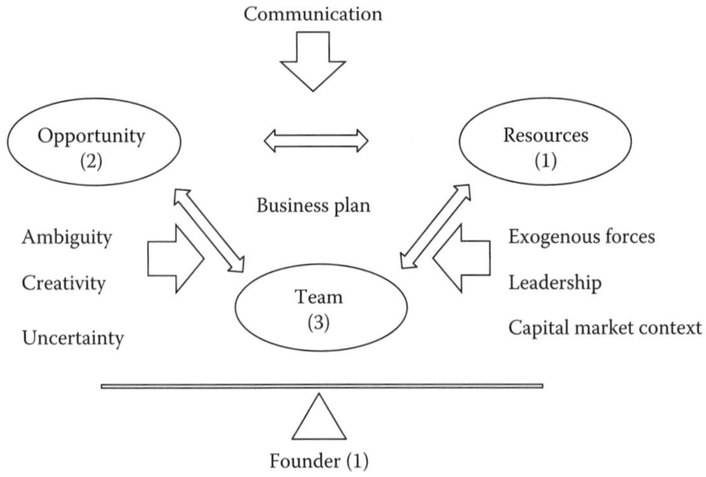

Figure 7.20 The Timmons model of the entrepreneurship process.

7.10 Entrepreneurial marketing

> Your most unhappy customers are your greatest source of learning.
>
> **Bill Gates**

Marketing is not about selling only. Relatively it is right because that is the eventual goal of marketing; however, it is more than just selling. It is about approaches in attracting the attention of target customers. One common aspect between entrepreneurship and marketing is the study of methods through which marketing concepts and principles could be applied in the field of entrepreneurship.

EM is a concept that was developed at the interface between two sciences, marketing and entrepreneurship. It came out in 1982 at a conference at University of Illinois, Chicago, sponsored by International Council for Small Business and American Marketing Association, two of the largest professional and academic associations in these fields (Hills et al. 2010). EM has come to describe the marketing activities of small and new ventures which they do not follow conventional methods in their marketing practices and their functions are so specific.

7.10.1 Marketing variables

Marketing is of crucial importance for the success or failure of an enterprise, as its success is eventually decided in the market, competing for the target customers. In funding, venture capitalists will usually look the way enterprise plans to enter the market, target groups to be addressed, and the approach by which firm's product or service offers will be communicated and distributed to potential customers. EM is difficult to calculate and is rather based on the entrepreneur's visionary and creative marketing ideas (Volkmann and Berg 2011). Innovative marketing is made up of six components given as six Ps of marketing variables as shown in Table 7.2.

EM refers, therefore, to an entrepreneurial approach to marketing functions; that is to say, it denotes the innovative, proactive, and risk-taking approach to the processes of creating, communicating and delivering value to customers.

For entrepreneurs and owner managers of small enterprise, marketing is a method or in other words, a tactic for catching opportunities. Moreover, marketing functions in these companies are highly dependent on available recourses, life cycle of both, company and its product/service, personality, knowledge and experience of owner manager, and degree of customer satisfaction (Odwyer et al. 2009).

Table 7.2 The six Ps of Marketing Variables

Product	Product enhancement: What is your real product? What should or should not you offer?
People	Customer focus: Who uses your product? What do they care about?
Price	Market focus: Can customers afford your product or service? How do they value it?
Place	Product distribution: How do customers get to your product? Where is it distributed? How is it delivered?
Production	Unique proposition: Can you do what you promise? Can you meet the market demand? Is your production flexible enough to meet changing market needs?
Promotion	Innovative marketing: How do you let people know what you have? How good does your promotion work?

7.10.2 Marketing activities

EM is difficult to calculate and is rather based on the entrepreneur´s creative marketing ideas. Marketing activities of small companies with limited access to funding and resources should be based on creativity and simplicity (e.g., budget or guerilla marketing). Guerrilla marketing is an alternative approach to advertise a product or a service. This strategy is more focused on helping small companies and entrepreneurs to advertise their products and services without spending a large amount of money. It involves word-of-mouth advertising, addressing consumers in their everyday work situation, for example via email, sticker and poster campaigns with electrostatic, self-adhesive materials, sending personal messages via Bluetooth, advertisements on cars, T-shirts, covered advertising, advertisements on sales receipts, projection of images, texts, or videos in public areas via beamer or laser (Volkmann and Berg 2011).

Viral marketing is another customer-based approach for EM which uses social networks such as Twitter and Facebook to gain brand awareness. The term was coined in 1997 (Phelps et al. 2004). However, the more common use of social networks has recently pushed this buzzword to a whole new level. The viral-marketing model is based on a similar concept to the spread of infectious diseases. Information about the product passes virally and efficiently from person to person at a low-cost marketing. It assumes that one starts with a seed of people who send a message by infecting their friends, where the expected number of new infectious people generated by each existing one is called the "reproduction rate." Viral marketing describes an interactive platform that encourages individuals to pass on a marketing message to others, creating the potential for exponential growth. It is especially attractive to smaller businesses or companies because viral marketing can be easily an economical alternative to traditional marketing and can propagate easily through various online channels including blogs, microblogs, posts, and so on.

The effective adoption of marketing technology may require that business owners develop competency in the area of web analytics in order to be able to unlock the value which is contained in that data. This relates to data generated across a range of platforms where customer engagement takes place—paid media, social media, and owned media (Chaffey and Ellis-Chadwick 2012).

7.11 Dimensions and determinants of technology entrepreneurship

The way to succeed is to double your failure rate.

Thomas J. Watson

As we are in this book concerned with technology-driven entrepreneurial businesses, we may define technology to encompass a wide range of issues, disciplines, and utilities. It ranges over a wide spectrum, from engineering and life sciences to virtual enterprises and can imply products, processes, and systems. Entrepreneurial business ventures in the technology area are highly dependent on three aspects:

- Opportunity recognition and innovative idea leading to smart design
- Relevant information leading to awareness, knowledge, and application
- Appropriate action leading to sustainable brand building

These three key aspects constitute a simple definition of the entrepreneurial dimension within which in a qualitative sense all decisions with regard to the business venture will be made. These three key terms are chosen on a qualitative basis as a result of the literature survey. Entrepreneurship goes hand in hand with "design" and relevant and timeous information which must result in appropriate activities. Entrepreneurship also requires very specific personal attributes: a propensity to accept calculated risk, drive, courage, and ability to make correct choices among a number of options. These are covered in the "action" dimension of the space (Winzker and Pretorius 2014) as shown in Figure 7.21. Once entrepreneurs have developed the idea, they must begin the process of assessing whether or not the idea is in fact a viable business opportunity. Entrepreneurs are able to create wealth by identifying opportunities and then developing competitive advantages to exploit them (Hitt et al. 2001).

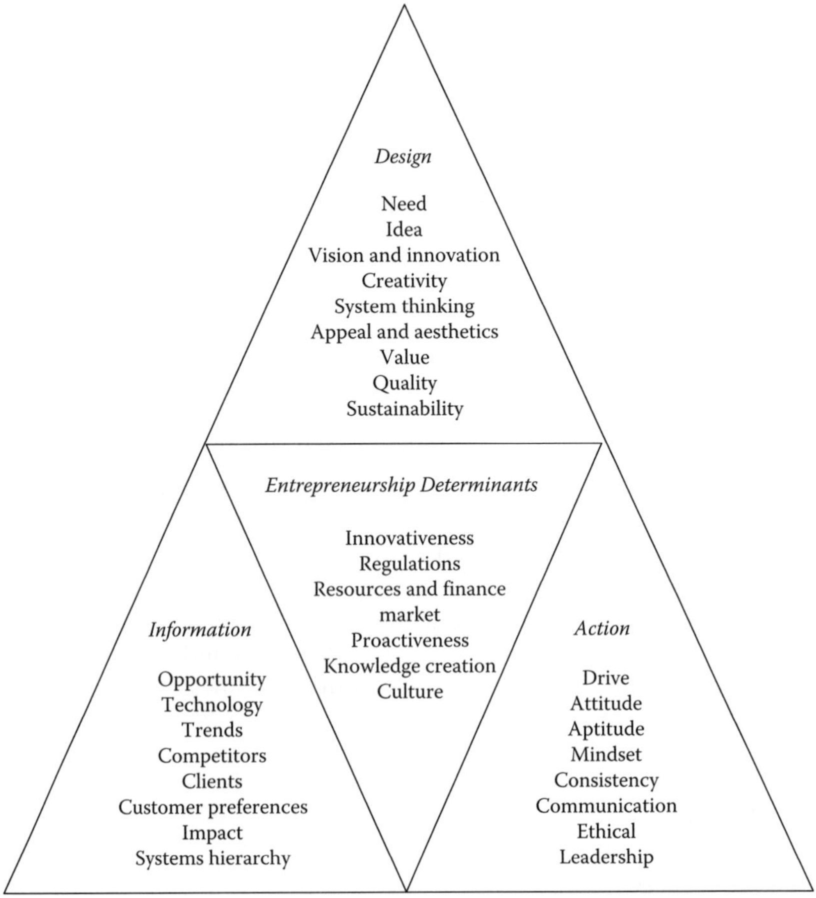

Figure 7.21 Interactive dimension of the entrepreneurial space.

As we move on to the third dimension "information" it becomes clear that the definition of the entrepreneur as creative or innovative is not sufficient. There are innovative thinkers who never get anything done. It is, therefore, necessary to move beyond the identification of opportunity to its pursuit. A significant characteristic of good entrepreneurs is a multi-staged commitment of resources with a minimum commitment at each stage or decision point. Entrepreneurial management requires learning to do a little more with a little less. Until one step has shown to be successful, resources will not be forthcoming.

Typical elements of risk-taking such as heavy borrowing, committing a large portion of assets to a course of action, or action in the face of uncertainty are associated with the risk-return trade-off. Risk can be managed by engaging in experiments, testing the markets, acquiring knowledge, and the use of networks. Interestingly, studies have shown that entrepreneurs perceive a business situation to be less risky than non-entrepreneurs.

Leadership in itself does not yield into entrepreneurial orientation. However, each type of leadership pose behaviors and talents that are connected to the entrepreneurship and such include creativity, innovation, vision, and proactive personality. These dimensions when integrated are vital to the success of any business venture.

Finally, the entrepreneurial determinants selected for the analysis (in shaded box) are chosen based on their relevance. Innovativeness is in some views intrinsically linked to entrepreneurship in that entrepreneurs create new combinations of resources by the fact of their entry into the market. In this context, innovativeness typically emphasizes the importance of technological leadership to the company, as well as improvement in its product lines. Regulations may include administrative burdens, bankruptcy regulations, intellectual property, safety and health guidelines, income taxes, and social security. The market indicator focuses on competitors by measuring whether the related activity is dominated by a few business groups or spread among many firms. The finance determinant includes access to funds, loans, credits, and stock market. Proactiveness describes the characteristic of entrepreneurial actions to predict future opportunities, in terms of technologies, products, markets, and consumer needs. The knowledge determinant includes investment in R&D, university/industry partnership, technology diffusion, and collaboration within the industry. The last determinant of culture covers risk attitudes in society, attitudes toward entrepreneurship and education.

7.12 Academic entrepreneurship

> If you are an entrepreneur, you have to think outside of the box.
>
> **Wayne Rogers**

7.12.1 Can entrepreneurship be taught?

> The entrepreneurial mystique, it is not magic, it is not mysterious, and it has nothing to do with the genes. It is a discipline. And, like any discipline, it can be learned.
>
> **Peter Drucker (1985)**

Given the widely accepted notion that entrepreneurial ventures are the key to innovation, productivity, and effective competition, the question whether entrepreneurship can be taught is obsolete (Kuratko 2005). This may be one of the most debated entrepreneurial questions out there, with a lot of varying opinions. Ronstadt (1987) posed the more relevant question regarding entrepreneurial education: What should be taught and how should it be taught?

It is becoming clear that entrepreneurship, or certain facets of it, can be taught. Business educator and professionals have evolved beyond the myth of the entrepreneurs are born not made. Gorman et al.(1997) indicated that entrepreneurship can be taught or at least encouraged by entrepreneurship education. Henry et al. (2005) also suggest that entrepreneurship is perceived as behavior pattern (which is thought) and therefore can be influenced from an early age through experiences, family, education, or cultural imprinting. Literature review on the entrepreneurship education pedagogy reveals that entrepreneurship education should take the action-learning or experience-oriented learning approach in order to increase the likelihood of effective entrepreneurship outcomes. The main challenge for entrepreneurship educators is to create the appropriate learning environment which reflects the life world of the entrepreneurs. Therefore, entrepreneurship education has increasingly adopted experiential approaches (Gibb 1993, 1996). Learning through experience, which combines experiences, perceptions, cognitions, and behaviors, is seen as the "innovative" alternative to traditional education. It emphasizes the central role that experience plays in the learning process (Rae and Carswell 2000; Sanja and Djula 2011).

In many engineering programs around the world, it is no longer sufficient to adequately train engineers with excellent left-brain skills: analysis, logical thinking, and quantitative thought. According to Dean Julio M. Ottino of the Robert R. McCormick School of Engineering and Applied Science at Northwestern University, solving problems is not enough. He states that there is no prize for solving correctly what may turn out to be the incorrect problem. It is important to acquire the skills to solve the correct problem behind the perceived problem, and this entails more than left-brain thinking alone. In fact, these right-brain skills, which include competitive differentiation, business adaptability, innovation and the development of a growth culture, and strategic thinking are the key competencies required to differentiate business in the future (Benade and Heunis 2005).

It is no surprise that majority of entrepreneurship courses are offered in business schools (Wilson 2008). However, it is questionable whether business schools are the most appropriate place to teach entrepreneurship: innovative and viable business ideas are more likely to arise from technical, scientific, and creative studies (EC 2008). Entrepreneurship needs to be expanded across the campus especially to the technology and science departments, where many innovative ideas and companies originate. Topics for undergraduate entrepreneurship courses could be introduction/principles of entrepreneurship, new venture creation/development, and entrepreneurship strategy. Courses under the introduction/principles of entrepreneurship topic area are courses providing an introduction to entrepreneurial attitudes, behaviors and processes. New venture creation and development courses also reflect the topic area title in that they focus on the processes of business entry and expansion (Sá et al. 2014).

The real challenge is to build transdisciplinary approaches, making entrepreneurship education accessible to all students, creating teams for the development and exploitation of business ideas, mixing students from economic and business studies with students from other faculties and with different backgrounds. Goldberg (2006) proposes that engineering programs need to educate entrepreneurial engineers. He makes the case that strong technical skills are not enough and that engineers should have the ability to communicate effectively, sell ideas, manage time, and recognize and properly evaluate opportunities.

Entrepreneurial learning should be effectively integrated into the curriculum, rather than only being offered as standalone programs or courses, in order to enhance students' mind-set and develop attitudes, skills and behaviors. Experiential learning and skilled teachers are the main key factors to entrepreneurship education. Plans and programs will not have any impact without effective educators to develop the necessary enthusiasm and understanding among students. Another key success factors for entrepreneurship education is the effective engagement of the private sector in facilitating entrepreneurship (UNCTAD 2009).

7.12.2 Impact of entrepreneurial education

Entrepreneurship has emerged over the last few decades as arguably the most potent economic force the world has ever experienced (Kuratko 2005). It has become fashionable to view entrepreneurship and entrepreneurship education as the panacea for stagnating or declining economic activity. Therefore, it has been maintained that the need for entrepreneurship education has never been greater, and the opportunities have never been so many.

Entrepreneurship education can be considered part of an innovation education continuum that ranges from the topic of creativity on one end, to new venture development and enterprise management on the other (Duval-Couetil and Dyrenfurth 2012). Using this basic framework, creativity and product development are considered the inputs or "innovation process" and the consequences of innovation, including entrepreneurship, intrapreneurship, and business/technology management, are the innovation "outcomes."

The idea of entrepreneurship education is not to create the next nontraditional successful entrepreneurs instantly. No one can guarantee that when entrepreneurship is taken as a major, the next Steve Jobs is created. Using milestones such as: identifying opportunities and how to capture those opportunities, how to build and manage team, how to write a realistic business plan, seeing how venture capitalists actually operate, and just observing successful entrepreneurs in action may at least show some of the pitfalls and how to avoid them (Larso et al. 2009).

The ultimate objective of enterprise and entrepreneurship education is to develop entrepreneurial and intrapreneurial effectiveness which students can attain to different degrees depending on variables such as their personality, prior learning, motivation, ability, and context (QAA 2012). Probst (2007) adds that in order to develop intrapreneurial skills, it is important to introduce students the skills and characteristics of intrapreneurs so that they can aspire to develop or enhance these characteristics. In the contest to motivate the students about enterprise and entrepreneurship, it is needed to think hard about what values the students are taught, and why. It is needed to be aware of the dangers of promoting the idea that making money is to be prized above everything else. Instead the emphasis should be on finding better approaches of accomplishing things and on thinking in new ways to solve real-life problems.

Through the creation of a mutual relationship between education, research, technology transfer, and business creation, educational institution can develop an entrepreneurship pipeline that can transform innovation into business innovation that can impact economic and community development. New business creation generates both jobs and revenue for companies in which engineers work; it is also the engine that maintains the economy. Engineering education, therefore, must teach engineers how to be entrepreneurially minded so they can be key influencers in creating new business and jobs. This emerging educational paradigm must include not only instruction in fundamentals of engineering, but also incorporate insight into the importance of customer awareness, an introduction to business principles, as well as a focus on societal needs and values. Figure 7.22 reflects the impact of entrepreneurship education on economic development.

By integrating entrepreneurship into engineering, students are likely to be more connected to their learning and thus are more likely to continue with their studies. However, students must also be able to see the relevance of their learning to their future careers. When Duval-Couetil and Wheadon (2013) interviewed engineering graduates they learned that having entrepreneurship experience on their resumes improved job prospects. In addition to aligning engineering education with workforce

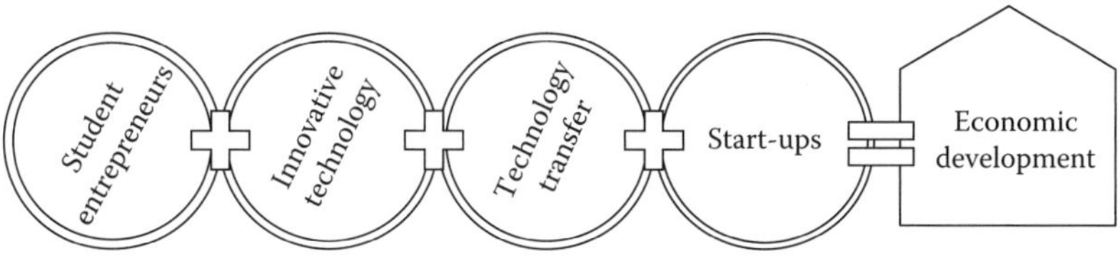

Figure 7.22 Impact of entrepreneurship education on economic development.

needs, the integration of entrepreneurship can prepare students to start their own companies based on their own innovations (Duval-Couetil 2013).

Also, higher education is uniquely positioned to play a leading role in supporting SD, especially by promoting student innovation-driven thinking and creative problem-solving toward solving social and environmental challenges. The creativity and innovation perspective is a new approach toward education of sustainability; therefore, this approach is a highly rewarding sphere for those students, who view sustainability as their inspiration and as a starting point for innovation-driven business planning (Brazdauskas 2015). Entrepreneurship education with a specific focus on sustainability could be one of the mechanisms that can be used to stimulate future entrepreneurial behavior in energy-related green sectors.

In their study "Impact of Entrepreneurship Education," Alberta Charney and Gary Libecap (2000) concluded that entrepreneurship education helps produce self-sufficient enterprising individuals, successful business leaders and champions of innovation. Their findings were based on a comparison of University of Arizona Berger Entrepreneurship Program graduates to other University of Arizona Business School Graduates (Kauffman Center 2001). Additionally, there was indication of a link between these entrepreneurship programs and an increased number of start-ups launched by students either during or closely following school. While the causal link is not clear, the Kauffman center explains, "Research indicates entrepreneurship program graduates are three times more likely to be involved in the creation of a new business venture than their non-entrepreneurship business counterparts."

Another high-impact approach involves creating intensive entrepreneurship programs and experiences for highly motivated students. Successful examples include the University of Texas at Austin's idea to product (I2P) competition, the NCIIA's E-Team program for launching student ventures, and a growing number of entrepreneurship-themed "living-learning" communities (combining student residence with curricular and extracurricular activities) at several universities (Inkelas et al. 2008).

7.12.3 Experiential entrepreneurship learning

Experiential learning in innovation and entrepreneurship has spread outside of business schools and moved into the fine arts, science, and engineering programs. It defines learning as the process whereby knowledge is created through the transformation of experience (Kolb et al. 2000). Entrepreneurship is taught most effectively using experiential methods (Duval-Couetil et al. 2015). Similar to other fields, entrepreneurship education is considered more effective if it includes a strong experiential component, requiring students to intellectually and physically engage in the learning process and reflect on their experiences (Kolb 1984). Therefore, entrepreneurship courses and learning modules typically include experiential learning activities to help students gain skills and confidence in a number of areas.

Entrepreneurship education according to Wilson (2008) should provide a combination of experiential learning, skill building, and mind-set shift. The goal of engineering education is to make every student innovation ready, ready to spot opportunities, envision possibilities, realize new ideas, learn and succeed. Developing a mind-set that fosters creativity, risk-taking, and motivation to succeed in the marketplace should start early in life. The mind-set concept focuses not just upon the notion of establishing a venture but upon the ability of an individual to cope with an unpredictable external environment and the associated entrepreneurial ways of doing, thinking, communicating, and organizing. At those early ages, science, technology, engineering, and financial literacy should be taught alongside business, math, social science, and the arts. Engineering is indispensable in solving technical problems; however, problem-solving alone is insufficient to produce new-to-the-world ideas for products and services. Figure 7.23 shows the key stages leading toward building entrepreneurial success.

Kolb and Fry (1974) view learning as an integrated process with each stage being mutually supportive of and feeding into the next. It is possible

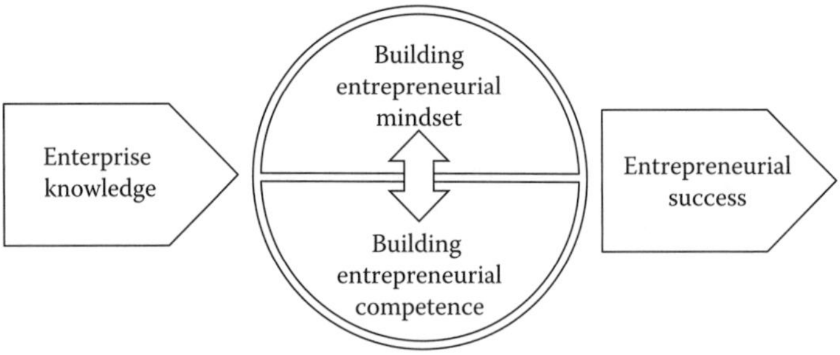

Figure 7.23 Building entrepreneurial mind-set and competence toward success.

to enter the cycle at any stage and follow it through its logical sequence. However, effective learning only occurs when a learner is able to execute all four stages of the model. Therefore, no one stage of the cycle is an effective as a learning procedure on its own.

7.12.4 Entrepreneurial curriculum building

> People believe there's an "entrepreneurial personality," but there's not. It's a set of skills you can learn.
>
> **Castaldo (2015)**

A typical approach to entrepreneurship education in a university or college setting is to leverage existing courses already offered to offer in addition to lectures, case studies, real-life projects, and new venture-based learning (see Figure 7.24). However, for real innovation to occur, extra steps need to be taken. Real innovation in education occurs when faculty and management think of new ways and new approaches of teaching, employing new experiences, and concentrating on learning models that are both engaging and expressive.

7.12.4.1 Business case writing

An effective way that familiarizes students to entrepreneurship is to use case studies that are designed to be integrated into existing engineering courses. These case studies are meant to demonstrate ways that

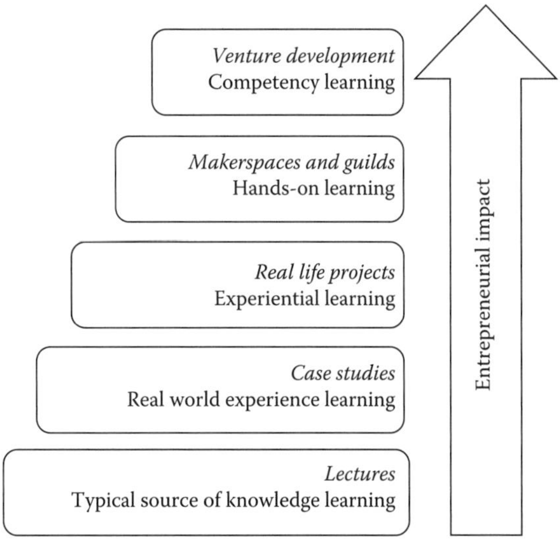

Figure 7.24 Entrepreneurial learning building in the undergraduate level.

entrepreneurs have capitalized on their knowledge of specific engineering topics covered in typical undergraduate courses to create successful business ventures. The aim is to continually showcase inspiring and successful engineering entrepreneurs and to provide standard exposures to principles of entrepreneurship throughout the curriculum. The ideal, long-term vision is to have at least one case study for each course. Case studies may help bridge the gap between theory and practice. They present the students with a real engineering scenario requiring application of a particular technical discipline while illustrating the often-critical nontechnical aspects of a problem. This helps motivate the students as to the relevance and importance of the subject matter and allows integration of important nontechnical aspects of the profession (including personal viewpoints, policies, ethical and moral considerations, business considerations, intellectual property, soft skills, etc.). Cases offer the opportunity for an engaging class period as opposed to the theory-plus-problem-solving pedagogy. Cases often involve situations that do not have clearly right or wrong answers. This may help students understand and develop a tolerance for uncertainty. Cases can offer many opportunities including defining a project, communicating a business case effectively in writing, comparing costs and advantages of alternative solutions to the problem, applying risk assessment techniques to mitigate problems, working out alternative solutions, and gaining support from key stakeholders.

7.12.4.2 Real-life projects

Working with real-life project assignments provides valuable information on enterprises, their culture, and processes in addition with strict requirements and timetables. Some projects may be the ones that students can spin off into a venture. However, entrepreneurship is not simply about preparing students to launch a venture; it is also about better equipping students to be intrapreneurs who innovatively contribute to existing organizations. The idea is to put students in the process of entrepreneurial learning in safe environment. There are several methods for this. One way is to encourage students to think of new ideas that can be turned into projects. It is necessary to blend these projects with engineering courses and/or in student activities. A project task that blends entrepreneurship in an engineering course is given in Section 7.16.3.

It is also necessary to develop conductive setup for entrepreneurship because students need the space to work on ideas, access to equipment, or seed funding to cover costs. It is important to engage students in competitions within the university and outside. Such efforts require an active engagement from the administration by providing student funds, facilitating resources, and creating competitions events for design projects and entrepreneurship initiatives.

7.12.4.3 Makers and guilds

Makerspaces inspire the new generation of makers, inventors, and entrepreneurs by offering resources and connections to explore new concepts and technologies, and learn to deliver market-ready ideas. At universities, makerspaces provide training to students and enhance the ability to deliver project-based and capstone classes in prototyping and manufacturing. They also connect university's innovation ecosystem to the world, serving as physical location for the university community to collaborate physically or virtually through advanced tools and technologies.

One movement that shows potential for a huge global impact and, over the past few years, has attracted an impressive number of followers is the "maker movement" (Lalande 2014). The emergence of locally based and globally networked makerspaces to harness and empower the creative and innovative energy of communities may indeed accelerate the pace of local entrepreneurship. In the author's opinion, community and university-based makerspaces and fab labs represent an inclusive and more appropriate model for promoting grassroots entrepreneurship and innovation rather than start-up incubators and accelerators. Makerspaces and fab labs integrated with local entrepreneurship ecosystems serve to educate, promote collaboration, problem-solving, and ignite creativity (Lalande 2016b).

On the other hand, Lalande is currently utilizing the dynamics of historical craft guilds that holds some potentially useful insights into organizing today's diverse and skilled creatives. He uses the term "creatives" more loosely than conventional definitions that restrict the term to occupations usually associated with creative vocations such as artists, sculptors, artisans, designers, architects, etc. Creatives include those self-identified "makers" who produce objects with the latest digital fabrication tools such as 3D printers. Such individuals may hold "day" jobs in what we might traditionally define as "noncreative" but they nonetheless strive for opportunities for creative expression. Creatives may also include what has been identified as "pro-ams," professional amateurs like passionate hobbyists, "garage" tinkerers, and even scientists who stray into more artistic endeavors, among others.

Guilds, according to Lalande, potentially represent such a novel value-creation and value-capture organizational experiment. The recent wave of lower cost, higher quality personal manufacturing tools may, over time, shift power back to the "owners" of the skills (e.g., the creatives) away from the owners of the capital. Guilds should be designed as an organizational experiment to promote and accelerate "collective" innovation and develop market opportunities for its members.

It is also worth considering the importance of having a "curator-entrepreneur" who could serve as the guild champion. The guild curator would play a vital role in identifying a particular niche for creative endeavor; attracting creatives to join and/or participate in guild activities; seeking market opportunities for guild members including business development; managing the guild's administrative requirements; harnessing the creative skills of the guild membership for specific project opportunities; establishing connections and facilitating self-organization among guild members; finding patrons and other means of financing guild activities; devising a platform of web-based resources and tools to encourage co-creation, collective and individual training and education (Lalande 2014).

7.12.5 University spin-off and venture development

Academic entrepreneurship by way of university spin-offs is an emerging field that focuses on the process of creating, discovering, and exploiting technological opportunities created by university education. The field of entrepreneurship and of university spin-offs, in particular, still lacks a widely accepted methodology or theoretical framework. Therefore, the deliberate and emergent dimensions of academic entrepreneurship need to interact and converge in building a cumulative body of knowledge and practice (Van Burg 2010). The latest trend around entrepreneurship and innovation is launching a venture; taking the designed idea and turning it into an operational company. To accelerate, incubators and accelerators are popping up everywhere, including in universities. These incubators and accelerators are fantastic and there has been many success stories coming out of these facilities (McLeod 2016).

Of particular interest to engineering programs trying to integrate the entrepreneurial mind-set, a combination of technical skills, business savvy, team building and management, and high-integrity leadership, is how to assess the methods by which they measure success in these programs. Most of the possible contents of entrepreneurship courses are relevant for students from all fields of studies. However, in order for the teaching to be tailored to the specific needs of different categories, more emphasis is placed on one aspect or another; for instance: entrepreneurship within science and technology studies is especially concerned with exploiting intellectual property, creating spin-off companies and venturing, and offers courses on issues such as management techniques, marketing, commercializing and selling of technology-based ideas, patenting and protecting technology-based ideas, financing and internationalizing high-tech ventures (EC 2008).

Today universities can decontextualize and contextualize design principles and solutions with their research findings and practices to realize venture development process. The process of abstracting solutions and their underlying principles from, for example, a first sample of new venture process through a university-based incubator involves "decontextualization." Similarly, applying a set of general principles to, for example, the creation of spin-offs in university implies adaptation to the institutional and regional setting of the university. Moving from top to bottom in Figure 7.25, knowledge becomes increasingly "contextualized," also in view of the rapidly diversifying nature of organizational, industrial, technological, and cultural settings (MacCormack and Verganti 2003).

The next step to design is to develop business plan, which serves as a simulation of the process of creating a new venture. This activity has been used widely because it draws on a wide range of skills. In an educational context, the business plan serves as a roadmap for the process of creating a business, encompassing: validation of the need for a particular product or service; analysis of the financial requirements, funding sources, and potential returns; substantiation of a marketing and distribution plan; and evidence of a team with the talent necessary to execute the plan (Wheadon and Duval-Couetil 2014). Integrating faculty and students from other disciplines, particularly business and the cross-pollination of expertise that provides, is also considered essential.

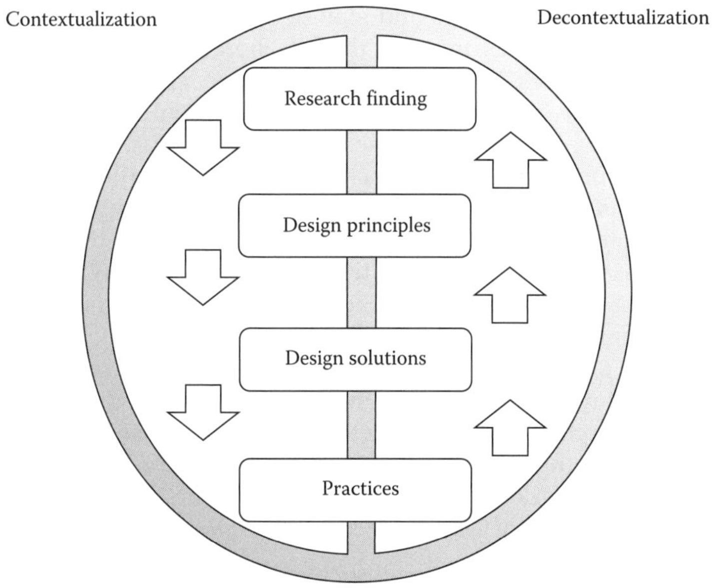

Figure 7.25 Faces of design that reflects the need to decontextualize and contextualize design principles.

7.13 *Role model case: A leading entrepreneurial engineer*

Opportunity is missed by most people because it is dressed in overalls, and looks like work.

Thomas Edison

This case is based on the following entrepreneurial research: Ensign, P. C. and N. P. Robinson. 2009. Growing from a monopoly situation: Case study of Med-Eng Systems Inc. SMEE Review August: 26–33

The case provides a glance into the company "Med-Eng Systems" under the leadership of an entrepreneurial engineer "Richard L'Abbe" who stood behind the company's innovative products "bomb suits and protective gear."

7.13.1 Med-Eng systems

Med-Eng was an Ottawa, Ontario-based firm incorporated in 1981 to manufacture a newly designed explosive disposal helmet. The firm began international activities in 1982. By the end of 2006, the firm had 450 employees and over $250 million in revenues, the bulk of which was outside of Canada. For this privately held venture, the key to developing cutting-edge products and staying on top was to "do the right things right every time, through the collaboration of its clients, vendors, and employees." This stakeholder-driven approach to product development and marketing led Med-Eng to be on the receiving end of numerous awards, for both the company's competitive strength and international posture. Superior technologies had also led to superior profit margins. Although at inception in 1981, the firm manufactured protective helmets solely, it went into the design and production of the Bomb Disposal Suit in 1992 when the company was thrown into a hostile market environment. Its approach to the market had not been used in this industry before Med-Eng developed its own suit. After developing a suit prototype, resulting into a less cumbersome, more comfortable suit, and helmet combination with two-way radio technology, Med-Eng staff proceeded to complete a three-month tour to clients in 45 countries. Involving potential clients in the design process helped Med-Eng develop a superior suit and created immediate buy-in among their customer base, most of whom placed orders within a relatively short period of time.

This type of gusto was a trademark of the Med-Eng culture, where management believed in the product so much that they actually tested their suits with live bombs. Richard L'Abbé was surely the only CEO in the world to "blow himself up" for his company. L'Abbé had tested his product 19 times with the use of explosives such as C4 and dynamite, often enough to destroy a car. His tests and the faith he displayed changed public opinion that bomb suits were "little more than just body bags." Terrorism and escalating conflict in many parts of the world led Med-Eng equipment to markets in over 120 countries worldwide, and an 85% global market share.

7.13.2 Richard L'Abbé, the entrepreneurial engineer

Richard L'Abbé graduated from the uOttawa, Canada, with a degree in mechanical engineering in 1979. He went to work briefly at Biokinetics (www.biokinetics.com), with which Med-Eng still maintained a relationship for product testing. The *Ottawa Business Journal* (CEO Profile, October 20, 2003) recounts that in 1980:

> He was busy studying the mechanics of bodily injury when the company received a contract from the Royal Canadian Mounted Police to design a bomb-disposal helmet. When the president of the company approached L'Abbé to ask for some fresh ideas, L'Abbé sat down for three hours and drew a basic blueprint of what he thought the prototype should look like. After seeing the drawing, his boss replied, "That's pretty cool, why don't you build it?"

In its inception year, Med-Eng was profitable. However, as in many start-ups, the realities of poor sales activity resulted in near catastrophic revenues. In 1982 Med-Eng, the company's total revenues were $4000… one helmet to the German Federal Police. A six-country promotional tour of Europe in September 1982, the company's first promotional trip abroad ever, resulted in their first major sale to "a seemingly rude and arrogant gathering of potential clients in France," six suit and helmet combos for $66,000. From that point on, things looked up in subsequent years. The First Gulf War led to "spectacular" sales in 1991.

L'Abbé credits an executive seminar he attended where Clayton Christensen spoke as the impetus for Med-Eng's disciplined approach to developing innovative new products. A few follow-up phone calls with the Harvard Business School professor, eminent for his insights into innovation and disruptive technology, furthered L'Abbé's resolve toward strategic planning and an entrepreneurial search for growth. And, inspired by another speaker, Dr. Ram Charan, Med-Eng went on to build a robust human resource strategy in an effort to hire only top performing staff.

7.13.3 Competition and new entrants

Med-Eng Systems produced most of its "gear" with a mix of sourced and in-house components and tried to integrate the ideas and concerns of stakeholders into the process. Not all products of firm could succeed in the market especially its riot-type protective gear. The number of companies offering riot-type protective gear exploded, with many firms offering suits of varying degrees of quality throughout the world.

Med-Eng's problems were further compounded as competitors chose to copy the company's superior product design. This was a sizable issue in countries where local authorities wanted to buy suits from local companies, and also had little respect for the concept of intellectual property. As a result, Med-Eng's designs were reproduced and sold throughout the globe illegally, under different brand names. Roughly seven or eight clones of Med-Eng's product emerged in the world market and the firm twice successfully challenged and won injunctions against companies that stole Med-Eng's product design.

7.13.4 Closing one door, opening another

The firm's sales in the industrial sector flopped and by the summer of 2004, Richard L'Abbé decided to "pull the plug" on selling cooling apparel to the industrial market. Coinciding with this let down was a spark—the US Army began to face a dilemma in Iraq: troops located there were having difficulty tolerating the extreme heat of the Middle Eastern climate while driving often unair-conditioned vehicles and wearing several pounds of thick personal body armor and other heavy gear that acted as insulators trapping body heat.

"This was expected to be a very profitable new market for Med-Eng," remarked L'Abbé, "we hoped to see sales of our cooling devices to the US Armed Forces go into the stratosphere." On the redeployment of resources, L'Abbé commented:

> So, despite an unsuccessful bid at selling cooling systems in the industrial market, we managed to acquire a new skill set that could be used more generally in the manufacture of other products.

7.13.5 The potential

As Med-Eng grew, those on both the outside and inside began to ask whether a company with revenues in excess of $250 million would be better served as a publicly held firm. L'Abbé and many others in the firm were convinced that the volatility of the company's revenues would disappoint analysts and create an element of instability. L'Abbé knew that because of their client base and trends in police and defense spending, revenues could not be steadied to the extent that investors would tolerate and that fickle investors would not appreciate the company's business structure. The firm's CEO was convinced that being privately held meant that the company could operate free of the rhetoric and pressures of investors looking for a steady return.

On top of that, thus far Med-Eng had been successful in getting enough private venture capital to grow. Med-Eng received approximately $12 million in venture capital needed to rid itself of the founding shareholders who were critical of CEO Richard L'Abbé. Less than 3 years later, Mr. L'Abbé was named "CEO of the Year" by the *Ottawa Business Journal* as revenues skyrocketed.

On October 30, 2005, Richard L'Abbé stepped aside as CEO of Med-Eng Systems. In 2008, the Canadian Venture Capital and Private Equity Association named Richard L'Abbé 'Entrepreneur of the Year' for the spectacular increase in shareholder equity realized by Med-Eng during the 25 years he managed the company.

7.13.6 *Devotion to education and community outreach*

Mr. L'Abbé has been visionary in his devotion and support to education and community outreach. He played a leading role in initiating and encouraging entrepreneurial activities at the University of Ottawa (uOttawa). The uOttawa Richard L'Abbé Maker Space is one of his lasting contributions to the University. This facility is a sandbox where students can design and develop projects. It is the first invent-build-play space at the University. It is a home to a lot of technologies such as 3D printing, virtual reality, the IoT, and wearables. Its goal is to further creativity, problem-solving skills, and interest in technology entrepreneurship. In addition, Mr. L'Abbe has used his vast experience to help students deliver better design and entrepreneurial ideas and projects.

In 2016, Richard L'Abbe was awarded an honorary doctorate by uOttawa for being a visionary Canadian entrepreneur and philanthropist (Figure 7.26).

7.13.7 *Case research questions*

- Does success breed success?
- Is entrepreneurial success a skill, or is it luck?
- Which is more crucial to the success of a start-up: the idea or the implementation?
- Who has a better chance of being a successful entrepreneur: an electrical engineer, mechanical engineer, computer engineer, computer scientist, or software developer?

Figure 7.26 Richard L'Abbé at the University of Ottawa in 2016.

- What stakeholder-driven approach means? How this approach benefited Med-Eng Systems?
- Why do ventures require dynamic leaders who understand vision, strategy, risk, and tactics? You may answer this question considering Mr. L'Abbé as an example.
- What skills are important for entrepreneurship and creative self-employment?

7.14 Knowledge acquisition

Attempting to answer the following questions involves acquisition of knowledge from this book and other books, documents, and the Internet.

- What is entrepreneurial mind-set?
- What contributes to the development of a successful entrepreneur?
- What is the relationship between design, innovation, and entrepreneurship?
- How does entrepreneurship change over time and how is it important?
- What are the twenty-first-century learning, innovation, and career skills?

- Describe the benefits and drawbacks of entrepreneurship.
- What are major strategic constraints and challenges confronted by entrepreneurs today?
- What are the objectives of innovation and intrapreneurship?
- What excites you about being an entrepreneur? What are your major concerns?
- How shall a person find a good business idea out?
- How do engineers view entrepreneurship?
- What is technology entrepreneurship?
- What is the role of ethics in technology entrepreneurship?
- How technology entrepreneurship differentiates from other entrepreneurship types?
- What is the relationship between technological innovation, entrepreneurship, and development?
- What is small-scale enterprise?
- What is project? What is business plan?
- What is opportunity analysis?
- What is market survey?
- What is environmental analysis?
- What is venture capital?
- Can academic research and entrepreneurship work together to solve problems?
- Do serial entrepreneurs succeed more than first-time entrepreneurs?
- Can entrepreneurship be taught?
- Does entrepreneurship education make people better entrepreneurs?
- What are the content standards for entrepreneurship education?
- What is KEEN and what are its outcomes?
- What are the factors responsible for becoming sustainopreneur or initiating green business?
- Is sustainopreneurship really affecting the businesses in a constructive way or it is just a marketing approach?
- Where do most entrepreneurs obtain their ideas from?
- Explain the forces that drive the growing of entrepreneurship.
- Who is more likely to get funded, a new entrepreneur or a tried and true one? Explain.
- What is the best academic research on entrepreneurship?

7.15 Knowledge possession

- How can entrepreneurs position themselves to succeed when the majority of small businesses fail?
- Ask five entrepreneurs what the term entrepreneurship means to them. Be prepared to present the commonalities and differences of these definitions to the class.

- What impact does entrepreneurship have on your local, state (or province), and national economies? Use evidence to back up your reasons.
- Research the policy statements of your local, state (or province), and national governments for their goals and objectives regarding the importance of entrepreneurship and ways of inspiring it.
- As an entrepreneur, how will you develop your technology and products, transfer or license them, and ensure commercial success?
- Speak to people from five different sectors and ask what entrepreneurship means to them and how their sector culture helps and/or hinders entrepreneurship.
- How sustainopreneurship has evolved from the past years and why it represents the need of future?
- What science and technology skills and competencies, and at what levels, are crucial for innovation-driven development, and how can policymakers ensure that education institutions respond to those needs?
- Starting own business, or venturing as it is often called, has its own advantages and disadvantages for consideration. Discuss that.
- To what degree do experiential activities contribute to higher entrepreneurial self-efficacy?
- How and to what extent do schools of engineering incorporate entrepreneurial practices in their design courses? You may investigate this question in regard to faculty or school.
- What is the incremental value of an engineering student taking more than one entrepreneurship course on their perceived knowledge and self-efficacy?
- What changes are required in engineering educational systems to provide students with the entrepreneurial knowledge and skills needed for the future? Which are the priorities? How can policymakers enable these changes?
- Why innovation and entrepreneurship are important in education system? Provide some recommendations about the best approaches of integrating the practices of innovation and entrepreneurship in the education system.

7.16 *Knowledge creation*

In the following tasks, collaborate with peers on learning or you may work with others outside the class to narrow down the objectives of each task. You may access class and online resources, and analyze data and information to create new ideas and balanced solutions. High-level digital tools may be used to develop multimedia presentations, simulations or animations, videos and visual displays, electronic portfolios, feasibility studies or consulting reports, apps or other computer program, website or recording, debates or innovation pitches, original piece of art, or well-researched and up-to-date reports that reflect the objectives of each task.

7.16.1 *Feasibility study on smart entrepreneurial library*

One of the world's first and most famous libraries, in Alexandria, Egypt, was frequently home some 2000 years ago to the self-starters and self-employed of that era. When you look back in history, they had philosophers and mathematicians and all sorts of folks who would get together and solve the problems of their time (Badger 2013). This old idea of the public library as coworking space now offers a modern answer for how these aging institutions could become more relevant two millennia after the original Alexandria library burned to the ground. Libraries meanwhile may be associated today with an outmoded product in paper books. But they also happen to have just about everything a twenty-first century innovator could need: Internet access, work space, reference materials, professional guidance.

In this task, conduct a feasibility study for an entrepreneurial pilot program to enable a network of coworking business incubators and start-up accelerators inside your community public library. Visit the library and outline its resources in terms of Internet access, work space, reference materials, and professional guidance.

7.16.2 *Feasibility study on smart agriculture farming*

IoT is flourishing in every industry. This concept took hold in some familiar places like home, cars, cities, and factories, but what about the agricultural industry? Can IoT bring a change in the way we produce food (Mohan 2015)? The IoT is set to push the future of farming to the next level. Smart agriculture is already becoming more commonplace among farmers, and high-tech farming is quickly becoming the standard thanks to agricultural drones and sensors (Meola 2016).

In this task, conduct a feasibility study for an entrepreneurial pilot program to enable farmers employing some high-tech farming techniques and technologies in order to improve the efficiency of their day-to-day work. Show how so seemingly different concepts, digital technology, and agriculture can refine precision farming application.

7.16.3 Project on monitoring and control of a photovoltaic power plant

The rapid evolution of renewable energy sources has led to the installation of many wind and photovoltaic (PV) systems all around the world. PV energy is a clean and renewable. With the rise of PV system installation, a real-time monitoring system is significant to evaluate and optimize all key parameters such as irradiance, output voltage, current, power, humidity, wind speed, panel and ambient temperature.

The PV power station often consists of PV array strings, storage batteries bank, power conditioning unit, and electrical loads appliances. In the operation of such station especially with large size (kilowatt or megawatt scales), the system performance should be carefully monitored and a proper decision must be taken in time. There is also, at present, considerable interest in the storage and dispatchability of PV energy, together with the need to manage power flows in real time.

The objective of this project is to develop an innovative utility-scale computer-based grid-connected data-acquisition system (DAQ) to monitor and control solar irradiance and PV power generation system's operational parameters and ultimately to improve utilization efficiency of solar energy. The innovative effort should join the sensor circuitry and signal display elements with the data acquisition and serial communication topics often taught in electronic instrumentation courses. This system should present a configuration of acquisition devices, including wired and wireless sensors distributed around the plant, which measure the required information. The developed system could be redesigned to be scaled to any number of measurement points. In addition, when a storage system is engaged in the plant, the system should also be able to monitor it.

This innovative project can be implemented as a capstone project or as an extra credit project assignment in several PBL-based courses within the domains of electronics or mechatronics. This project may present a design experience that addresses numerous facets of a signal acquisition system by merging design credits for two undergraduate, senior-level courses. For electrical engineering students, the emphasis is given to the apparatus instrumentation and computerized data acquisition and control, while a detailed description of the electronics and mechanical design and test results will be required from mechanical and mechatronics engineering students.

The overall goals of this project are to (1) create a considerable, design-driven learning experience for engineering seniors; (2) increase student interest by attaching a renewable energy context to an instrumentation project that would otherwise be generic; (3) enhance creativity and innovation notions with the domain of mechatronics; and (4) develop entrepreneurial mind-set skills with this interdisciplinary field of study.

The DAQ and the monitor program should be very flexible to new additions and changes. In addition, the designed system should accomplish the following additional objectives:

- Acquire measured data with high accuracy and speed.
- Process and analyze measured data for immediate use.
- Display the raw (measured) and processed (analyzed) data in graphical and numerical forms.

Students interested in participating are asked to organize themselves into groups consisting of three to four individuals. Each group has to pitch a proposal in an effort to convince the customer (in this case, the instructor and other students) that their design is the best and most cost-effective solution. To do so, each group has to provide supporting evidence that their design is in fact the best. The evidence is required to be compiled into a product proposal that includes the following:

- Define the project idea (project name, target clientele, and limitation) (maximum 200 words).
- Describe the team's strengths and weakness in carrying out the project (maximum one page).
- Define the context of the project (target market, competition, potential income, sources of information) (maximum one page).
- Describe the offer (description of the product, innovativeness, sale price) (maximum one page).
- Develop the communication and action plan (methods selected, cost, production, and advertising) (maximum one page).
- Develop the human resources and financing plan (tasks and funds) (maximum one page).
- Provide a list of required components and materials.
- Conduct cost analysis (including the profit and the team members' fees).
- Show system design and simulation (using well-known simulation software).
- Implement and show technical details and testing procedure.
- Demonstrate the layout of the printed board (polychlorinated biphenyl: PCB) and packaging schematic delivery time.
- Display the generated digital data in a computer.

Each group has to develop their bill of materials, cost analysis, and testing plan based on an initial customer order of 1000 systems. Each system is required to operate in the range of 25°C–100°C. The temperature sensing element had to be a thermistor, which is a variable resistor whose resistance value varies significantly with temperature. Students should provide a block diagram for the various components of the acquisition system.

7.16.4 Debate on engineers to entrepreneurs

With an increasingly technical and research-based business world, engineers are needed for their expertise in the fields of science, mathematics, and technology. In addition, engineers are extremely useful in today's environment of constant innovation because of their creativity and rationality that, coupled with a distinctive competence in a specific field, make for a great entrepreneur (McLaughlin 2015). On the other hand, engineering challenges can be much interesting and the mind-set required to

Objective	Introducing an open-ended debate in the classroom to help students understand argument on the concepts of engineering and entrepreneurship.
Time	15 min for debate and 15 min for review
Level	For and against
Learning outcomes	Make an argument about a particular opinion; evaluate the arguments of peers; and understand the concept of counterarguments.
Capabilities demonstrated	Developing skills on public speaking, research, teamwork, critical thinking, communication, and professional judgment
Arrangement	Students are organized into two position groups of three speaking in a specific order. Three argue for an opinion and three argue against. One or two students might each work on the opening and closing statements while the group is investigating the subject; however, the entire group should revise the statements. Each group should read an opening and closing statement for the debate.
Ideas for the topic	Debate both opinions: engineers are well suited to becoming successful entrepreneurs; and engineers are not well suited to becoming successful entrepreneurs. Consider creative engineers in history. Debate around terms like value creation, risk-taking, radical ideas, innovation promotion, commercial and social opportunities.
Assessment	Indicate what you consider the best arguments in favor of the topic. How were they substantiated? Identify arguments that are based on poor facts, not ethical and/or sustainable or not well substantiated?

successfully meet them is not naturally focused on risks, opportunities, and commercial value creation (Goffin and Carter 2011).

7.16.5 Piece of art on marketing a new technology venture

Digital tools and technology continue to disrupt the world of marketing. Engagement on new platforms, such as messaging services and social media is reshaping the future of marketing. This pattern shift represents the next opportunity to change the way consumers and brands engage with one another.

> Key Point: There is a well-known story of two shoe entrepreneurs who go to China to do market research. They find, to their surprise, that no one in China wears shoes. One entrepreneur comes home and laments that there is no market of shoe-wearers to sell to. The other entrepreneur returns with a smile on his face, knowing that if he can access it the market is enormous (Felser 2011). Be optimistic!

For this task, propose an approach for marketing a new technology venture, which combines viral-marketing tools with old-fashioned mass media in a way that yields far more predictable results than "purely" viral approaches such as word-of-mouth marketing. You may use metaphors and symbols, as long as those are clearly communicated or explained in supplementary materials.

7.16.6 Poster on the responsibility of government

What is the responsibility of government in entrepreneurship? To what extent should it help engage people in entrepreneurial activities? Should it simply get out of the way and leave the market to reward or punish inappropriate behavior? How can governments develop logical and effective policies for entrepreneurship education? How can other stakeholders be engaged? What specific steps should the government take or what steps have it taken that should be reversed? Create a poster that collectively answers the above questions and possibly beyond.

7.16.7 Developing an entrepreneurship course

List the content that you believe is necessary for a technology entrepreneurship introductory course. Include content for lectures, guest speakers who will talk about their experiences as entrepreneurs or in a start-up, case study, building product project with makerspace experience. Do you believe that ethics and social responsibility should be part of an entrepreneurship course? Justify your design argument for the course content.

7.16.8 Video contest on supporting employee intrapreneurs

> Intrapreneurs often feel isolated and disengaged from their colleagues and need to have avenues for support and collaboration that make them feel engaged and able to drive value.
>
> **(Firrier 2014)**

Intrapreneurs are a real channel to generate and execute new thinking approaches within a corporate environment. For this task, form a team and work together in a collaborative environment to formulate a combined message through a 3-min video to encourage intrapreneurial efforts within an organization to create cultural change of accepting new ideas and thinking; engage existing employees; and attract new and high-potential individuals into the organization. Try to explore a wide range of approaches and models for organizations to support and improve the effectiveness of intrapreneurs. This may include incentives, training, and various types of support.

References

Acs, Z. J., and D. B. Audretsch. 1990. *Innovation and Small Firms*. Cambridge, MA: MIT Press.

Amabile, T. 1993. Motivational synergy: Toward new conceptualizations of intrinsic and extrinsic motivation in the workplace. *Human Resource Management Review* 3(3): 185–201.

Anastakis, D. 2008. The people's tycoon: Henry Ford and the American century. *Business History Review* 82(1): 119–126.

Aparicio, A. C., and A. M. Ruiz-Teran. 2007 Tradition and innovation in teaching structural design in civil engineering. *Journal of Professional Issues in Engineering Education and Practice* 133(4): 340–349.

Badger, E. 2013. Start-up incubators. http://smartcitystudio.com/.

Bailetti, T. 2012. Technology entrepreneurship: Overview, definition, and distinctive aspects. *Technology Innovation Management Review* 2(2): 5–12. https://timreview.ca/article/520.

Baker, G. 2013. Entrepreneurship: It's in our DNA. http://www.entrepreneurialengineers.com/news/20/16/Entrepreneurship-It-s-in-our-DNA.

Benade, S., and C. Heunis. 2005. Management development: More right brain skills required. *Management Today* 21(6): 54–58.

Bennett, A. 2014. A historical perspective on entrepreneurship. http://www.canberraentrepreneur.com/musings/historical-perspective-entrepreneurship/.

Bhidé, A. 2009. *The Venturesome Economy: How Innovation Sustains Prosperity in a More Connected World?* Princeton, NJ: Princeton University Press.
Brazdauskas, M. 2015. Promoting student innovation-driven thinking and creative problem solving for sustainability and corporate social responsibility. *Journal of Creativity and Business* 1: 75–87.
Brown, J. S., A. Collins, and P. Duguid. 1989. Situated cognition and the culture of learning. *Educational Researcher* 18(1): 32–42.
Bury, K. 2016. Entrepreneur ecosystem: Bringing balance to the entrepreneurship ecosystem in MENA. https://www.entrepreneur.com/article/275400.
Busse, W. 2011. Successful entrepreneurs of the 19th century. http://www.brighthub.com/office/entrepreneurs/articles/74678.aspx.
Carlsson, B. 1989. The evolution of manufacturing technology and its impact on industrial structure: An international study. *Small Business Economics* 1(1): 21–37.
Carlsson, B. 1992. The rise of small business: Causes and consequences. In Adams, W. J. (Ed.), *Singular Europe: Economy and Polity of the European Community after 1992*. Ann Arbor, MI: University of Michigan Press, pp. 145–169.
Carlsson, B., P. Braunerhjelm, M. McKelvey, C. Olofsson, L. Persson, and H. Ylinenpää. 2013. The evolving domain of entrepreneurship research. *Small Business Economics* 41(4): 913–930.
Caruso, S.M. 2017. 5 historical figures who mastered the art of the hustle. GA Blog. https://generalassembly/blog/art-of-the-hustle/.
Castaldo, J. 2015. There's no such thing as a born entrepreneur. www.canadianbusiness.com/innovation/theres-no-such-thing-as-a-born-entrepreneur/.
Chaffey, D., and F. Ellis-Chadwick. 2012. *Digital Marketing Strategy, Implementation and Practice*. Harlow, UK: Pearson Education.
Charney, A., and G. Libecap. 2000. The impact of entrepreneurship education: An evaluation of the berger entrepreneurship program at the University of Arizona, 1985–1999. Social Science Research Network.
Cogliser, C. C., and K. H. Brigham. 2004. The intersection of leadership and entrepreneurship: Mutual lessons to be learned. *Leadership Quarterly* 15(6): 771–799.
Collins, A., J. Brown, and S. E. Newman. 1989. Cognitive apprenticeship: Teaching the crafts of reading, writing, and mathematics. In Resnick, L. (Ed.), *Knowing, Learning and Instruction: Essays in Honour of Robert Glaser*. Mahwah, NJ: Erlbaum.
Crawford, M. 2012. From engineer to entrepreneur. ASME. https://www.asme.org/career-education/articles/entrepreneurship/from-engineer-to-entrepreneur.
Currie, G., A. Humphreys, D. Ucbasaran, and S. McManus. 2008. Entrepreneurial leadership in the English public sector: Paradox or possibility? *Public Administration* 86(4): 987–1008.
Diesendorf, M. 2000. Sustainability and sustainable development. In Dunphy, D., J. Benveniste, A. Griffiths, and P. Sutton (Eds.), *Sustainability: The Corporate Challenge of the 21st Century*. Sydney, NSW: Allen and Unwin, chapter 2, pp. 19–37.
Dorf, R. C., and T. H. Byers. 2007. *Technology Ventures: From Idea to Enterprise*. Second edition. New York: McGraw-Hill.
Drucker, P. E. 1985. *Innovation and Entrepreneurship*. New York: Harper and Row.

Duval-Couetil, N. 2013. Assessing the impact of entrepreneurship education programs: Challenges and approaches. *Journal of Small Business Management* 51: 394–409.

Duval-Couetil, N., and M. Dyrenfurth. 2012. Teaching students to be innovators: Examining competencies and approaches across disciplines. *International Journal of Innovation Science* 4(3): 11.

Duval-Couetil, N., A. Shartrand, and T. Reed-Rhoads. 2015. The role of entrepreneurship program models and experiential activities on engineering student outcomes. *Advances in Engineering Education* 5(1): 1–27.

Duval-Couetil, N., and J. Wheadon. 2013. The value of entrepreneurship to recent engineering graduates: A qualitative perspective. *IEEE Frontiers in Education Conference,* October 23–26, pp. 114–120.

EC. 2008. Entrepreneurship in higher education, especially within non-business studies-Final Report of the Expert Group, European Commission. http://ec.europa.eu/enterprise/entrepreneurship/support_measures/training_education/entr_highed.pdf.

Eckhardt, J. T., and S. A. Shane. 2003. Opportunities and entrepreneurship. *Journal of Management* 29: 333–349.

Edward, D. 2012.The motivations of catering companies to deliver food to elderly people. Laurea University of Applied Sciences. http://www.theseus.fi/bitstream/handle/10024/53104/Donkor_Edward%20Final%20Thesis%20pdf.pdf?sequence=1. Entrepreneur. 2017. Henry Ford: the man who taught America to drive. https://www.entrepreneur.com/article/197524.

Eisenmann, T. 2013. Entrepreneurship: A working definition. *HBR blog network-Harvard Business Review.* http://blogs.hbr.org/hbsfaculty/2013/01/what-is-entrepreneurship.html.

FE. 2013. Steve Jobs. Famous Entrepreneurs. http://www.famous-entrepreneurs.com/steve-jobs.

Felser, J. 2011. An Engineer's guide to applying appropriate technology. 2011 AHS Capstone Projects. Paper 21. http://digitalcommons.olin.edu/ahs_capstone_2011/21.

Firrier, A. 2014. Activities to support your employee intrapreneurs. http://www.innovationmanagement.se/2014/12/08/supporting-your-intrapreneurs-part-ii/.

Fleischmann, F. 2006. Entrepreneurship as emancipation: The history of an idea. Free University of Berlin, Berlin, Germany, July 12. https://www.entrepreneurship.de/wp-content/uploads/2011/01/Entrepreneurship_as_Emanzipation.pdf.

Fuller, A., and L. Unwin. 2010. Change and continuity in apprenticeship: The resilience of a model of learning. *Journal of Education and Work* 25(5): 405–416.

Gale, T. 2007. Intrapreneurship. http://www.encyclopedia.com/social-sciences-and-law/economics-business-and-labor/businesses-and-occupations/intrapreneurship.

Gale Encyclopedia of U.S. Economic History. 2000. Firestone, Harvey Samuel. http://www.encyclopedia.com/people/social-sciences-and-law/business-leaders/harvey-samuel-firestone.

Gibb, A. A. 1993. The enterprise culture and education: Understanding education and its links with small business, entrepreneurship and wider education goals. *International Small Business Journal* 11(3): 11–34.

Gibb, A. A. 1996. Entrepreneurship and small business management: Can we afford to neglect them in the 21st century business school? *British Journal of Management* 7: 309–321.

Goldberg, D. E. 2006. *The Entrepreneurial Engineer*. Hoboken, NJ: John Wiley & Sons.

Gordon, J. S. 2014. Entrepreneurship in American history. *Imprimis* 43(2). https://imprimis.hillsdale.edu/entrepreneurship-in-american-history/

Gorman, G. 1997. Some research perspectives on entrepreneurship education, Enterprise education and education of small business management: A ten year literature review. *International Small Business Journal* 15(3): 56–78.

Hamel, G. 2002. *Leading the Revolution*. Revised edition. Boston, MA: Harvard Business School Press.

Haller, H.E. 2016. Steve Jobs; the ultimate intrapreneur. Intrapreneurship Institute http://www.intrapreneurshipinstitute.com/intrapreneurship-case-studies/steve-jobs-the-ultimate-intrapreneur-and-entrepreneur/.

Henry, C., F. Hill, and C. Leitch. 2005. Entrepreneurship education and training: Can entrepreneurship be taught: Part I–II? *Education and Training* 47(2): 98–111.

Herbert, R. F., and A. H. Link. 1982. *The Entrepreneur: Mainstream Views and Radical Critiques*. New York: Praeger Publishers.

Hills, G. E., C. M. Hultman, S. Kraus, and R. Schulte. 2010. History, theory and evidence of entrepreneurial marketing: An overview. *International Journal of Entrepreneurship and Innovation Management* 11(1): 3–18.

Hisrich, R. D. 2005. The Entrepreneurial Perspective, White Paper. http://highered.mheducation.com/sites/dl/free/0070922926/388147/01_Hisrich_ch_01.pdf.

Hisrich, R. D., and M. Peters. 2002. *Entrepreneurship*. New York: McGraw-Hill.

Hitt, M. A., R. D. Ireland, S. M. Camp, and D. L. Sexton. 2001. Strategic entrepreneurship: Entrepreneurial Strategies for wealth creation. *Strategic Management Journal* 22: 479–449.

Hockerts, K., and R. Wüstenhagen. 2010. Greening Goliaths versus emerging Davids—Theorizing about the role of incumbents and new entrants in sustainable entrepreneurship. *Journal of Business Venturing* 25(5): 481–492.

Inkelas, K. K., K. Szelényi, M. Soldner, and A. M. Brower. 2008. National study of living-learning programs: 2007 Report of Findings. http://drum.lib.umd.edu/bitstream/1903/8392/1/2007 NSLLP Final Report.pdf.

Jemmell, R. M. 2012. Socio-cognitive foundations of entrepreneurial venturing. PhD Thesis, Weatherhead School of Management, Case Western Reserve University, Cleveland, OH.

Jones, G., and R. D. Wadhwanim. 2007. Entrepreneurial theory and the history of globalization. Business and Economic History on Line. http://www.thebhc.org/sites/default/files/joneswadhwani.pdf..

Kauffman Center. 2001. The growth and advancement of entrepreneurship in higher education: An Environmental Scan of College Initiatives. Prepared by the Kauffman Center for Entrepreneurial Leadership Staff. http://www.unm.edu/~asalazar/Kauffman/Entrep_research/College_Scan.pdf

Kemeny, T. 2010. Does foreign direct investment drive technological upgrading? *World Development* 38(11): 1543–1554.

Kihlstrom, R. E., and J. J. Laffont. 1979. General equilibrium entrepreneurial theory of firm formation based on risk aversion. *Journal of Political Economy* 87: 719–748.

Kolb, D. 1984. *Experiential Learning: Experience as the Source of Learning and Development*. Englewood Cliffs, NJ: Prentice Hall.

Kolb, D. A., R. E. Boyatzis, and C. Mainemelis. 2000. Experiential learning theory: Previous research and new directions. In Sternberg, R. J., and L. F. Zhang (Eds.), *Perspectives on Cognitive, Learning, and Thinking Styles*. Mahwah, NJ: Lawrence Erlbaum.

Kolb, D. A., and R. E. Fry. 1974. Toward an applied theory of experiential learning. MIT Alfred P. Sloan School of Management, Boston, MA.

Kriewall, T. J., and K. Mekemson. 2010. Instilling the entrepreneurial mindset into engineering understanding. *The Journal of Engineering Entrepreneurship* 1(1): 5–19.

Kuratko, D. F. 2005. The emergence of entrepreneurship education: Development, trends, and challenges. *Entrepreneurship Theory and Practice* 29: 577–598.

Lalande, L. 2014. The maker movement and guilds 2.0—Part 1. University of Ottawa, Ottawa, ON. https://medium.com/@luclalande/the-maker-movement-and-guilds-2-0-part-1-c07a42674f8c#.rmv0yhboa.

Lalande, L. 2016a. Apprenticeships: An old-new approach to entrepreneurship education. https://medium.com/@luclalande/apprenticeships-an-old-new-approach-to-entrepreneurship-education-7d2367ff0a14#.premgn17r.

Lalande, L. 2016b. Entrepreneurship, innovation and appropriate technologies for international development—A new initiative at the University of Ottawa. https://medium.com/@luclalande/entrepreneurship-innovation-and-appropriate-technologies-for-international-development-2afb9143c1d8#.mqe4d5o7r.

Lans, T. 2009. Entrepreneurial competence in agriculture: Characterization, identification, development and the role of the work environment. In Education and Competence Studies. Dissertation, Wageningen University, Wageningen, the Netherlands.

Larso, D., Yulianto Yulianto, S. Rustiadi, and L. Aldianto. 2009. Developing technopreneurship program at the Center for Innovation, Entrepreneurship, and Leadership (CIEL). School of Business and Management, Bandung Institute of Technology (ITB), Jawa Barat, Indonesia. *PICMET 2009 Proceedings*, August 2–6, Portland, OR.

Lindsay, E., R. Munt, H. Rogers, D. Scott, and K. Sullivan. 2008. Making students engineers. *Engineering Education Journal of the Higher Education Academy Engineering Subject Centre* 3(2): 28–36.

LTBN. 2007. Entrepreneur's hall of fame: Thomas Alva Edison. http://www.ltbn.com/hall_of_fame/Edison.html.

Lumsdaine, E., and M. Lumsdaine. 1995. *Creative Problem Solving*. New York: McGraw-Hill.

MacCormack, A., and R. Verganti. 2003. Managing the sources of uncertainty: Matching process and context in software development. *Journal of Product Innovation Management* 20(3): 217–232.

Mason, C., and R. Brown. 2014. Entrepreneurial ecosystem and growth oriented entrepreneurship. Background paper prepared for the workshop organised by the OECD LEED Programme and the Dutch Ministry of Economic Affairs on Entrepreneurial Ecosystems and Growth Oriented Entrepreneurship. https://www.oecd.org/cfe/leed/Entrepreneurial-ecosystems.pdf.

McCarthy, P. 2002. *Henry Ford: Building Cars for Everyone*. Berkeley Heights, NJ: Enslow Publishers.

McClelland, D. C. 1987. Characteristics of successful entrepreneurs. *The Journal of Creative Behaviour* 31(3): 219–233.

McCormick, B., and B. Folsom. 2003. A survey of business historians on America's greatest entrepreneurs. *Business History Review* 77(4): 703–716.

McLeod, J. 2016. Entrepreneurial apprenticeship Coop 2:00: StudIoT. University of Ottawa, Ottawa, ON. https://medium.com/@StudIoTeHub/entrepreneurial-apprenticeship-co-op-2-0-studiot-ae1ca52bd9c6#.w8vdhsauc.

Meola, A. 2016. Why IoT, big data and smart farming are the future of agriculture. http://www.businessinsider.com/internet-of-things-smart-agriculture-2016-10.

Meredith, G. G., R. E. Nelson, and R. A. Neck. 1991. *The Practice of Entrepreneurship.* Lagos, Nigeria: University Press.

Meyer, M. 2003. Academic entrepreneurs or entrepreneurial academics? Research-based ventures and public support mechanisms. *R&D Management* 33(2): 107–115.

Mohan, N. 2015. IoT and agriculture: The revolution begins. https://medium.com/@Keleno/iot-and-agriculture-the-revolution-begins-e6e07399d231#.25s3l1gsv.

Nandy, S., and S. Kumar. 2014. Women entrepreneurship in 21st century India. *Global Journal of Finance and Management* 6(9): 967–976.

Nidumolu, R., C. K. Prahalad, and M. R. Rangaswami. 2009. Why sustainability is now the key driver of innovation? *Harvard Business Review*, September, pp. 1–10.

Obaji, N. O., and M. U. Olugu. 2014. The role of government policy in entrepreneurship development. *Science Journal of Business and Management* 2(4): 109–115.

Odeigah, R. 2012. Introduction to entrepreneurship development. https://fredericodeigah.wordpress.com/2012/10/12/introduction-to-entrepreneurship-development/.

Odwyer, M., A. Gilmore, and D. Carson. 2009. Innovative marketing in SMEs. *European Journal of Marketing* 43(1/2): 42–61.

OEERE. 2000. From invention to innovation. Office of Energy Efficiency and Renewable Energy, US Department of Energy, Washington, DC.

Pech, R. M. 2015. Achieving the innovative edge in technology, engineering design, and entrepreneurship. *Journal of Innovation and Entrepreneurship* 5: 6.

Pech, R. M. 2016. Achieving the innovative edge in technology, engineering design, and entrepreneurship. *Journal of Innovation and Entrepreneurship* 5(6): 1–18.

Petersen, O. G., W. M. Jordan, and R. Radharamanan. 2012. Proposed KEEN initiative framework for entrepreneurial mindedness in engineering education. American Society for Engineering Education.

Phelps, J. E., R. Lewis, L. Mobilio, D. Perry, and N. Raman. 2004. Viral marketing or electronic word-of-mouth advertising: Examining consumer responses and motivations to pass along e-mail. *Journal of Advertising Research* 44(4): 333–348.

Pinchot, G. 1985. *Intrapreneuring: Why You Don't Have to Leave the Corporation to Become an Entrepreneur.* New York: Harper and Row.

Pinchot, P. 2016. Why intrapreneuring is suddenly happening again. http://www.pinchot.com/intrapreneuring/.

Price, R. W. 2011. What is entrepreneurial management? Global Entrepreneurship Institute. https://news.gcase.org/2011/10/24/what-is-entrepreneurial-management/.

Probst, H. 2007. 2 Inspire: Increasing intrapreneurial skills through pedagogy, increases innovation, retention and employability. The higher Education Academy, Sheffield Hallam University, Sheffield, UK. http://www.shu.ac.uk/research/hsc/ourexpertise/2-inspire-increasingintrapreneurial-skills-through-pedagogy-increases-innovation-reten.

Pushprofile. 2016. History of entrepreneurship. http://pushprofile.com/history/history-of-entrepreneurship/.

QAA. 2012. Enterprise and entrepreneurship education: Guidance for UK higher education providers. The quality assurance agency for higher education. QAA's publications. http://www.qaa.ac.uk/en/Publications/Documents/enterprise-entrepreneurship-guidance.pdf.

Rae, D., and M. Carswell. 2000. Using a life story approach in researching entrepreneurial learning: The development of conceptual model and its implications in the design of learning experiences. *Education and Training* 42(4/5): 200–207.

Ricketts, M. 2008. Theories of entrepreneurship: Historical development and critical assessment. In Basu, A., M. Casson, N. Wadeson, and B. Yeung (Eds.), *The Oxford Handbook of Entrepreneurship*. Oxford: Oxford University Press, pp. 33–58.

Ronstadt, R. 1987. The educated entrepreneurs: A new era of entrepreneurial education is beginning. *American Journal of Small Business* 11(4): 37–53.

Russell, J. 2013. *Thrivability: Breaking Through to a World That Works*. Devon, UK: Triarchy Press.

Sá, C., A. Kretz, and K. Sigurdson. 2014. *The State of Entrepreneurship Education in Ontario's Colleges and Universities*. Toronto, ON: Higher Education Quality Council of Ontario.

Sanja, P., and B. Djula. 2011. Fitting Kolb's learning style theory to entrepreneurship learning aims and contents. *International Journal of Business Research* 11(2): 216–223.

Sarasvathy, S. D. 2001. What makes entrepreneurs entrepreneurial? White Paper. The Darden Graduate School of Business Administration, University of Virginia, Charlottesville, VA. http://www.effectuation.org/sites/default/files/documents/what-makes-entrepreneurs-entrepreneurial-sarasvathy.pdf.

Schaper, M. 2010. Understanding the green entrepreneur. In Schaper, M. (Ed.), *Making Ecopreneurs—Developing Sustainable Entrepreneurship,* Second edition. Burlington, VT: Ashgate, pp. 7–20.

Schumpeter, J. A. 1934. *The Theory of Economic Development*. Cambridge, MA: Harvard University Press.

Schumpeter, J. 1947. The creative response in economic history. *The Journal of Economic History* 7(2): 149–159.

Schumpeter, J. 1978. *Can Capitalism Survive?* New York: Harper and Row.

Schwartz, S. 2017. Biography of Nikola Tesla. https://www.thoughtco.com/nikola-tesla-1779840.

Sellers, K. D. 2014. From startup to success: An entrepreneurial case study. Undergraduate Honors Thesis, East Tennessee State University, Johnson City, TN.

Seshadri, D. V. R., and A. Tripathy. 2006. Innovation through Intrapreneurship: The road less travelled. *Vikapala* 31(1): 17–29.

Shane, S. 2000. Prior knowledge and the discovery of entrepreneurial opportunities. *Organization Science* 11(4): 448–469.

Shane, S. 2003. *A General Theory of Entrepreneurship: The Individual-Opportunity Nexus*. Cheltenham, UK: Edward Elgar Publishing.

Shepherd, D. A., and H. Patzelt. 2011. The new field of sustainable entrepreneurship: Studying entrepreneurial action linking "What is to be sustained" with "What is to be developed." *Entrepreneurship: Theory and Practice* 35(1): 137–163.

Stam, E. 2014. The Dutch entrepreneurial ecosystem. http://ssrn.com/abstract=2473475; http://dx.doi.org/10.2139/ssrn.2473475.

Stam, E. 2015. Entrepreneurial ecosystems and regional policy: A sympathetic critique. *European Planning Studies* 23(9): 1759–1769.

Stam, E., and B. Spigel. 2016. Entrepreneurial ecosystems. In: Blackburn, R., D. J. De Clercq, J. Heinonen, and Z. Wang (Eds.), *Handbook for Entrepreneurship and Small Business*. London: SAGE.

Stevenson, H. H., and D. E. Gumpert. 1985. The heart of entrepreneurship. *Harvard Business Review* 63(2): 85–94.

Stevenson, H. H., M. J. Roberts, and J. I. Grousbeck. 1985. *New Business Ventures and the Entrepreneur*. Burr Ridge, IL: Richard D. Irwin, pp. 16–23.

Stopford, J. M., and C. W. F. Baden-Fuller. 1990. Corporate rejuvenation. *Journal of Management Studies* 27(4): 399–415.

Tillery, F., and F. Young. 2009. Sustainability entrepreneurs: Can they be the true wealth generators of the future? *Greener Management International* 55: 79–92.

Timmons, J. A. 1978. Characteristics and role demands of entrepreneurship. *American Journal of Small Business* 3(1): 5–17.

Toftoy, C., and J. Chatterjee. 2005. The intrapreneurial revolution: Now is the time for action. http://sbaer.uca.edu/research/icsb/2005/paper192.pdf.

Turner, J. 2012. 4 reasons why engineers make great entrepreneurs. Careerealism. http://www.careerealism.com/great-entrepreneurs-engineers/.

Tushman, M. L., and P. Anderson. 1968. Technological discontinuities and organizational environments. *Administrative Science Quarterly* 31: 439–465.

UNCTAD. 2009. Entrepreneurship education, innovation and capacity-building in developing countries. *United Nations Conference on Trade and Development*, Geneva, Switzerland. http://unctad.org/en/docs/ciimem1d9_en.pdf.

Van Burg, E. 2010. Creating spin-off designing entrepreneurship conducive universities. PhD Thesis, Eindhoven University of Technology, Eindhoven, the Netherlands.

Volkmann, V., and H. Berg. 2011. Entrepreneurial marketing. Guest Lecture Kosice. http://www.ekf.tuke.sk/files/utorok.pdf.

Von Hippel, E. 2005. *Democratizing Innovation*. Cambridge, MA: MIT Press.

Wheadon, J. D., and N. Duval-Couetil. 2014. Business plan development activities as a pedagogical tool in entrepreneurship education. *The Journal of Engineering Entrepreneurship* 5(1): 31–48.

Wilson, K. 2008. Entrepreneurship education in Europe. In Potter, J. (Ed.), *Entrepreneurship and Higher Education*. Paris, France: OECD Publications.

Winzker, D. H., and L. Pretorius. 2014. A technology management model for entrepreneurial business ventures. Proceedings of *PICMET '14. Infrastructure and Service Integration*, pp. 1201–1212.

part three

The Design Landscape

chapter eight

Engineering design

Engineers make stuff and fix stuff.

Nelson (2012)

8.1 Objectives

- Focus on design theory or design science (what design is).
- Provide a historical perspective of engineering design.
- Define foundational concepts of design such as creativity, innovation, enterprise, ethics, and sustainability.
- Explore engineering design and its relation to innovation and entrepreneurship.
- Know about types of design and explore the importance of engineering design.
- Explore the philosophy of design and its transdisciplinary factor.
- Discuss engineering design language and the required communication tools.
- Define and discuss the concept of design problem.
- Understand factors such as human, technical, and environmental that influence design including design purpose and setting.
- Understand the role of engineers in an entrepreneurial context.
- Know the types, characteristics, advantages, and disadvantages of various design theories.
- Understand the concepts of systems engineering (SE) and concurrent engineering (CE).
- Realize V-design approach in project life cycle development.
- Describe the notion of deterministic design (DD).
- Distinguish between different systematic design (SD) models.
- Learn about standards and codes in engineering design.
- Highlight the scope of human factors engineering (HFE).
- Explore the impact of habits of mind (HoM) in modern engineering design.
- Discuss the role and impact of design in engineering education.
- Learn about experiential design learning.

- Realize engineering design throughout an interlinking case of mechatronic system design.
- Provide three end of chapter pedagogical knowledge strategies, namely knowledge acquisition, knowledge possession, and knowledge creation to help understand the above topics and generate new and alternative ideas and solutions.

8.2 Historical perspective

> Those who do not remember the past are condemned to repeat it.
>
> **George Santayana**

History is full with examples of people who achieved great goals by means of their intelligence or determined will, but seldom without passion. History is also full with people who never accomplished much of anything because they could never bring themselves away from mind inactivity. Designers belong to the first category that is passionate about conceiving, planning, creating, and executing ideas.

8.2.1 Early history

The forefathers of engineers such as practical artists and craftsmen advanced their work primarily by trial and error. Yet, playing blended with creativity produced many marvelous designs. Many ancient monuments cannot fail to provoke appreciation. The appreciation is exemplified in the tag "engineer" itself. It originated in the eleventh century from the Latin word *ingeniator*, meaning one with *ingenium*, the ingenious one. The name, used for builders of ingenious fortifications or makers of ingenious devices, was closely related to the notion of imagination, which was confined in the old meaning of "engine" until the word was taken over by the Industrial Revolution and its technologies. Leonardo da Vinci bore the official title of *Ingegnere Generale*. His notebooks reveal that some renaissance engineers questioning what works and why (Gill 1966; Finch 1978; Grafton and Alberti 2000).

Around the year 1400, Filippo Brunelleschi (1377–1446), the Italian architect, engineer, and sculptor, won a prestigious opportunity to design and build the cupola (dome) of the new cathedral for the city of Florence. Until that time, however, buildings were not really engineered at all. The craft was then known as artisanship, and basically involved using well-understood principles and trial-and-error methods of building. This was not good enough for Brunelleschi. In the environment of artisanship, the artisan simply starts building or manufacturing the product. When insurmountable problems are encountered, the entire project is junked and started over again. This contributed to making engineers extremely conservative; innovation was rarely encouraged and often discouraged because of its implied risks (Kocabiyik, 2004).

After Brunelleschi, design remained largely unchanged for hundreds of years. This was relatively due to the enormous success of the method and partly due to the highly qualitative nature of design. Science and technology began to take hold and lead to a variety of exciting discoveries that allowed all kinds of new products to be developed. No one perceived how poorly designed the products were because they were in whatever form they took far greater to their predecessors (Salustri 2005).

8.2.2 Scientific revolution

The first phase of modern engineering emerged in the scientific revolution. Galileo's 1638 *Two New Sciences,* a canonical text of early modern science which seeks methodical explanations and adopts a scientific approach to practical problems, is a landmark regarded by many engineering historians as the beginning of structural analysis, mathematical representation, and design of building structures. "Two New Sciences" is the basic of engineering sciences that today's engineering students study as strength of materials and dynamics. Each was motivated by practical interest. This book begins with a discussion of design in the great arsenal of Venice.

Galileo's work was published by an English press in 1665, where it entered the tradition begun by Francis Bacon, who was Galileo's contemporary, and subsequently developed in monumental fashion by Newton in "Principia" in 1686. Francis Bacon, too, plays an important role in the origins of design research, because his project was to begin a great instauration of learning that would lead to our ability to command nature in action, where nature would be molded by art and human ministry in the creation of "artificial things." Bacon's project is clearly a design project. And perhaps it is the design project (Buchanan 1999).

This phase of engineering lasted through the first Industrial Revolution, when machines, increasingly powered by steam engines, started to replace muscles in most production. While pulling off the revolution, traditional artisans transformed themselves to modern professionals. The French, who were more rationalistic oriented, spearheaded civil engineering with emphasis on mathematics and developed university engineering education under the sponsorship of their government. The British, more empirically oriented, pioneered mechanical engineering and autonomous professional societies under the "laissez-faire" attitude of their government (Armytage 1976; Benvenuto 1991).

8.2.3 *Apprenticeship to scientific university education*

As discussed in Chapter 7, apprenticeship is a system of learning by doing. It is a formal, on-the-job training program through which a novice learns craft, trade, or vocation under the guidance of a master practitioner. Historically, most American engineers began as apprentices on canal and railway projects. Around the 1850s some schools started following the French model, the "polytechnics." Learning institutions started reducing shop learning hours and adding more basic science in the classroom. After World War I, the Europeans brought their concepts on engineering education to the United States. However, substantial change came after World War II. The Cold War after the 1950s brought the arms and space race and therefore more interest in engineering.

Gradually, practical thinking became scientific in addition to intuitive, as engineers developed mathematical analysis and controlled experiments. Technical training shifted from apprenticeship to university education. Information flowed more quickly in organized meetings and journal publications as professional societies emerged (Armytage 1976; Buchanan 1985; Benvenuto 1991). This trend was only reinforced after World War II. The United States attributed much of their victory to the ability of the American scientists to develop technologies that were superior to those of their enemies in that war. The fact was that most of those scientists were in fact engineers. After the war, engineers decided to focus their energies on teaching and researching in the scientific, quantifiable areas of engineering: analysis and manufacturing. They redesigned their university curricula to suit this goal.

During the period immediately following the Grinter Report (1955), engineering curricula swung from a practical base to a scientific base with more emphasis on theoretical approaches and less emphasis on the "machinery" of engineering (Sheppard and Jenison 1997). By the late 1960s, the weight had swung severely toward science. The very engineers—like Theodor von Karman, Hungarian-American mathematician, aerospace engineer, and physicist who was active primarily in the fields of aeronautics and astronautics—who advocated so heavily for science in the 1930s protested the swing was too far away from engineering design.

8.2.4 *By the 1980s and later*

By the 1980s, hands-on skills declined enormously. The result was that design was treated poorly and therefore almost neglected. Disappointment with this lack of skills further fueled shift back to laboratory and design skills. Engineering schools were criticized for offering too few practical and hands-on courses. Students were not sufficiently schooled in teamwork and team approaches to problem-solving. There was too much compartmentalization of engineering disciplines, and there is insufficient drilling in both written and oral communication. Other criticisms have to do with retention; too many students become discouraged in the first few terms of an engineering curriculum and because of inadequate exposure to engineering and engineering design, many switch out of engineering (Sheppard and Jenison 1997). By the late 1980s, capstone courses had become common, and it also had become clear to engineering educators that design education could not begin during students' senior years. Today, most programs have some kind of design course at the freshman or sophomore level to introduce students to the process of solving real problems in the face of uncertainty. These recent trends have been positive changes for engineering and technology students and engineering education. Students often find design classes to be fun and challenging, and apply themselves more in these classes than they do in traditional engineering science lecture courses (Newcomer 1999).

The 1990s brought significant changes in engineering education. Freshman and capstone design, lab experiences throughout the curriculum started becoming standard. Universities attended more to industry concerns, causing even greater shift away from science to the "hands-on" and applied work. Universities are increasingly introducing design-intensive curricula and expecting to accomplish many objectives, including introducing students to real-world problems; engaging students with multiple learning styles; teaching teaming and improving communication skills; teaching creativity, innovation, and habit of lifelong learning; and integrating knowledge from science and business considerations.

8.3 Design explained

> Design is not just what it looks like and feels like.
> Design is how it works.
>
> **Steve Jobs**

8.3.1 Design defined

The word "design" takes on a variety of noun and verb meanings. In its noun form, dictionaries suggest concepts of sketch, drawing, plan, pattern, intention or purpose, or the art of producing them. In its verb form, the same dictionaries suggest elements of definition involving representing an artifact, system or society, or the fixing of its look, function, or purpose. The word "design" therefore has meanings ranging from the abstract conception of something to the actual plans and processes required to achieve it. The concept of design as a way of making sense of things has been the subject of many studies (Krippendorff 1989) as has the design thinking (DT) process itself (Brown 2008, 2009).

Design is defined in the dictionary in different ways as listed in Figure 8.1. The verb "design" comes from the Latin word *designare*, which means to *specify*, as in pointing out what to do. The modern sense of design is held to have originated in the Renaissance, when architect and builder functions came to be two separated functions. Similarly, the noun "design" comes from *signum*, which is not so much in the modern sense of root "sign" (as in symbol, mark; semantics, semiotics, etc.) as is sometimes claimed (Kocabiyik 2004).

Design is one of the oldest endeavors among intellectual and technological pursuits. It is a field of innovation; its core is the creation of something new and unique. Basic characteristics in the nature of design are as follows:

- Design is a formal professional endeavor requiring specific knowledge, skills, and abilities (Adams 2015).
- Design is rational involving logical reasoning, mathematical analysis, computer simulation, laboratory experiments and field trials, etc. (Adams 2015).
- Design requires inquiry into the stakeholder's requirements and expectations, available design techniques, previous design solutions, past design failures and successes, etc. (Adams 2015).
- Design is naturally integrative, not separative (Owen 1988), interactive; and requires a transdisciplinary team.
- Design is intellectually soft, intuitive, informal, and cookbooky (Simon 1996).
- Design is iterative. Artifacts are analyzed with respect to functional and nonfunctional requirements, constraints, and cost. Revisions are based on experience and feedback mechanisms (Adams 2015).
- Design requires value judgments. Courses of action and selection from competing solutions are based on experience and criteria provided by the system's stakeholders (Adams 2015).

To conceive of fashion in the mind
To formulate a plan for
To plan out in a systematic, usually graphic form
To create or contrive for a particular purpose or effect
To create or execute in an artistic or highly skilled manner

Figure 8.1 Dictionary definitions of design.

Professionally, managers, engineers, architects, scientists, etc. all act designerly in the context of industry while they conceive and plan out in the mind, and devise for a specific function or end. Design is also at the heart of professional training as schools get their students ready to meet the needs of life.

Academically, design is in humanities (literature, history, philosophy, mathematics, etc.), in sciences (natural, mathematical, behavioral, physical, economical sciences, etc.), in engineering (electrical, civil, chemical, human engineering, etc.), and in arts in the means of personal expression and research context (Kocabiyik 2004).

The study of engineering design is a discipline within the broader field of engineering. Design theory (or design science) and design methodology represent two academic subjects within the discipline of engineering design that each have their own unique features and contents. Both design theory (what design is) and design methodology (how to do design) will be broadly discussed in Chapters 8 and 9.

8.3.2 *Philosophy of design*

The philosophy of design has been experiencing a dramatic evolution since the positivist scientization of design introduced by the modern movement of design, in the early 1920s. It then witnessed the backlash of the 1970s, against the science-inspired design methodologies and the claim that the epistemology of science was in disarray and had little to offer to an epistemology of design, that there were forms of knowledge peculiar to the awareness and ability of the designer, and that we should rather concentrate on the "designerly" ways of knowing, thinking, and acting. Important contributions to this debate have been developing recently in the field of information systems, where the evolution of systems design has been described as incorporating four categories: design as functional analysis, design as problem-solving, design as problem setting, and design as emergent evolutionary learning (de Figueiredo 2008).

Design as functional analysis assumes requirements to be fully available at the outset, so that the designer just needs to analyze the problem and deductively proceed to the solution, following a path closely inspired by the traditional basic sciences (de Figueiredo and Cunha 2007).

Design as problem-solving resolves complex, namely organizational, problems by simplifying them to a level where they can still satisfy a minimal set of criteria leading to their rational solution (de Figueiredo and Cunha 2007). This category of design is inspired by Herbert Simon's concept of "bounded rationality" (Simon 1973), which reflects an epistemological standing closer to some popular visions of the social sciences.

Design as problem setting views design as a systemic activity needing the discovery and possible negotiation of unstated goals, implications, and criteria before a problem can be formulated and, subsequently, solved (Simon 1973).

Problem-solving is important but problem finding is arguably more so. In purely pragmatic terms, if a customer knows exactly what his or her problem is, he or she can probably find the solution on his or her own. Problem finding requires learners to ask questions, to investigate, and to check and cross-check. They will need to reframe problems to see if they are dealing with a symptom or an underlying cause. Teachers who are problem finders tend to ask questions to which they genuinely do not know the answers. They are comfortable with not having tightly structured tasks to offer their students and happy to live with the uncertainty of not knowing quite which way a project will develop (de Figueiredo 2008).

Design as emergent, evolutionary, learning sees design as the convergence of problem and solution in an emergent process of learning about a situation and then planning short-term partial goals that emerge as the process progresses (Suchman 1987). Aspects of the solution are thus explored in conjunction with those of problem understanding: not only the problem is unclear at the start of the process, but the goals of the design are also ill-defined (Gasson 2004).

Today, nearly every discipline has been converted into a science. The borderlines between the pure or epistemic sciences, on the one hand, and the action sciences or applied science, on the other hand, have become fuzzy. Thus, all disciplines have more or less theoretical, empirical, and practical issues as well. Any given science can act as an ancillary discipline to any other science. While practical design seems to be only a matter of technology, the study of possibly alternative design is a task for the technological sciences. Yet today, design is done in a scientific and computer-aided way as never before. The thinking in alternatives requires that the practical design has become not only a practical task but also a scientific task (Kornwatchs 2016).

8.3.3 Design paradigms

According to the dictionary, a paradigm is an example, a pattern, or a model. When attempting to characterize the major patterns which operate within the world of design today, three, in particular, seem to each be characterized by specific discourses and values (see Figure 8.2) and to be practiced by large numbers of designers and other professionals. Technology-driven design, sustainable design, and human-centered design are major movements which usually lead to distinguishably different results despite operating within the same legal, regulatory, contextual, and economic constraints (Giacomin 2012). The different core discourses based on technical novelty, planetary impact, or human meaning lead to notable differences in the resulting product, system, or service.

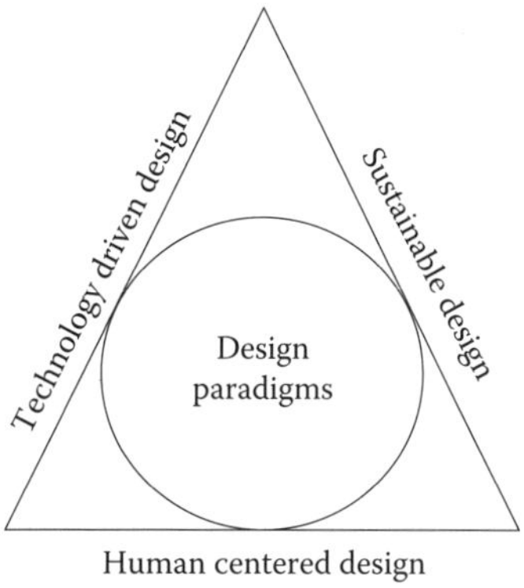

Figure 8.2 Three major design paradigms.

Technology is the imagination of the end user and developer communities alike. It had always been a powerful tool of human being to rule over his or her environment and to meet his or her needs in the most effective way. By implementing latest technologies in consumer products, integration of unnecessary features turns out to be right with the excuse of being novel. With the effects of the overall transformations in technology in the past decades, the rate of technological obsolescence shortens and new forms of interactions are integrated into products. Many products that are designed with a technology-driven approach surpass the needs and requirements of users, and simple products began to accommodate added and often unnecessary functions. As a result, these rapid changes toward complexity in interfaces require users to involve in a continuous learning process to use everyday products, leading to the adaptation difficulties and deficiencies in product use (Gultiken 2004).

Human-centered design has its roots in semiscientific fields such as ergonomics, computer science, and artificial intelligence (Giacomin 2012). The echoes of this past can be noted in international standards such as ISO 9241-210 "Ergonomics of human-centered system interaction" which describes human-centered design as "an approach to systems design and development that aims to make interactive systems more usable by focusing on the use of the system and applying human factors/ergonomics and usability knowledge and techniques." Such engineering-based approaches address well the needs of the users of tools since tools have predetermined functions.

Human-centered design is defined by Krippendorff (1989) as follows: "an approach to design and research that takes seriously the proposition that behavior and understanding go hand-in-glove, that the use of artifacts is inseparable from how users conceive of them and engage with them in their world. The proposition may be stated more concisely:

Humans do not respond to the physical qualities of things but to what they mean to them."

Many businesses are shifting their emphasis away from matters of technology and manufacture, moving instead toward a growing preoccupation with how their products, systems, or services are perceived and experienced by the consumer. The commercial imperative of this shift is demonstrated by statistical analysis such as the work of Eric Von Hippel (2007) of the MIT Business School who has noted that "70%–80% of new product development (NPD) that fails does so not for lack of advanced technology, but because of a failure to understand users' needs."

Sustainable design has emerged as a guiding paradigm in the creation of a new kind of built environment: one that meets the needs of the present without compromising the ability of future generations to meet their own needs. This topic will be discussed in detail in Chapter 10.

8.3.4 *Types of design*

Design is broad and integrative with complex structuring in various professions. Today, design is reflected in various specializations and disciplines that may sometimes work together (Dhillon 1985). These types can be treated as a hierarchy, in the sense that each type informs other types in the process of development.

- Engineering design
- Product design
- Interface design
- Visual design

Design is the "central creative process" of engineering (Brzustowski 2004). It is, in fact, the "essence of engineering." The Canadian Academy of Engineering (CAE 1999) states, "Engineering is a profession concerned with the creation of new and improved systems, processes, and products to serve human needs." The central focus of engineering is design, an art entailing the exercise of ingenuity, imagination, knowledge, skill, discipline, and judgment based on experience.

Engineering design is a distinguished discipline since it (1) synthesizes new information for product realization; (2) establishes quality through defining functionality, materialization, and appearance of artifacts; and (3) influences the technological, economic, and marketing aspects of production (Horvath 2001). It facilitates the creation of new products, processes, software, systems, and organizations through which engineering contributes to society by satisfying its needs and aspirations.

Engineering design is concerned with applying various techniques and scientific principles to the development and analysis of basic functional features of systems, devices, and service. Engineering design as stated by Dym et al. (2005) is as follows: "a systematic, intelligent process in which designers generate, evaluate, and specify concepts for devices, systems, or processes whose form and function achieve clients' objectives or users' needs while satisfying a specified set of constraints." In the context of engineering design, creativity is important, but it is not design! Design problems do reflect the hard fact that the designer has many constraints that may positively or negatively affect the outcome of the design; for example, the designer has a client to satisfy and for whose benefit the item/artifact and/or project is being developed (Akili 2015).

Product design is related with those items that are manufactured and ultimately to be sold to consumers. It needs to take into account how the item will perform its planned functionality (engineering design) in an efficient, safe, and reliable manner. The goal of product design is to generate and prioritize functionality that could possibly deliver value to users in correspondence with the product's desired purpose.

Interface design is concerned with the processes for desired transformation and adaptation of a product. The goal is to translate the functionality conveyed by the product designer and formulate the way the user operates and experiences the functionality of the product. It is usually concerned with the design of components, tools, and equipment. It focuses on anticipating what users might need or expect to do and ensuring that the interface has features that are easy to access, understand, and use to facilitate those needs. Interface design may be the design of new products or it may be the modification or expansion of existing ones.

Visual design is concerned with the appearance features of an item. It reflects personal expression (artistic), concrete (realism), or abstract. The goal of visual design is to ensure that the product conveys a perception of quality and draws the proper emotional response from its users. Although visual design is the most aesthetic and subjective design type, it is also the most perceptible one.

In some respects, product design is narrower than engineering design. Engineering designers work with product and interface designers. Engineering designers are responsible for applying various techniques and scientific principles to the development and analysis of basic functional features of systems, devices, and services (Kocabiyik 2004). The practical question that may be asked by developers is how much attention to give each of these types of design, especially when their general hierarchy is recognized. The answer basically boils down to how much usability friction users can be expected to tolerate (on the interface front) and how central the notions of quality and emotion are to the product's value proposition (on the visual front) at any given release point (Hendrickson 2012).

8.3.5 The transdisciplinary factor

Design is generally regarded to be the central or differentiating activity of engineering. To design is to invent, says Ferguson (1992). Invention and creativity are inherent in the professional title of engineer which has its origins in the "one who uses ingenuity," "to bring about or make manifest" and the Greek root word *techne* meaning "the arts of the mind" (Freeman-Bell and Balkwill 1996; Watts 2001).

Engineering is a profoundly creative process. A most sophisticated description is that engineering is about design under constraint. The engineer designs devices, components, subsystems, and systems and, to create a successful design, in the sense that it leads directly or indirectly to an improvement in our quality of life, must work within the constraints provided by technical, economic, business, political, social, and ethical issues (NAE 2004). Engineering design affects almost all areas of life, uses the laws of science, and requires professional integrity and responsibility. It is part of human nature and is at the intersection of engineering science, economics, politics, psychology, engineering technology, visual art, and industrial design as shown in Figure 8.3 (Gopsill et al. 2013; Gopsill 2014). It is a transdisciplinary and highly collaborative exercise.

The Canadian Engineering Accreditation Board (CEAB 2004) describes engineering design as integration of mathematics, basic sciences, engineering sciences, and complementary studies in developing elements, systems, and processes to meet specific needs. It is a creative, iterative, and often open-ended process subject to constraints that may be governed by standards or legislation to varying degrees depending on the discipline. These constraints may relate to economic, health, safety, environmental, social, or other pertinent interdisciplinary factors. Obviously, successful engineering design requires a broad cross section of knowledge, skills, and attitudes.

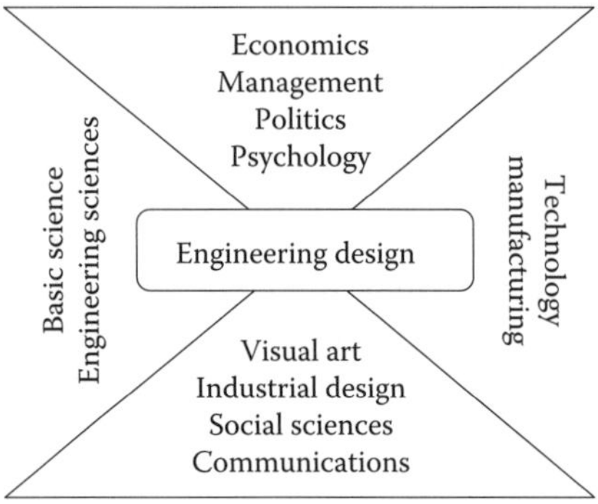

Figure 8.3 The transdisciplinary engineering design.

Engineering accreditation bodies have also recognized the need for design and multidisciplinary capabilities. Engineering is often a tactile, visual, verbal, cerebral, and physical activity. A formal definition of engineering design is found in the curriculum guidelines of the Accreditation Board for Engineering and Technology (ABET). The ABET definition states that engineering design is the process of devising a system, component, or process to meet desired needs. It is a decision-making iterative process, in which the basic sciences, mathematics, and engineering sciences are applied to optimally convert resources to meet a desired objective (Haik and Shahin 2011). The ABET (2006) specifies that "an ability to design a system, component, or process to meet desired needs within realistic constraints" and "an ability to function on multidisciplinary teams" are two of the eleven key elements (engineering programs must demonstrate) which are present in their new graduates.

Engineering design is essentially an issue of cost and performance trade-offs. The key element of performance is time to market. In the academic world, the design requires engineering analysis and synthesis. It requires breaking the problem into practicable parts, understanding their behaviors, and integrating partial solutions into a functional system.

8.4 Visualization in design

Design is more than applied knowledge.

Armand Hatchuel
École des Mines

8.4.1 Design thinking

DT is a methodology used by designers to solve complex problems, and find desirable solutions for clients. A design mind-set is not problem focused, but it is solution focused and action oriented toward creating a preferred future. DT draws upon logic, imagination, intuition, and systemic reasoning, to explore possibilities of what could be and to create desired outcomes that benefit the end user (the customer) (Naiman 2017). DT is a methodology that imbues the full spectrum of innovation activities with a human-centered design ethos.

DT is a human-centric, holistic approach to problem-solving and business thinking that employs empathy, ideation, prototyping, and experimentation to solve real-world issues. DT is popular among educators and social entrepreneurs for social innovation because it approaches problem-solving from the point of view of the end user and calls for creative solutions by developing a deep understanding of unmet needs within the context and constraints of a particular situation. Some designers are picking up the skills when working in close collaboration with other domain specialists in the field of engineering, economics, and social sciences (Mootee 2013).

DT, which stresses that products be created to respond empathetically to a problem, consists of four key components (Maxey 2012):

- Empathy: Imagining a products use from the perspective of the customer.
- Collaboration: Being open to bringing in opinions from a variety of sources to better understand and solve problems.
- Integrative thinking: Refusing to make compromises that jeopardize the effectiveness of a product.
- Experimentalism: Using creative thinking to solve larger issues with products rather than making incremental upgrades.

DT processes combine empathy for the context of a problem, creativity in the generation of insights and solutions, and rationality in analyzing and fitting various solutions to the problem context. They can be powerful ways of creating new possibilities in education (AITSL 2014). It is regarded as a system of three overlapping spaces including viability, desirability, and feasibility, where innovation increases when all three perspectives are addressed (Chasanidou et al. 2015). For example, Thomas Edison created the electric light bulb and then wrapped an entire industry around it. The light bulb is most often thought of as his signature invention, but Edison understood that the bulb was little more than a parlor trick without a system of electric power generation and transmission to make it truly useful. So he created that, too. Thus, Edison's genius lay in his ability to conceive of a fully developed marketplace, not simply a discrete device. He was able to envision how people would want to use what he made, and he engineered toward that insight. Edison's approach was an early example of DT (Brown 2008).

8.4.2 Design problem

The word problem is used in a variety of contexts, for example, understanding a confusing phenomenon is a problem, how to find a better approach to do something is a problem, the best way to design or build something is a problem, and how to create a creative work may be a problem.

Jonassen (2004) has focused on three types of problems. These include story problems, troubleshooting problems, case and system, and policy analysis problems. Story problems are the most commonly used and extensively researched kink of problems. These are the types of problems frequently found at the back of textbooks. Students identify key words from the problem description (the story) and select or adapt an appropriate solution methodology.

Troubleshooting problems are, probably, the most commonly found problems in the engineering domain. It is concerned with identifying faults in some nonfunctional systems and repairing or replacing components to return the systems to their operational condition. Effective and efficient troubleshooting requires three kinds of knowledge and skills: system knowledge (knowledge of how the system operates), procedural knowledge (problem-solving and test procedures), and strategic knowledge (how to apply the procedures). These knowledge and skills are integrated and honed by the trouble shooter's own experiences.

Case problems are usually found everywhere except the classroom, probably, because they are ill structured and complex and, therefore, difficult to assess. They are very often used in domains where knowledge is less hierarchical and linear or sequential in nature (after some level foundation knowledge and a set of basic principles have been acquired by the learner). This is unlike medicine, where the knowledge is more encyclopedic (Perrenet et al. 2000). Very often case problems have no opportunity to actually be implemented, and tend to focus on the thought processes rather than the creation of an actual product.

Design problems are not restricted to the story, troubleshooting, or case problems, because they incorporate new designs, and they are practical and achievable in a way. They can be a blend of all the above three, but tending toward cases, which have a possibility of an achievable result. Design problems and case problems are similar in that there is no single correct answer. The range of problem types is shown in Figure 8.4.

Design problems are usually unclearly defined compared to analysis problems. Unlike an analysis problem, a design problem often begins as an abstract idea in the mind of the designer. Creating a clear definition of a design problem is harder than defining an analysis problem where a design problem evolves through a series of processes as an understanding of the problem is developed. According to Brown and Chandrasekaran (1989), design problems can be classified into three classes as presented in Table 8.1.

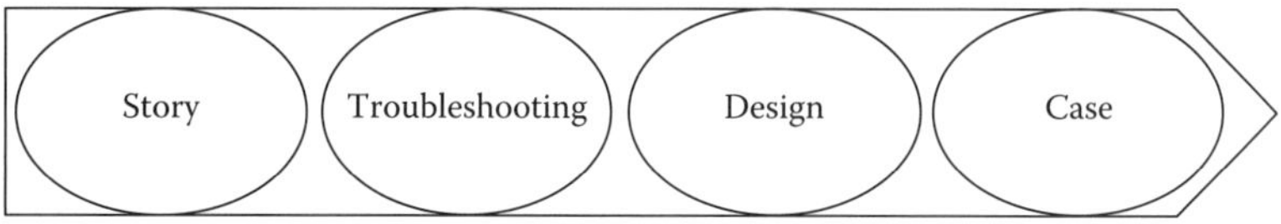

Figure 8.4 Range of problem types.

Table 8.1 Classes of design problems

Class 1	Open-ended, nonroutine creative activities where the goals are ill structured, and there is no effective design plan specifying the sequence of actions to take in producing a design model
Class 2	Existing, well-developed design and decomposition plans (e.g., designing a new product)
Class 3	Routine where design and decomposition plans are known as well as customary actions taken to deal with failures (e.g., writing a computer program)

8.5 Engineering design communication

> Design is really an act of communication, which means having a deep understanding of the person with whom the designer is communicating.
>
> **Donald A. Norman**

8.5.1 Design tools and methods

Since the early days of engineering design, the challenge of communicating designs to other people has been manifest (Allen 2015). Communication is an essential part of any design process (Clarkson and Eckert 2005). Therefore, engineering design communication (EDC) is fundamental to almost all engineering design activities as it provides the ability for knowledge and information to be shared between engineers. It is part of "what we do." This communication contains a great deal of rationale relating to the evolution of product development and is essential for understanding "why the product is the way it is" (Gopsill et al. 2013). Today, EDC plays an important role in the coordination of tasks between designers and engineering project teams.

Since the early days of engineering design, the challenge of communicating to other people has been obvious. In their most basic form, engineering drawing is the major tool used to communicate information about a design to others who will be engaged in producing or realizing the design. What is the best way to characterize an engineering design? Tools and methods have always played a major role in that process, because design is all about manipulating the physical environment. Tools are roughly related to the various design stages. Methods are instructions of how to go about doing something, and can involve tools. For instance, brainstorming is a method that involves pens and papers as tools.

In earlier times, engineering designs were described on paper drawings. The oldest and most powerful conceptual tool is, and likely will remain, the pen(cil). Consequently, it is the most widely used tool in any conceptual work. The pen enables designers to quickly, with minimal cost and effort, try out ideas, communicate these ideas, change, and either discard or refine them. It is thus ideally suited for the early stages of design, when ideas are quick and plentiful (Knörig 2008). A good sketching tool must be barely noticeable, fast and easy, yet at the same time highly expressive. It is hard to imagine a computer-based tool that is as ready to hand as pen and paper, but the computer could add the ability to make animated or even interactive sketches, more suitable for designing interactions. However, designers make very creative and opportunistic use of the material available to them for sketching their ideas.

Early engineering drawings were similar to artistic drawings than technical drawings in that they could be communicated to users without involving any set of rules in order to understand and eventually produce the designs. One of the famous talents of the past, Leonardo da Vinci created many inventions on drawing that demonstrates the early artistic-centric EDC. Many of Leonardo's inventions were not realized in his time due to a variety of limitations in prototyping, simulation, and manufacturing knowledge during his life. Today, most of the drawings are in bound volumes kept in museums and libraries. Many of those documents contain written explanations that help one understands the objective of the drawings.

The language of engineering drawings has evolved over the years into a specific methodology that certain skills and training are required to understand them. While paper and pencil were initially used to communicate design ideas, a special set of tools and aids is developed over time to make the design process faster and more accurate, a theme that continues to evolve. Schematics used to be drawn by hand, but most are now prepared using schematic editors of CAD programs that run on engineering workstations. CAD is a set of methods and tools to assist product designers in creating a geometrical representation of the artifacts. It helps to feed information between teams, organizations, and subsequent design steps including computer-aided engineering (CAE) for drafting and modeling designs and CAM for managing manufacturing processes, all by using a computer system. This challenge proposes that engineering design data be represented in a textual, human-readable language. In a typical CAD setting, the computer primarily serves as a precise drafting and visualization tool, enabling the designer to view the emerging geometry from several angles and in different projections.

Digitization supports sharing and cooperation. The digital file can easily be shared with others without scanning. It enhances the ability to sketch faster and make edits to explore more design options. It influences existing assets by use of existing digital files like photos and CAD files for reference. A digital representation also makes it possible to do several analytical tasks. Figure 8.5 shows a general flow of design process and the required tools.

Today, educators at engineering schools are experiencing new pressure to change the way they teach design-related courses in order to equip their students with the ability to interact with CAD/CAM/CAE systems and have the knowledge of their fundamental principles (McMahon and Browne 1998).

8.5.2 Virtual and augmented reality

Virtual reality (VR) is an emerging computer technology of simulating or replicating to increase the realism and impact of simulations by placing the user in the center of an interactive 3D environment, complete with spatialized sound, haptic feedback, and eventually olfactory and taste feedback as well. VR has the potential to further revolutionize the engineering design cycle by providing engineers with a new window into the computer world. With the appearance of VR software and peripherals, companies have started building VR applications to solve real problems (Bell and Fogler 1997). With traditional CAD tools, it is not possible to view models with natural stereoscopic vision.

VR application in engineering includes the use of 3D modeling tools and visualization techniques as part of the design process. It assists engineers to observe their projects in 3D and acquire a better understanding of how it operates. In addition, they can identify any faults or possible risks before employment. VR has the ability to show details of an engineering product to maintain the illusion. This means high-end graphics and videos. Currently, car manufacturers use VR for prototyping purposes during the design process. This assists them to produce numerous versions which are then tested. Also, it eliminates the need to make physical prototypes and speeds up the development stage.

In addition to VR, augmented reality (AR) is an enhanced VR, where users can see and experience the world around them with the addition of computer simulation. It is a technology that layers computer-generated enhancements atop an existing reality in order to make it more meaningful through the ability to interact with it. Current AR uses a device like a

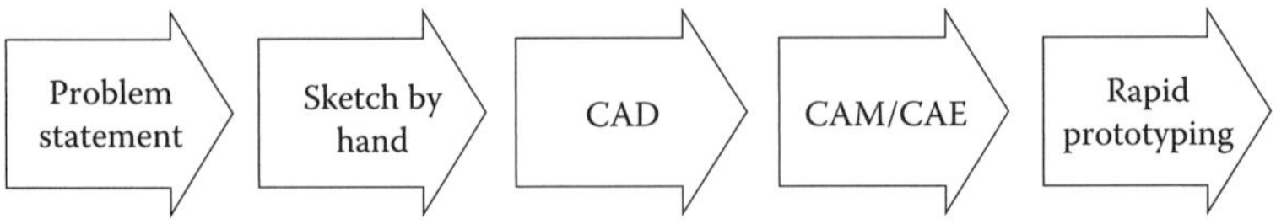

Figure 8.5 General flow of design process and the required tools.

smartphone or tablet to capture the surrounding environment via camera and place a digital effect. This effect is only viewable on the device being used to capture the surrounding area (Gonzalez 2016). The key feature of the VR and AR environments is that it combines VR-based interaction with functional behavior simulation. The VR tool allows users to conduct functional evaluation and usability test before engaging in the costly and time-consuming process of building physical prototypes. VR tools have also been applied within the manufacturing domain.

AR and VR both leverage some of the same types of technology, and they each exist to serve the user with an enhanced or enriched experience. Their environments can support product design through achieving reductions in development time and increasing customer's satisfaction. Previous VR applications within engineering design have been primarily restricted to use by engineers and designers, for example, for illustrative purposes, and for the study of user behavior and interaction with products. In product design, it is important (as with capturing customer needs) to accurately transform customer needs into the actual forms and functions of the products. Furthermore, it is essential for the people involved in the development process to visualize the design as effortlessly as possible and to gain a comprehensive understanding of the functional behavior of the product. Traditional physical prototyping as well as CAD software cannot reflect the functional behavior of the product. A VR environment was reported to satisfy such requirement through the case of design evaluation of a digital consumer product (Park et al. 2008). Dukic et al. (2007) detailed a case study focusing on the verification of visual demands in car assembly work using virtual tools. Computer mannequins were created for the analysis of the ergonomics of assembly operations. This use of VR within the product development process allowed the early identification of ergonomic problems using only virtual mock-ups. Cappelli et al. (2007) detailed a virtual environment for disassembly that could accept a virtual CAD assembly prototype as an input and produce necessary information for the identification of the disassembly path.

8.5.3 Social media support

A designer constantly needs to present the state of the accomplished work to team members, the client, or other stakeholders. For review purposes, design tools should allow a quick and direct presentation function for discussion, including the ability to explain the design details. Designers need to communicate with engineers and other production people, and the design tool should be able to export documentation in a suitable format.

Today, engineers spend a significant portion of their time developing and communicating as they fill in the gaps left by formal documentation and processes. Communication remains an equal partner with design as students and engineers write problem statements, scripts for interviewing users, minutes of meetings, memos to clients and faculty, progress reports, and proposals. For oral communication experience, they conduct meetings, run focus groups, hold design reviews, and give formal presentations (Hirsch et al. 2001).

Social media has been developed significantly over the past decade and the tools are becoming increasingly central to the digital lives of consumers and, to a lesser extent, businesses (Boyd and Ellison 2007). Furthermore, media tools generally support synchronous and asynchronous communication, which has benefits in enabling communications to continue independent of users' schedules, time differences, and location (Gopsill et al. 2013). For example, Facebook allows users to upload photographs and enables them to comment. Thus, there is potential in applying such technology within the engineering design context.

With the advent of social media and associated technologies that better support communication within a given community, it is contended that a social media approach could be a key significance to EDC in relating rationale and understanding behind engineering work, records, and project management (Gopsill 2014).

8.6 Design science and theories

Bad design is smoke, while good design is a mirror.

Juan-Carlos Fernandez

8.6.1 Design science

Design science has long undertaken issues to support the practice of design engineering, including understanding the complexity of the products, understanding the people who design them and those who use them, and understanding the process of designing, together with the organization around the process. The field of design science crosses discipline boundaries. This understanding is built upon knowledge: from within the engineering domain, of modeling products, and human behavior in design; for example, understanding better the process of creativity, thus establishing the discipline of design research (Papalambros 2015). Design science involves developing design knowledge, both knowledge of design and knowledge for design (Horvath 2001).

Design science in the future is likely to be transdisciplinary, not only in borrowing research methods or theories from other fields and applying them to design problems as we do today, but also in impacting research beyond design to facilitate the new generation of products (systems/services/digital), processes, and people (Papalambros 2015).

Design theories aim to lead a designer to good solutions to a design problem. To develop a scientific theory of design, it is necessary to discuss what makes a design methodology scientific. In essence, an activity can be considered scientific if it is based on a set of rules which meet specific criteria. In order to make design more scientific, a set of rules or principles must be developed, which guide its practice in order to improve the results of the process (Dyas 2005). The literature provides a wide range of design theories; however, it has been noticed that existing design theories are either process or product oriented (Evbuomwan et al. 1996). In this regard, several theories will be investigated.

8.6.2 Axiomatic design: Process-oriented design theory

The axiomatic design (AD) method is a mapping of one set of variables to another. It is a type of design specification that is obtained by examining the customers' needs and expressing them as a list of attributes. These attributes are mapped into a set of functional requirements. AD has two axioms. In mathematics, axiom is a proposition that is assumed to be true without proof for the sake of studying the consequences that follow from it. Axiom 1 (independence axiom) states that the independence of functional requirements should be maintained in the design (Suh 1990). Axiom 2 (information axiom) states that the information content of designs should be minimized; among designs that satisfy function requirements, the design with the minimum information content has the highest probability of success and thus should be chosen.

AD is applied not only in product development, but also in many other applications. Through the systematic approach and the consideration of independence axiom and information axiom, even highly complex projects can be mastered reducing the complexity in the design task (Rauch et al. 2016). In addition to product design, AD applies to all designs, including hardware, software, materials, manufacturing, and organizations. It is also an appropriate tool for the design of nonengineering systems such as business plans and organizations (Martin and Kar 2002). Moreover, instead of prioritizing system requirements and focusing on most important factors, AD considers the system as a whole and incorporates all requirements, even the least important, in the system.

AD operates with a model of the design process that uses state spaces to describe different steps in generating design concepts. According to AD, the world of design has four domains: customer domain with characteristic vector of customer attributes, functional domain with characteristic vector of functional requirements, physical domain with characteristic vector of design parameters, and process domain with characteristic vector of process variables, as seen in Figure 8.6. The domain on the left represents "what we want to achieve" and the domain on the right corresponds to "how we want to achieve it" (Suh 1995).

The AD methodology begins with the identification of customer needs and the conversion of these needs into a set of one or more high-level functional requirements. The goal is to develop the minimum set of independently achieved requirements that completely characterize the desired functions of the design (Suh 1990). In AD, the same design is represented in each space by a vector of different variables. Customer attributes are the customer needs and wants that the completed design must fulfill. Functional requirements are the variables that describe the intended behavior of the device. Design parameters are the physical characteristics of a particular design that has been specified through the design process. Process variables are the variables of the processes that will result in the physical design described by the set of the design parameters.

In product development, the designer following the AD process produces a detailed description of what functions the object is to perform, a description of the object that will realize those functions, and a description of how this object will be produced. Whether the design solution is a tangible product, service, or a process, designers typically understand their customers need, define the problem they aim to solve, create a solution from among many, and check the design based on customer needs.

8.6.3 *Concept knowledge theory*

The concept knowledge (C-K) theory is a design theory based on the distinction between concept and knowledge, as its name suggests. C-K is a cognitive theory that has been initially proposed by Hatchuel (1996) and developed by Hatchuel and Weil (1999). The C-K theory offers a formal framework that interprets existing design theories as special cases of a

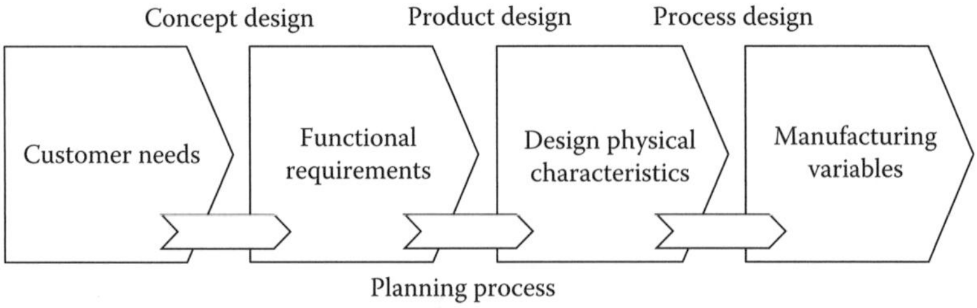

Figure 8.6 Design according to AD.

unified model of reasoning. The core idea of this framework is to separate concept and knowledge in two different spaces, and to keep in mind that the object of study never has invariant definitions and properties (Hatchuel et al. 2004). Since its introduction, the principles of C-K theory have been industrially applied several times in order to model and support industrial design processes (Hooge et al. 2012).

The C-K theory, at the core of its scope, integrates creative thinking and innovation. It makes use of two spaces: (1) K—the knowledge space is a space of propositions that have a logical status for a designer and (2) C—the concept space is a space containing concepts that are propositions, or groups of propositions that have no logical status (e.g., are undecidable propositions) in K. This means that when a concept is formulated, it is impossible to prove that it is a proposition in K. Design is defined as a process that generates concepts from an existing concept or transforms a concept into knowledge, for example, propositions in K (Hatchuel et al. 2004).

A concept is defined as a proposition that is neither true nor false. It might emerge from market needs, that is, when a technical or market requirement is not satisfied by existing solutions/technologies. Basically, the concept is equal to idea. Concepts are gathered on the C-space.

A central point in the C-K theory is the dependences between what is known (and hence, what can be used as a resource for the design process) and what is yet to exist (a set of variants for a seed project with innovative elements). The claim of the theory is that this conceptive reasoning process is defining the essential characteristic of design and it is fundamentally different from the usual processes prevalent in formal sciences (e.g., deductive or inductive processes).

Knowledge is defined as the group of propositions with known logical status (we know whether they are true or false): all that we (or the designer) know belongs to this. Knowledge is contained in the K-space. When a concept/idea is tested in reality, we come to know whether the proposition is true or not (if the idea is feasible). Consequently, the proposition becomes part of the knowledge and leaves the concept space, resulting in an expansion of the K-space.

Figure 8.7 shows the function of the C-K theory. It shows that a new idea/concept can be generated from existing knowledge. In an education environment, students need some preparation at the lowest level (basic knowledge) before they can properly apply this knowledge (application of knowledge level). Additionally, they need both basic knowledge and some experience in its application before being able to judge and critique the knowledge (critical analysis). Extending knowledge beyond what is received, creating new knowledge, making inferences, and transferring knowledge to usefulness in new areas of application will be the highest of knowledge space.

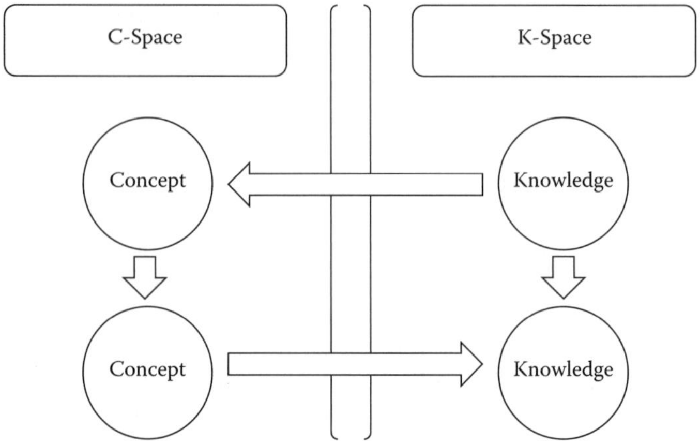

Figure 8.7 C-K design theory.

When an idea is conceived based on another idea, there is a concept expansion. It is important for concept expansion to support, incentive, visualize this process, and leave behind all consideration such as feasibility or other limitation. This expansion can be guided by other frameworks/tools (in 2011).

8.6.4 Systematic design

SD refers to a process of design that considers not only at the problem that needs to be overcome, but also at the surrounding environment, and other systems that are linked to the problem. As such, SD is the basis for a lot of appropriate technology. SD does not only apply to technological design, but applies to architecture and planning, and broader social system design.

The SD process model aims at making it easier to find an optimal design for a product-to-be. To that end, it is necessary to encompass the broadest range of solutions, that is, to search for solutions in a structured, systematic way. The breadth-first top-down strategy is adopted, which means first finding the largest possible number of abstract solutions (breadth-first) and then more concrete ones (top-down). The reasons are that each solution on an abstract level represents a set of different concrete solutions, and that it is more costly and time-consuming to develop, describe, and evaluate concrete solutions than abstract ones (Motte 2008).

The SD process models are organized as problems to solve, following the archetype: understanding the problem, generating solutions, evaluating and choosing solutions, and implementing. The models allow for iterations between every step, and the problem-solving process is also repetitive: the first step can, for example, be considered as a problem in itself and be decomposed accordingly.

As an ongoing basis for engineering design process models, the stage-gate process model implies that the engineering design process must be much more integrated in the product development process. The SD process models were mainly built around the technical system; the engineering design process model must be set toward the goals behind product development: developing a product-to-be in accordance with production, marketing, and corporate strategy. This is the domain of integrated product development (IPD), where the current models still use the systematic engineering design process models as a basis for the engineering design activity, with activities decomposed by the concretization level of the product (Motte 2008). Chapter 9 is largely based on SD.

The well-known German systematic model (Pahl and Beitz 1984) distinguishes three stages for any design process: the functional, conceptual, and embodiment design phases which can often overlap. In the German approach, the three phases are only experiential scheme that can be useful in many engineering cases. Figure 8.8 shows the hierarchical nature of process-based design.

The first phase requires understanding the opportunity by dealing with the identification of customer needs to define the problem to be addressed, where various tools like customer survey interviews are used to understand the customer needs.

The next phase is conceptual design by dealing with finding solutions to solve the problem. Various techniques such as functions modeling and concept engineering are used in this phase. The concept engineering is generally considered as the final outcome of developing a concept, where the concept generation and concept selection activities are done.

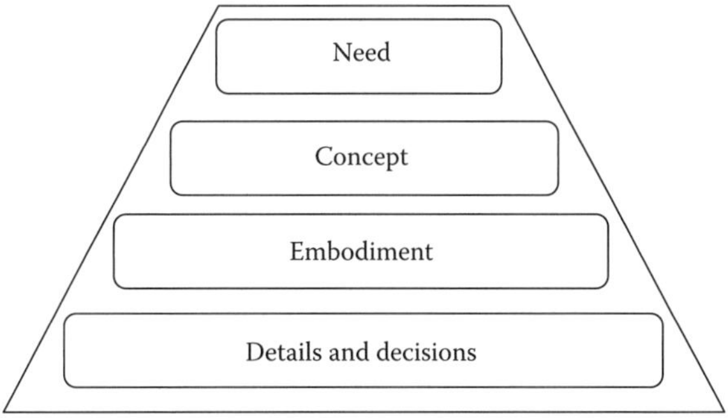

Figure 8.8 Hierarchical nature of process-based design.

The third phase is implementing the concept which deals with the embodiment design. The embodiment design takes the abstract conceptual design from the second phase and molds it into a system that can actually be produced. This phase finalizes material selection, dimensions, and tolerances and generates product details and decisions. Every step in the design process generates design information which reduces the uncertainty. Final design decisions should be justified by mathematical and physical proofs.

Product-oriented design theory is based on some specific properties explicitly required from the product to be designed. Therefore, product-based theory is in fact specification theory. Suh (1990) defines two universal product attributes. These specifications only form new functional requirements that could be added to the primary functional requirements used to build Suh's matrix. The same could be said from others (Matchett and Briggs 1966). Evolutionary design (Hybs and Gero 1992) is an interesting attempt to mix process and product but it is basically a problem-solving theory where problems are discovered progressively.

Traditional engineering design methods (EDMs) are based on a bottom-up approach. Beginning with a set of known elements, design engineers create the product or system by a combination of system elements. However, it is unlikely that the functional need will be met on the first attempt unless the system is simple. After determining the product's performance and deviation from what is required, the elements and their combination are altered and the performance is determined again. The bottom-up process is iterative, with the number of iterations (and design process efficiency) determined by the experience and creativity of the designer, and the complexity of the product or system (Blanchard and Fabrycky 1990).

SE methodologies are more directed, and based on a top-down approach to design. The key idea is that large complicated problems can be reduced to a set of smaller problems that are easier to solve. Requirements at the interfaces ensure that the solutions to these smaller problems will form a coherent solution. While some iterations may inevitably occur, these should be less than is typical for bottom-up methodologies. SE also considers the entire life cycle of the design. Manufacturing, sale, distribution, service, maintenance, and, finally, disposal must be considered in addition to the actual use of the product. In the top-down approach, the requirements are always satisfied through every step of the design process because it is an inherent part of the methodology, whereas in the bottom-up approach the methodology provides no assurance that the requirements are always satisfied (Blanchard and Fabrycky 1998).

8.6.5 Modular design

Modular design (MD) allows managing, developing, and organizing complex systems (such as a software program, an electronic circuit, or a mechanical system) as a set of distinct components that can be developed independently and then assembled together. During design, different modules may be built by separate developers. Designing modules simultaneously reduces overall time to market for a product, therefore maximizing sales and revenue.

Modularity requires grouping functionally similar parts into subassemblies, which then can be put together to form the product. Effective modularization can be achieved only if interfaces are standard (Chiu 2011).

Modularization can improve and support many aspects of a system such as manufacturing operations. In practice, approaches for MD are diversified. However, the focuses of these MDs can be classified into three groups: function oriented, structure oriented, and recycle oriented. Function-oriented MD makes several components working together as a module to ensure the operation of the system. Structure-oriented MD includes components with proper structures to reduce the cost and difficulty of manufacturing or assembling. Recycle-oriented MD is to make recycling process easier or cheaper by considering material properties or life cycle of the product in the design phase.

8.6.6 Design for X

Design for X (DfX) refers to the usage of a recognized methodology to optimize a specific aspect of a design. The variable X represents an area of focus. The design strategy usually proposes an approach and corresponding method that may help to generate and apply technical knowledge in order to manage, improve, or probably to invent particular characteristics of a product.

Chiu (2011) presents the DfX methods using two organizing themes [design for efficiency (DfE) and green design (GD)] to show their complementary nature. In addition, the author categorizes the DfX methods using three ranges of perception: product scope, system scope, and ecosystem scope. In this context, the efficiency is defined as the ratio of the effective or useful design process output (e.g., designed artifact and the process itself) to the total input to the design process and the designed artifact (e.g., information and materials). The DfX concepts relating to efficiency may be perceived in two ranges: product scope and system scope. On this basis, it is possible to group design for manufacturing (DfM), design for assembly (DfA), design for variety (DfV), design for quality (DfQ), design for reliability (DfR), design for disassembly (DfD), design for maintainability

(DfM), design for supportability (DfS), and design for obsolescence (DfO) within the product scope. The system scope covers design for supply chain (DfSC), design for logistics (DfL), and design for network. Figure 8.9 presents the overall structure of categorization.

The main purpose of DfE is expressed as reducing cost and lead time of a product while sustaining or improving its quality. DfE concept is divided into two ranges of perception: product scope and system scope. The product scope focuses on the product aspects, which enable efficiencies at the shop floor within a company (e.g., altering the design of a product to reduce machining time). The system scope concentrates on the integration and coordination of the value chain starting with the design stage and ending with the delivery and maintenance system (Chiu 2011).

GD is practicing engineering with the inclusion of natural system as a fundamental consideration (Ogot and Kremer 2004). On the basis of our review, we group design for sustainability (DfS), DfE, and design for life cycle (DfLC) under GD-related DfX concepts. The ultimate purpose of GD is to design a product, which will have minimum negative environmental impact during its life cycle.

8.6.7 *Robust design*

Robust design (RD) is to choose settings for product or process parameters to reduce the variation of that product or process response from the products or process target. RD seems to be well in line with the sustainability challenges in all disciplines. Thus, it is time to enhance RD to again focus on preventing losses to the society and thereby contribute to sustainability.

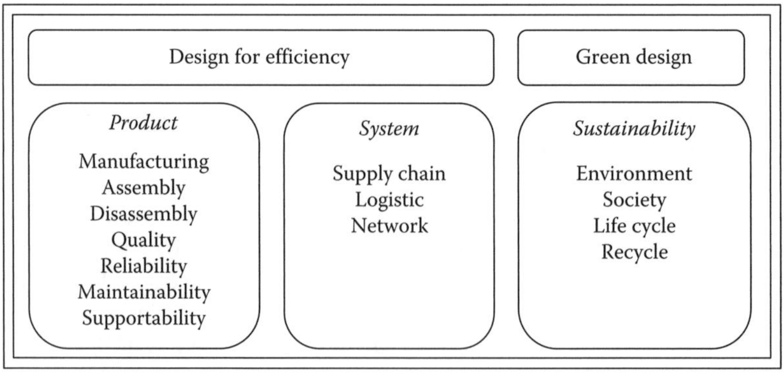

Figure 8.9 DfX domain.

Since the early 1990s, aligned with the boom in SD initiatives, environmental concerns have been interspersed in the development and manufacturing of products. The term "green product" is widely used with regard to research and practices of manufacturing "environmentally friendly" products (Baumann et al. 2002).

In order to relate RD to sustainable product development, three main concepts or models underlying this framework are reviewed. It can be said that each of them answers to a fundamental question related to RD. Thus, why is it important to deal with variation? The answer can be understood through the concept of a quadratic loss function. What causes unwanted variation? The concept of noise factors, often within a product or process (P)-diagram (Figure 8.10), is fundamental to answering this question. The concept of off-line and online quality control contributes to an answer to when should actions be taken to reduce unwanted variation.

The desired levels of product characteristics are usually referred to as target values (Kacker 1985). However, sources of unwanted variation might cause characteristics to deviate from their target values. These sources are often referred to as noise factors in RD studies. One way to conceptually analyze the noise factor and their influence on a product or process is by the use of the P-diagram (Figure 8.10) relating the input to a system (signal factor) to a desired response, simultaneously considering control and noise factors (Phadke 1989). In other words, signal factors as inputs are related to the response variables, as the output, considering both noise and control factors. Later versions of the P-diagram also add as an output various error states, that is, undesired outputs (Davis 2006).

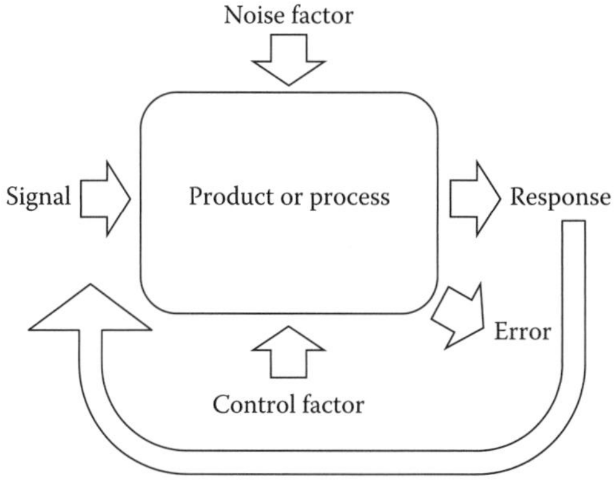

Figure 8.10 Product (P)-diagram.

Noise factors are often categorized as outer disturbances, manufacturing disturbances, and inner disturbances (Taguchi and Wu 1979). Efforts for quality control are often divided into two categories based on the point in a product life cycle (PLC) when they are applied. Online efforts are applied during production and off-line efforts in the design of products and manufacturing processes. Online and off-line efforts can also be related to their ability to reduce variation.

8.6.8 Concurrent design

By the 1970s, one idea had emerged from all the theory and research in design: concurrent design (CD), which is integrated design of products and processes, including manufacture and support. Today, it represents a key improvement in product development. CD requires technical and organizational solutions. It allows activity overlapping, information transfer within cross-functional integrated teams, and broadly experienced leadership. The extent to which two activities can be effectively overlapped depends on the relationship between those activities (Prasad 1996; Yassine et al. 1999). The essence of CD is the myriad of interactions that occur at the interfaces among all of the members of a design team and all their tools.

The foundations of CD were built on the concepts of DfM and DfA (Kusiak and Larson 2009). Research in CD focuses on developing these multidisciplinary design tools, communication technologies, and management processes. The design of complex engineering systems may then be coordinated across disciplines, both inside single companies and across a complex network of companies that includes suppliers, for instance, of individual components, manufacturers of turbine systems, and developers of entire wind plants. After selecting the best modules, the details that constitute the modules will be worked out.

According to *NASA Systems Engineering Handbook*, CD involves all elements related to design of complex technical systems including people and tools as well as organizational processes and facilities (NASA 2007). It involves designing a product through collaboration among multidisciplinary product developers associated with the entire PLC, including preliminary design, detailed design, manufacturing, assembly, testing, quality control, and product services (Shen et al. 2008).

8.7 Methodologies and approaches for product development

A product development process may be defined as the series of steps or activities that an enterprise engages to conceive, design, and commercialize a product. The conventional product development process employs a design-build-break philosophy. The sequentially executed product development process often results in a prolonged lead time and an elevated product cost (Chang et al. 2016). Today, there are several methodologies for NPD in current use, including SE, CE, V-cycle model, model-based design (MBD), and DD. Definitions of both SE and CE may lead an observer to conclude that they occupy the same territory and could even be synonymous. Both forms of engineering have the same objective in developing and introducing new products (Gardiner 1995).

8.7.1 SE approach

SE as a term was used as early as in the 1940s by Bell Telephone Laboratories. SE was developed to cope with the growing complexities of designing and developing large-scale telecommunications and military systems during the aftermath of World War II. SE has its roots in the electronics, military, and aerospace industries and it provides a systematic approach to the design and development of complex engineering systems (M'pherson 1980).

SE is a transdisciplinary field of engineering that focuses on how complex engineering projects should be designed and managed over their life cycles. SE approaches develop the philosophy of integrated systems with emphasis on the interplay between tools and techniques of different disciplines. These approaches have the four characteristics: holistic, transdisciplinary, integrated/value driven, and long term/life cycle oriented. The approach is holistic in that it considers the full technical system, including any number of performance criteria, as well as potentially nontechnical concerns related to human factors or societal impacts. SE work is transdisciplinary, involving engineering, natural, computational, and even social sciences. It is also integrated and value driven by considering the needs and interests of all stakeholders including customers. Finally, SE is focused on the long-term life cycle of the system. Figure 8.11 shows the key characteristics of SE and associated large-scale complex technical systems.

On the whole system dimension, SE applies a top-down, bottom-up approach as shown in Figure 8.12, establishing cost and performance parameters and a conceptual design at the system level, then working down through the systems hierarchy, allocating requirements to each level in such a manner as to ensure that the final components of the system will combine to meet the cost and effectiveness targets. Throughout this top-down design, optimization, and specification of the system, the entire system life cycle is taken into account; manufacturing processes, operating procedures and costs, maintainability, logistic support, phase

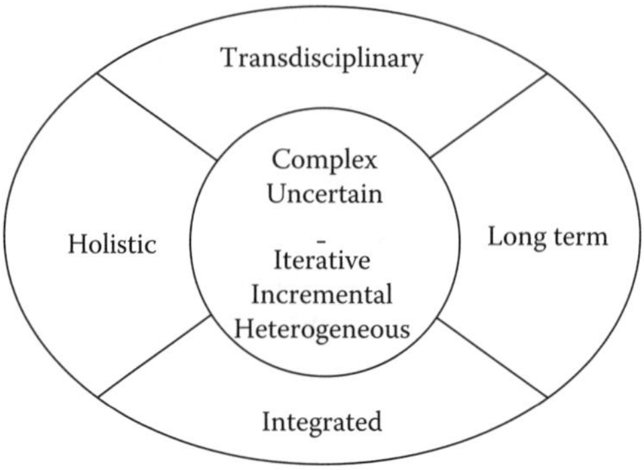

Figure 8.11 Key characteristics of SE and associated large-scale complex technical systems.

out, and disposal are all considered from the very earliest phases of the design and specification process (Colbome and Steyn 1998).

8.7.2 CE approach

CE, also known as simultaneous engineering, may be defined as a systematic approach to the combined, simultaneous design of technical systems, products, and processes, including manufacturing technical support. It is a process of designing and developing products, in which the various stages run simultaneously, rather than sequentially. It shortens product development time and as well as the time to market, leading to improved productivity and reduced costs. CE seeks for better products

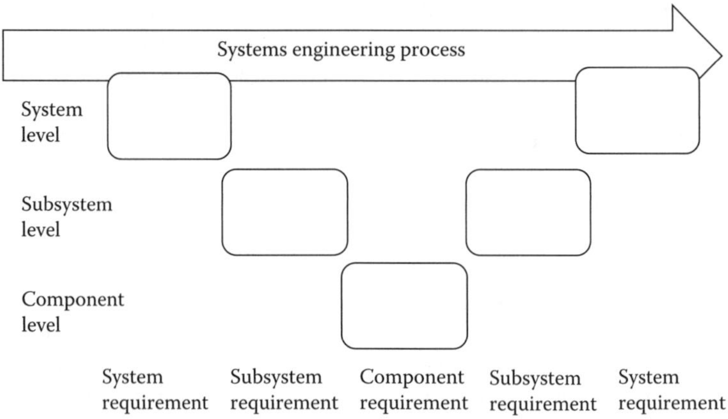

Figure 8.12 Top-down, bottom-up system design.

by refining the different design processes inside the entire development process. It considers all the elements, starting from the design and incorporating manufacturing and support, quality, cost, schedule, and customer needs.

The concept of CE is not new. According to Ziemke and Spann (1993), CE was commonly used in the development of the US weapon and transportation arsenal in the World War II era. After this period, many US and Western producers forgot this good way of engineering as corporations grew, products became more complex, and greater specialization of the work force took place. Its redevelopment began in the 1980s with the coining of the term "computer supported cooperative work" by Grief and Cashman in 1984 (Shen et al. 2008). Shortly thereafter, the concept of CE was developed and defined as "a systematic approach to the integrated, concurrent design of products and their related processes, including manufacture and support" (Turino 1992; Shen et al. 2008). In particular, CE grew out of the American automobile industry's efforts to emulate the Japanese approach to product development. According to Hartley (1992), many Japanese companies have been using the basic elements of CE successfully for over 30 years. They did not call it CE, but the success of Japanese manufacturing, particularly in the automotive industry, created strong competitive pressure on American and European automobile manufacturers.

One of the key early contributions of the field was to bring the downstream considerations of manufacturing and product services upstream into the preliminary design consideration process (Haskins et al. 2010). Therefore, CE primarily emphasizes two traits: a multidisciplinary approach to the overall design process, and the use of technologies and processes that build communication across organizational teams involved in the design process.

CE is also an engineering management approach and a business strategy which enables the integrated development of products and processes with the goal of completing the entire cycle in a shorter time (Smith and Reinertsen 1995). CE which is an effective concept to guide product development is a fundamental improvement to traditional product development approach. CE is a nonlinear product or project design approach during which all phases of manufacturing operate at the same time. Both product and process design run in parallel and occur in the same time frame. It removes the need to have multiple design reworks, by creating an environment for designing a product right the first time round.

The CE approach is based on five key elements: a process, a multidisciplinary team, an integrated design model, a facility, and a software infrastructure (Bandecchi et al. 1999). The design process is carried out in a series of sessions in which all designers and specialists take part.

It is an iterative process that addresses all aspects of the system design quickly and completely. A fundamental part of the CE approach is to create a highly motivated, interdisciplinary team that performs the design work in real time. The design process is "model-driven" using information derived from the collection and integration of the tools used by each specialist. Software tools for the generation of the model, documentation, and storage are the infrastructure to implement CE.

The core of CE is a process of task integration. It is more focused on the synergy of time. By using the right technology and techniques, CE can improve the processes of product development, adopt new quality concepts to meet the changing needs of users, and use new computer-aided tools (e.g., quality function deployment; design for manufacture and assembly; failure modes and effects analysis; DfM and DfS; and Pugh concept selection matrix to ensure early problem discovery, early decision-making, work structuring, teamwork affinity, and knowledge leveraging). Figure 8.13 shows the seven Ts that influence the CE.

CE is the ability to implement parallel design and analysis in which safety, manufacturability, serviceability, marketability, and compliance matters are reflected early on and during the process. The implementation of CE addresses three main areas: people, process, and technology. It involves major organizational changes because it requires the integration of people, business methods, and technology and is dependent on cross-functional working and teamwork rather than the traditional hierarchical organization. Collaboration rather than individual effort is required and information sharing is the key to success.

In the typical (serial) engineering approach, finishing all the physical manufacture of a prototype is a requirement before realizing any test, but CE allows the ability to implement parallel design and analysis at the same time before the real unfolding. Safety, manufacturability, serviceability, marketability, and compliance issues are considered early on and during the process. This interdisciplinary approach emphasizes the use of cross-functional equipment and allows employees to work in all the

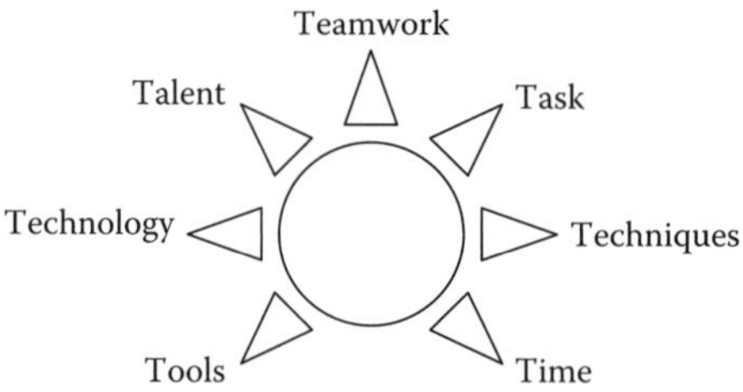

Figure 8.13 The seven "Ts" that influence CE.

aspects of a project from the beginning. CE is possible through the application of modern CAD, DfM, DfA, DfS, DfR, and DfE software.

An important aspect to the realization of the CE process is to assemble a capable team to carry out the task. All disciplines that are involved in hardware and software configurations should be engaged. Typical engineering disciplines that may be represented on a CE team are system hardware or software design, operations, manufacturing and assembly, machining, and safety. As shown in Figure 8.14, consideration of each of these disciplines, coupled with the user requirements, is important to a successful integrated design. To do this, all configurations under consideration must be communicated to all team members. This is where computer-aided modeling and simulations, VR, and graphical CAE analysis techniques come into action.

8.7.3 The V-cycle development model

In terms of management processes, there are various methods within SE to guide the design life cycle, including the waterfall, spiral, and Agile models (Haskins et al. 2010). The waterfall model is a sequential development model where requirement should be clear before going to the next phase of design and each phase of development proceeds in order without any overlapping. Agile stands for moving quickly. However, the emergence of all above design life cycle management processes is the design V (validation and verification)-model, which represents the different phases

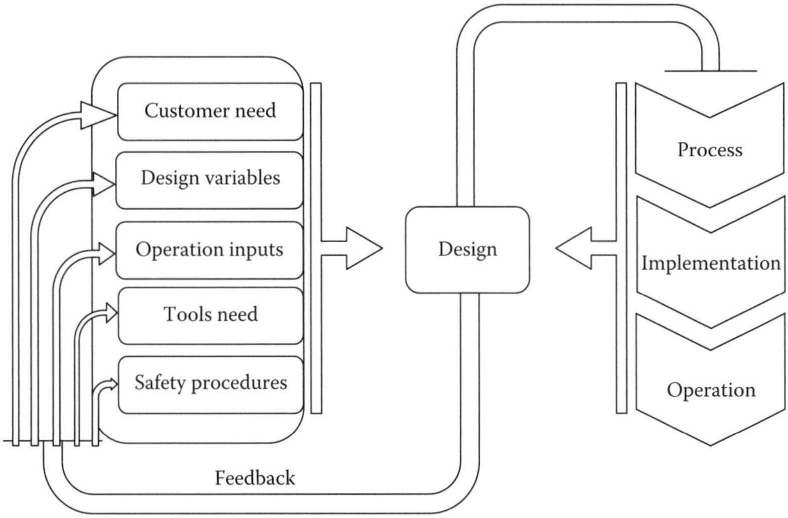

Figure 8.14 CE integrated design.

in design and the coordination activities that should occur across them at each step (Haskins et al. 2010). Verification is the process of proving that each product meets its specification. Validation is the process of demonstrating that the product satisfies the user needs, regardless of what the system specification requires. Figure 8.15 shows an overview project development within the V-model.

The V-model is an adapted version of the common waterfall method. It involves a sequential evolution of plans, specifications, and products that are baselined and put under configuration management. The V-model is a sequential path of implementation of processes. Each phase must be completed before the next phase begins. Product testing is organized in parallel with the corresponding phase of development. Unlike the waterfall method, there is no prohibition against doing detailed work early in the cycle. In the initial stage, hardware and software feasibility models are required.

The V-model represents a sequence of steps in a project life cycle development. It describes the activities to be performed and the results

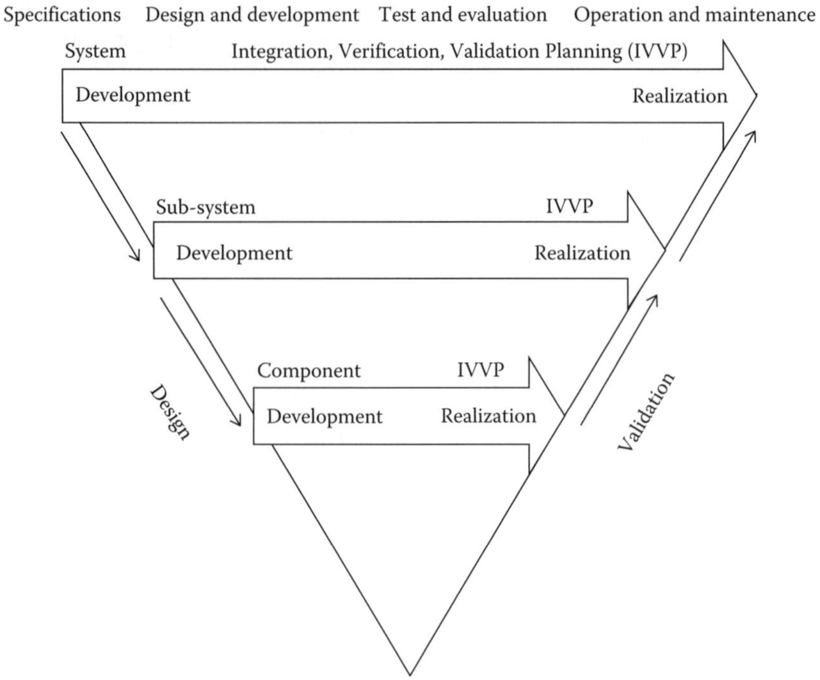

Figure 8.15 Design V-model for project development. (From Haskins, C. et al. *INCOSE Systems Engineering Handbook: A Guide for System Life Cycle Processes* and *Activities*, Vol. 3.2, INCOSE-TP-2003-002-03.2, International Council on Systems Engineering, San Diego, CA, 2010.)

generated during product development. The left side of the V-model represents the decomposition of requirements and creation of system specifications. The right side represents the integration of parts and their validation. This developmental process is balanced and depends on the verification from the previous steps before moving forward. At each level, there is a direct correspondence between activities on the both sides of the model. At the bottom level on the V-model, the tasks break into three parallel efforts: operations, hardware development, and software development. The systems engineer in charge must be capable to direct significant skills between these different activities.

Beyond design life cycle models, there are various processes and associated tools to support the design of large-scale complex systems including team management (organizational structure), product structure management, workflow and process management, design change management, visualization-based collaborative workspace (emphasizing visualization decision-support tools), and integration interface management (Shen et al. 2008).

Finally, the V-model should be used for small- to medium-sized projects where requirements are clearly defined and fixed. It should be chosen when sufficient technical resources are available with needed technical expertise. In the V-model, developer and tester work in parallel. Moreover, the model provides guidance for the planning and realization of projects.

8.7.4 MBD methodology

MBD is a development workflow based on system model with different levels of abstraction as illustrated in Figure 8.16. MBD provides an efficient approach for establishing a general structure for communication throughout the design process while supporting the V-cycle model. This simulation-based approach offers a better understanding of design options and trade-offs than traditional hardware prototype-based design techniques, enabling designers to optimize their design to meet predefined performance criteria.

Currently, educators are adopting project-based learning (PBL) to engage their students more in the learning process. MBD is well suited for PBL. Thanks to MATLAB®/Simulink®'s automatic code generation capabilities and support for low-cost hardware connectivity (such as Lego, Arduino, Raspberry Pi, and Beagle Board support packages), where students can influence the advantages of simulation and are better introduced to the implementation phases.

MBD allows various system solutions to be assessed, without significant time-consuming and often costly investment in physical system prototyping. In particular, it provides designers of complex motion systems like mechatronic systems an innovative and cost-effective methodology

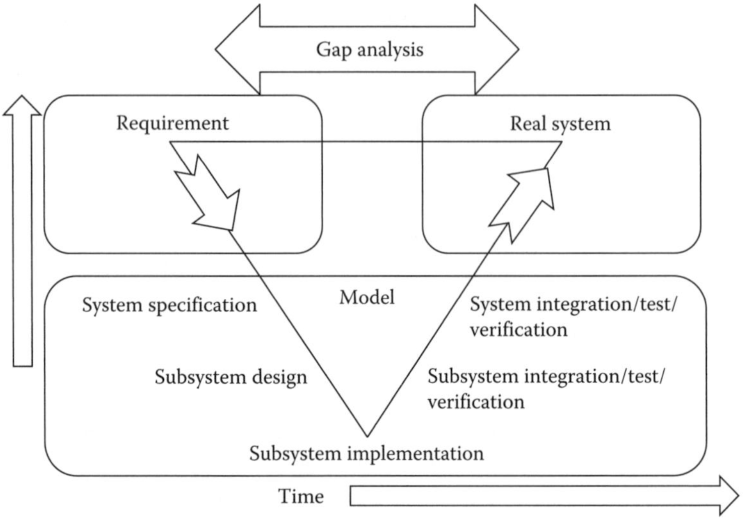

Figure 8.16 MBD scenario.

for movement control. Precise modeling allows the systems designer to choose components that will best support the desired application. The iterative process of modeling and testing must be repeated until the outputs of model and system match the parameters of interest.

8.7.5 DD approach

DD is a process that seeks to minimize unknowns, and to map out a solution path and implementation plan. It has its origins in precision engineering (attention to detail and requirement of wide variety of knowledge including measurement, fabrication, and control issues) (Slocum 1995) and axiomatics (Suh 1990), but it uses a more relaxed format that encourages designers to feel free to think hair-raising unstructured thoughts, which preserves the fun of design (Graham et al. 2007). The key to DD is the funneling of creativity by means of continuous risk assessment and systematic collection, creation and analysis of design information. It significantly reduces risk and redundancy, resulting in simple, high-performance, cost-efficient, and reliable products.

DD merges qualities of the scientific method with the business focus of risk assessment and countermeasures, and schedules. The procedure of idea generation and selection that is used to produce strategies is repeated again, at a different level. Idea development is a sequence of three stages: strategies, concepts, and modules. At each step of creating (strategy, concept, and modules), a deterministic process occurs. Individuals create (and write down their ideas), peer-review evaluation process (PREP), and then brainstorm. The so-called scientific method generally includes the typical steps shown in Figure 8.17. The initial step in any design process is to identify the functional requirements. To meet these requirements, a set of possible strategies is generated. At this stage, there is room for creative thinking and one may come up with many ideas. The next step is to develop concepts to implement it. Freethinking allows for the generation of many ideas, and based on the previously stated design principles, it is possible to make a deterministic choice of the best concept. Once a concept has been decided, modules that are the building blocks of a given concept will be generated.

Each stage of development occurs in three phases that make up PREP as shown in Figure 8.18. The key to deterministic design is the funneling of creativity by means of continuous risk assessment and systematic collection, creation and analysis of design information. Individual thought constitutes the first phase during each of the three stages of developing ideas. During the second phase in developing ideas, a peer-review process is employed, where (N) people circulate their milestone reports to the other ($N - 1$) people for comments. Brainstorming is the third phase, which helps teams solve personal creativity deadlocks and helps to ensure nothing has been overlooked (Graham and Slocum 2005).

During the individual thought phase individuals are required to not only think wild and free, but to address essential issues. Individuals independently develop ideas for the stage of development, and team members gather around a table and pass their ideas to their neighbor to be silently reviewed. The process continues until each individual has his or her work reviewed by every other team member, and then team members discuss the ideas they have reviewed to synthesize ideas into

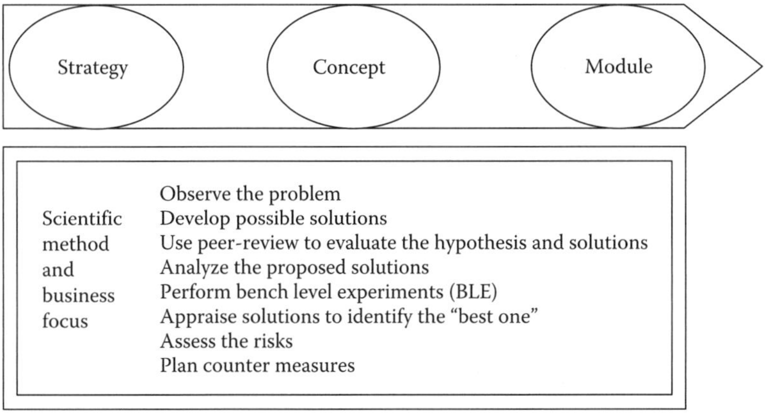

Figure 8.17 Stages of idea development with a platform of scientific method and business focus.

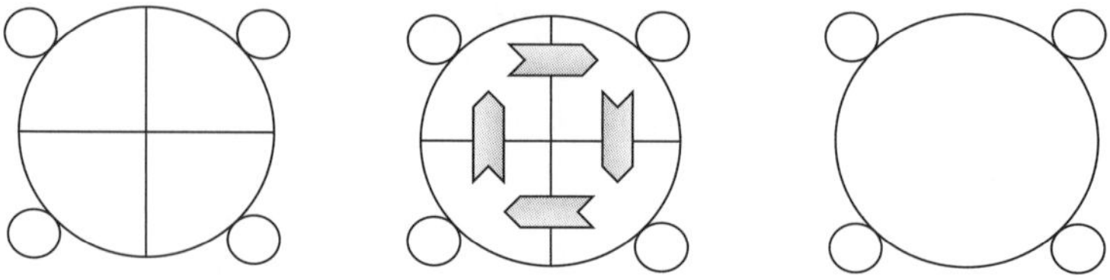

Figure 8.18 PREP.

the "best" idea. During discussion, weighted selection occurs to determine the top idea, while also identifying strong points from all ideas and how the most favorable characteristics can be incorporated in the top idea (Pugh 1996).

Weighting is decided based on factors such as objectives, customer requirements, and interests of stakeholders. PREP is thus especially useful for diverse teams of designers with members from various cultures, races, genders, physical disabilities, and personalities because it first empowers people to contribute without fear of confrontation that often occurs in group brainstorming sessions. Furthermore, it provides a written record of who first came up with a concept, which can be valuable for assessing promotions or for inventorship (Graham et al. 2007).

8.8 Standards and codes in engineering design

Standards are documents that describe the important features of a product, service, or system. Design standard encapsulates what has become accepted best practice for the design of particular types of product. Almost all the engineering products are made by meeting the requirements of a series of guidelines including standards, codes, specifications, and technical regulations (Table 8.2). Also, all the engineering products implicitly use components prepared based on these guidelines and/or explicitly designed by considering one or more of the recognized guidelines. For example, ISO 5840 is applicable to cardiovascular implants. These implants can be used in medical setting only after meeting the various requirements of the standard.

The purpose of developing and adhering to standards is to ensure minimum performance, to assure safe operation of systems, to reduce cost by allowing manufacturers use standard parts, to make sure that the product/system/process is consistent and repeatable, to simplify maintenance and repair, and to provide for interfacing with other standard-compliant equipment.

Table 8.2 Brief description of standardization guidelines

Standard	Document approved by a recognized body that provides, for common and repeated use, rules, guidelines, or characteristics for products or related processes and production methods, with which compliance is not mandatory. It may also deal exclusively with terminology, symbols, packaging, marking, or labeling requirements as they apply to a product, process, or production method.
Code	Any set of standards set forth and enforced by a local government agency for the protection of public safety, health, etc. as in the structural safety of buildings (building code), health requirements for plumbing, ventilation, etc. (sanitary or health code), and the specifications for fire escapes or exits (fire code).
Specification	An explicit set of requirements to be satisfied by a material, design, product, or service. Standards may be referenced or included in specifications.
Technical regulation	Document which lays down product characteristics or their related processes and production methods, including the applicable administrative provisions, with which compliance is mandatory. It may also include terminology, symbols, packaging, marking, or labeling requirements as they apply to a product, process, or production method.

Standardization is concerned with the use of common components, products, or processes to satisfy heterogeneous needs. It necessitates designing an overly robust product or the use of a robust process. Tarondeau (1998) argues that standardization results in higher productivity, larger lot sizes, decrease in the number of reference points to be managed, decrease in the stock level, and the reduction of complexity of a manufacturing system.

Relevant standards should be consulted early in the design process and since it is reasonable to assume many companies will want their products to be accepted in the global market. Standardizing design work also involves defining and implementing "best practices" for each design environment. Standardizing does not suppress speed or creativity.

8.9 Human factors engineering

> Design is not just what it looks like and feels like.
> Design is how it works.
>
> **Steve Jobs**

HFE, also known as comfort design, functional design, ergonomics, or human engineering, is the practice of designing products, systems, or processes to take proper account of the interaction between them and the people who use or operate them. HFE is used to designate equally a body of knowledge, process, and profession. As a body of knowledge, HFE is a gathering of data and principles about human characteristics, capabilities, and limitations in relation to machines, jobs, and environments. As a process, it refers to the design of machines, work methods, and environments to take into account the safety, comfort, and productiveness of human users and operators. As a profession, it includes a range of scientists and engineers from several disciplines that are concerned with individuals and small groups at work. The field has seen contributions from numerous disciplines, such as psychology, engineering, biomechanics, industrial design, physiology, and anthropometry. In essence, it is the study of designing equipment and devices that fit the human body and its cognitive abilities. The discipline contributes to the design and evaluation of organizations, tasks, jobs and equipment, environments, and products and systems. It focuses on the inherent characteristics, needs, abilities, and limitations of people and the development of sustainable and safe working cultures (OGP 2011).

Two general premises characterize the approach of the HFE in practical design work. The first is that the engineer must solve the problems of integrating humans into machine systems by rigorous scientific methods and not rely on logic, intuition, or common sense. In the past, the typical engineer tended either to ignore the complex and unpredictable nature of human behavior or to deal with it summarily with educated guesses. Human-factors engineers have tried to show that with appropriate techniques it is possible to identify human–machine mismatches and that it is usually possible to find workable solutions to these mismatches through the use of methods developed in the behavioral sciences (Holstein and Chapanis 2016).

HFE is a "sociotechnical" approach to systems design. It recognizes that any complex technological system that involves people is critically dependent on the organizational and social context in which it operates. HFE is a multidisciplinary approach to engineering that focuses on the integration of several elements as illustrated in Figure 8.19. A focus on the integration between the above elements is the unique and often critical perspective that HFE brings to the development of sociotechnical systems.

Applications of HFE have been made to simple devices such as highway signs, telephone sets, hand tools, and stoves, and to a host of modern, sophisticated complexes such as data processing systems, automated factories and warehouses, robots, and space vehicles. The modern push-button telephone handset gives a good example of a simple device that has required a great deal of HFE. The layout of the keys in the four rows

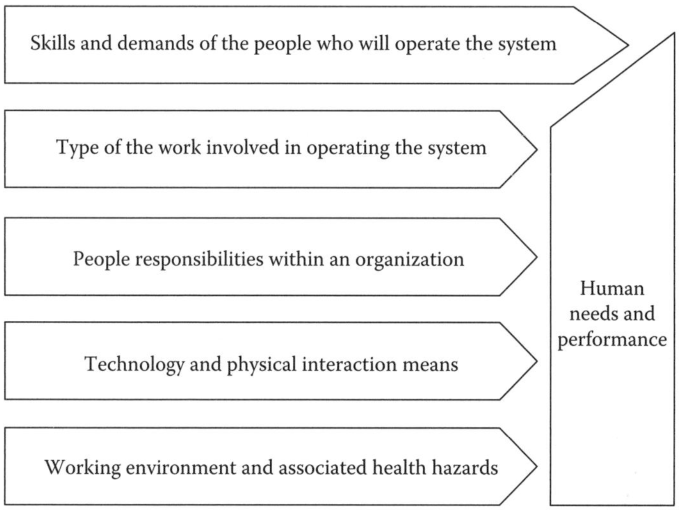

Figure 8.19 Scope of HFE.

of three buttons, for example, was selected only after extensive tests on a variety of arrangements: circular, two vertical rows of five buttons, two horizontal rows of five, and a diagonal pattern; the arrangement of the numerals and letters on the keys, in the order of left to right and from top to bottom, was chosen as superior to other arrangements such as that used on many desk calculators, in which the numbers increase from bottom to top (Holstein and Chapanis 2016).

8.10 HoM for modern engineering design

Where there is a will there is a way.

Proverb

Engineering HoM allows engineers to regularly find solutions to problems or improvements to ways of doing things. Modern engineering practice is dependent on the development of a mind-set and skills necessary to transcend disciplinary limitations in solving problems.

8.10.1 Systems thinking

Systems thinking (ST) means seeing whole, systems and parts, and how they connect, by examining the linkages and interactions, pattern sniffing, and recognizing interdependencies. It is a framework for seeing interrelationships and repeated events rather than things. ST enables people to

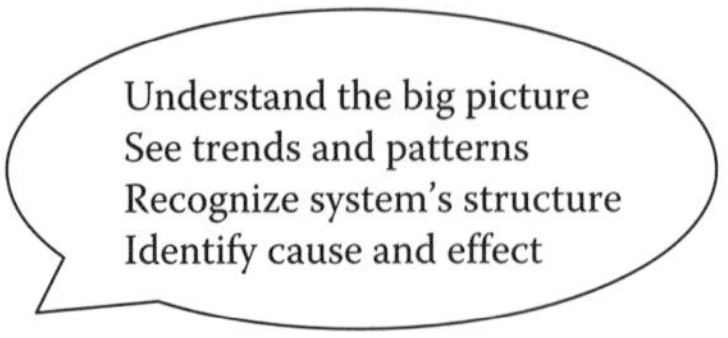

Figure 8.20 Numerous habits used by systems thinkers in solving problems.

look at problems in new ways which leads to new solutions. Figure 8.20 shows numerous habits used by systems thinkers in solving problems.

ST is an approach which facilitates the integration of people, purpose, process, and performance because it is a framework for seeing and working with the whole(s), rather than only the individual part, and for seeing the interrelationships between parts (Senge et al. 2008). Engineering leaders in industries therefore need to develop themselves in succession plans on how to be better systems thinkers.

8.10.2 Creativity

Creativity is an integral part of the engineering design process, its presence often being the major influence on the impact of a product. It is the ability to use novel approaches for generating, investigating, and representing ideas. Creativity is a quality that is highly valued, but not always well appreciated. Thus, creativity is the ability to see connections and relationships and to think in intuitive, nonverbal, and visual terms. The creative process is very similar in all fields including engineering. The creative work usually builds new relationships and new situations.

Design is a planned activity that requires creative thinking in response to difficult circumstances. Essentially, the design process is a problem-solving process, and the designer, just like the laboratory scientist, will be most successful if the problem is approached in a systematic manner.

Within industry, creativity does not necessarily equate to success; however, without some form of innovative NPD, the long-term failure is a certainty. In order for firms to increase their organizational creativity, therefore resulting in enhanced innovation, the creative process of individuals must be considered within the design process (Bharadwaj and Menon 2000).

8.10.3 Optimism

Optimism comes from the Latin word *optimus*, meaning "best," which describes how an optimistic person is always looking for the best in any situation and expecting good things to happen. Optimism is the tendency to believe, expect, or hope that things will turn out well (Beattie 2017). It is a trait that should become more common, judging by Winston Churchill's famous quote that "a pessimist sees the difficulty in every opportunity; an optimist sees the opportunity in every difficulty." Engineers, as a general rule, believe that things can always be realized and improved. Just because it has not been done yet, it does not mean it cannot be done. Good ideas can come from anywhere and engineering is based on the premise that everyone is capable of designing something new or different.

8.10.4 Collaboration

The dictionary defines collaboration as working together toward a shared goal (especially an intellectual endeavor). Engineering successes are built through collaboration and communication. The complexity of current engineering design demands the collaboration of specialists. Collaboration involves both communication between and coordination among members of a design team. Currently, design collaboration is carried out through the use of schedules, specifications, and drawings which only capture the end results of the design process (Favela et al. 1993).

For students working on design projects, teamwork is essential and collaboration can bring about solutions to problems that no individual working alone could ever have imagined. In workplace, the best engineers are willing to work with others. They are skilled at listening to stakeholders, thinking independently, and then sharing ideas.

8.10.5 Ethical criteria

Ethical considerations, reflections, and actions can make a difference in whether a particular innovation, design, or research finding will have a helpful or harmful impact on society. In education, the mission of a plan in ethics and global awareness can be summarized in the following four points (Stephan 2004): develop student awareness of the ethical implications of their work, foster in students an understanding of the ethical standards in their professions, improve students' judgment of ethical conduct, and encourage students to put their knowledge of ethics into action. Ethical aspects can be of significance in design processes. In particular, it shows two ways in which ethical issues may emerge in engineering design processes, for example, first in formulating requirements, specifications, and design criteria and second in assessing trade-offs between criteria and in making decisions on what constitute acceptable trade-offs (Van de Poel 2001).

8.10.6 Social criteria

Engineering design, as practiced by engineers in the workplace, is a highly social, highly iterative process where, often, no single right answer exists. Engineering relies more on social involvement, cooperation, and collaboration to get its work done than is the case in many other fields. Dym (1994) described the context in which engineering design occurs as necessarily open ended, suggesting a plethora of acceptable (though not optimal) solutions, and ill structured, indicating that solutions cannot be found routinely by simply applying mathematical formulas in a structured way. Similarly, in their analysis of engineering design, Jonassen et al. (2006) described the inherently ill-defined context in which engineering design problems are embedded, alluding to the vagueness of goals, implicit constraints, and availability of multiple solutions and paths to reach solutions.

Engineering design criteria expanded again when economics began to explicitly affect design decisions in the late nineteenth century (Lesser 1945). Economic criteria are now an integral part of modern engineering design and good contemporary design is not just the "technically best," but the best at a given cost. For example, a relatively straightforward application of engineering techniques allows an engineer to specify the wall thickness needed in a pipe to withstand a given internal pressure. But the ready availability of low-cost piping in standard stock sizes provides the engineer with the choice between low-cost piping that is stronger than it needs to be and piping of exactly the required strength but which is more expensive since it must be custom-made.

This development of engineering design criteria from purely technical considerations to include economics did not require any slighting of the technical criteria. The latter still had to be satisfied, but engineers had to acquire new skills in cost analysis in order to generate creative engineering designs that met these broader criteria. Thus, it is not unusual for new design criteria to evolve from changes in the social environment within which engineers operate. In the past several decades, vastly increased public sensitivity to the impacts of technology has fostered the concept that engineering should consider societal concerns as part of the design process.

To develop a broad knowledge of engineering, students must also understand the critical role engineering plays in society: that is, it has a long history, is relevant to today's problems and opportunities, and is influenced by cultures and societies. Important to this is that engineers generate technological solutions that are intended to solve society's problems, but that may sometimes have negative consequences.

8.11 *The design entrepreneur*

> Effective design is a combination of innovation, integrity, co-creation and hassle-free offerings for customers.
>
> **Anuvat Chalermchai**
> *COTTO*

Innovation, technology, and commercialization are central to offer new products or services that meet customer needs. Innovation is the act of creating something new and worthwhile and entrepreneurship is the act of carrying an innovation to market in a commercial manner. Innovation and innovative design play a major role within entrepreneurship, but it is upstream from the commercialization stage, which is a major component of entrepreneurship (IPENZ 2002).

8.11.1 *Design entrepreneurship*

Design entrepreneurship began more than a century ago. Today, it is the answer to the question: What is next (Heller and Talarico 2016). The conceptualization, production, and marketing of a design idea are a critical case for a company's growth and survival. Simultaneously, design is a key strategic activity in many firms because new products contribute incessantly destroying the old one and define new competencies and qualifications in the market place. The amount of contribution of a designer in a value creation process has been a controversial issue for a long time; on the other hand, there is a very large pool of empirical studies promoting design and its result as added value in the literature (Gunes 2012).

Design entrepreneurship is about creating business and new opportunities by the help of design. It is a natural outgrowth of the typical design practice, yet it is not limited by creating viable concepts but by marketing their intellectual rights (Heller and Talarico 2008). It means that to motivate industrial design activity to be more entrepreneurial for to take a product from concept to market which require giving the designers crucial and extra insights about the total product development process. Design entrepreneurship is the collection of correct skills and abilities to develop the right ideas and market them as the successful design products (Gunes 2012).

8.11.2 *Innovation, entrepreneurship, and design*

In the context of entrepreneurship and innovation, design processes tend to be as much emerging as deliberate in nature (Hargadon and Douglas

2001). Innovation and entrepreneurship are essential components of the skill set that engineering graduates entering the modern competitive worldwide workplace must possess. To be entrepreneurs, among the substances necessary that individuals have to possess are entrepreneurial skills. Thus, the early development of potential entrepreneurs through entrepreneurship education should be conducted to coach students with this substance. Entrepreneurship education teaches engineering students in all disciplines the knowledge and attitudes that are required to identify opportunities and bring them to life.

Figure 8.21 demonstrates the interrelationship between innovation, engineering design, and entrepreneurship. It is cyclical because innovations need to be designed and engineered to be commercially viable. Companies often retain a competitive advantage when they make profit and reinvest a proportion of their profit to continue being innovative. As an example, see the case of Apple, Inc.—first established the iMac, then the iPod, the iPhone, and then the iPad. The "i" brand continued with the design path. This example demonstrates that innovation needs suitable design for the purpose of commercialization.

This intrinsic motivation to accumulate knowledge in a specific field of interest makes entrepreneurs knowledge agents for innovation in their field. Consequently, students who desire to become entrepreneurs have to learn how to accumulate the necessary knowledge in a specific field so to become knowledge agents for innovation. The field of entrepreneurship education has been divided as to whether entrepreneurship can be taught or not. Those who favor it as an independent academic discipline see it as a distinctive, if not unique, component of the free enterprise system. A second consideration is that entrepreneurship contains specific knowledge, concepts, and theories that apply in a reasonable manner across the discipline.

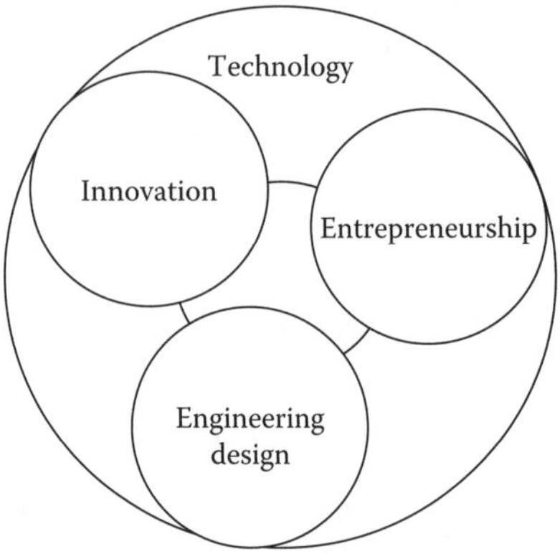

Figure 8.21 Interrelationship between innovation, its design, and entrepreneurship.

One of the prime concerns that technologists have is that their innovation is a means of increasing the quality of life—the ease, comfort, safety and, in many cases, the enjoyment of life. A number of key figures have made contributions that may rightfully claim to have changed the world. Usually for the better, although at times, there were those who experienced unexpected negative consequences, too. The Swede Alfred Nobel in the 1860s, for example, devised an explosive that was safe to handle for the purpose of better exploiting minerals, such as coal to fuel the Industrial Revolution. And then by the 1880s, he stood shocked in his later years to discover that his innovation, a dry and portable explosive which he called "dynamite," could equally be deployed to advance nations politically through inflicting death and war. He became reclusive in the last decade of his life and left his huge fortune to his household servants and the larger portion to a prize in his name that would celebrate scientific achievement—specifically medicine, physics, chemistry, literature, and ostensibly peace—and later economics (Pech 2015).

8.11.3 Engineer, entrepreneur, and design entrepreneur

The designer has the human instinct that every problem could be transformed to a design problem which may enclose a design revolution solution (Pilliton 2009). A great number of design solutions have changed the world in different areas such as water, well-being, energy, education, mobility, playing, enterprises, etc. Perhaps, the most basic differentiator between the perceived role of the designer and the engineer is that though both work to a brief, the designer prefers a loosely defined brief and will often even attempt to transcend it, whereas the engineer prefers tightly controlled parameters to design.

Innovative designs owe their origins to many sources—the chance intersection of knowledge and circumstances, necessity, dictate, genius, and unique childhood experiences (Shavinina 2003).

But innovations may also come from methods that are learnable (Liu and Boyle 2004; Liu and Schönwetter 2004). As a learnable process, innovative engineering design fits within a larger framework of engineering design.

Engineers, entrepreneurs, and design entrepreneurs share different traits. In broad terms, engineers realize ideas and utilize design elements and tools to create an identity that differentiates a product, service, or entity from its competition. They then communicate that entity through models to manufacturers. By contrast, entrepreneurs generate and execute creative business ideas. An entrepreneur is one who is willing to taking a risk and turns an idea into a profitable innovation. The common trait that engineers and entrepreneurs possess is that both are problem solvers. Figure 8.22 shows various traits of engineers, entrepreneurs, and design entrepreneurs.

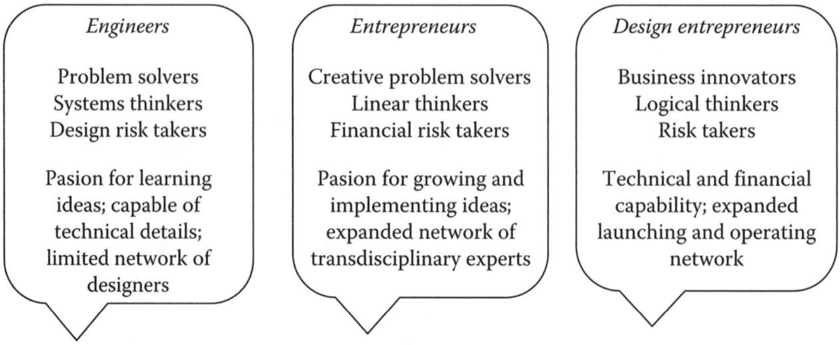

Figure 8.22 Traits of engineers, entrepreneurs, and design entrepreneurs.

There are several traits that engineers need to develop to become successful entrepreneurs. In an academic teaching system that is still predominately knowledge based (linear thinking), many nondesign educational programs actually eradicate creativity (nonlinear thinking). Students are afraid of failure because they have been taught from a young age that there is a right and wrong answer (Trummer and Lieras 2012). Design education differs in that students are taught to live with ambiguity and to navigate complexity, as well as to routinely combine creative and analytical thinking. Even though these two paths of education are significantly different, design education should capitalize and provide supplemental business education for their students along with opportunities to experience, apply, and develop these skills in collaboration with other disciplines (Anderson 2014). By doing this, engineers will be able to combine the languages of business and DT and formulate the most logical and creative ways to become successful design entrepreneurs.

The transition from engineers to successful entrepreneur also requires engineers to work in a more transdisciplinary way to form partnerships with other professionals, entrepreneurs, and specialists from a diverse set of backgrounds. When engineers know how to build wide and diverse networks, it increases their visibility, stimulates their business idea, links them to potential investors, finds them employees, and connects them to potential customers. Figure 8.23 shows the conceptual view of role of future engineers in industry. Virtual innovation hub includes engineering entrepreneurs and other critical entrepreneurial functions, for example, marketing, finance, and project management. Those hubs are responsible for feeding innovations into industry, who then secure the best virtual design/manufacturing/service factories to produce and deliver required products and services to customers.

Engineers need business education and resources to become entrepreneurs. They must be self-sufficient, self-sustaining lifelong learners responsible for expanding their knowledge, experience, and skills as

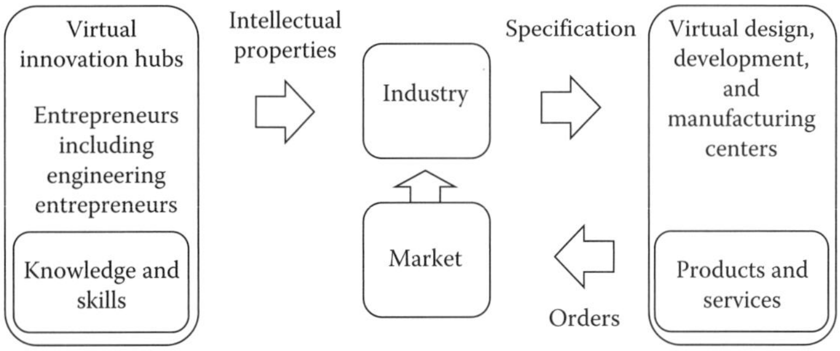

Figure 8.23 Conceptual view of role of future engineers in industry.

entrepreneurial engineers. Engineers must be able to effectively apply entrepreneurial skills as they join teams. The only outstanding question that remains is when should an engineer be taught these skills? For sure, it would be more beneficial for students to learn these studies earlier in their education.

8.11.4 Design practice, thinking, and leadership

Leadership has long been identified as a key driver for innovation and design. The role of the design leader is to raise awareness of the design process and the positive effects design generates (Topalian 2012) such as competitive advantage, build and improve image, better return on investment, and customer satisfaction (Lockwood 2008). Gloppen (2009) states that the strategic function of design leadership is related to the vision for how design might be used within the organization to reach the collective goals.

According to Alnelind and Alvén (2014), design leadership is when a person at a leading position having enough design knowledge and skills to lead, motivate, and support designers in their process of developing new products for the organization. They also have to have skills in motivating coworkers to be part of the design process within the organization and contribute creatively, as well as establish design-related visions and strategic direction for the organization.

DT is a recent concept entering the literature with much promise. The notion of "design" as a "way of thinking" was introduced in 1969 (Simon 1996), and adopted for a business process in the 1990s (Richard 1992). DT is described as a successful tool for business seeking innovation, exploring greater synergies between business strategies and product and service innovation (Martin 2011).

The process of DT consists of five steps, namely, "empathize," "define," "ideate," "prototype," and "test." Communications with end users are given paramount importance in DT, since observations of the responses of individuals yield valuable clues for new products and services. In the "empathize" process, an ethnographic approach based on a field work is adopted in order to derive insights into potential needs. In the "test" process, simple prototypes are utilized to verify business hypothesis through communications with potential users. It is rational to contact with customers in the early stage of business development since ideas born in meeting room or laboratory do not often meet the true needs of customers. A similar principle is observed in famous "build, measure, and learn" cycle in lean start-up (Ries 2011; Suzuki 2016).

In a design environment, designers ask questions such as: How do we make something beautiful and usable? How does it mechanically go together? How do we reflect the brand? Leaders ask questions such as: What are we trying to achieve and why? How do we accomplish our goals? What people and resources do we need to make it happen? By weaving together the leadership process with the design process, Horth and Buchner (2016) have identified six innovative thinking skills to create a healthy environment for innovation and design as shown in Table 8.3. Using these skills, organizations are able to create something that is useful and desirable—whether it is a breakthrough technology, a valuable service, or a fresh solution to an old problem. Each of these skills shifts understanding of a situation and opens the door for new approaches and solutions.

Table 8.3 Innovative thinking skills to create a healthy environment for innovation and design

Thinking skills	Description
Paying attention	Paying attention is the ability to notice what has gone unnoticed. It is about looking more deeply at a situation, being a clear-eyed observer, perceiving details, seeing new patterns, and grasping the situation.
Personalizing	For innovative thinking, personalizing is a twofold process: tapping into own broad scope of knowledge and experience, and understanding customer in a deep, personal way.
Imaging	Imagery is a very good way to take it in and make sense of it. Pictures, stories, impressions, and metaphors are powerful tools for describing situations, constructing ideas, and communicating effectively.
Serious play	Innovation requires bending some rules, branching out, and having some fun.
Collaborative inquiry	Innovations are rarely made by a "lone genius." Insights come through thoughtful, nonjudgmental sharing of ideas.
Crafting	Innovation requires thinking and seeing the whole as inclusive of opposition and open to a third (or fourth or fifth) solution.

Successful companies, such as Tesla, Apple, and Google, innovate continuously, enabled by design practice, design culture, and design leadership. These firms connect with customer needs and emotions to create exciting products and services through collaborative team design and development processes.

8.12 *The "what" of learning in design*

> No pleasure, no learning. No learning, no pleasure.
>
> **Wang Ken**
> *Song of Joy*

8.12.1 *The challenge of teaching as design science*

> Everyone designs who devises courses of action aimed at changing existing situations into desired ones.
>
> **Herbert Simon**

Teaching is changing. It is no longer simply about passing on knowledge to the next generation. Teachers in the twenty-first century, in all educational sectors, have to cope with an ever-changing cultural and technological environment. Teaching is now a design science. Like other design professionals such as architects, engineers, and programmers, teachers have to work out creative and evidence-based ways of improving what they do (Laurilland 2012). To develop an effective pedagogical pattern, the teacher-designer should have a good knowledge of different learning approaches, such as learning through acquisition, through inquiry, through discussion, through practice, and through collaboration (Rapanta 2014).

Unlike historical treatment of engineering as an intuitive blend of design, art, and scientific analysis, engineering has increasingly been taught as an applied science since World War II. Engineering educators suggest that the creative or soft issues in engineering design (such as aesthetics and cultural appropriateness) have become "effete and marginal" and that engineers would rather be called scientists than artists or designers (Schon 1983, 1991; Ferguson 1992; Gelernter 1998). Engineering started swinging back toward a design emphasis in the early 1980s with efforts for innovative transformation in engineering education. It may be the right time to revisit the historical roots of engineering and realize that design is a fundamental aspect of engineering.

Implanting design knowledge (design science and design theory) throughout the curriculum, learning engineering skills and using exercises to build certainty, and learning skills to solve open-ended and ill-defined problems are few of the challenges of modern engineering education. The availability of online resources has transformed the learning environment for students, expanding access to knowledge, information, data, and people exponentially. Whether it is solving new problems or finding new solutions to old problems, design is at the core of what engineers really do. As such, design knowledge should be an aspect of teaching the fundamentals of engineering. A thorough sense should be innovated and new tools and methods should be exploited to create design-based engineering curriculum by being willing to rethink current teaching methods and reinforcing engineering fundamentals and other skills. Although such transition to design-integrated curricula is challenging, the potential for long-term advantage is enormous.

The knowledge needed to design, in education as in other fields, is actionable knowledge: that is, knowledge that is sufficient to inform action in the world. The key challenge is to make academic institutions more hospitable environments for design to build design capacity among all staff (lecturers, professional staff, and managers), and also to help students become more capable, self-managing participants in the processes that complete and enact designs for learning (Goodyear 2015).

Design phenomenology is the subfield of the scientific study of design that focuses on the nature of the products of the design process. This goes beyond the form and configuration of designed artifacts (Cross's original definition). Indeed, one of the goals of design phenomenology, when higher education is the application area, is to identify the scope of what can and should be designed.

Many design practices are oriented to the (eventual) creation of material products: simple or complex. In recent years, service design has become more widely practiced (e.g., in the design of public services) (Boland and Collopy 2004; Meroni and Sangiorgi 2011). Design for learning is a hybrid, involving mixtures of service, product, and space design. This hybridity is accompanied by a need for a more complex knowledge base for design than is sometimes found in discussions of knowledge for university teaching.

Another challenge comes from an industry perspective where Nicolai (1998) complains that there is a tendency for engineering courses to produce researchers and scientists, not engineers, who lack the ability to define a problem and develop solutions. Nicolai cites this as the primary engineering design ability. He also criticizes a shortage of realistic, open-ended problems, stating that industry places a higher value on engineering design than universities do.

The need to incorporate design into teaching is becoming more evident and most especially in blended learning higher education contexts in which instructors are asked to integrate new digital technologies in their existing teaching practice in order to achieve high-order learning goals. This book gives us an idea of what it takes to plan a teaching activity using any types of tools and resources in a way that ensures effective learning (Rapanta 2014).

8.12.2 *How to teach the "what" of engineering design?*

Design typically results in the creation of specifications of some kind, rather than directly in a finished product. It produces blueprints, plans, and sketches: inscriptions of various types that guide the creation of an imagined end product. Etymologically, design relates to both "making marks" (drawing) and "marking out" (designating; giving significance to) (Krippendorff 2006).

There are two primary reasons to include a design component in the majority of engineering science classes: design provides a necessary link from theoretical material to its application in the real world, and design presents a clear opportunity for achieving other student learning objectives beyond the development of analytical skills and engineering knowledge. If a goal is to have students graduate being both competent and comfortable applying precise engineering approaches to new technical problems, then it is needed to teach traditional material as tools for solving new problems. Learning requires emphasis, and traditional engineering science classes should introduce applications through design projects. Both PBL and experiential learning have positive outcomes for design learning. To achieve such goal, the following instructional approaches should be implemented in the process of learning.

8.12.2.1 *Knowledge transfer*

Transfer means taking knowledge and skills learned in one context and apply them in another. Teaching for transfer aims to increase transfer within a discipline which enables the students to transfer their knowledge to other areas in the future. Features that support transfer include engaging learners in challenging tasks, teaching using case studies, and priming student motivation (NRC 2012).

8.12.2.2 *Scientific design method*

The scientific design method (SDM) and the EDM (or engineering design process as described in Chapter 9) are typically helpful to guide a project, investigation, or innovation. The goal of the SDM is to help students understand the EDM in the context of a project or any other learning task. The SDM usually starts with asking a question which is equivalent to defining the problem in the EDM. This is followed by conducting background research for both SDM and EDM (Cowen 2013). Instead of developing a prototype as in the EDM, test with an experiment satisfies the need in the SDM. The data are then analyzed and a conclusion is drawn.

Using case studies as a way to assist, students develop necessary skills to meet the requirement of SDM. Setting goals for each case will aid in determining the scale and scope of the case that should be used. It is needed to reevaluate the tools used to support student learning such as lab exercises and textbooks. Laboratories usually reinforce basic concepts but by avoiding cookbook labs and include problems with multiple possible solutions.

Students are encouraged to keep up their level of understanding of the analysis techniques taught in the theoretical courses, and, in that sense, stimulate their motivation and interest in these courses (Daems et al. 2003). Along with helping students develop and learn to apply analytical skills, case studies present the opportunity to help students develop other necessary skills as well. They can motivate students by providing high-profile outlet for their technical creativity, developing business skills, and developing a better understanding of ethics and professionalism for life-long learning throughout their careers.

While working on cases, students should learn that using standards saves their time and provides them with a set of technically valid, industry-approved test procedures (Goldberg 2012). Students must be prepared for engineering practice through a curriculum culminating in a major design experience based on the knowledge and skills acquired in earlier course work and incorporating appropriate engineering standards and multiple realistic constraints (ABET 2013).

8.12.2.3 Online learning library

Effective communication sharpens student thinking and can uncover flaws in their designs. For students, communication is an essential component of the engineering design process. It is central to each of the steps of the process. Students must learn to document their work and to communicate effectively with one another as well as with their "client," sharing information about the progress that they are making at each step of the design process. This communication is often documented in their engineering design notebook and includes the design requirements, constraints, prioritization of the design goals, safety considerations, ideas from the brainstorming session, key mathematics and science concepts that influence the design, rationale for design decisions, detailed illustrations and specification of the prototype, design decisions or changes made while building the prototype, and data collected during testing (Ross and Bayles 2007).

Instructors may develop their own online learning libraries. Tools for constructing, analyzing, testing, and sharing such libraries are available from various resources including academic institution learning centers as well as from social and professional media. Internet hosting services such as slideshare.com and YouTube provide a framework for creating such libraries where Powerpoint slide, PDF file, or movie can be embedded and shared easily. The author has developed two web-based learning resources to showcase his student projects and activities: www.g9toengineering.com and www.greenengineers.ca.

8.12.3 Piloting engineering design

How can we create a lesson or a case study that uses engineering design? You may start in collaboration, working with a like-minded colleague who teaches the same subject or level. Also, you may collaborate with many teachers using a web support system like Maker Space (makerspace.com). Once collaborators have been identified, begin with the three interlinked areas of engineering design found in Appendix I of NGSS: define, design, and optimize (2013):

- *Question/define*: Attend to a broad range of considerations in criteria and constraints for problems of social and global significance.
- *Analyze/design*: Break a major problem into smaller problems that can be solved separately.
- *Optimize*: Prioritize criteria, consider trade-offs, and assess social and environmental impacts as a complex solution is tested and refined.

Designing a perfect engineering design unit or lesson may not be attainable. However, what we can do is use a particular design, test it in practice, and improve it as necessary. In other words, piloting a particular design activity will result in the need to improve it for the second iteration (much like engineering design practice). Also, it is unlikely that the unit a teacher designs and implements in his or her building will be identical to the one employed by other teachers. One's building, classroom, and community are unique, and so is the project one creates. The notion of creating a new unit from scratch can be daunting. Start with the "define" stage and move toward the "design" stage, both of which should be completed before the students are ready for the project. The "optimize" stage is implemented while the students are working on and completing their projects. The process of optimization is intended to improve the project and start over again, moving through the wheel (Turner Jr et al. 2016).

Experiential hands-on skills, communication skills, teamwork and leadership skills, project management skills, and creative problem-solving ability are all assets that students must have when they graduate and these can be developed through the use of design projects. Numerous pedagogic strategies can be used to integrate real examples into the classroom. These include teaching with case studies or with investigative cases.

Cases provide a context-rich opportunity for students to learn about real problems and to think critically about designing potential solutions to these problems. They constitute a kind of virtual apprenticeship, in which students can apply ethical principles to actual situations and discuss the outcomes with each other and with a faculty mentor. The case method is the best pedagogy to teach design, engineering, managerial, and engineering ethics (Gorman et al. 2000).

8.13 Interlinking case: Mechatronic system design

> A scientist discovers that which exists. An engineer creates that which never was.
>
> **Theodore von Karman**

This case considers steps for understanding the potential of adopting methodologies and approaches for product development as discussed in Section 8.6, with the three interlinked areas of engineering design described in Section 8.12.3. It reviews methods as applied to related technical systems and illustrates how these methods can be combined in a system framework to meet the research, design, and development needs.

8.13.1 Question/define: Piezoelectric energy harvesting system

Energy harvesting is the process of obtaining a small amount of energy from an existing environment. Mechanical energy is one of the most pervasive energies that can be reused in our surroundings. The sources of mechanical energy can be a shaking structure, a moving object, and vibration induced by flowing air or water. The energies related to movement by the flow of air and water are the wind and hydroelectric energy, respectively.

Mechanical waste energies can be harvested by using vibration-to-electricity conversion. Harvesters can be employed as battery rechargers in various environments, such as industries, houses, and military. The most distinguished characteristic of this kind of waste energy harvesting is initially identified for low power generations.

The word "piezoelectricity" is derived from the Greek word *piezein*, which means to "squeeze" or "press." Piezoelectric materials are considered as one of the existing smart materials due to their special characteristic of producing voltage upon being subjected to stress. In addition, this effect also works in the reverse direction, in which applying voltage across the material will produce stress within the material. This phenomenon is known as the piezoelectric effect.

The piezoelectric material converts mechanical strain into electric current or voltage. It is based on the fundamental structure of a crystal lattice. Certain crystalline structures have a charge balance with negative and positive polarization, which neutralize along the imaginary polar axis. When this charge balance is perturbed with external stress onto the crystal mesh, the energy is transferred by electric charge carriers creating a current in the crystal. Conversely, with the piezoelectric effect an external charge input will create an unbalance in the neutral charge state causing mechanical stress (Calio et al. 2014). Figure 8.24 shows a typical power harvesting system for self-powered sensor nodes and microsensors. It includes an external energy source, a transducer to convert energy from external energy to electric power, a harvesting circuit to optimize the harvesting efficiency, and a storage battery or a load circuit.

Since the mechanical vibration of a piezoelectric element generates an alternating voltage across its electrodes, most of the proposed electrical circuits include an ac–dc converter to provide the electrical energy to its storage device.

8.13.2 *Analyze/design: Piezoelectric wind tunnel energy*

A lot of waste energy of heat, vibration, wind energy, and natural wind energy can be harvested into useful energy by using technology of piezoelectric devices. Metropolitan cities around the world are building more underground subway systems in order to minimize the number of vehicles on the roads. A new technology of piezoelectric wind energy system is introduced to utilize the wind power inside the tunnel. The features of the proposed topology are as follows:

- Delivering power at high voltage levels
- Starting operation at low cut-in speeds of about 1 m/s
- Robust structure for operation at high-speed wind flows

The proposed topology consists of a two-blade system with permanent magnets (PMs) and a piezoelectric beam with a PM proof mass, which interacts with the PMs in the fan to harvest wind power as shown in Figure 8.25.

The choice of the storage device for the harvested energy should consider first the estimated power budget of the whole system against the desired level of service continuity. The devices used more often in energy harvesting applications are supercapacitors and rechargeable batteries (Sudevalayam and Kulkarni 2008).

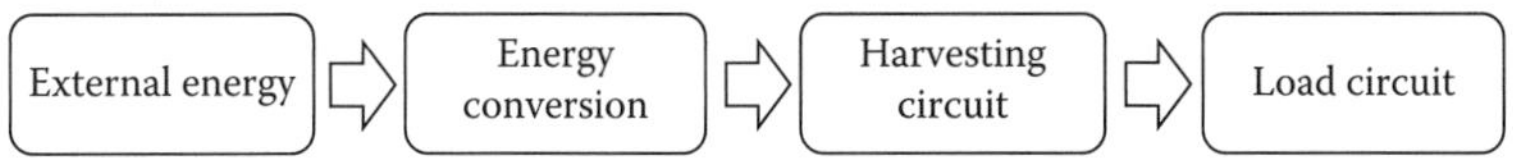

Figure 8.24 Typical power harvesting system.

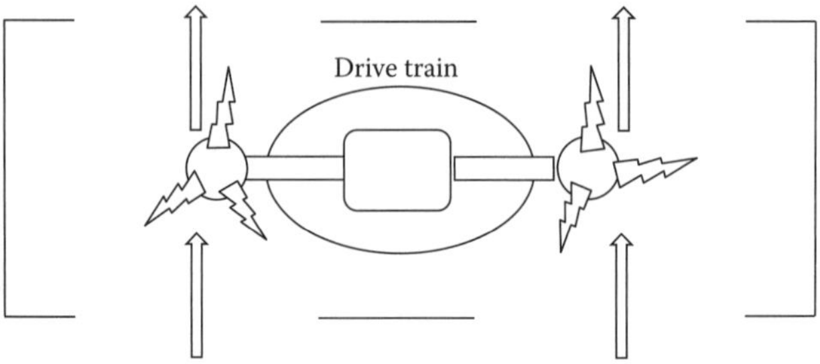

Figure 8.25 Wind turbine system.

It is proposed to design a novel wind tunnel energy harvester, for example, inside a subway tunnel. With a train speed of 200 km/h and a wind speed of 15 m/s, energy will be produced based on the specifications of the wind turbine as shown in Figure 8.26. The design approach will develop the fundamental nature of design abstractions as a key engineering tool. Design includes problem description, organization of resources, synthesis of ideas, construction, testing, and evaluation. These steps are needed to develop the wind turbine under consideration. A fundamental understanding of how systems are integrated from components and subsystems is necessary to view a problem and its environment from an overall perspective. Finally, evaluation can only be effected through the application of metrics.

At the turbine level, manufacturers can design the main configuration for each subsystem according to related standards, including blades,

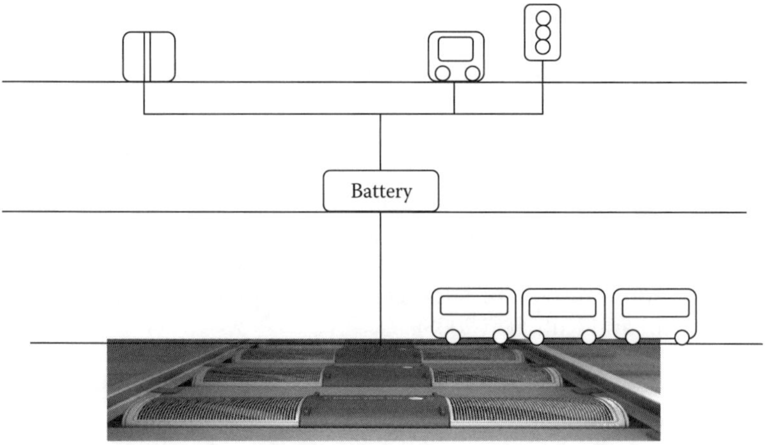

Figure 8.26 Layout of wind energy generated by subway trains.

drivetrain, and tower but also secondary systems for pitch, yaw, power conversion and quality, braking, control strategy, operating modes—fixed speed and variable speed—and interaction between the wind turbine and the grid.

The blade system configuration includes a number of rotors; diameter; tilt; and key airfoil design variables such as tip speed ratio, rotor solidity, and maximum chord, taper, trailing edge configuration, and tip design.

For the drivetrain, the configurations depend on the characteristics of the system: direct drive or a gearbox. Also, it depends on the number of generators, type of generators (dc, synchronous, or induction), and all the aspects of a particular design within a configuration such as the gear ratio, type of generator, and machine rating and speed.

As an example of highly complex mechatronic system (physical, electronics, and configuration), it is necessary to look at the system of a modern wind turbine. Wind energy system design results on sizing components and listing specifications of various subsystems. Component design inputs, which include configuration, geometries, materials, assembly, and auxiliary features such as sensors and actuators, are usually fed through the system requirements.

8.13.3 Optimize

Designing mechatronic systems is challenging. Even so, mechatronic systems are considered as one of the most important means of innovation in several areas including automobile, aerospace, and renewable energy. One well-known consideration when designing a mechatronic system is the choice of the most appropriate systematic methodology that may effectively help in guiding and organizing the design process.

Traditional product development methodologies, even ones with a mechatronic design focus, rely on treating the individual domains separately and only integrating them at a point in time rather far into the detailed design phase of the development process. In order to solve problems that arise in later design stages, the design engineers need to backtrack to earlier stages and in the worst case repeat substantial parts of the work. This is both time and cost inefficient (Malmquist 2014).

Optimization is commonly used in multi-interdisciplinary engineering in order to find the most possible design. Novel mechatronic design methodologies to enable design of better products, in terms of development cost, size, and sustainability, are presented in this section. The methodologies rely on using optimization approaches to efficiently determine the potential of a system concept, described by combining a number of component models—including motors, transmissions, and structural ones—from a component library and specifying their parameters.

8.13.3.1 V-cycle development

The V-cycle model was first developed by the German federal administration to regulate a software development process in 1997. However, after some adoption and modification the V-model has been suggested by the Association of German Engineers (VDI) committee A127 as the "VDI guideline 2206: design methodology for mechatronic systems," which is today often used as a basis for applied engineering design methodologies (Vielhaber et al. 2010). V-cycle is one of the most widely cited mechatronic design methods. It relies on treating the involved engineering disciplines separately, hence only integrating them in late design stages. It is easy to see why this generally results in suboptimal products. The name of the method refers to that the design is verified/validated against specification and requirements resulting from earlier phases at each level of granularity (Malmquist 2014).

A top-down approach is recommended, where modularization with precise component interfaces is arranged in order to focus on the overall system design. The design cycle starts with conceptual stage which consists of system requirements specification as shown in Figure 8.27. The

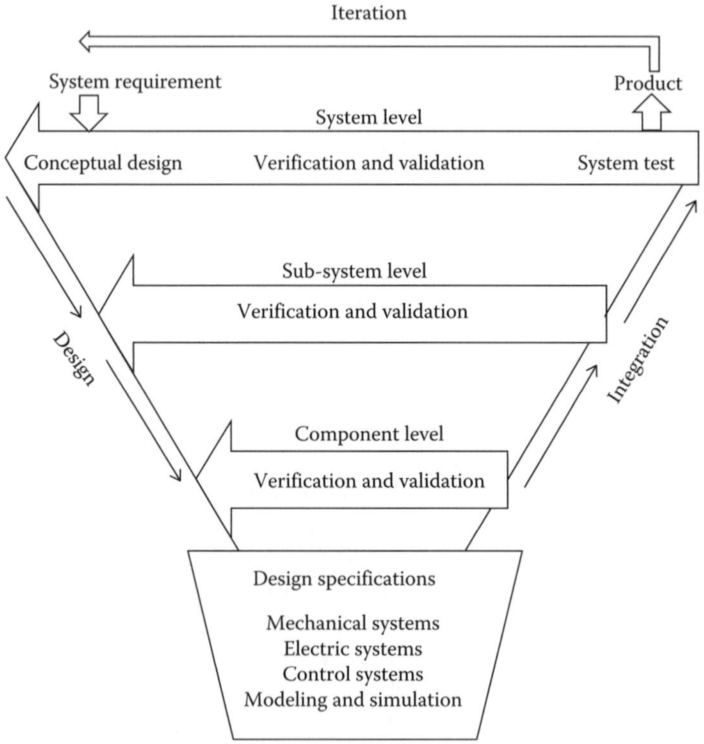

Figure 8.27 The V-cycle development for a mechatronic system.

main goal at the system level is to produce a cross-domain solution principle, which thoroughly describes the main physical and logical operating characteristic of the system or product, which is going to be developed. Therefore, the planning and task clarification phase and the conceptual design phase of the design process are carried out at the system level (Abd. Rahman et al. 2007). The result is a design specification and a formulated set of desired measurable behaviors of the future product, which introduces the quality measure into the design process.

At the subsystem level, the solution principle is divided into the respective domains involved. As for the solution principle of the piezoelectric brake, since it consists of the mechanical and electrical operating principles only, the design process of the brake is divided into two main engineering domains, which are mechanical engineering and electrical engineering. After the division, the embodiment design phase proceeds separately in the respective domains (VDI 2004). The milestones of the subsystem level after completing the embodiment design phase are a rough layout design, which consisted of a general arrangement and spatial compatibility of the respective mechanical and electrical systems as well as the preliminary parts list for reference purposes when doing the detailed design phase in the component level.

On the component level, a further concretization of the parts in each system is carried out. In terms of Pahl and Beitz (2006), this constitutes the detailed design phase which involves activities like detailed design of the parts, detailed calculation, parts analysis, etc. At the end of the detailed design phase, the complete design of the mechanical parts as well as the electric parts will be produced. Next, all of these parts are documented. Testing is conducted later on to validate and verify all these parts to ensure that they have been designed according to the specification prescribed in the requirements list. The next task is the first integration process of the parts. This first integration process is known as the subsystem integration, which is done in the respective domains.

On the integration stage, all domain-specific models and solutions will be integrated into overall system and all interactions will be investigated and verified. Once this is accomplished, the system, for the first time, will be tested as a unified whole to determine whether it meets its technical requirements, specified at the beginning. All engineering techniques such as CAD, CAM, and other CAE system can be engaged. Creating 3D models or running different kind of calculations is generated. While the V-model just proposes the integration on different levels of the system, the actual design process is strongly driven by so-called integration stages, which happen regularly and integrate the design status of different parts in order to test the whole system behavior and properties.

These stages somehow form iterations of the V-model, though they are mainly restricted to the integration phase, for example the adjusting of the subsystems and not the requirements development. By that, the data of the software can be seen as part of the detailed design of the V-model (bottom) or as part of the integration of the V-model (right top), depending on whether you see the V-model as a model for the levels of the mechatronic product or as a time-oriented process model.

A final product is not produced with one single cycle. Depending on the complexity of the design problem, the main cycle may be divided into sub-cycles. Initial requirements definition and design problem formulation are the starting points of the first cycle. The product is a full system specification and this is also an input for the next cycle. The last cycle will produce tested and verified final product.

8.13.3.2 Deterministic design

In the design of systems subject to uncertainties, both DD and non-DD approaches are very popular. In a DD approach, normally, the size of the storage or backup/auxiliary components is determined based on a suitable worst-case scenario to achieve a predefined level of reliability of power supply, while the remaining components are optimized for minimizing the system cost (Maheri et al. 2012). By adopting this approach, the multiobjective optimization problem with two main objectives of reliability and cost is reduced to a simple single-objective optimization problem with the objective of cost only. After sizing the storage or backup/auxiliary components, a simple single-objective optimization search can be employed to find the optimum size of the renewable components. In a deterministic approach, all calculations are based on the averaged values and the stochastic nature of demand load and renewable resources.

For systems of this kind, normally, first the battery bank is sized for a defined worst-case scenario and then the rest of the system is sized within an optimization problem with the objective of minimizing the cost subject to the constraint of having the yearly averaged renewable power greater than the yearly averaged load with a reasonable margin of safety.

8.13.4 Case research questions

- Categorize three mechatronic products of your choice according to their physical, electronics, and configuration complexity.
- What are advantages and disadvantages of the V-cycle model and when to use it?
- How can mechanical, electrical, control, and software engineering disciplines be treated concurrently in order to evaluate the potential of a mechatronic system without increasing the design task to a complex level?

8.14 Knowledge acquisition

Attempting to answer the following questions involves acquisition of knowledge from this book and other books, documents, and the Internet.

- What is design and what is a designer?
- What is the difference between ST and DT?
- What does design mean in an engineering context?
- What are the qualifications of a designer?
- Why is technology-driven design necessary?
- Why is human-driven design the most necessary model in the design paradigm?
- What are the advantages of sustainable design?
- How do augmented and virtual realities differ? How do they work together?
- How exactly do engineers turn science into reality? How do engineers think and act?
- What specific things do engineers design, help to manufacture or build, or help to operate and maintain?
- How to become a design entrepreneur?
- How do you go about being an entrepreneur as an engineer?
- Why the minds of engineers matter?
- Why engineering design needs entrepreneurship and why entrepreneurship needs design.
- Compare the engineering design process and the scientific method.
- What are the elements of engineering design?
- What are the most common design theories and methods?
- What is a system? What is SE?
- Why is estimation important for engineering design?
- What is CE or manufacturing? Why is it important?
- Who invented CE? Why?
- When is CE used? What are some examples?
- Why do companies adopt CE methods?
- What is the purpose of establishing standards, codes, specifications, and technical regulations?
- Which is more exhilarating to you, the process of coming up with a new design or witnessing the product that comes out of it?
- What makes someone a design entrepreneur? Why are not more designers starting their own companies?
- Can design be taught? If possible, how to teach it? And who can teach it?
- How best can the education system develop learners who think and act like engineers?

8.15 *Knowledge possession*

Attempting to answer the following open-ended "not explicitly expressed" questions may require research and investigation beyond the scope of this book, mostly by engaging in conversation, class discussion, and Internet-based research.

- What makes a good and successful design tool? Frame general criteria that a good tool should follow.
- What is the role of design in the new product process? How has that role changed in recent years? Is design becoming more or less important?
- How important are design skills for design leaders, in particular when the concept of design leadership is applied to leaders without design skills? In a world where design leads the path to competitive advantage, could a leader bring an organization to the forefront without design leadership skills?
- What does DT entail and how might it be used in furthering innovation?
- What practical strategies can be used to evaluate the infusion of engineering design into technology education learning activities?
- What features of the engineering design process can be identified within the context of technology education learning activities, where engineering design is the focus for curriculum?

8.16 *Knowledge creation*

In the following tasks, collaborate with peers on learning or you may work with others outside the class to narrow down the objectives of each task. You may access class and online resources, and analyze data and information to create new ideas and balanced solutions. High-level digital tools may be used to develop multimedia presentations, simulations or animations, videos and visual displays, electronic portfolios, feasibility studies or consulting reports, apps or other computer program, website or recording, debates or innovation pitches, original piece of art, or well-researched and up-to-date reports that reflect the objectives of each task.

8.16.1 *Piece of art on understanding the value of designing before building*

To perform any project properly and efficiently, appropriate planning is required and this is the reality to take into consideration. This task is intended to produce piece of art or an image that provides brief guidelines for the development of authentic engineering design learning approach, to describe instructional strategies for introducing engineering design experiences to high school students. The information is intended to be useful in planning, organizing, and implementing the infusion of engineering design challenges in high school Science, Technology, Engineering, Art, Math (STEAM) courses.

8.16.2 Design of virtual resource on DT

The objective of this task is to develop an online resource on DT to help students and others who want to generate ideas about this important notion. The instructor may set an outline for the online resource and allocate the resource tasks to groups of three to five students, each. The tasks may be developed around five major themes: empathy, collaboration, integrative thinking, and experimentalism. Each group should adopt a theme and prepare its task in PowerPoint format. At the end, all tasks should be combined into one file for the purpose of developing an online learning resource.

8.16.3 Design competition on a smart popsicle bridge

The goal is to build a bridge structure that—in addition to being strong and looks professionally smart—autonomously detects boat traffic and opens the bridge deck to allow passage. The system must also control vehicles, pedestrian, and bike traffic traveling over the bridge and close the bridge deck after the boat has passed.

Students work in teams to design and build their own bridge out of glue and popsicle sticks. They test their bridges using weights, evaluate their results, and present their findings to the class.

8.16.3.1 Objectives

- Know about civil engineering, structural engineering and design, planning and construction, sensing and control engineering, and working in groups.
- Know how bridges are engineered to withstand weight, while being durable, and in some cases aesthetically pleasing.

8.16.3.2 Types of bridges

In general, there are six main types of bridges including arch, beam, cable-stayed, cantilever, suspension, and truss.

8.16.3.3 *Competition requirements*

- The bridge must be able to span 16 in., about 5 in. wide and about 6 in. high.
- No more than 50% of stick surface can be glued to another stick.
- The bridge must have room for a toy car to pass through the bridge on a roadway.
- The width of the bridge must be between 3 and 5 in.

The activity can be set up as a competition between student groups. There may be different categories under which the final bridges created and built by the students are evaluated.

- The first category might be the overall best bridge as specified in the problem description: the bridge with the highest applied load to bridge weight ratio that meets all project constraints should win this category. All bridges would be weighed and then weight applied at the mid-span until failure of the bridge.
- The second category might be the most professional-looking bridge with aesthetic appeal.
- The third category goes to the team whose bridge reflects the most creative idea for spanning two distances.
- The fourth category defines the technology used to make the bridge autonomous.

8.16.3.4 *Planning stage*

- Team members (two to five) discuss the problem.
- Discuss, develop, and agree on a design for your bridge.
- Think about what patterns might be the strongest.
- Draw the generated design to show art of making plans and how the bridge is to be made and how it will work and look.
- Present your design to the class. You may choose to revise your plan after you receive feedback from the class.
- Build the structure of the bridge. In the process, you may need to revise the design and generate another sketch.
- Test the bridge to see if it can withstand the required weight for at least one full minute.

8.16.3.5 *Evaluation*

A worksheet to evaluate team's results should be developed. The sheet may consider the following questions:

- Did you succeed in building a bridge that held the required weight? If not, why did it fail?

- Did you decide to revise your original design while in the building phase? Why?
- How many Popsicle sticks did the team end up using? Did this number differ from the initial plan? If so, what changed?
- Do you think that engineers have to adapt their original plans during the building of systems or products? Why might they?
- If you had to do it all over again, how would your planned design change? Why?
- Do you think you would have been able to complete this project easier if you were working alone? Explain.
- What sort of trade-offs do you think engineers make between functionality, safety, and aesthetics when building a real bridge?

8.16.3.6 SE and CE

The prediction and consideration of a product's downstream behavior early in the design stage is a typical characteristic of CE. Unlike sequential engineering, more diverse design objectives are exercised when working with CE.

- What are the requirements for a bridge?
- What kinds of engineers build bridges?
- What is the procedure for designing and constructing a bridge?
- Did you notice any opportunity to implement SE and/or CE in the bridge project?
- What does a bridge system engineer do?

8.16.4 Feasibility study on designing super grid

Fossil fuels continue to increase in price and the recent Japanese tsunami has given doubts about the future of electricity needs. Accordingly, entrepreneurs in China, South Korea, Russia, and Japan have signed a memorandum of understanding that seeks to create the Asian Super Grid. It will transmit electrical power from renewable sources from areas in the world that are best able to produce it to consumers in other parts of the world (Hanley 2016). For this task, a team of three to four students will investigate the topic to provide a brief design of a regional interconnected super grid as a way to tap distributed renewable energy sources and transition to sustainable, low-carbon energy that will benefit the entire region. The team should develop a technical feasibility study on the challenges of generating and transmitting electricity across long distances.

8.16.5 Debate on scientific process and design process

It is often claimed that the scientific process is quite opposite to the design process, mainly based on the former's analysis of existing phenomena in order to develop a theory, while the design process is an act of synthesis that creates something new in the world (Mottel and Bjärnemo 2011). In real life, the distinction between science and engineering is not always clear. Scientists often do some engineering work, and engineers often apply scientific process.

Objective	Introducing an open-ended debate in the classroom to help students understand argument on the concepts of scientific process and design process
Time	15 min for debate and 15 min for review
Format	For and against
Learning outcomes	Make an argument about a particular opinion, evaluate the arguments of peers, and understand the concept of counterarguments
Capabilities demonstrated	Developing skills on public speaking, research, teamwork, critical thinking, communication, and professional judgment
Arrangement	Students are organized into two position groups of three speaking in a specific order. Three argue for an opinion and three argue against. One or two students might each work on the opening and closing statements while the group is investigating the subject; however, the entire group should revise the statements. Each group should read an opening and closing statement for the debate.
Ideas for the topic	Scientists are practitioners who use experience and other practical skills as much as other professionals. In design process theory, it should take into account the similarities with the scientific process and not be defined in opposition to it. The C-K theory, by defining design as the creation of knowledge, does that (Motte and Bjärnemo 2011). Which process should you follow for your engineering project, scientific or design?
Assessment	Indicate what you consider the best three arguments in favor of the topic. How were they substantiated? Identify arguments that are based on poor facts, not ethical and/or sustainable or not well substantiated.

8.16.6 *Poster on design for reuse*

Design for reuse, whereby current artifacts are designed with a specific emphasis on promoting, extracting, and enhancing reusable knowledge elements, has been shown in previous studies to have the most significant impact on the realization of "reuse" related benefits (Duffy and Ferns 1999). MD is a product structuring principle whereby products are developed with distinct modules for rapid product development, efficient upgrades, and possible reuse (of the physical modules) (Meehan et al. 2007). In this case task, explore MD and its relation to reuse.

8.16.7 Piece of art on MBD in automotive industry

An MBD is useful in embedded systems like automotive software due to the fact that development in these systems is driven by the evolutionary development of automotive systems, dealing with the iterated combination of new functions into a large amount of existing functionality from pervious system versions. In this task, collaborate with peers or you may work with others outside the class to narrow down the objectives of this task. You may access online resources to create a piece of art that reflects the above subject.

8.16.8 Video contest on how to become a design entrepreneur

The current role of the engineer designer is changing, more than ever before. Designers are becoming more valued for their naturally entrepreneurial mind-set. Designers contribute new and alternative ideas to solve problems. In this task, collaborate with peers or you may work with others outside the class to narrow down the objectives of this task. You may access online resources to create a 3-min video that reflects the above subject. In this regard, you may clearly answer the following three questions:

- How does the mind-set apply if a designer is working in house or in an organization?
- Can engineer designers be successful entrepreneurs?
- What stops designers from becoming entrepreneurs?

References

Abd. Rahman, R., U. Pulm, and R. Stetter. 2007. Systematic mechatronic design of a piezo electric brake. *International Conference on Engineering Design*, Cite Des Sciences et de L'industrie, Paris, France, August 28–31.

ABET. 2006. 2005–2006 criteria for accrediting engineering programs. Accreditation Board for Engineering and Technology, Baltimore, MD. www.abet.org.

ABET Website. 2013. http://www.abet.org/engineering-criteria-2012-2013.

Adams, K. M. 2015. *Non-functional Requirements in Systems Analysis and Design*. Cham, Switzerland: Springer International Publishing.

AITSL. 2014. Learning through doing: Introduction to design thinking. Australian Institute for Teaching and School Leadership, Melbourne, VIC. http://www.aitsl.edu.au/docs/default-source/partnerships-documents/aitsl---learning-through-doing---introduction-to-design-thinking_v4.pdf?sfvrsn=2.

Akili, W. 2015. Perspectives on engineering design learning: Realities, challenges, and recommendations. *IEEE Frontiers in Education Conference*, IEEE Computer Society, Washington, DC, pp. 1–7, October 21–24.

Allen, R. 2015. The history of engineering design tools: Engineering 1.0. http://blogs.solidworks.com/solidworksblog/2015/01/the-history-of-engineering-design-tools-engineering-1-0.html

Alnelind, J., and C. Alvén. 2014. Design leadership skills questioning the difference between design leadership and generic leadership in SME manufacturing organisations. Linnaeus University, Växjö, Sweden.

Anderson, D. 2014. Graphic design and entrepreneurship: The rise of the design entrepreneur. http://www.creatividadysociedad.com/articulos/22/05_Anderson.pdf.

Armytage, W. H. G. 1976. *A Social History of Engineering*. London: Faber and Faber.

Bandecchi, M., B. Melton, and F. Ongaro. 1999. Concurrent engineering applied to space mission assessment and design. *EESA Bulletin*, September.

Baumann, H., F. Boons, and A. Bragd. 2002. Mapping the green product development field: Engineering, policy and business perspectives. *Journal of Cleaner Production* 10(5): 409–425.

Beattie, L. 2017. Optimism and the power of positive thinking. http://www.sparkpeople.com/resource/wellness_articles.asp?id=835.

Bell, J. T., and H. S. Fogler. 1997. The application of virtual reality to chemical engineering education. Proceedings of the *1997 International Conference on Simulation in Engineering Education* (ICSEE '97), Phoenix, AZ, January 12–15.

Benvenuto, E. 1991. *An Introduction to the History of Structural Mechanics*. New York: Springer-Verlag.

Bharadwaj, S., and A. Menon. 2000. Making innovation happen in organizations: Individual creativity mechanisms, organizational creativity mechanisms or both? *Journal of Product Innovation Management* 17(6): 424–434.

Blanchard, B., and S. Fabrycky. 1998. *Systems Engineering and Analysis*. Third edition. Englewood Cliffs, NJ: Prentice Hall.

Boland, R., and F. Collopy (Eds.). 2004. *Managing as Designing*. Stanford, CA: Stanford University Press.

Boyd, D. M., and N. B. Ellison. 2007. Social network sites: Definition, history, and scholarship. *Journal of Computer-Mediated Communication* 13: 210–230.

Brown, T. 2008. Design thinking. *Harvard Business Review*, June, 1–10.

Brown, T. 2009. *Change By Design: How Design Thinking Transforms Organizations and Inspires Innovation*. New York: HarperCollins Publishers.

Brown, D. C., and B. Chandrasekaran. 1989. *Design Problem Solving: Knowledge Structures and Control Strategies*. San Mateo, CA: Morgan Kaufman.

Brzustowski, T. 2004. Science and engineering research Canada: Design engineering and the innovation strategy. Presentation at the *2004 CDEN Conference*, McGill University, Montreal, QC.

Buchanan, R. A. 1985. The rise of scientific engineering in Britain. *British Journal for the History of Science* 18: 218–233.

Buchanan, R. 1999. Design research and the new learning. MIT. http://www.ida.liu.se/~steho87/desres/buchanan.pdf.

CAE. 1999. Evolution of engineering education in Canada. Canadian Academy of Engineering, Ottawa, ON.

Calio, R., U. B. Rongala, D. Camboni, M. Milazzo, C. Stefanini, G. de Petris, and C. M. Oddo. 2014. Piezoelectric energy harvesting solutions. *Sensors* 14: 4755–4790.

Cappelli, F., M. Delogu, M. Pierini, and F. Schiavone. 2007. Design for disassembly: A methodology for identifying the optimal disassembly sequence. *Journal of Engineering Design* 18(6): 563–575.

CEAB. 2004. Accreditation criteria and procedures. Canadian Engineering Accreditation Board, Canadian Council of Professional Engineers, Ottawa, ON. http://www.ccpe.ca/e/files/report_ceab.pdf.

Chang, K.-H., J. Silva, and I. Bryant. 2016. Concurrent design and manufacturing for mechanical systems. *Concurrent Engineering Research and Application (CERA) Journal* 6(2): 131–144.

Chasanidou, D., A. A. Gasparini, and E. Lee. 2015. Design thinking methods and tools for innovation. In Marcus, A. (Ed.), *DUXU 2015, Part I, LNCS 9186*, pp. 12–23. Cham, Switzerland: Springer International Publishing.

Chiu, M.-C. 2011. Investigation of the applicability of design for X tools during design concept evolution: A literature review. *International Journal Product Development* 13(2): 132–167.

Clarkson, J., and C. Eckert. 2005. *Design Process Improvement: A Review of Current Practice*. London: Springer Verlag.

Colbome, P. A., and Steyn, H. 1998. System engineering and concurrent engineering: What's it for manufacturers of mature industrial products? *The South African Journal of Industrial Engineering* 9: 21–35.

Corbett, A. 2005. Experiential learning within the process of opportunity identification and exploitation. *Entrepreneurship Theory and Practice* 29(4): 473–492.

Cowen, A. 2013. Understanding the engineering design process: Teachers embracing engineering design. http://www.sciencebuddies.org/blog/2013/09/understanding-the-engineering-design-process.php.

Daems, W., B. De Smedt, P. Vanassche, G. Gielen, W. Sansen, and H. De Man. 2003. PeopleMover: An example of interdisciplinary project-based education in electrical engineering. *IEEE Transactions on Education* 46(1): 157–167.

Davis, T. P. 2006. Science, engineering, and statistics. *Applied Stochastic Models in Business and Industry* 22: 401–430.

de Figueiredo, A. D. 2008. Toward an epistemology of engineering. Workshop on Philosophy and Engineering. The Royal Academy of Engineering, London, November 10–12.

de Figueiredo, A. D., and P. R. Cunha. 2007. Action research and design in information systems: Two faces of a single coin. In Kock, N. (Ed.), *Information Systems Action Research: An Applied View of Emerging Concepts and Methods*. Laredo, TX: Springer.

Dhillon, B. S., 1985. *Quality Control, Reliability, and Engineering Design*. New York: Marcel Dekker.

Duffy, A. H. B., and A. F. Ferns. 1999. An analysis of design re-use benefits. *12th International Conference on Engineering Design*, Munich, Germany, August 24–26.

Dukic, T., M. Ronnang, and M. Christmansson. 2007. Evaluation of ergonomics in a virtual manufacturing process. *Journal of Engineering Design* 18(2): 125–137.

Dyas, S. P. 2005. A unified theory of engineering design. MSc Thesis, Texas A&M University, College Station, TX.

Dym, C. L. 1994. *Engineering Design: A Synthesis of Views*. Cambridge: Cambridge University Press.

Dym, C. L., A. M. Agogino, O. Eris, D. D. Frey, and L. J. Leifer. 2005. Engineering design thinking, teaching, and learning. *Journal of Engineering Education* 94(1): 103–120.

Evbuomwan, N. F., S. Sivaloganathan, and A. Jebb. 1996. A survey of design philosophies, models, methods and systems. *Journal of Engineering Manufacture* 210: 301–320.

Favela, J., A. Wong, and A. Chakravarthy. 1993. Supporting collaborative engineering design. *Engineering with Computers* 9(3): 125–132.

Ferguson, E. S. 1992. *Engineering and the Mind's Eye*. London: MIT Press.

Finch, J. K. 1978. *Engineering Classics*. Kensington, MD: Cedar Press.

Freeman-Bell, G., and J. Balkwill. 1996. *Management in Engineering: Principles and Practice*. Second edition. New York: Prentice Hall.

Gardiner, G. S. 1995. Systems engineering and concurrent engineering: Synonymous, complementary or contrary? *IEEE Colloquium on Systems Engineering for Profit*, London, March 22, 6/1–6/4.

Gasson, S. 2004. Organizational "problem-solving" and theories of social cognition. Working Paper. http://www.cis.drexel.edu/faculty/gasson/Research/Proble m-Solving.html.

Gelernter, D. 1998. *Machine Beauty: Elegance and the Heart of Technology*. New York: Basic Books.

Giacomin, J. 2012. What is human centred design? P and D Design 2012, Human Centred Design Institute, Brunel University, Uxbridge, UK. http://hcdi.brunel.ac.uk/files/What%20is%20Human%20Centred%20Design.pdf.

Gille, B. 1966. *Engineers of the Renaissance*. Cambridge: MIT Press.

Gloppen, J. 2009. Perspectives on design leadership and design thinking and how they relate to European service industries. *Design Management Journal* 4 (1): 33–47.

Goldberg, J. R. 2012. Standards in capstone design courses and the engineering curriculum. *IEEE PULSE*, September/October, pp. 42–44.

Gonzalez, C. 2016. Does virtual reality have a future in engineering? http://machinedesign.com/blog/does-virtual-reality-have-future-engineering.

Goodyear, P. 2015. *Teaching as Design. HERDSA Review of Higher Education*, Vol. 2. www.herdsa.org.au/herdsa-review-higher-education-vol-2/27-50.

Gopsill, J. A. 2014. A social media approach to support engineering design communication. PhD Thesis, Department of Mechanical Engineering, University of Bath, Bath.

Gopsill, J. A., H. C. McAlpine, and J. H. Ben. 2013. A social media framework to support engineering design communication. *Advanced Engineering Informatics* 27(4): 580–597.

Gorman, M. G., M. M. Mehalik, and P. H. 2000. *Werhane, Ethical and Environmental Challenges to Engineering*. Englewood Cliffs, NJ: Prentice-Hall.

Grafton, A., and L. B. Alberti. 2000. *Master Builder of the Italian Renaissance*. New York: Hill and Wang.

Graham, M. M., and A. H. Slocum. 2005. Product development by deterministic design. *First Annual CDIO*, Kingston, ON.

Graham, M. M., A. H. Slocum, and R. M. Sanchez. 2007. Teaching high school students and college freshmen product development by deterministic design with PREP. *Journal of Mechanical Design* 129: 677–681.

Grinter, L. E. 1955. Report on evaluation of engineering education. American Society of Engineering Education (ASEE), Washington, DC.

Gultiken, P. 2004. The negative effects of technology-driven product design on user-product integration and product usability. MSc Thesis, The Graduate School of Natural and Applied Sciences of Middle East Technical University, Çankaya/Ankara, Turkey.

Gunes, S. 2012. Design entrepreneurship in product design education. *Social and Behavioral Sciences* 51: 64–68.

Haik, Y., and T. Shahin. 2011. *Engineering Design Process*. Second edition. Stamford, CT: Cengage Learning.

Hanley, S. 2016. China, Japan, Russia, and South Korea plan renewable energy super grid. https://cleantechnica.com/2016/09/21/china-japan-russia-south-korea-plan-renewable-energy-super-grid/.

Hargadon, A. B., and Y. Douglas. 2001. When innovations meet institutions: Edison and the design of the electric light. *Administrative Science Quarterly* 46(3): 476–592.

Hartley, J. R. 1992. *Concurrent Engineering*. Cambridge, MA: Productivity Press.

Haskins, C., K. Forsberg, M. Krueger, D. Walden, and R. Hamelin. 2010. *INCOSE Systems Engineering Handbook: A Guide for System Life Cycle Processes and Activities*, Vol. 3.2. INCOSE-TP-2003-002-03.2. San Diego, CA: International Council on Systems Engineering.

Hatchuel, A. 1996. *Les théories de la conception*. Paris, France.

Hatchuel, A., P. Le Masson, and B. Weil. 2004. C-K theory in practice: Lessons from industrial applications. Proceedings of *International Design Conference*, Dubrovnik, Croatia.

Hatchuel, A., and B. Weil. 1999. Pour une théorie unifiée de la conception, Axiomatiques et processus collectifs, Paris, CGS Ecole des Mines/GIS cognition-CNRS 1–27.

Heller, S., and L. Talarico. 2008. *The Design Entrepreneur: Turning Graphic Design into Goods that Sell*. Beverly, MA: Quayside Publishing.

Heller, S., and L. Talarico. 2016. *Becoming a Design Entrepreneur: How to Launch your Design-Driven Ventures?* New York: Allworth Press.

Hendrickson, M. 2012. Three types of design. https://medium.com/@markymark/three-types-of-design-3623c3243aa6#.ms5hh04kb.

Hirsch, P. L., B. L. Shwom, C. Yarnoff, J. C. Anderson, D. M. Kelso, G. B. Olson, and J. E. Colgate. 2001. Engineering design and communication: The case for interdisciplinary collaboration. *International Journal of Engineering Education* 17(4 and 5): 342–348.

Holstein, W. K., and A. Chapanis. 2016. Human-factors engineering. Encyclopaedia Britannica. https://www.britannica.com/topic/human-factors-engineering.

Hooge, S., M. Agogu´e, and T. Gillier. 2012. A new methodology for advanced engineering design: Lessons from experimenting C-K theory driven tools. *International Design Conference*—Design 2012, Dubrovnik, Croatia.

Horth, D., and D. Buchner. 2016. Innovation leadership: How to use innovation to lead effectively, work collaboratively, and drive results. White Paper, Centre for Creative Leadership, Greensboro, NC. http://insights.ccl.org/wp-content/uploads/2015/04/InnovationLeadership.pdf.

Horvath, I. 2001. A contemporary survey of scientific research into engineering design. In Culley, S. et al. (Eds.), *International Conference on Engineering Design*, IMechE, Glasgow, Scotland, August 21–23, pp. 13–20.

Hybs, I., and J. S. Gero. 1992. An evolutionary process model of design. *Design Studies* 13(3): 273–290.

IN. 2011. CK theory. Innovation Now. https://innovatingnow.wordpress.com/2011/05/05/ck-theory/.

IPENZ. 2002. The drive for innovator and entrepreneurs: School governance and technology education. The Institution of Professional Engineers New Zealand (IPENZ), IPENZ Informatory Note Seven, Wellington, New Zealand.

Jonassen, D. H. 2004. *Learning to Solve Problems: An Instructional Design Guide.* San Francisco, CA: John Wiley & Sons.

Jonassen, D. H., J. Strobel, and C. B. Lee. 2006. Everyday problem solving in engineering: Lessons for engineering educators. *Journal of Engineering Education* 95(2): 1–14.

Knörig, A. 2008. Design tools design: How to design tools for designers, and a proposal of two new tools for the design of physical interactions? Master of Design (Interface Design) Thesis, University of Applied Sciences Potsdam, Potsdam.

Kocabiyik, E. 2004. Engineering concepts in industrial product design with a case study of bicycle design. Master of Industrial Design, Izmir Institute of Technology, Izmir, Turkey.

Kornwatchs, K. 2016. Design technoscience—What's up with responsibility? Proceeding of the *Design 14th International Design Conference,* Cavtat, Dubrovnik, Croatia, May, 16–19.

Krippendorff, K. 1989. On the essential contents of artifacts or on the proposition that "design is making sense (of things)". *Design Issues* 5(2): 9–39.

Krippendorff, K. 2006. *The Semantic Turn: A New Foundation for Design.* Boca Raton, FL: CRC Press.

Kusiak, A., and N. Larson. 2009. Concurrent engineering. In Sage, A. P., and W. B. Rouse (Eds.), *Handbook of Systems Engineering and Management.* Second edition. Hoboken, NJ: John Wiley & Sons.

Laurilland, D. 2012. *Teaching as a Design Science: Building Pedagogical Patterns for Learning and Technology.* Third edition. New York: Routledge.

Liu, S., and L. M. Boyle. 2004. Engineering design: Perspectives, challenges and recent advances. *Journal of Engineering Design* 20(1): 1–16.

Liu, Z., and D. J. Schönwetter. 2004. Teaching creativity in engineering. *International Journal of Engineering Education* 20(5): 801–808.

Lockwood, T. 2008. Design value: A framework for measurement. In Lockwood, T., and T. Walton (Eds.), *Building Design Strategy: Using Design to Achieve Key Business Objectives.* New York: Allworth Press.

Maheri, R., R. Bagali, C. Chilak, F. Scheller, and A. I. Mustafa. 2012. Deterministic versus non-deterministic design of hybrid renewable energy systems. *2nd International Symposium on Environment-Friendly Energies and Applications (EFEA),* Northumbria University, Newcastle upon Tyne, Tyne and Wear, UK.

Malmquist, D. 2014. A tool for holistic optimization of mechatronic design concepts. Doctoral Thesis. KTH Royal Institute of Technology, School of Industrial Engineering and Management, Department of Machine Design, Stockholm, Sweden.

Martin, R. 2011. The innovation catalysts. *Harvard Business Review* 89(6): 82–87.

Martin, S. B., and A. K. Kar. 2002. Axiomatic design for the development of enterprise level e-commerce strategies. Proceedings of the *ICAD2002, Second International Conference on Axiomatic Design,* MIT, Cambridge, MA, June 10–11.

Matchett, E., and A. H. Briggs. 1966. Practical design based on method (fundamental design method). In Gregory, S. A. (Ed.), *Design Method*. London: Butterworth, pp. 183–199.
Maxey, K. 2012. Product concept design: Past, present, and future. http://www.engineering.com/DesignerEdge/DesignerEdgeArticles/ArticleID/4986/Product-Concept-Design-Past-Present-and-Future.aspx.
McMahon, C., and J. Browne. 1998. *CAD/CAM: Principles, Practice, and Manufacturing Management*. New York: Prentice Hall.
Meehan, J. S., A. H. B. Duffy, and I. Whitfield. 2007. Supporting design for reuse with modular design. Concurrent engineering. *Research and Applications* 15(2): 141–155.
Meroni, A., and D. Sangiorgi (Eds.). 2011. *Design for Services*. Aldershot, UK: Gower Publishing Ltd.
Mootee, I. 2013. Design thinking for strategic innovation. https://blog.hubspot.com/agency/excerpt-design-thinking-for-strategic-innovation#sm.0001g8omj81djwekow2i989ia21sj.
Motte, D. 2008. A review of the fundamentals of systematic engineering design process models. *International Design Conference*, Dubrovnik, Croatia, May 19–22.
Motte, D., and R. Bjärnemo. 2011. A note on the debate on scientific process vs. design process. *International Conference on Engineering Design*, Technical University of Denmark, Kgs. Lyngby, Denmark, August 15–18.
M'pherson, P. K. 1980. Systems engineering: An approach to whole-system design. *The Radio and Electronic Engineer* 50(1): 545–558.
NAE. 2004. The engineer of 2020: Visions of engineering in the new century. National Academy of Engineering, Washington, DC.
Naiman, L. 2017. Design thinking as a strategy for innovation. http://www.creativityatwork.com/design-thinking-strategy-for-innovation/.
Nelson, E. 2012. What makes an engineer an engineer? Structures Congress, Fort Worth, TX, pp. 1152–1159.
Newcomer, J. L. 1999. Design: The future of engineering and engineering technology education. *IEEE Frontiers in Education Conference*, San Juan, Puerto Rico, November 10–13.
NGSS. 2013. Next generation science standards. Next Generation Science Standards. www.nextgenscience.org.
Nicolai, L. M. 1998. Viewpoint: An industry view of engineering design education. *Journal of Engineering Education* 14(1): 7–13.
Ogot, M., and G. E. Kremer. 2004. *Engineering Design: A Practical Guide*. Victoria, BC: Trafford Publishing.
OGP. 2011. Human factors engineering in projects. International Association of Oil and Gas Producers, Report 454.
Owen, C. L. 1988. Design education and research for the 21st century. Illinois Institute of Technology, Chicago, IL. http://www.id.iit.edu/papers/Owen_singapore88.pdf.
Pahl, G., and W. Beitz. 2006. *Engineering Design: A Systematic Approach*. Second edition. New York: Springer-Verlag.
Papalambros, P. Y. 2015. Design science: Why, what and how. *Design Science* 1: e1.
Park, H., J. S. Son, and K. H. Lee. 2008. Design evaluation of digital consumer products using virtual reality-based functional behaviour simulation. *Journal of Engineering Design* 19(4): 359–375.

Pech, R. M. 2015. Achieving the innovative edge in technology, engineering design, and entrepreneurship. *Journal of Innovation and Entrepreneurship* 5: 6.
Perrenet, C., P. A. J. Bouhuijs, and J. G. M. M. Smits. 2000. The suitability of problem-based learning for engineering education: Theory and practice. *Teaching in Higher Education* 5(3): 345–358.
Phadke, M. S. 1989. *Quality Engineering Using Robust Design*. Upper Saddle River, NJ: PTR Prentice-Hall.
Pilliton, E. 2009. *Design Revolution*. New York: Thames and Hudson.
Prasad, B. 1996. *Concurrent Engineering Fundamentals: Integrated Product and Process Organization*. Upper Saddle River, NJ: Prentice Hall.
Pugh, S. 1996. *Creating Innovative Products Using Total Design*. Clausing, D., and R. Andrade (Eds.), First edition. Boston, MA: Addison Wesley Longman.
Rapanta, C. 2014. Book review: Teaching as a design science. Learning and Teaching in Higher Education. *Gulf Perspectives* 11(1). http://lthe.zu.ac.ae.
Rauch, E., D. T. Matt, and P. Dalasega. 2016. Application of axiomatic design in manufacturing system design: A literature review. *Procedia CIRP* 53: 1–7.
Richard, B. 1992. Wicked problems in design thinking. *Design Issues* 8(2): 5–21.
Ries, E. 2011. *The Lean Startup: How Today's Entrepreneurs Use Continuous Innovation to Create Radically Successful Businesses*. New York: Crown Publishing Group.
Ross, J. M., and T. M. Bayles. 2007. Implementing the INSPIRES curriculum: The role of professional development. *The National Symposium on Professional Development in Engineering and Technology Education*, Dallas, TX. http://conferences.illinoisstate.edu/NSA/papers/RossBayless.pdf.
Salustri. A. 2005. An extremely abbreviated history of engineering design. http://deed.ryerson.ca/~fil/t/history0.html.
Schon, D. A. 1983. *The Reflective Practitioner: How Professionals Think in Action?* First edition. London: Ashgate Publishing.
Schon, D. A. 1991. *The Reflective Practitioner: How Professionals Think in Action*. Second edition. London: Ashgate Publishing.
Senge, P., B. Smith, N. Kruschwitz, J. Laur, and S. Schley. 2008. *The Necessary Revolution: How Individuals and Organisations Are Working Together to Create a Sustainable World*. New York: Doubleday.
Shavinina, L. V. 2003. Understanding innovation: Introduction to some important issues. In Shavinina, L. V. (Ed.), *International Handbook of Innovation*. Oxford: Elsevier Science, pp. 3–14.
Shen, W., Q. Hao, and W. Li. 2008. Computer supported collaborative design: Retrospective and perspective. *Computers in Industry* 59(9): 855–862.
Sheppard, S., and R. Jenison. 1997. Examples of freshman design education. *International Journal of Engineering Education* 13(4): 248–261.
Simon, H. A. 1973. The structure of ill-structured problems. *Artificial Intelligence* 4: 181–201.
Simon, H. 1996. *The Sciences of the Artificial*. Third edition. London: MIT Press.
Slocum, A. H. 1995. *Precision Machine Design*. Dearborn, MI: Society of Manufacturing Engineers.
Smith, P. S., and D. G. Reinertsen. 1995. *Developing Products in Half the Time*. New York: Van Nostrand Reinhold.
Stephan, K. 2004. Can engineering ethics be taught? *IEEE Technology and Society Magazine*, Spring, pp. 5, 8.
Suchman, L. 1987. *Plans and Situated Action*. New York: Cambridge University Press.

Sudevalayam, S., and P. Kulkarni. 2008. Energy harvesting sensor nodes: Survey and implications. Department of Computer Science and Engineering (CSE), Indian Institute of Technology Bombay (IITB), Tech. Rep. IITB/CSE/2008/December/19. http://www.cse.iitb. ac.in/internal/techreports/reports/TR-CSE-2008-19.pdf.

Suh, N. P. 1990. *The Principles of Design*. London: Oxford University Press.

Suh, N. P. 1995. Designing-in of quality through axiomatic design. *IEEE Transactions on Reliability* 44(2): 256–264.

Suzuki, K. 2016. Entrepreneurship education based on design thinking and technology commercialization in Japanese universities. *IEEE 5th International Congress on Advanced Applied Informatics,* Kumamoto, Japan, July 10–14, pp. 779–764.

Taguchi, G., and Y. Wu. 1979. Introduction to off-line quality control. Central Japan Quality Control Association, Nagoya, Japan.

Tarondeau, J. C. 1998. *Stratégie Industrielle,* Second edition. Collection Gestion, Vuibert.

Topalian, A. 2012. Frontline roles for design leaders in the multiverses of business. *Design Management Journal* 7(1): 29–39.

Trummer, J. A., and S. Lieras. 2012. Reflections on design education in a changing world. *Design Management Review* 23(4): 14–22.

Turino, J. 1992. *Managing Concurrent Engineering.* New York: Van Nostrand Reinhold.

Turner Jr, K. L., M. K. Kettle, and S. Bober. 2016. Engineering design for engineering design: Benefits, models, and examples from practice. *Inquiry in Education* 8(2): 1–12.

van de Poel, I. R. 2001. Investigating ethical issues on engineering design. *Science and Engineering Ethics* 7(3): 425–442.

VDI. 2004. *Guideline 2223: Systematic Embodiment Design of Technical Products.* Düsseldorf, Germany: Association of German Engineers.

Vielhaber, M., D. Bergsjö, and A. Ćatić. 2010. Mechatronics systems engineering: Theory and automotive practice. *International Design Conference-Design 2010,* Dubrovnik, Croatia, May 17–20.

Von Hippel, E. 2007. An emerging hotbed of user-centered innovation, breakthrough ideas for 2007. *Harvard Business Review,* Article R0702A, February.

Watts, M. 2001. *Heidegger: A Beginner's Guide.* London: Hodder and Stoughton.

Yassine, A. A., K. R. Chelst, and D. R. Falkenburg. 1999. A decision analytic framework for evaluating concurrent engineering. *IEEE Transactions on Engineering Management* 46(2): 144–157.

Ziemke, M. C., and M. S. Spann. 1993. *Concurrent Engineering's Roots in the World War II Era. Concurrent Engineering: Contemporary Issues and Modern Design Tools.* London: Chapman and Hall.

chapter nine

Engineering product design and development

> A good scientist is a person with original ideas. A good engineer is a person who makes a design that works with as few original ideas as possible.
>
> **Freeman Dyson**

9.1 *Learning objectives*

- Focus upon product development (PD) with emphasis on design methodology and process. More precisely, how does engineering design methodology execute the technical processes of the design stage of the system development life cycle?
- Understand new product development (NPD) processes and how product innovation happens within companies.
- Differentiate between design theory, design methodology, and design process.
- Identify the major stages associated with design and development of products.
- Describe the function and details of each design process stage.
- Discuss the phases and corresponding steps of each PD stage.
- Analyze the various phases of design stage including brief, conceptual, embodiment, and detailed design (DD).
- Explain how layout, configuration, parts, topologies, and features are used to create design.
- Know about elements of DD including design for manufacturing (DfM), design for assembly (DfA), design for operability (DfO), design for maintainability (DfM), and design for environment (DfE).
- Outline the additional tools used in DD including analysis, modeling, simulation, and optimization.
- Highlight the impact of iteration in the product design process.
- Describe the PD life cycle and identify the need of management.
- Discuss the various aspects and challenges of product design education and highlight the need for transdisciplinary curriculum.

- Introduce Bloom's taxonomy through a case that outlines the design of an enhanced wind turbine and describes the engineering design steps followed.
- Provide three end of chapter pedagogical knowledge strategies, namely knowledge acquisition, knowledge possession, and knowledge creation to help understand the above topics and generate new and alternative ideas and solutions.

9.2 Historical perspective

The engineer has been, and is, a maker of history.

James Kip Finch

9.2.1 The relevance of artisanship

For most of history, when people needed a particular object, they either created it themselves or found someone to make it for them. Individuals may have specialized in their production, such as shoemakers and carpenters, but their output was still largely unique creations (Kuen Chang 2016).

Around the year 1400, Filippo Brunelleschi (1377–1446), the Italian architect and engineer, won a prestigious opportunity to design and build the cupola (dome) of the new cathedral for the city of Florence. He was especially worried that his contemporaries would try to steal his ideas, so he devised a way of designing the cathedral that would ensure the exact nature of the structure would remain hidden until it was too late. Until that time, however, buildings were not really engineered at all. The craft was then known as "artisanship," which means a person skilled in making a product by hand but involving well-understood principles and trial and error methods of building. Brunelleschi began by keeping a journal in which he sketched and described individual ideas for features and components of the cathedral from both architectural and civil engineering perspectives. Slowly he pieced together an overall concept for the cathedral, which he described in a single master plan, of which there was only one copy, and that he guarded carefully. Then, Brunelleschi did something new. He knew he would have to subcontract the construction of the building materials to other people, but he did not want to show them the master plan. So he created a large collection of individual drawings. Each drawing specified only a few components of the cathedral's structure. He then distributed the drawings to the various manufacturers who delivered their products to various off-site locations (Salustri 2005). Brunelleschi decided to build without scaffolding in such a way that it supported itself as it progressed. Completed by 1436, the dome of the cathedral remains one of the most famous features and tallest building of the Florentine skyline. The 45 m wide dome represents the central place of geometrical coordination and the rebirth of ancient models in Renaissance architecture, and miracle of design and engineering.

9.2.2 Moving closer to historical origins

At the beginning of the twentieth century research companies like Thomas Edison's Menlo Park R&D lab pioneered a product concept design (PCD) cycle that was based on experimentation. Edison's approach was remarkably successful. He and his assistants created products like the modern light bulb, the phonograph, and an electrical grid based on dc current. Around the same time Henry Ford perfected the assembly line by introducing conveyor belts to the production cycle. The introduction of a consistent, unrelenting conveyor transformed the world economy by laying the groundwork of the factory-based model (Maxey 2012).

Throughout the twentieth century, along with balancing the needs of the user and manufacturer, differences in politics and culture were evident in the design of objects. A rising consumer culture in the post–World War II period meant that manufactured goods doubled as a cultural proxy, intertwining national pride and economic reinvention (Kuen Chang 2016). New ideas about the way products would be developed catalyzed a second phase of product innovation. Individuals like Vannevar Bush believed that a structured collaboration between universities and industry would create better products. Bush's ideas for the development of products meant that administrators of large-scale projects would be essential to successful innovation. These administrators would become the conduit through which fundamental research was translated to industry (Maxey 2012).

Companies like Xerox found their stride during the period after World War II. Having changed the business world with the introduction of its 813 copier in 1963, Xerox set out to revolutionize PCD. In 1970 Xerox opened the Physical Activity Research Centre (PARC). Headed by Jack Goldman, a professor at the University of Washington, PARC created an explosion of new products, driven by collaborations between Stanford University students and PARC scientists. Among the products produced through this collaboration were the graphical user interface, laser printers, the Xerox Alto, and the Ethernet local area network (LAN) (Maxey 2012).

Today, by integrating the Internet and rapid prototyping with DT it is possible to realize that PDC is moving closer to its historical origins where a single individual can create something that changes the way people interact with the world. The Internet is enabling inventors and designers to realize their work quickly. For example, companies like "Invention Partner", "Quirky" and "Edison Nation" are examples of online resources available for innovators. They receive ideas that have the potential for licensing and production, if they fit with their product line, and connect the realized products to market. On the other hand, 3D printing makes it possible for product designers to visualize and share their concepts more economically at an initial stage.

9.3 Product development

> In the beginner's mind there are many possibilities,
> in the expert's mind there are few.
>
> **Shunryu Suzuki**

9.3.1 Product defined

A product is an object or service that is the result of design activity. It may be physical or in virtual or cyber form. Every product is made at a cost and each is offered at a price. A product needs to be functionally able to do what it is supposed to, and do it with a good quality. With trends, time, and change in parts, the product should lend itself to adjustment to make it more appropriate and maintain its value.

Products according to ISO 9000:2000 are results of a set of consisted or interacting activities which transform inputs into outputs, and include services, software, hardware, and processed materials, also known as goods and services, or as artifacts and processes. Some of these have a substantial contribution from engineering, technical systems, and technical processes (Eder and Hosnedl 2007).

In design theory of the early and middle decades of the twentieth century, products were often understood from an external perspective. The focus was on the form, function, materials, manner of production, and use of products. This is why form and function emerged so large in theoretical discussions of both graphic and industrial design, and why materials, tools, and techniques figured so prominently in the early phases of design education. With the move away from visual symbols and things as the focus of attention, designers and design theorists have tried to understand products from the inside, not physically inside, but inside the experience of the human beings that make and use them in situated social and cultural environments. While form, function, materials, and manner of production continue to be significant, there have been an opportunity for new understanding through an investigation of what makes a product useful, usable, and desirable (Buchanan 2001). Figure 9.1 shows the external and internal product perspectives.

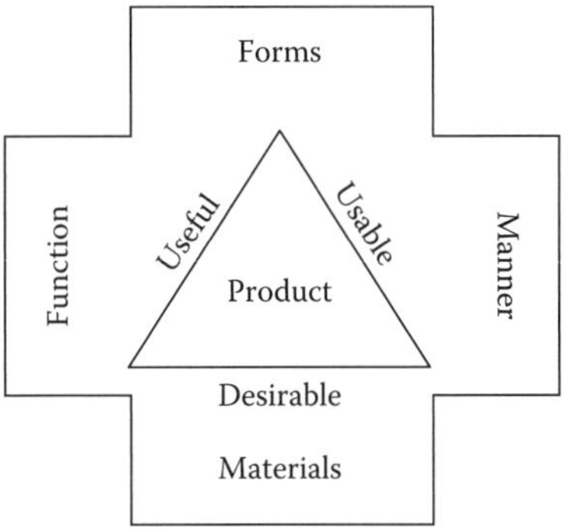

Figure 9.1 External and internal product perspectives.

9.3.2 *Product design and development*

It is widely accepted that both engineering and design are concerned with the evolution and development of products, systems, and technologies. Engineering is the creative process of turning abstract ideas into physical representations. PD appears as a key process of business and sustainability. However, the complex nature and uncertain environment which describe PD lead organizations toward collaboration in order to share risks, reduce costs and time to market, improve quality, and benefit from complementary knowledge and competence throughout the development process (Littler et al. 1995).

Design is one of the most important stages in PD because system performance greatly depends on the quality of design (Suh 1999). In the design context, the emphasis lies on how to build a physical object that realizes a certain function. This function is often depicted in terms of a list of specifications which the object to be designed must meet (Kroes 2002). The design process is usually iteration loop that follows the same basic steps over and over. For PD, there are different ways to represent the process, from the instant the initial idea comes to mind, or the job gets designated, through the design and implementation stages, until the product is in the market for sale.

Product design is the solving of a design problem from the initial idea to the final product design. Before manufacturing, every part has to be detailed with, geometry, material, and manufacturing. The design activity is cyclic or iterative, whereas analysis problem-solving is primarily sequential. In most cases, it involves a redesign of a previous product and can range from innovative to routine.

Product design is concerned with the efficient and effective generation and development of ideas through a process that leads to new products (Morris 2009). The design function includes engineering design (mechanical, electrical, software, control, etc.) and industrial design (aesthetics, ergonomics, user interfaces, etc.). The goal of the product design is to create a product that fulfills its desired functions, looks good, can be produced economically, and is sustainable. This may be accomplished by the designer with approaches that may bring embodiment design to the conceptual phase.

Engineering design is and was dealing with the future. This is one of the important arguments to shorten the PD time. Other arguments are the cost or competitors or technologies. Markets (customers, subsuppliers, custom and law situations, exchange rates, etc.) are changing with increasing speed. All of this leads to pressure on PD to be fulfilled in shorter time (faster, parallelized), robust (reduced risk), and transparent (for operation as well as management) in different dimensions (quality, cost, etc.) (Papalambros 2015).

9.3.3 Design methodology and design process

Design is an adaptable and progressing process, not very organized or a specific procedure one should follow on every project to ensure realization. Design methodology which is strongly process based aims at the improvement of design processes particularly by exploiting scientific techniques. A methodology is a body of knowledge comprising the principles, guidelines, systematic analysis, best practices, organization, and processes. There are a number of features that should be possessed by each and every design methodology. The features are noticeable elements characteristic of each and every successful engineering design endeavor. Therefore, methodology is a much broader concept than a process.

The design process may be seen as a set of successive cycle of design information where the design problem cycle is gradually transformed into a solution cycle. The process should start with a clear product scope and requirement. It should be flexible enough to meet the needs of both a particular organization and a particular project. While an effective design process does provide a framework within which work is carried out, there is always a need for innovation along the way when designing products.

Achieving optimal design solutions requires an effective design process that provides a framework within which designers can consistently deliver high-quality work. To the greatest degree possible within the constraints of a particular PD effort, this should be a user-centered design process, but such constraints also require that it be a flexible process (Gabriel-Petit 2010).

9.3.4 Axiomatic design methodology

There are seven major historical design methodologies that were developed along the past few decades. These include the seven-phase methodology by Morris Asimow (1962), the eight-stage model by Nigel Cross (2008), the four-phase model by Michael J. French (1998), the four-phase, six-step model by Vladimir Hubka and W. E. Eder (1995), the total design activity model by Stuart Pugh (1991), the four phases made up of seven stages by the Association of German Engineers (VDI 1987), the four main phases model by Pahl et al. (2011), and the ADM model by Nam P. Suh which will be discussed in this section (Suh 1997).

Axiomatic design methodology (ADM) in particular provides a systems-based framework for design that permits design alternatives to be evaluated based on quantitative analysis, eliminating the need for messy qualitative and cost-based models (Adams 2015). ADM was developed by Professor Nam P. Suh while at the Massachusetts Institute of Technology. Professor Suh's design framework is founded upon two axioms of systems theory, that he titles the independence axiom and the information axiom. Suh uses these axioms, in conjunction with the concept of domains to develop a framework where customer attributes are transformed into process variables in a completed design. The basic idea of an axiomatic design framework was envisioned by Dr. Suh in the mid-1970s and was first published in 1990 (Suh 1997) and updated in 2001 (Suh 2001).

A key concept in ADM is that of four domains: the customer domain, the functional domain, the physical domain, and the production domain as shown in Figure 8.6. The domain on the left relative to the domain on the right represents "what the designer wants to achieve," whereas the domain on the right represents the design solution, or "how the designer proposes to satisfy the problem." The customer domain is characterized by customer needs or the attributes the customer is looking for in a product, service, or system. In the functional domain, the designer formally specifies customer needs in terms of functional requirements. In order to satisfy these requirements, design parameters are conceived in the physical domain (Suh 1999).

A key concept of ADM is the independence axiom. The independence axiom states: maintain the independence of the functional requirements (Suh 2005). Simply stated, each functional requirement should be satisfied without affecting any other functional requirement. During the conceptualization process the functional requirements are transformed from the functional domain where they state "what," to the physical domain where they will be met by "how."

9.3.5 Typical steps in the engineering design process

According to Dictionary.com, the primary and long-standing sense of the noun process is "a series of progressive and interdependent steps by which an end is attained" or, more particularly, "a series of operations performed in the making of a product." Therefore, a process is holistic in nature and is devised with a specific goal in mind. A process is a change, procedure, course of events, taking place over a period of time, in which an object transforms from one state to a preferably more desirable state, generally called a transformation process. The smallest steps in a process are called operations. A technical process is that part of a transformation process performed mainly by or with the help of outputs delivered by a technical system.

While different organizations employ PD processes that are similar in many respects, their processes also differ in certain ways. The process usually begins with a problem and concludes with a solution, but the internal steps may vary. It is not a linear process. Successful design and problem solving require going back and forth between the main steps in the process. This is called an "open-ended" design process because when it starts to solve a problem, there is no idea about the best solution to meet the requirements. The process is iterative and may begin at, and return to, any step. In general, design process involves repeating the same steps down from open concepts to details. Once the developers learn the fundamentals of a process, they may easily apply it over and over again as the process evolves.

Among the fundamental elements of the design process is the establishment of objectives and criteria, synthesis, analysis, construction, testing, and evaluation. Furthermore, it is essential to include a variety of realistic constraints, such as economic factors, safety, reliability, aesthetics, ethics, and social impact (Haik and Shahin 2011). Typical steps in the engineering design process provide a framework on which tasks and decisions have to be executed in the various phases of the design project. The process is a series of interdependent and frequently overlapping activities that transform an idea into a prototype and on to a marketable product. Figure 9.2 shows the main project development steps as perceived by various investigators. However, in this chapter, four main stages namely idea, design, implementation, use and operation are adopted as shown in Figure 9.3. Each of the above stages incorporates multiple phases of activities.

Different factors affect the performance of design and PD process. These factors are related either to the product's macro- or micro-economic, technological, political, social, and environmental factors; user-friendliness, lifespan, size; and to the collaboration process

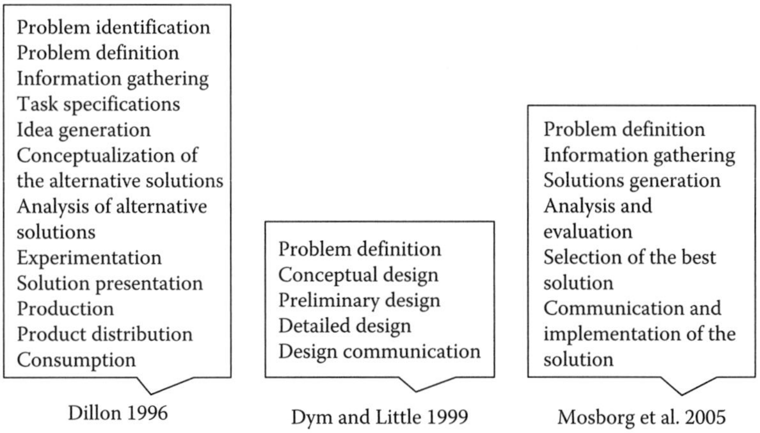

Figure 9.2 Questions from problem statement.

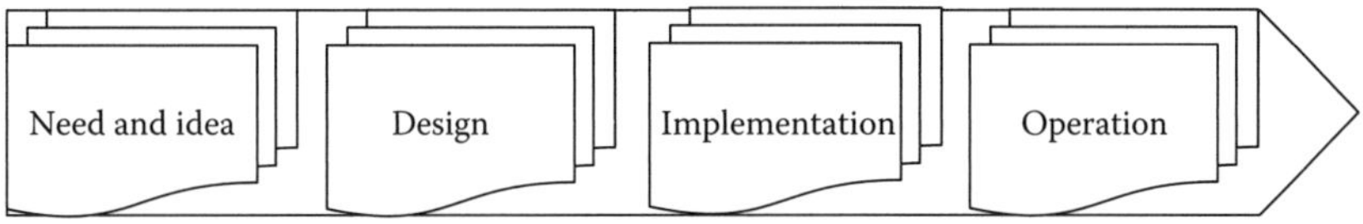

Figure 9.3 Stages of engineering design and PD.

incorporated with feedback and assessment of each stage of development as shown in Figure 9.4.

9.4 *Stage 1: Need and idea*

> If you generate one idea, it is probably a poor one. If you generate twenty ideas, you may have a good one.
>
> **David G. Ullman**

9.4.1 *Customer requirement*

Engineering design idea always happens as an answer to a human need. Prior to developing a clear definition statement for a design problem, it is needed to recognize the need for a new product, system, or service. The primary source of ideas is mostly customers, the motivating market force in the design of products and services. Simply, a customer requirement is

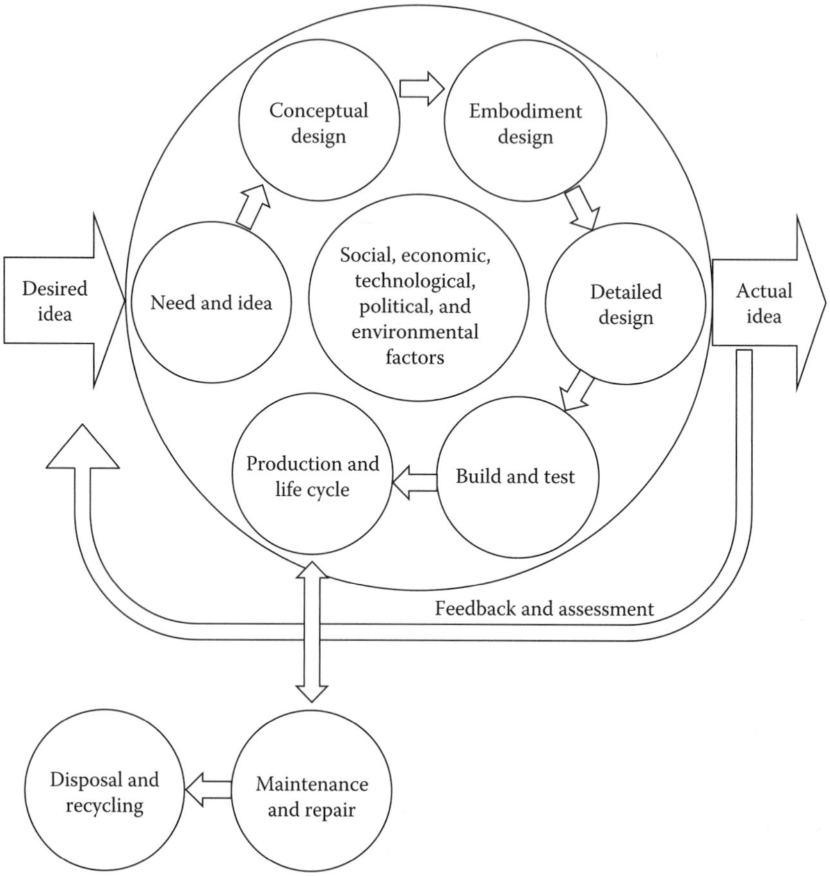

Figure 9.4 A typical PLC with emphasis on the development process.

a statement that specifies what a product should do and look like, but it does not specify how it should do it.

Finding an idea for a product requires identification of people's need or market's need (customers). The customers may be people, institutions, or other companies. In some circumstances, the customer and designer may be one and the same person. The needs of customers are often multiple and complex. However, establishing these needs is the most difficult aspect of market led design, since customers do not always know what they want. Therefore, the customer requirements represent the system's actual requirements.

Industry's survival depends on producing a product that people will buy and can be manufactured and sold at a profit. Ultimately, consumers establish a need, because they will purchase and use a product that they perceive as meeting a need for comfort, health, recreation, transportation, shelter, and so on. Likewise, members of community decide whether they need safe drinking water, roads and highways, libraries, schools, fire protection, and so on. The perceived need, however, may not be the real need. Before delving into the details of producing a solution, it is needed to make sure that enough information is available to generate a clear, unambiguous problem definition that addresses the real need. This initial phase of understanding the opportunity deals with the identification of customer needs which define the problem to be addressed using tools like customer survey interviews.

For the industry, competitors are another source of ideas. A company realizes by observing its competitors' products and services and the success rate of these products and services. This includes looking at product design, pricing scheme, and other aspects of the operation. Another way of using competitors' ideas is to buy a competitor's new product and study its design features. Using a process called reverse engineering a company's engineers gently disassemble the product and analyze its parts and structures. Product design ideas are also produced by a company's R&D department, whose role is to develop product and process innovation. Other sources of ideas are part suppliers, the company's employees, and new technological developments.

9.4.2 *Problem definition*

Once an idea is established for a product, it should be described by writing a problem statement. The problem statement must answer several questions as shown in Figure 9.5. Often, the results of the activities in this design step determine how the design problem is decomposed into smaller and more manageable design tasks. Sometimes not enough information is known yet about the problem and decomposition occurs later in the design process.

Problem definition is the preliminary thoughts about the new product, process, or service incorporating an understanding of user needs. It is a translation of technical ideas for meeting user needs into a preliminary design; initial calculations and drawings that demonstrate theoretical

What is the problem or need?
Who has the problem or need?
Why is it necessary to solve the problem?
What are the desired features?
What are the constraints?
What is the criterion of evaluating design?

Figure 9.5 Questions from problem statement.

validity of product definition. At this stage lots of time should be allocated and very little money should be spent. Critical thoughts and review of literature in depth should be the theme of this phase. All assumption should be challenged (OEERE 2000).

9.5 Stage 2: Engineering design process

Iterate, iterate, iterate!

The understanding of the design process is important for the teaching of design, the improvement of products, and the efficiency of engineering-based companies; it is also the foundation on which a lot of design research is conducted. The design process is cyclic or iterative in nature, whereas analysis problem-solving is primarily sequential. It is labor-intensive process culminating in the proposal of a product or process (Howard et al. 2007). There are several ways of describing the design process, of which three main phases have been identified as shown in Figure 9.6. During each phase, the design engineer can focus on a portion of the problem and the decisions taken are based on the accuracy of the forecasts made and the quality of criteria used.

If the problem-solving team has been successful in clearly defining constraints and criteria for solving the problem, they are well prepared for the stage of finding a solution. All three design phases require considerable level of system knowledge, procedural knowledge, and strategic knowledge. The traditional approach, however, tends to place more emphasize on application of procedural knowledge (Jonassen 2004).

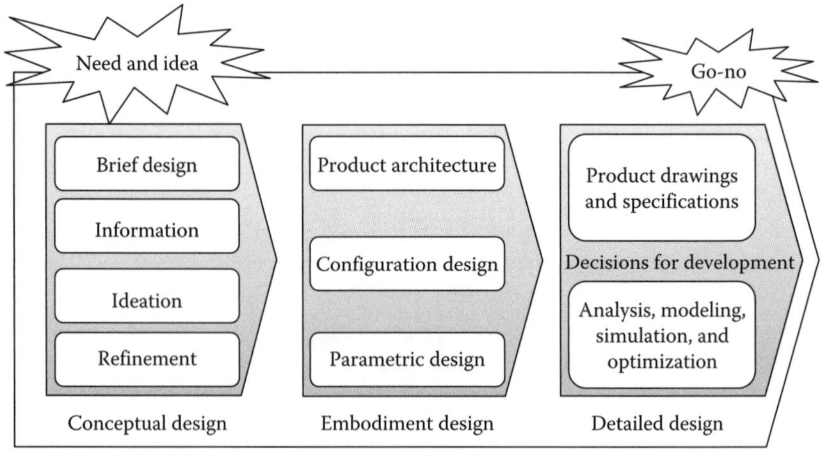

Figure 9.6 Phases in engineering design stage.

9.5.1 Conceptual design

Product conceptual design is the heart of product innovation and development, directly impacting DD in the latter phase of design. It is the very early phase of the product design process in which technical drawings or prototypes are the dominant tools. Conceptual design is conducted to determine a feasible system-level design baseline for a new concept. It involves the process of developing a research idea into a realistic design. Conceptual design emphasizes principles of structural design and material selection. It is as much about investigating requirements as it is about investigating design.

Conceptual design is a series of methodical, planned, and targeted design activities from analyzing needs of users to generating conceptual products, and expressing an evolving process from crude to refined, from fuzzy to clear and from abstract to concrete (Deng et al. 2002). It is common that as much as 60%–80% of the total product cost is committed already during the concept stage of the design process. Accordingly, a design methodology focusing on this early concept stage, front-loading certain design activities to it, has a much larger impact potential than the ones focusing mainly on the DD (Ullman 2010).

A concept is an idea that is sufficiently developed to evaluate the basic principles that govern its behavior. Concepts are means for providing function including how the product might be used and maintained over its lifetime. The main outputs are the conceptual designs, design specifications, project schedule, cost estimate, design review, and proposal report.

The concept generation phase should impede early judgments. Often creative ideas develop slowly and require time to proceed in an explicit manner. Thus, the concept generation phase should not be hampered by critical judgment at the initial level. This phase often includes the creation of high-quality surfaces to define the products shape. It can be time-consuming, depending on the level of expertise of the investigator. Figure 9.7 shows the path of concept generation starting from customer need.

9.5.1.1 Design brief

Design brief is about figuring out the objectives of the project as well as the starting point of the corresponding tasks. It is a formal document which describes what is to be designed and for which target markets. It develops from an analysis of the need or problem. The main purpose of design brief is to illustrate the design solution for a new product or service and how the solution is devised. It provides clear guidance on how the product is developed and details the design principles and the procedures to achieve

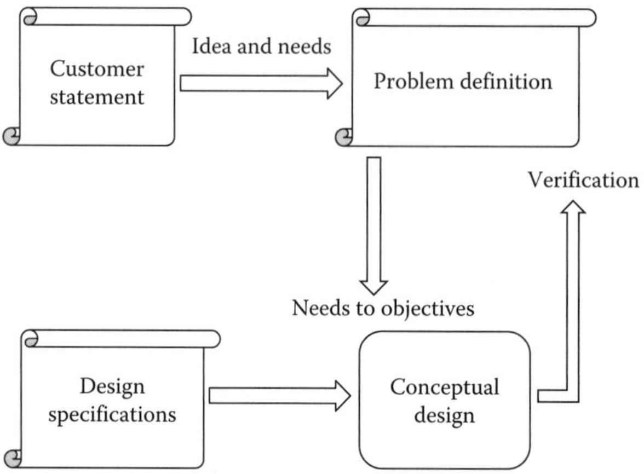

Figure 9.7 Path from customer need to conceptual design.

these principles. It is like a business plan for a specific project. Figure 9.8 shows the essential elements of a good design brief.

9.5.1.2 *Information and background research*

In this step relevant information should be gathered and background research on the problem should be carried out. The context of design challenge can be enhanced through the Internet, patent, standards and codes, textbook, handbook, and catalog research; literature published by vendors and suppliers; company reports; trade journals; product user manuals, user interviews; and other techniques such as web forums and online courses. In the following activities, the design teams utilize this knowledge about the problem to generate product design ideas. This usually takes some time as it should involve the collaboration of all

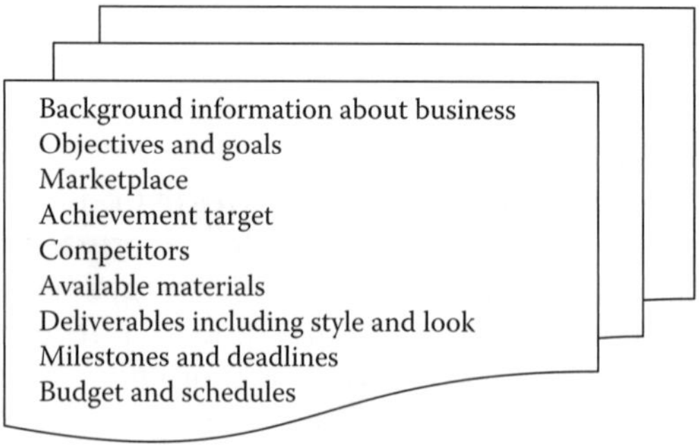

Figure 9.8 Essential elements of design brief.

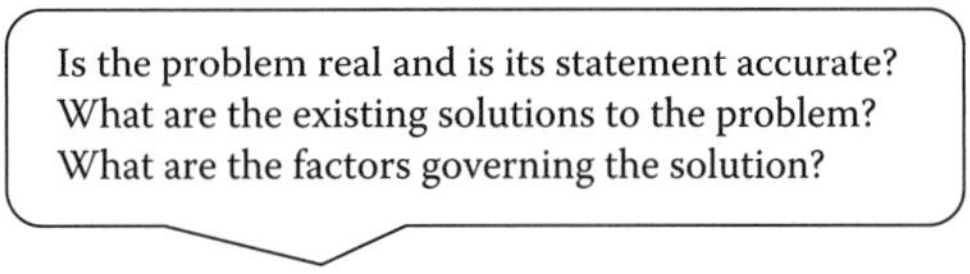

Figure 9.9 Questions that may be asked during research step of design.

members of the design team. Questions that may be asked at this stage are shown in Figure 9.9.

Background research is important for design projects to learn from the experience of others rather than stumble around and repeat their mistakes. To make a background research plan, a roadmap of the research topics should be outlined:

- Gaps in the literature
- State of science
- Target user or customer
- Existing products that solve similar problems
- Development and operation of the product
- Network of experienced people

9.5.1.3 Ideation

To get a good idea, get lots of ideas.

Ideation is an important activity in the design process. Ideation involves rapidly generating many different possible workflows and user interface design solutions and capturing them by sketching designs on easel pads or whiteboards or in notebooks (Gabriel-Petit 2010). During this process, the first kind of representation (e.g., freehand sketches and rough physical models) serves designers to exteriorize and visualize their design intentions or to communicate with each other. Later on, designers employ a second type of representation (e.g., digital 3D models, drawings and images) to better communicate asynchronously. At the end of the process, a third kind of representation is reached [e.g., detailed technical drawings and rapid prototyping (RP) models] to communicate exact and definitive information to build the product.

Ideation often happens not on a computer but rather through sketches on paper or mock-ups using soft materials, steering away from the exactness of digital representations and the inconsistencies of interfaces. Therefore, computers are limited to represent already designed ideas. In the early phase of design, where ideas are still not clear, traditional pen-and-paper sketches and physical models remain the tools of choice to do ideation because they are intuitive, are direct, and allow ambiguous, abstract, and imprecise representations (Dorta 2008).

During ideation, everyone on the design team has an opportunity to communicate requirements and constraints. The keyword here is creative thinking which is critical for concept generation and PD. The process of creative thinking can be viewed as a step to move from an unstructured idea to a well-structured one, from an implicit to an explicit design. Figure 9.10 shows the questions that might stimulate the flow of ideas.

9.5.1.4 *Brainstorming*

Brainstorming is one of the most powerful methods in the innovation toolbox. In the 1950s Alex Osborn (1953) advocated "brainstorming" (Figure 9.11) as a group interaction technique that produces more and better ideas. In brainstorming, quantity rather than quality of ideas is emphasized, criticism is forbidden, wild ideas and "free-wheeling" are welcomed, modification, combination, and improvement of ideas are sought.

Brainstorming is a group problem-solving process in an open form without criticism. It is a technique to generate ideas in a nonthreatening atmosphere. Coming up with many possible solutions is a powerful way to begin a project. As important as is the initial idea, it is not really an innovation until it has undergone further development. As knowledge flows toward realization and possible commercialization, the idea gains substance and intellectual capital. The process generally goes through several phases including the following:

Which new ideas may arise from combining purposes or functions?
What design elements from another product can be applied to new product?
What do certain ideas have in common?
How to refine ideas?
What differentiates ideas? What to add? What to eliminate?
How to improve ideas?
Is a certain component necessary?
Is there a new way to use it?

Figure 9.10 Questions that might stimulate the flow of ideas.

Rule 1: Postpone and with hold your judgment of ideas
Rule 2: Encourage wild and exaggerated ideas
Rule 3: Quantity counts at this stage, not quality
Rule 4: Build on the ideas put forward by others
Rule 5: Every person and every idea has equal worth

Figure 9.11 Classical brainstorming.

- Research in which basic concepts are tested
- Development, in which the elements of practicality and economy are alloyed with the fundamental concepts
- Demonstration, in which the best ideas are tested in pilot form or in full scale
- Commercialization, through manufacture, sales, licensing, and/or other steps

9.5.1.5 Refinement and concept evaluation

Evaluations (assessments) and decisions (selections) that are made in the conceptual phase of the design process are crucial, since a decision made in this phase will act as a watershed for future activities of the process (Derelöv 2009). Evaluation involves comparison and decision-making. Evaluation is based on feasibility of design or readiness of technology. In Figure 9.12 the conditions outlined above are formalized into a basic process for the evaluation and decision-making.

As the expression indicates, evaluation and decision-making comprise two separate parts: one evaluation part and one decision-making part. Even if evaluation and decision-making are often closely associated with each other when discussing design issues, it is important to keep in mind that they are, in theory, two widely different activities (Derelöv 2009).

One quantitative technique, called the Pugh method, helps engineers in design decisions by establishing a procedure to choose the best design from the considered designs. This method is also known as decision-matrix or Pugh concept selection. Pugh matrix invented by Stuart Pugh (1981) can be used whenever there is need to decide among a number of alternatives. This is a widely accepted method for comparing concepts that are not refined enough for direct comparison with engineering requirements. Steps involved in this method are shown in Table 9.1.

9.5.2 Embodiment design

The embodiment design phase takes the abstract conceptual path, chosen in the conceptual design phase, and mold it into a system that can actually be produced. When a conceptual design is completed, a set of concepts is

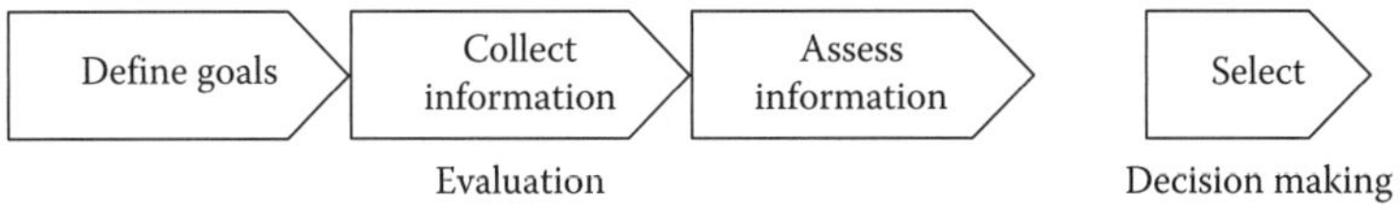

Figure 9.12 Outline of a basic evaluation and decision-making process.

Table 9.1 Pugh concept

Steps	Detailed
1	Develop criteria for comparison: make a list of the criteria to compare between different designs. Each criterion should be an objectively quantifiable measure. The criteria can be identified by examining customer needs and generating a corresponding set of engineering requirements.
2	Select alternatives for comparison: the alternatives refer to the alternate ideas developed during concept generation (brainstorming). All concepts should be compared at the same level of generalization.
3	Generate scores: establish weights factors for each criterion. A number between 1 and 10 can be chosen for each criterion.
4	Compute the total score: calculate each design score.

evaluated to produce a single concept or a small set of concepts for further development. Embodiment design starts to finalize product architecture, quantify important design parameters, material and process selection, and determine form and shape of parts that will satisfy required functions.

The most time absorbing part of the design process is, in general, embodiment design: going from idea to realization. Kesselring (1954) was the first to refer to embodiment design and introduced a set of principles: minimum manufacturing costs, minimum requirements, minimum of weight, minimum losses, and optimal handling. These principles are often calculated at the end of the design process and are typically used as verification. Embodiment design is a design phase in which, starting from the concept of a technical product, the design is developed in accordance with technical and economic criteria and in the light of further information, to the point where subsequent DD can lead directly to PD.

The embodiment phase is the bridge between the conceptual phase and the DD phase of the design process stage. Its input is an outline sketch and project documentation as well as design requirements. The output is a detailed scheme drawing accompanied by documentation including calculations, dimensions and tolerances, and proposed materials. In addition, it includes shape, style, size, and general appearance. Importantly, embodiment design provides both technical solutions and characteristics of the final product that meets the need and demand of the customer. Figure 9.13 shows the domain of embodiment design.

9.5.2.1 Product architecture

Product architecture is the plan step of the product physical components to facilitate the product required function. This concept, with respect to product design, is identical with the layout, configuration, or topology of functions and embodiments. This process affects most aspects of product

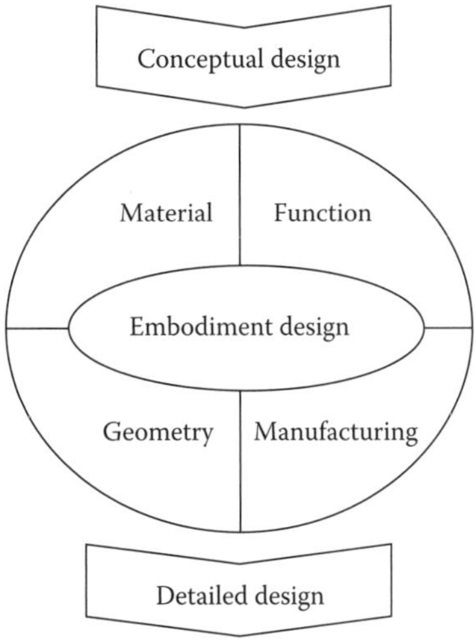

Figure 9.13 Domain of embodiment design.

design and manufacturing including product design, product structure, and customization. Usually, the outcomes of architecture decisions are implemented during assembly.

Architecture emerges informally during the concept generation phase and becomes an explicit concern during configuration or layout design (Ulrich and Eppinger 2000). Given the geometric and spatial constraints associated with form solutions, a large number of design issues arise during this stage such as part number and complexity, manufacturing and assembly, product family variety, standardization, serviceability, and industrial design.

The basic layout and the architecture of the product are established by defining the basic building blocks of the product in terms of function and interfaces. These basic building blocks are also known as chunks. Therefore, product architecture identifies the product in terms of chunks, or the functional systems, and how these systems are arranged to work together. Each chunk is an assembly of components that would accomplish a certain function. It is concerned with dividing the overall system into small subsystems and modules. This step determines the organization of the physical components that perform the functional duties of the system.

There are two styles of product architecture. First is the modular architecture and second is the integral architecture. In the first style, each function is delivered by an independent element; the building blocks implement only one or a few intended functions, and the interactions between the building blocks are well defined. In the case of integral architecture, functions are shared by physical elements; and the operation of a function is carried out by only one or few building blocks often leading interfaces between the building blocks. Typical product architecture contains a blend of both the modular and the integral architecture.

Ullman (2001) states that: engineers generally work from the function of a system, to the architecture of an assembly, to the shape of parts. The goal of this architecture design task is ultimately to create a spatial arrangement of components and assemblies although initial steps may include manipulation (chunking/partitioning) of a functional model to impose some desired modularity and/or integrality early on. Most products have both characteristics, modular and integral. In general, the product architecture can be established in four-step processes:

- Create a schematic diagram of the product.
- Cluster the elements of the schematic.
- Create a rough geometric such as 2D drawings and RP.
- Identify the fundamental and incidental interactions given by geometric arrangement, and physical implementation of functional elements.

9.5.2.2 *Configuration design*

Configuration design is a form of design where a set of predefined components is given and an assembly of selected components is sought that satisfies a set of requirements and obeys a set of constraints. Shape and general dimensions of components are established in this step. According to Mittal and Frayman (1989), the configuration of an artifact is a set of interconnected components that are chosen from predefined sets of component types called the catalog of component types. Specifically, a component is described by a set of properties, ports for connecting it to other components, constraints at each port that describe the components that can be connected at that port, and other structural constraints.

The configuration design consists of the following three constituent tasks: selection of components, allocation of components, and interfacing of components. The shape and the general dimensions of components are realized although the exact dimensions and tolerances would be finally established during parametric design in a later step. This step represents the beginning of the manufacturing process. The configuration design strongly depends on the availability of the materials and production procedures that would be used to develop the product.

9.5.2.3 *Parametric design*

Parametric can be defined as any set of physical properties whose values establish the characteristics of a system. Parametric design is a common framework which is primarily concerned with design variables; the development of solutions for specific values and attributes of various design elements that exist in the configuration design. It starts with information from the configuration design process and aims to establish the exact dimensions and tolerances of the product as well as to assess if the design is successful or not.

Parametric design is also about setting the dimensions and tolerances so as to enhance quality and performance and optimize the cost of manufacturing. Some aspects of parametric design include design for reliability, robust design, and tolerance. Permissible tolerances must be placed on dimensions to control the acceptable variations in the dimension of a product.

Engineering simulation, with its ability to design, prototype, and test products in the low-risk virtual world, has provided many companies with a fast, cost-effective way to create robust product designs (Boucher 2008). These advanced solutions have an enormous potential to support parametric analyses in which certain design parameters are modified and the effect of these variations is studied across the entire design in an iterative process. By understanding the impact of each small change, the speed of the PD process can be increased by a large factor (ANSYS 2011).

9.5.3 Detailed design

DD of the product is the last design activity before implementation begins. It is the phase wherein the necessary engineering is done for every component of the product. Tolerances, materials, and finishes are defined, and the design is documented with product overall layout and operational flows. In this phase, the design reaches a state where it has the complete engineering description of a tested product. Engineers work closely with PD team to ensure designs take full advantage of technology opportunities and observe all technical constraints.

Alternatively, interactive prototypes can be developed, with either low-or high-fidelity interactivity that shows how the product should behave. However, when developing physical prototypes, it is still needed to write specifications to fill the details that have not been implemented in the prototype to ensure development team does not make any wrong assumptions. The working model is often less than full scale, inexpensively and roughly constructed, and need not function optimally. It is meant to test the most basic operating parameters and to aid in the design of an engineering prototype. This is an actual working form of a product which is used to gather data on operating performance and production requirements (OEERE 2000).

9.5.3.1 *Design for manufacturing*

DfA is a design method of a product for ease of assembly. It is a series of guidelines to follow in order to produce a product easily and profitably. The guidelines focus on design simplification which means reducing the number of parts of the product. A simple product reduces the material, overhead and labor, and overall part production cost; shortens the PD cycle; and minimizes the complexity of manufacturing. By using interchangeable parts it is possible to build a great range of flexible products with less inventory and significantly lower cost. Figure 9.14 shows the guidelines for DfM.

Computer technology is used by designers and manufacturers in DfM. It can be used in estimating and reducing the total number of parts of the product. It helps in identifying and designing multipurpose parts to be multifunctional wherever possible.

9.5.3.2 *Design for assembly*

The aim of DfA is to simplify the product so that the cost of assembly is reduced. It means that components are designed in order to optimize proper assembly and function. It supports the analysis and design of products for ease of assembly. The impact of DfA is visible throughout the overall design and manufacturing process. For example, DfA reduces the number of parts of a product and this shortens the assembly time, which leads to a decrease in costs. It is a tool used to assist in product design that will end up productions at a minimum cost, focusing on the number of parts, handling and ease of assembly. In general, it encourages modular design, design parts for retrieval, handling, and component symmetry for insertion. DfA helps the designer to focus on the relationship between the features of a design and its components and the effort and resources necessary to assemble these components into the desired product.

Simplify operation
Minimize part count
Adopt modular design and avoid tools
Design parts for multiple products
Create modular assemblies
Simplify and reduce the number of manufacturing processes
Standardize parts and materials

Figure 9.14 Guidelines for DfM.

9.5.3.3 Design for operability

DfO takes into consideration the needs of the operator and user of the product or system. This means a competitive system must have a reasonable operational cost and an appropriate operational value because of the high economic impact of design in operability.

The complexity of the product that is presented to the operator including the array of components must be of concern in design. The various components comprising the product should be clearly labeled for easy identification. Those items that must be manipulated in order to connect the product (cable connectors, patch panels, switches) should be arranged in a simple and logical order and should be plainly marked (Suh 1999). Mock-ups and prototypes under realistic conditions are very helpful in uncovering potential problems early in the design process.

9.5.3.4 Design for maintainability

Maintainability is a design parameter that defines the degree to which a product allows safe, quick, and easy replacement of its component parts. Maintenance can thus be an important consideration in the long-term effectiveness of a product. It is the probability that, when a specified maintenance action is taken, a failed product will be restored to operable conditions in a specified downtime. Thus, design features that will accelerate maintenance will enhance maintainability.

DfM means inclusion in the design of those features that can be conceived to assist in the maintenance process. Specific features include the degree of accessibility for product replacement, facilities for fault isolation, special tools or test product requirements, the level of servicing skills required, servicing documentation requirements, and spare part stocking requirements.

DfM is a significant aspect of any system lifecycle. It is the degree to which a product allows safe, quick, and easy replacement of its component parts. It is embodied in the design of the product. There are several approaches to evaluate the maintainability of a product at the design stage. They are maintainability design checklists, maintainability evaluation using physical mock-ups, maintainability evaluation using digital mock-ups and virtual reality, and maintainability evaluation using quantitative approaches (Ding 2009).

Modularity improves maintainability, but carries cost penalties. Few operational products can be perfectly reliable while meeting other product trade-offs such as rust.

Critical rating factors include grouping of components by electrical function, use of integral fault indication for basic modules; components or functional assemblies removable without interruption of permanent electrical connections; elimination of tool requirements for mechanical disassembly; direct access to removable assemblies; products commonality; and identification of replaceable components (Suh 1999).

9.5.3.5 Design for environment

DfE is a way to scientifically consider design performance with respect to environmen, health, and safety over the product life cycle (PLC). It blends environmental aspects into product design to enhance performance throughout its lifecycle. The idea behind DfE is to ensure that all relevant environmental considerations and constraints are integrated into product design and realization process. The driving force behind DfE includes customers and governmental agencies, who are all stake holders in the environmental well-being.

Incorporating a DfE process that fits into the existing PD process has significant potential to help manufacturing firms achieve their environmental objectives. It bears in mind the potential environmental impact throughout the life cycle of the product including emission of harmful substances, excessive use of energy or nonrenewable energy sources. It also considers the life cycle of the materials from extraction to disposal. In this way the designers do not create just a product but a whole life cycle.

DfE involves investigating the impact of quantities like temperature, shock and vibration, radiation, and more. Temperature is a powerful agent for electrical, chemical, and physical deterioration for two basic reasons. The physical properties of almost all known materials are modified by changes in temperature, temperature gradients, and temperature extremes. The rate of most chemical reactions is influenced by the temperature of the reactants. Shock and vibration can harmfully flex leads and interconnects, dislodge parts or foreign particles into bearings, pumps, and electronics, cause acoustical and electrical noise, and lead to structural instabilities. Protection against the effect of electromagnetic radiation has become a sophisticated engineering field of electromagnetic compatibility design. When exposure cannot be avoided, shielding and filtering are important protective measures (Suh 1999).

Virtually all methods for reducing the impact of an engineering design are predicted on intentionally designing to reduce, reuse, and recycle mass and energy. These are called the three Rs (reduce, reuse, recycle) and their order of preference is indicated in terms of design. Reduce means try to design a system that requires a reduced amount of mass and energy. Reuse means try to design a system that can be reused as many times as possible. Recycle means plan to use material in a different form (McCahan et al. 2015).

9.5.3.6 Design for excellent

Design for eXcellencet (DfX) is a program and toolbox for proactively including end-user experience in the PD phase. It emphasizes the consideration of all design goals and related constraints in the early product design stage (Sangarappillai and Peter 2001). By considering all goals and constraints early, developers can produce better products. The product will enter the marketplace earlier because an inherently simpler product is designed correctly the first time without the introduction of problems, delays, and changes of orders.

DfX guidelines implementation has led to enormous benefits including simplification of products, reduction of assembly and manufacturing costs, improvement of quality, and reduction of time to market. Environmental concerns required that disassembly and recycling issues should be considered during the product design phases. The effort to reduce total lifecycle cost for a product through design innovation is becoming an essential part of the current manufacturing industry (Kuo et al. 2001).

9.5.3.7 Design for sustainable mass customization

Mass customization is defined as design and manufacture of customized products at mass production efficiency and speed (Anderson and Pine 1997). It is a business strategy that emerged due to an increasing market demand for individually customized products. However, demands for products which are sustainable have also increased. It uses flexible design processes and manufacturing systems to produce a variety of customized products at a lower cost than standardized mass-production systems; it can provide customers with products capable of fulfilling most of their individual needs. Automation, happening in mass customizing production systems, may lead to shorter lead times and may reduce costs if savings generated overcome initial investments in automating equipment. Also, modular product design as an enabler of mass customization has a positive impact on product recovery at its end of life. The added value of customization must be balanced among the product costs, manufacturing costs, and PD times.

The question is whether manufacturing customized products is really compatible with sustainability. It is known that achieving sustainability in mass customization implies a number of challenges which are different from those in achieving sustainability in mass production and in relation to product design, manufacturing, logistics, reuse, recycling, etc.

In order to define design for sustainable mass customization (DfSMC), DfX guideline, each of the DfX guidelines depicted in Table 9.2, should be studied individually in order to take those useful requirements that will be integrated to achieve the proposed DfSMC guideline.

Table 9.2 DfX guidelines studied toward DfSMC guideline

DfA	Design for assembly	DfEr	Design for ergonomics
DfD	Design for disassembly	DfC	Design for cost
DfM	Design for manufacturing	DfR	Design for reliability
DfSS	Design for six sigma	DfSTM	Design for short time to market
DfT	Design for testing	DfS	Design for safety
DfQ	Design for quality	DfMR	Design for minimum risk
DfSC&L	Design for supply chain and logistics	DfMN	Design for maintenance
DfR	Design for recycling		

Source: Osorio, J. et al., Design for sustainable mass-customization: Design guidelines for sustainable mass-customized products, The *20th International Conference on Engineering Technology*, 1–9, 2014.

9.5.4 *Tools in DD: Modeling, simulation, and optimization*

Design analysis involves a systematic step-by-step examination of all phases of the design of a particular item in relation to the function it performs. Designers often have the inclination to jump directly working on a solution without going through the analysis.

Analysis is the core of being an engineer; it is what differentiates an engineer from others. Engineering analysis helps in making evaluations and guides the design process. A design project without analysis is like a sports team without a coach! So what is engineering analysis? Basically, it is the breaking down of an object, system, or problem, into its fundamental parts to understand their relationships to each other and to outside elements. Figure 9.15 describes engineering analysis as the integration of

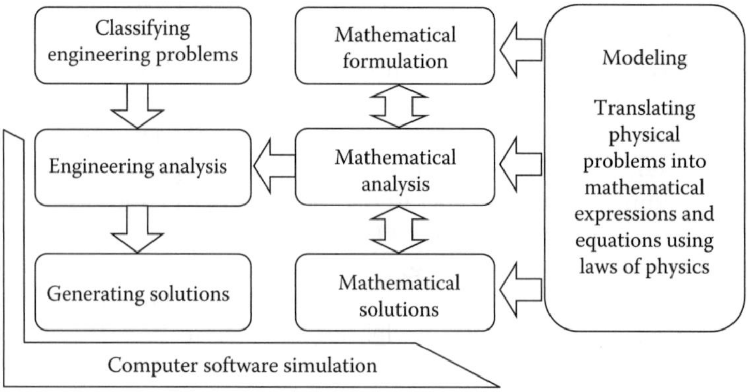

Figure 9.15 Integration of mathematical modeling and computer simulation in engineering analysis.

mathematical modeling and computer simulation guided by engineering problems under considerations.

One of the most fundamental tools used in engineering analysis is a model. Gass and Harris (1996) define a model as an idealized representation, an abstract and simplified description, of a real-world situation that is to be studied and/or analyzed. The basic function of a model is to transform a number of known input variables into a number of output variables, whose values are sought. Models are created because they are easier to manipulate than the real-world system or because they provide enhanced insight into the behavior of the real-world system.

Mathematical modeling is a practice involving the translation of physical (engineering) situations into mathematical forms with empirical formulas, algebraic equations and formulas from textbooks and handbooks, differential and integral equations with appropriate conditions fit the specific problems, and numerical solutions such as finite element method or finite difference method.

Modeling and simulation enables designers to test whether design specifications are met by using virtual rather than physical experiments. Simulation is the process of exercising a model for a particular instantiation of the system and specific set of inputs in order to predict the system response. Engineers usually use models for thinking, communicating, predicting, and controlling of design detailed. A simple mathematical model often helps improve a conceptual design. In embodiment design, full-scale modeling approach is needed. Modeling represents as a proof-of-concept model, prototype model, and scale model.

The use of virtual prototypes shortens the design cycle and reduces its cost. The introduction of CAD and CAE has had a major impact on the ability to produce better products. Simulation-based design is a process in which simulation is the primary means of design evaluation and verification.

Most CAD systems are effective at producing precise and accurate renderings of well-defined designs. They also include detailed analysis packages for obtaining simulation feedback on the performance of a design. However, very few CAD packages address the desires of a designer during the initial conceptual trial phase of design when the problem is still imprecise, and the solution has not yet been settled. CAD modeling can be divided into three categories: wireframe, surface, and solid modeling. Solid modeling provides better visualization together with a number of other key features, including fast 3D rendering, rounding, and history control.

Success or failure of engineering projects depends on the type of engineering analysis used to evaluate the design. This involves modeling, experimentation with different materials, and fastening techniques, shapes, and other things that need to be done before actual construction of the final design is undertaken. At this point, the designer begins to develop models and prototypes that represent the desired idea.

9.6 Stage 3: Implementation

> I love taking an idea to a prototype and then to a product that millions of people use.
>
> **Susan Wojcicki**

Once the DD has been completed and approved, it needs to be implemented and produced. Depending on the nature of the problem being solved, the solution to the problem could vary wildly and the implementation could also vary. It could consist of using a new process that was designed earlier, or it could consist of a manufacturing plan and producing of some physical object. This stage of PD involves several phases as shown in Figure 9.16. In all phases in this stage, one may find that potential solution is flawed and have to back up to a previous step to get a workable solution with proper testing.

9.6.1 Prototyping

Prototyping is the early phase of implementation of a new product. It is the first fully operational production of the complete design including test and evaluation of solutions. A prototype is a full-scale working model, technically and visually. The purpose of the prototype is to verify that the design meets all the customer requirements and performance criteria. Usually, hand built, the prototype must conform as closely as possible to the design standards for the final full-production product. Broad testing of the prototype provides the needed information for reliability and robustness of the design. It will also verify the environmental, safety, health, and legal requirements. Many designs are complex enough that modeling and calculations are not sufficient enough. In general, prototypes allow designers a chance to examine the design more closely and even test it before producing the final product.

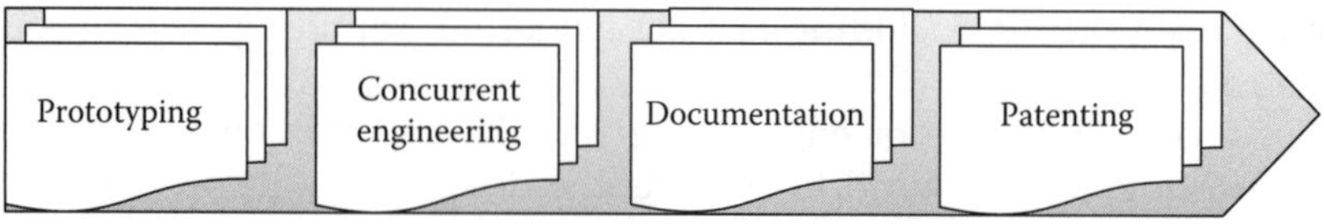

Figure 9.16 Phases of implementation and test stage of PD.

Prototyping of physical systems has benefitted greatly from the emerging field of RP. The main enabling technology behind time compression engineering is 3D CAD modeling. RP may be defined as an automated process which allows solid physical parts to be made directly from computer data in a short time. It acts as the manufacturing middle to link up the CAD process and manufacturing processes. If different design and manufacturing activities are carried out concurrently, it is possible to reduce the overall PD time. CE environments have evolved considerably to integrate 3D modeling with CAM, CAE, and rapid prototyping and manufacturing (RP&M).

In software engineering design, a prototype is an abridged version of an algorithm used to test the code for user interface, suitability, and appeal.

9.6.2 *Implementing concurrent engineering*

As discussed in Chapter 8, CE may be defined as the design of the complete lifecycle of a product simultaneously using a product design team and engineering and production tools. The definition stresses the importance of two key factors, people and equipment, with an emphasis on their interdependence upon each other.

The CE approach, as shown in Figure 9.17, seeks to detail the design while simultaneously developing production capability, field-support capability, and quality. The methodology features multifunctional teams that apply tools in the form of algorithms, software, and techniques to achieve concurrency in product, process, tool, and system design. CE in general addresses three main areas including people, process, and technology.

CE brings together interdisciplinary teams in which product developers from different functions work together and in parallel from the start of the project with the desire to get things right as quickly as possible. Such a transdisciplinary team may contain representatives of different functions such as electrical engineering, mechanical engineering, system engineering, manufacturing and production, maintainability and testability, computing and layout, and others.

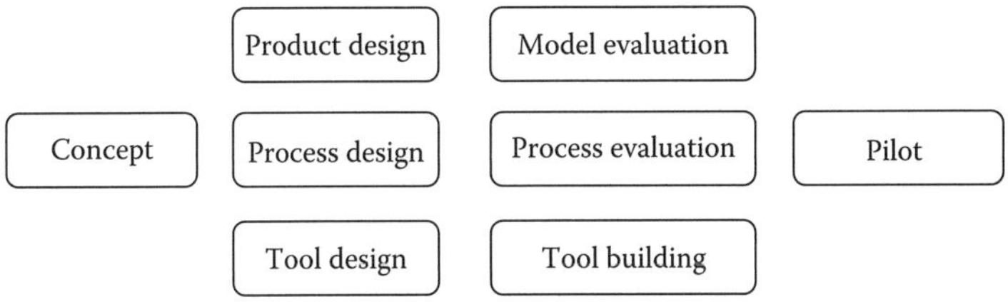

Figure 9.17 Implementing concurrent engineering in PD.

Processes for CE include DfM, DfA, DfQ, design for the life cycle, and design for cost (DfC). The main advantage for implementing CE is the extensive development of computer-aided engineering (CAE). Some examples of possible technologies include CAD/CAM/CAE systems, RP, rapid manufacturing, as well as technologies that enable the presentation of product design in virtual context, namely virtual reality which is a relatively new technology that involves the use of computers to create a digital prototype.

In CE, team spirit and leadership is crucial to the whole PD processes. As an approach CE significantly decreases time to market, enables faster PD, improves the quality of new designs, lowers work in progress, results in fewer engineering change orders, and increases productivity.

The steps of analysis, verification, design evaluation and review associated with product design may be undertaken using computing equipment and by building prototypes for experimental testing. RP&M is an effective tool that allows the CE design team to undertake engineering analysis and verification within the product and process cycle, shortening the time-to-market cycle. Specifically RP&M offer the CE team the advantages of visualization, verification, iteration, and optimization.

9.6.3 Documentation and communication

One of the most important activities in design is documenting and communicating the design information Design documentation is a vital part of the implementation stage since it forms the majority of the results. Its purpose is to communicate information in a way that is feasible and reliable. The quality of the documentation will largely determine whether or not a design project is successful. Written documentation provides the "glue" that stabilizes components and unifies the project. Decent design documentation does not just specify all the design details but can also communicate the high-level story, connect together the full picture, and get stakeholders excited about the vision of the project. It clearly defines information structure and creates trust and provides consistency for future iterations of the design thinking.

Communication assures coordination of effort across stakeholders. It may take many forms: written, oral, both written and oral (presentation), and graphical (drawing and pictures). The best design documentation provides the client a unified design language, a framework for talking about the design, and a platform for improving the design over time. Static documentation is quickly becoming a thing of the past. Currently, clients are looking for interactive, functional (prototype with notations as necessary) documentation techniques and delivery mechanisms. Providing design documentation marks the beginning of the client's journey, not the end. Finally, documentation and prototypes are mutually exclusive. Both are valuable in conveying the bigger picture and have a significant place in the design process.

9.6.4 *Intellectual property*

As discussed in Chapter 3, it is necessary to create some form of intellectual property (design registration, copyright, patent, etc.) for the final product. Design registration only protects the external appearance of a product. Copyright protects the work from being copied. It does not protect the conceptual content of the work. On the other hand, patents protect concepts, methods of manufacture, and the way a product operates. If an original solution to a design problem is developed, part of the implementation stage may include applying for a patent on the solution. In order for a design to be patentable it must meet several basic criteria including being novel and useful. Novelty means the design must not have been disclosed to the public prior to the application for the patent. It is a basic requirement in any examination as to matter and is an undisputed condition of patentability.

With a start-up operation, a patent will be considered an asset by investors and lenders. When intellectual property is included in the business plan, it will be easier to attract funding. For a company already in the business, the decision on whether to proceed with a patent application will primarily be a business priority.

9.6.5 *Iteration and development support*

Iteration in design has different meanings, ranging from simple task repetition to heuristic reasoning processes. Iteration is a process of repeating over and over, in a loop, in order to come as close to an ideal solution as possible, given time, resources and technology (McCahen et al. 2015). Ulrich and Eppinger (2000) formally define iteration as repeating an already completed task to incorporate new information (such as performing analysis followed by design revision, then repeating the analysis on the revised design).

Iterative design is an approach based on a continual process of prototyping, testing, analyzing, and improving a product. Changes and modifications are carried out based on the outcomes of testing of the most recent iteration cycle. Iterative design can be used at any phase of the design process, including when the product has already been in the market and the manufacturer is aiming to improve it. Importantly, the earlier in a product's life cycle to implement the iterative design, the more cost-effective the approach will be.

Designers know that iteration does not just take place at the end of the process; it happens during every stage and phase of the product development process. The process is not linear; it is common to jump from one step to another. Sometimes a designer may jump back and forth between steps several times before ever moving onto the next step. The goal is to create the best design possible by keeping improving it. The iterative process of design refinement should continue throughout PD as developers realize they need additional design details.

In industry the process will be more variable and unpredictable. It will be dictated by the flow of the project, changing technologies and circumstances, and input from the user. However, in school projects, a three-round sequence of iteration steps to reach an optimal solution may be suggested (McCahen et al. 2015). Figure 9.18 shows a three-round iteration process and the iteration details.

In iteration 1 the designer generates a large number of solutions. Two techniques are used: benchmarking existing designs and brainstorming. During brainstorming a group of design engineers and other participants generate as many proposals for solutions as possible within a given period of time. At this stage, solutions are evaluated against functions, objectives,

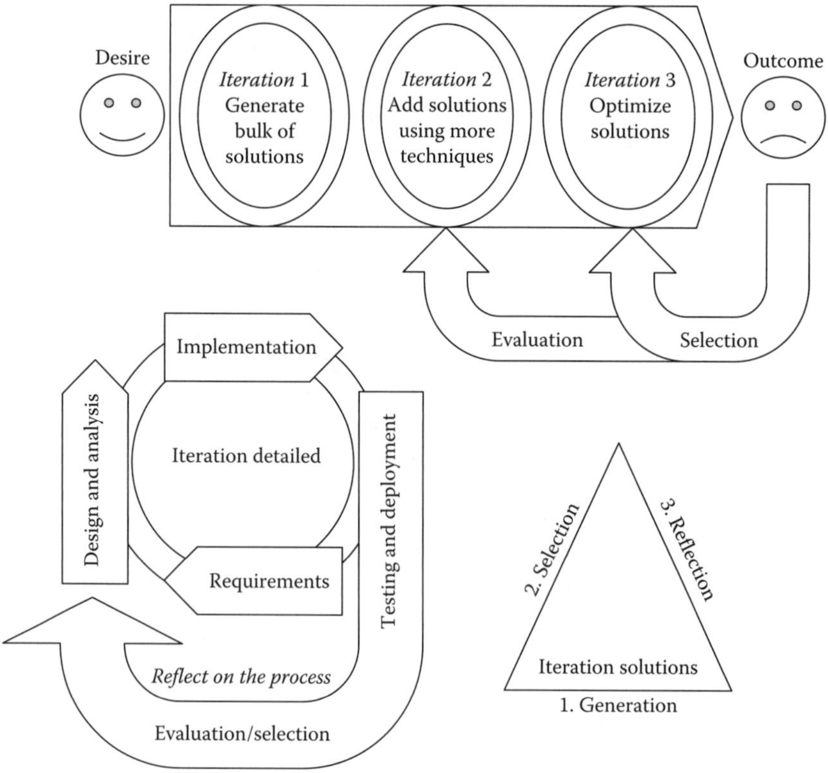

Figure 9.18 A multiround iteration process.

and constraints. If the designer feels that the solutions fall short of expectations, then the round should be repeated, including idea generation.

In iteration 2, the designer uses comparative techniques and deeper analysis to increase number of solutions and then reduce the number severely during the selection activity. Solutions are evaluated against functions, objectives, and constraints. Interpretation may depend on user wishes. Solutions are reduced but not to a single one. That will happen in iteration 3.

Iteration 3 will be left with a single design solution to move forward for DD work. In a typical product design, the bulk of the cost and time are incurred after iteration 3.

Assuming that design is by its nature an iterative and generative process, how should we understand waste in design? Waste has been characterized by Koskela and Huovila (1997) in terms of minimizing what is unnecessary for task completion and value generation. Consequently, that iteration is wasteful which can be eliminated without loss of value or causing failure to complete the project. Precisely what iteration can be thus eliminated is a matter for empirical research. Informal surveys of design teams have revealed estimates as high as 50% of design time spent on needless (negative) iteration.

9.7 *Stage 4: Sell and PD life cycle*

Great companies are built on great products.

Elon Musk

9.7.1 *New product development*

NPD is the full process of bringing a new product to market. It is described as the transformation of a market opportunity into a product available for sale (Krishnan and Ulrich 2001). It is a multiphase process that involves several steps; however, the number of phases and their descriptions vary from model to model. In this context, a related notion is the time for which a consumer uses the purchased product before it is replaced by a new one. This can be called period of ownership.

Conceptually, there are four key PD stages as shown in Figure 9.19. Each of the above stages can be broken down even further into several controllable tasks and milestones. These may be further expanded to accommodate various concepts of PD as shown in Figure 9.20 which outlines the stages and tasks of NPD. Product concepts provide detailed versions of new product ideas. The market need or opportunity for a new product idea can be attributed to advances in technology that provide an opportunity to improve existing products; need to improve existing

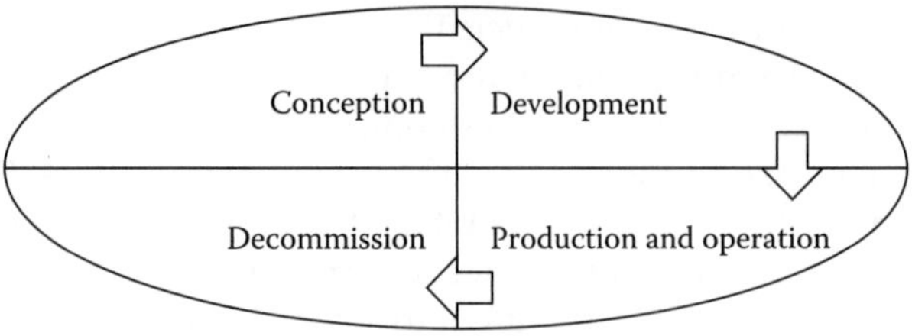

Figure 9.19 Key PD stages.

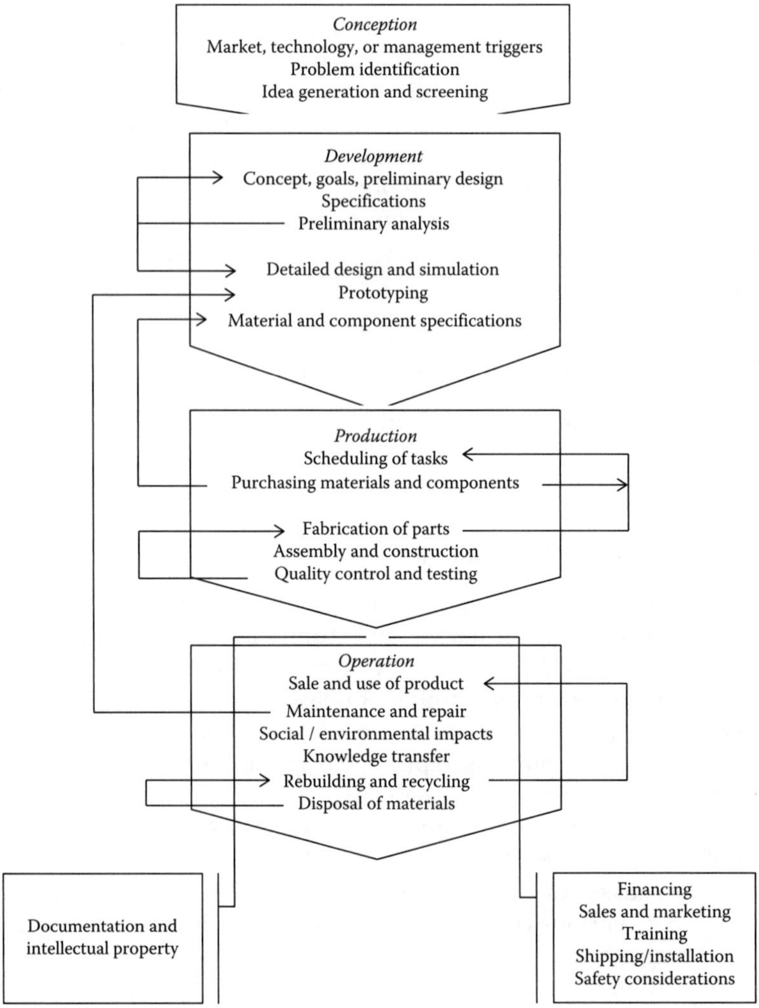

Figure 9.20 Stages and tasks of NPD.

product in response to competitor actions; and firm inspiration for development. The new product drivers generate a constant flow of new product ideas. There is also a constant screening of ideas to decide which ones to pursue further.

The development stage is concerned with the design specifications of the product to arrive at characteristics that may provide the desired product attributes determined in the conception stage. During this stage, the product may still be just an idea in the process of being manufactured or not yet ready for sale.

The production processes, in general, transforms tangible inputs (materials and assemblies) and intangible inputs (ideas and knowledge) into goods or services. Technically, a standard production stage involves few major steps including scheduling of tasks; providing materials and components; and fabrication of parts, assembly and construction, quality control and testing.

In addition to use of product, maintenance, and repair, the operation stage involves marketing and product support. Marketing deals with issues such as the target market, product positioning, penetration and long-run pricing, sales, shares, profit goals, logistics, sales promotion, competition, warranty, channels of distribution, and so on.

Finally, when customers buy a physical product, they expect availability of technical support, maintenance, spare parts, training, and upgrades to ensure satisfactory operation of the product.

9.7.2 Product life cycle

The PLC is a key concept in marketing and planning. It defines the phases a product goes through from when it was first thought of until it is removed from the market. It describes the sequence of phases over which a product is developed, brought to market, and eventually removed from the market. It considers the useful life of a product which is the age beyond which the product is considered to be unfitting for further use due to its incompetence to function satisfactorily. The concept is based on a simple biological analogy of phases over a product's life.

The cycle consists of four major phases: introduction, growth, maturity, and decline as shown in Figure 9.21. Products in the introductory phase are not well defined and neither is their market. In this phase, a business is trying to build market acceptance for a new product. The product is still new and the customer acceptance as well as sales is low. This phase involves introducing new product to customers. In order to further sales, the focus lies heavily on promotion. At this phase, there is a significant cash outflow, as the company is spending to support the product.

In the growth phase of the PLC, the market has accepted the product and sales start to rise. The company builds market share to maximize

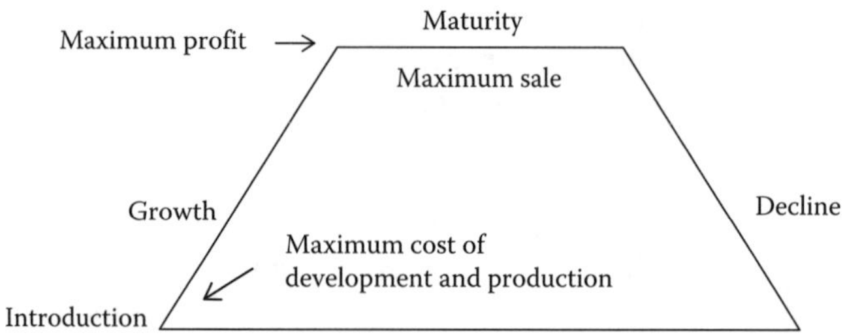

Figure 9.21 Phases of the PLC.

sales. Buyers have become familiar with the product and are ready to buy it. So new buyers join the market and former buyers come back to buy again. The promotion is still a key component for further growth. There can still be a cash outflow since the company is investing in more fixed assets and capital to support the growth of sales.

The third phase is that of maturity, where demand levels off and there are usually no design changes. There are many competitors, so the primary task is to defend market share where the market becomes saturated. The product is widely acceptable, sales are stable, profit is high and risk is low.

At a certain point of time, the product enters the decline phase, probably because of competitive new technology, superior product design, or market saturation. At this point, the product may not fulfill the current needs of the customers and accordingly sales start to decline. Profits dry up because of narrow profit margins and decreasing sales.

The first two phases of the life cycle may together be called the early phases of the PLC because the product is still being improved and refined, and the market is still in the process of being grown. The last two phases of the life cycle can be referred to as the later phases because the product and market are both well defined.

The duration of the PLC depends on the market. In some cases, a product may last for decades, while other products may have a life span of less than a year.

9.7.3 Technology life cycle

It is important to note that products are not equal to technologies (although these two terms are often confounded). A product is based on multiple technologies and a technology can form the basis for multiple products (Beck 2013).

In general, the development of a new technology tracks a typical S-shaped curve (Figure 3.8). In its early phase, the progress is limited by the lack of sophisticated ideas. A single good idea may generate many other good ideas, and the rate of progress will be exponential. Gradually the growth becomes linear when the progress is concerned with filling the gaps between the ideas. It is during this time when the commercial exploitation thrives. But with time the technology begins to run dry and increased improvements come with more complexity. This matured technology grows slowly and approaches a limit asymptotically. The success of a technology-based company lies in its capabilities of recognizing when the core technology on which the company's products are based begins to mature and through an active R&D program, transfer to another technology growth curve which offers better potentials.

9.7.4 Product life cycle management

Product life cycle management (PLCM) addresses the management of all stages of PLC, from conception until disposal. The goals of PLCM are to reduce time to market, improve product quality, reduce prototyping costs, identify potential sales opportunities, and lessen environmental impacts at end-of-life. PLCM is a strategy for managing products more effectively by linking product design with operations. From the business perspective, all systems that are operational in the field are an asset that provides many opportunities. From the technical perspective, the operational life of products is quite a challenge because systems keep evolving to fit in the market and to benefit from technical capabilities.

When looking at PLCM, many concepts come into play including development, financing, marketing, and manufacturing. For product companies, the PLCM concept takes into consideration the entire vision of effectively managing and connecting all information related to the process and production data needed to design, produce, validate, support, maintain, and dispose manufactured goods (PLMinfo 2016).

Often PLCM is used as an enabling framework to help connect, organize, control, manage, track, consolidate, and centralize all the mission-critical information that affects a product. Just as important, PLCM offers a process to streamline collaboration and communication between product stakeholders, engineering, design, manufacturing, quality, and other key disciplines. PLCM helps track information related to safety and control of components especially in aerospace, automotive, medical device, military, and nuclear industries. PLCM got its start in automotive, aerospace, and other industries that build very large, very complex products and systems. It was designed to provide everyone involved with an up-to-date view of every product throughout its lifecycle so people could make the right decisions and take the proper actions. It was these industries that led the way in the discipline of configuration management, which evolved into electronic data management systems, which then further evolve to product data management (PLMinfo 2016).

9.8 The how of learning in design

> The future belongs to a different kind of person with a different kind of mind: artists, inventors, storytellers-creative and holistic "right-brain" thinkers whose abilities mark the fault line between who gets ahead and who doesn't.
>
> **Daniel Pink**

9.8.1 The challenge

Design is an extremely wide subject, covering the whole range of disciplines within engineering. It is at the heart of what engineers do, and draws together all the skills and knowledge that we seek to foster in our students into an activity that has perhaps the greatest effect on society (McLaren 2008). Given the growing emphasis on design of consumer products, engineering educators have been challenged to introduce customer-driven design strategies to students and future engineers. Since product and engineering design activities are often not easily decoupled, a challenge has been to find and communicate a pattern that presents the steps in designing most products and processes. Related challenges in product design education arise because most of the idea generation and screening steps are qualitative in nature, and the feasibility analysis step often involves experimentation and possible consumer testing.

In the education system, the responsibility of educators is to train students and design professionals that can participate in and lead PD teams. In classrooms, as in industry community, discussion of engineering and management issues related to PD must be fostered. The activity of designing products requires basic skills in marketing, engineering analysis, process technology, manufacturing management, organizational behavior, and industrial design. These skills must be augmented by an ability to synthesize and organize. In order to properly educate design professionals who will create world-class products, faculty from diverse disciplines must work together in ways that are unfamiliar within the organization of the university (Eppinger et al. 1990). Education should focus on conditioning of the designer and offer all kind of tools for the all design stages so the motivation stays on right level.

9.8.2 The CDIO initiative: Design-build experience

Design is widely recognized as the core activity in engineering education, which integrates the subject-specific technical content with the needs of customers and business. Many approaches to design teaching recognize the above. Most prominent among curriculum models that follow this pattern is the CDIO initiative. The CDIO website (www.cdio.org) introduces the initiative as follows: The CDIO initiative is an innovative educational framework for producing the next generation of engineers. It provides students with an education stressing engineering fundamentals set in the context of conceiving-designing-implementing-operating real-world systems and products.

The CDIO initiative was developed with input from academics, industry, engineers, and students. It is universally adaptable for all engineering schools. CDIO initiative collaborators throughout the world have adopted CDIO as the framework of their curricular planning and outcome–based assessment.

The initiative grew out of collaboration between MIT and three engineering departments in Sweden, and has now expanded to a network of more than 100 partners worldwide.

The approach emphasizes the need to teach engineering fundamentals (which will be discipline specific) integrated with personal and professional skills, interpersonal skills, and product and system building.

Many engineering design modules require students to design and build some sort of project, mostly in teams. This is a significant element of the CDIO initiative. The term design-build denotes a range of engineering activities central to the process of project delivery system. Design-build experiences are structured and sequenced to promote early success in engineering practice. Iteration of design-build experiences and increasing levels of design complexity strengthen students' understanding of the product development process. Design-build also provides a strong foundation upon which to build deeper conceptual understanding of transdisciplinary knowledge and skills. The emphasis on building projects and implementing design processes in real-world contexts gives students opportunities to make connections between the various learning subjects and their future professional interests.

Two or more design-build experiences in the curriculum are required. The early-stage design-build experiences are often in the form of a set kit, with limited parts and options, from which the students have to construct a machine or object that meets certain design objectives. Within the CDIO scheme, this tends to address the Implement Operate aspects, and the Conceive-Design parts are predefined within the module. Senior design-build experiences, occurring later in the course, tend to address at least the Design-Implement Operate aspects, and possibly also the Conceive part (McLaren 2008).

9.8.3 Approaches to the teaching of product design

There are many approaches to the teaching of design, and each of them has a place in engineering education. These approaches range from the traditional to the truly innovative, encompassing tasks based on individual study and scholarship, to those that require all the skills of group work, management, logistics, and communication (McLaren 2008).

A product is no longer just a physical object. It has become a flow of interconnected experiences: brand image, design, function, interaction, communication, sharing, and content; each small part contributing to the success of the whole. Solving complex problems of this type requires the integration of many disciplines and is based on a reflective and broad system of knowledge.

Not only complex problems need to be solved but solutions must ensure long-term sustainability of design choices (Rittel and Webber 1973). The problem-solving approach needed to resolve complex issues must provide a wider view than what standard disciplinary problem-solving methods offer. Transdisciplinarity raises the question of not only problem solution but problem choice (Klein 2004).

Figure 9.22 shows how any educational system can arrange a program to teach engineering product design. The cognition, motivation, inspiration, and creativity are main properties of the product design program which should translate into design courses and engineering domain courses which may promote design through student projects.

Creativity is one of the major education reforms in the world nowadays. Advanced education puts more attention on the inspiration of student creative thinking ability (Shaheen 2010). Creativity can be developed and stimulated through training and learning activities and creative thinking teaching practices can promote students' learning motivation and cultivate their thinking ability (Adams 2006). In order to gain a creative design experience

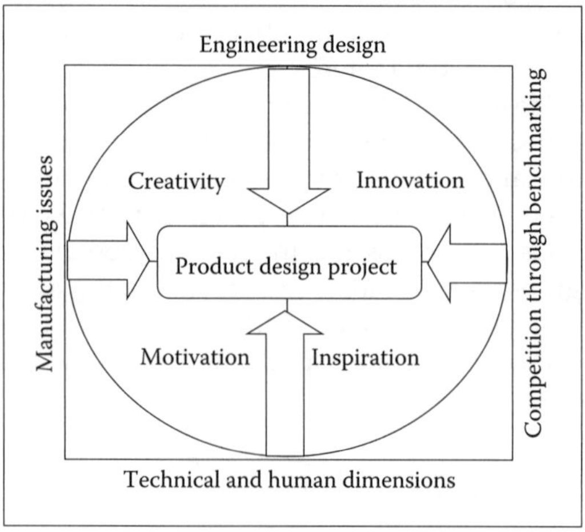

Figure **9.22** Product design education domain.

in which students experience intense connections between engineering and science as well as allow for the time to engage in iterative design, the design project must involve certain level of business and management instructional work. The following is a brief summary of principles that appear to guide successful experiences for student projects in product design.

- Design projects in engineering classes should involve approaches that highlight key engineering learning goals and allow for flexibility in the selection of goals. These projects should involve transdisciplinary collaboration to a certain level and encourage design activities that can be divided into multiple collaborating tasks. They should also require reflective performances rather than just the construction of prototypes or demonstrations of operation.
- Engagement of students in the pedagogy of designiettes: These are glimpses, snapshots, small-scale, short turnaround, and well-scoped design problems that provide a significant design experience. While most engineering programs around the world introduce design at distinct points in a curriculum, such as freshman and capstone design courses, designiettes pedagogy is where design is integrated across courses, semesters, years, and extra-curricular activities. This pedagogy, or framework, may be implemented in whole or in part in any engineering program (Wood et al. 2012). Designiettes help foster a culture of design, and enable the introduction of multidisciplinary design challenges across all core courses in each semester of study. They are ideal for class student competitions.
- 3D computer modeling, CAE analysis, material selection, prototyping, product costing, PLCM, and documentation should be available to students in order to develop a robust essence of the functional and aesthetic principles of design as well as strong understanding of the technical and business aspects of PD.
- Specially designed products which can serve as examples of PD projects in an integrated and comprehensible process should be used as learning examples in classes, so the path from initial idea to the final product can be followed.
- Initiation of transdisciplinary learning studios and/or labs to operate as an interface between the university and industry to deliver solutions to technical problems and to mentor selected students in some real projects as a way to leverage their learning.

9.8.4 Bloom's taxonomy

Perhaps the most widely accepted classification scheme for educational goals is the Bloom's taxonomy of educational objectives (Armstrong 2016). It was created in 1956 under the leadership of educational psychologist

Dr. Benjamin Bloom (1913–1999) who was interested in improving student learning by promoting higher forms of thinking in education, such as analyzing and evaluating concepts, processes, procedures, and principles, rather than just remembering facts. It is most often used when designing educational, training, and learning processes (Bloom et al. 1956).

9.8.4.1 Six major skill levels

Bloom's taxonomy is a classification of thinking organized by level of complexity. It gives teachers and students an opportunity to learn and practice a range of thinking and provides a simple structure for many kinds of questions and thinking (Ullah 2012). This taxonomy employs a hierarchical scale to articulate the level of expertise required to achieve each measurable student outcome. Based on the target type, there are three different taxonomies, one each for cognitive (knowledge-based) goals, psychomotor (skills-based) goals, and affective (heart-based feelings) goals (Qamar et al. 2016). The framework elaborated by Bloom and his collaborators consisted of six major categories as shown in Table 9.3 (Bloom and Krathwohl 1984).

Table 9.3 Bloom's six major skill levels

Lower level skills	Knowledge	Recall of specifics and universals, methods, and processes, or the recall of a pattern, structure, or setting; repeating memorized information
	Comprehension	Type of understanding or apprehension such that the individual knows what is being communicated; paraphrasing text, explaining concepts in jargon-free terms
	Application	Use of abstractions in particular and concrete situations; applying course material to solve straightforward problems; most undergraduate engineering courses focus on this level
Higher level skills	Analysis	Breakdown of a communication into its constituent elements such that the relative hierarchy of ideas is made clear; solving complex problems, developing process models and simulations, troubleshooting equipment and system problems
	Synthesis	Placing together of elements and parts so as to form a whole; designing experiments, devices, processes, and products
	Evaluation	Judgments about the value of materials and methods for specified purposes; choosing from among alternatives and justifying the choice, optimizing processes, making judgments about the environmental impact of engineering decisions, resolving ethical dilemmas

Bloom's taxonomy provides a way to organize thinking skills into six levels, from the most basic to the higher order levels of thinking. As discussed previously, designiettes may reflect Bloom's taxonomy benefit by engaging students in a cycle that advances from a stage of merely acquiring information to analyzing and ultimately synthesizing information to apply what they have learned in different situations (Telenko et al. 2014). These challenges combine problem clarification, concept generation and prototyping with subject content from curricula such as biology, chemistry, thermodynamics, mathematics, software, controls, etc.

9.8.4.2 Taxonomy revisited

In the 1990s, Lorin Anderson (a former student of Bloom) revisited the taxonomy. The names and subcategories of the six major categories were changed from noun to verb forms. Some subcategories were reorganized. The knowledge which is a product of thinking was replaced with the word remembering instead. Comprehension became understanding and synthesis was renamed creating in order to better signify the type of the thinking described by each category. The most suited taxonomy for teaching of engineering product design is shown in Figure 9.23.

Bloom's taxonomy can be useful for course design because the different levels can help move students through the process of learning from the

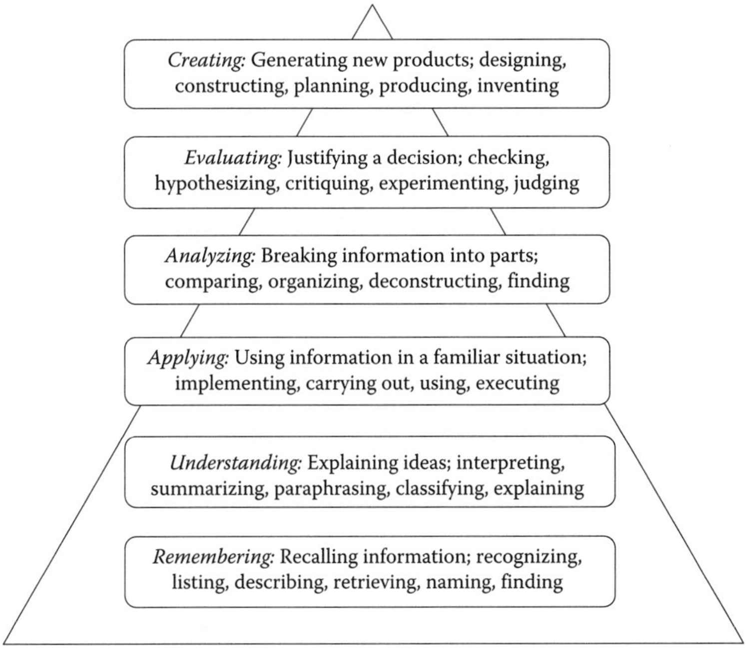

Figure 9.23 Bloom's revised taxonomy.

most fundamental remembering and understanding to the more complex evaluating and creating (Forehand 2010). Bloom's Taxonomy offers a significant framework for educators to use to focus on higher order thinking. By providing a pyramid of levels, this taxonomy can assist in designing performance tasks, shaping questions for discussing with students, and providing feedback on student work. Using the components of Bloom's taxonomy is a useful tool when it comes to writing objectives and identifying how well students comprehend a concept. Based on this taxonomy, a design case is presented in Section 9.9.

9.9 Bloom's taxonomy case: Designing a wind turbine

He who learns but does not think is lost.

Chinese proverb

Concerns regarding energy and environment are prevalent today. Such concerns sometimes become opportunities in that they lead to new inventions and innovations, often developed through EDM. In this case study, the EDM of a particular new product is described, which addresses both energy and environmental concerns by increasing the efficiency and enhancing performance.

A wind turbine was selected for this case in particular because it is a typical mechatronics renewable energy product of interdisciplinary engineering design. It is applicable to many engineering disciplines, and addresses the typical sustainability area of energy efficiency. This case study describes the development of a wind turbine; detailing the EDP and highlighting where appropriate health, safety and environmental considerations. The new product considered here is a performance-enhanced contra-rotating wind turbine. Conventional wind turbines have the potential to increase energy efficiency, but the enhancements described in this case study increase the system's efficiency further and producing less noise. The focus of this case study is mainly on redesigning two subsystems including rotor as well as the electric generator. Figure 9.24 shows the decomposed wind turbine with the major parts labeled. By the way, the exercise of labeling the picture with part names also belongs to the "understanding" level in Bloom's taxonomy. Figure 9.25 shows the functional decomposition of the wind turbine under consideration.

The educational purpose is to provide an aid to instructors for teaching engineering students and others the importance of a holistic approach to engineering design. This case activity is suitable not only for engineering students, but also for others, for example, students in related and

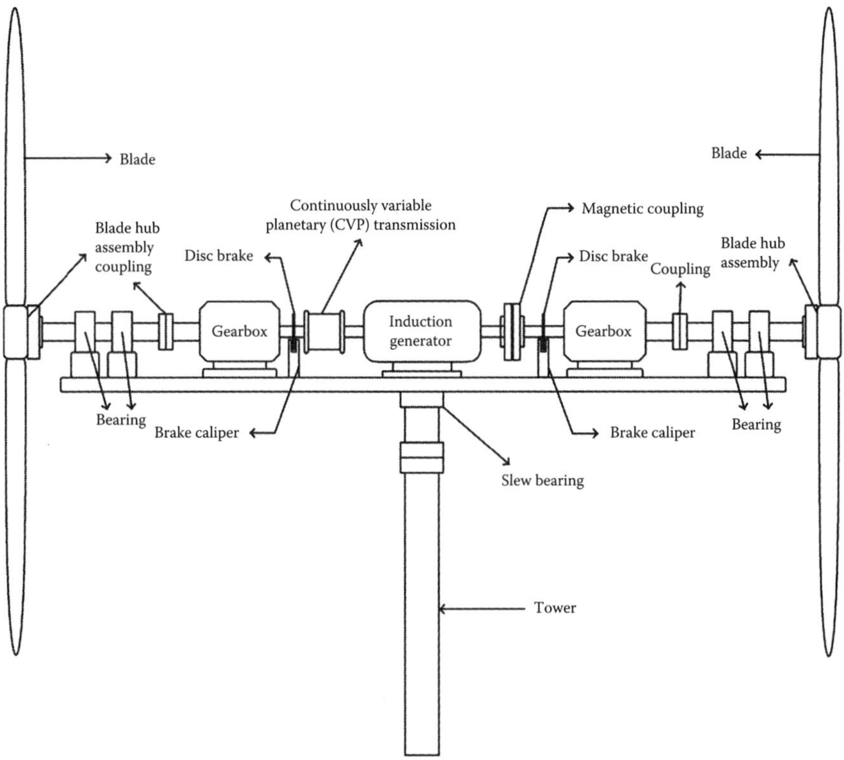

Figure 9.24 Contra-rotating small wind energy converter (From Habash, R.W.Y. et al. 2011. Experimental wind tunnel investigation of a contra-rotating small wind energy converter, *ISRN Mechanical Engineering*, 1–10, https://www.hindawi.com/journals/isrn/2011/828739/.)

complementary disciplines such as business and environmental studies, as well as practicing engineers and technologists, managers, and senior executives. The case is particularly suited for courses with significant coverage of engineering product design. Bloom's taxonomy is considered as a guideline and approach to teach students product design as shown in Figure 9.26.

9.9.1 Remembering

Remembering skill may be defined as the ability of students to recollect previously discussed or learned knowledge from their memory (Anderson and Krathwohl 2001). Verb examples that represent intellectual activity on this level include the following: define, describe, list, match, select, label, and recognize. Usually, students have the ability to keep in their mind certain information and later to remember and recall it, often with little adjustment.

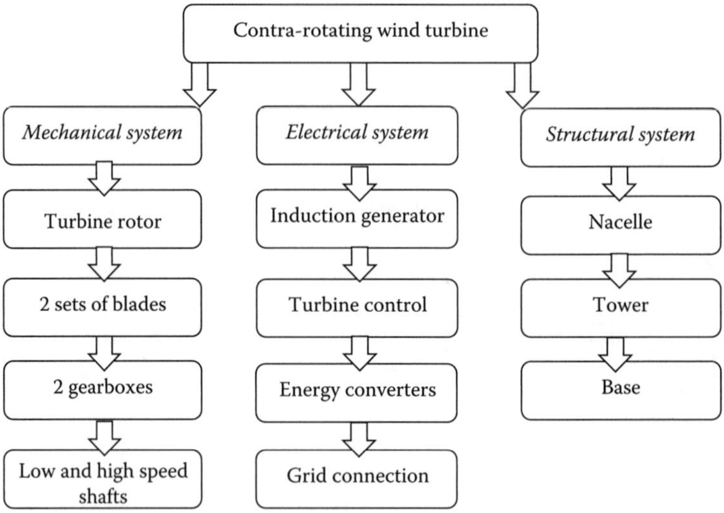

Figure 9.25 Physical decomposition tree diagram of the wind turbine.

- All students work through the remembering and understanding stages and select at least one activity from each other level.
- All students work through first two levels and then select activities from any other level.
- Some students work at lower level while others work at higher levels.
- All students select activities from any level.
- Some activities are tagged "essential" while others are "optional".
- A thinking process singled out for particular attention.
- Some students work through the lower levels and then design their own activities at the higher levels.
- All students write their own activities from the taxonomy.

Figure 9.26 Practical Bloom's possible approach within a class (From Ullah, L., *Project Based Learning, Common Core and Bloom's Revised Taxonomy: Putting it All Together,* Ullah Ventures, LLC, https://www.questar.org/services/rse-tasc-ii/presentations/instruction/Revised-Taxonomy-and-Project-Based-Learning.pdf, 2012.)

In the product design case study under consideration, students are given definitions and explanations at the beginning of the presentation for some technical terms that would be used throughout the case study. Some of these are engineering design terms (semantic, graphical, analytical, and physical); type of design problem (selection, configuration, parametric, original, redesign, and variant); concept generation technique (brainstorming, brainwriting, morphological method, etc.), design cycle (project definition and planning, specifications development, concept generation and selection, and PD), and more (Qamar et al. 2016).

Evaluation at this stage may be done through an activity like a quiz, a test, or an assignment in which the students would be asked to write the definitions of some of the above-mentioned technical terms used in the design process, with real-world examples related to renewable energy in general and wind turbine technology in particular. Figure 9.27 shows a student assignment on the first stage of remembering.

9.9.2 Understanding

Understanding skill refers to the capability of explaining information and concepts. Some familiar verbs related to this skill are as follows: explain, interpret, classify, illustrate, identify, summarize, and others (Anderson and Krathwohl 2001). Understanding is directly related to comprehension which refers to those objectives, behaviors, or responses that characterize an understanding of the accurate message contained in a communication, without essentially connecting it to other material. In this regard, the instructor could explain a design problem in which different types of designs are involved. Attainment of this skill could be done through a quiz in which the students are asked to list and briefly explain the different types of design problems applicable in a given example product.

Related to the case study, assume that the design team has been given the task of developing a new wind turbine. This mechatronics machine should generate more and produce less noise. Briefly explain the different types of design problems involved. A typical answer could be as follows: since this is a new product, original design work will definitely be needed. Later, configuration design needs to be carried out to come up with an optimum layout for the various parts of the wind turbine. To properly size and dimension each subsystem of the wind turbine, such as for

- Define renewable energy.
- List several sources of leading renewable energy sources that can be used to produce electricity.
- Select one technology from the list, wind turbine system for example, draw its block diagram and label its parts.
- Survey five students to see how they think about wind turbines, their types, sizes, and their impact on society.

Useful words: define, describe, list, match, select, label, and recognize.

Figure 9.27 Student's assignment on the first stage of remembering.

converting energy, parametric design is needed. Systems such as blades, gearbox, generator, controller, and tower would require selection design. Ingenuity and inventiveness could be incorporated into the project by trying to modify existing systems (such as blades or generator) rather than starting the design from scratch. The problem then becomes a redesign rather than an original design. Figure 9.28 shows a student assignment on the second stage of understanding.

9.9.3 *Applying*

Applying is the ability to use previously learned knowledge in situations which are either new, or contain new elements, as compared to the situation in which the abstraction was learned. Verb examples that represent intellectual activity on this level include practice, prepare, resolve, discover, implement, test, use, and others. Physical and functional decomposition of a real product may be a good example of a classroom activity to develop this skill. The instructor would explain this technique using wind turbine example, going from physical to functional decomposition and illustrating the use of tree diagrams in this regard. The students are asked to apply this technique to real wind turbine. As a first exercise, students are required to correctly label the view diagram of a standard wind turbine. They are then asked to arrange all subassembly/part names in a meaningful tree diagram structure representing the physical decomposition.

A typical wind turbine has only a few primary components: nacelle, generator, rotor, blades, gearbox, control and monitoring system, yaw, and tower. The focus of the development is first to modify the rotor system

- What are the advantages and disadvantages of using wind energy?
- Explain why some wind turbines are large and others are small.
- How do you get power from the electric grid? Draw a block diagram of power grid from power plant to user places showing major subsystems.
- Classify the power in the grid according to voltage levels. Identify major components of the grid.
- Connect a wind farm of five wind turbines, each one megawatt to the electric grid.
- Identify by using a graph the five-connected wind turbines and the way they are connected to the grid.

Useful words: explain, interpret, illustrate, classify, identify, summarize.

Figure 9.28 Student's assignment on the second stage of understanding.

into a contra-rotating system to make the turbine more aerodynamic and consequently more efficient. Second is to build a high-performance induction generator. However, it is recognized that modifications to the blade may impact the design of other components, and that efficiency gains may be attained by paying careful attention to them. Figure 9.29 shows a student assignment on the third stage of applying.

9.9.4 *Analyzing*

Analyzing is the breakdown of material into its constituent parts and detection of the relationship of the parts and of the way they are organized. Verb examples that represent intellectual activity on this level include the following: analyze, categorize, appraise, calculate, examine, outline, compare, discriminate, and reconcile.

Analyzing level exercises helps students in developing higher order skills and working knowledge of the material, as the classroom activity is to ask student groups to analyze the subsystem and components of the physical decomposition of the wind turbine, and to construct a functional decomposition. Students have to identify the functions associated with each subsystem or component, and to arrange all functions and subfunctions into a meaningful tree diagram.

Design is obviously an open-ended activity. Different student groups can come up with different ways of developing the physical decomposition and, consequently, the functional decomposition would also be different. Figure 9.30 shows a student assignment on the fourth stage of analyzing.

- Describe four different types of wind turbines.
- What types of electric generators are used in wind turbines?
- What types of gearboxes are used in wind turbines?
- Prepare a tree diagram for a typical and contra-rotating wind turbine and label their main components.
- What is the impact of implementing contra-rotating concept on turbine output power?
- What is the environmental impact of implementing a contra-rotating system?
- What are the drawbacks of implementing a contra-rotating concept?
- Do you think the contra-rotating concept is the best value for money?

Useful words: practice, prepare, resolve, discover, implement, test, use.

Figure 9.29 Student's assignment on the third stage of applying.

- Make a puzzle for wind turbine components.
- Compare vertical and horizontal wind turbines.
- Categorize wind turbines in terms of their rotor system orientation.
- Examine the function of every major components of the wind turbine.
- Compare the performance of wind turbines based on their height.

Useful words: analyze, categorize, appraise, calculate, examine, outline, compare, discriminate, and reconcile

Figure 9.30 Student's assignment on the fourth stage of analyzing.

9.9.5 *Evaluating*

Evaluating is making of judgments about the value, for some purpose, of ideas, works, solutions, methods, material, etc. It involves the use of criteria and standards for appraising the extent to which particulars are accurate, effective, or satisfying. It may be quantitative or qualitative. Verb examples that may represent intellectual activity on this level include the following: assess, conclude, estimate, predict, defend, justify, support, evaluate, criticize, and value.

In Bloom's revised taxonomy, evaluation was moved from level 6 to level 5. The rationale was that creation is the highest level in thinking activity, and many things need to be analyzed and evaluated before creating something new. This idea will be discussed again in the next section. Figure 9.31 shows a student assignment on the fifth stage of evaluating.

- Do you think that using contra-rotating wind turbine over a conventional one to supply electricity is a good idea?
- Assess changes would you recommend to the contra-rotating turbine to make it more sustainable.
- Justify the use of wind energy in farms.
- Evaluate the usage of wind energy in California. Explain your answer taking into consideration the water crises as well as environmental impact.
- Criticize the implementation of wind turbines close to residential areas.

Useful words: assess, conclude, estimate, predict, defend, justify, support, evaluate, criticize and value

Figure 9.31 Student's assignment on the fifth stage of evaluating.

9.9.6 *Creating*

In Bloom's revised taxonomy, creating is the highest level of skill, and would need the development of all other skill levels to successfully achieve this. Among the several essential requirements of higher education, the most prominent one is to kindle and develop the creativity of students (Kukk and Heikkinen 2015). Termed as "synthesis" in the earlier version, creating is simply the process of creating something new. It usually involves verbs such as propose, design, create, formulate, generate, build, develop, and others (Anderson and Krathwohl 2001). Synthesizing may be defined as the putting together of elements and parts so as to form a whole. This is the category in the cognitive domain that clearly calls for creative behavior on the part of the learner, but within the limits set by the framework.

This skill is a natural and obvious requirement in a course or assignment such as product design. This skill is needed in both the concept generation and PD stages of design. Using a function-based design approach, student groups have to use the functions and subfunctions identified earlier to generate identified concepts for an improved design of the wind turbine for concept generation. Using the technique of brainstorming, students may come up with at least three concepts for each lowest level subfunction. The morphological method is then used to combine subfunction solutions into concepts for each major function. Evaluation of existing products is necessary for the generation or creation of new concepts. These concepts have to be evaluated again to select the best or most optimum solution. This evaluation-creation-evaluation process is required in both the concept generation and product development (PD) stages. Figure 9.32 shows a student assignment on the last stage of creating.

9.9.7 *Case research questions*

- Do you think that passing traditional tests, according to Bloom taxonomy, can get students to the level of creation? What teachers should do to achieve that level?
- Examine the energy sources that currently supply electricity to a rural community. Think of an alternative energy plan that is more sustainable. Explain your reasons.
- Do you believe that using wind energy to supply energy to the rural community is a good idea? Explain your answer taking into consideration the environmental impact, views of the community, and government policies.

- The present case focuses on the design of a new wind turbine. Investigate the design process for wind turbines, and describe the concerns associated with the manufacturing of a wind turbine, and what measures are typically taken to address them?
- Is there design concern related to contra-rotating wind turbine under consideration in the case study? Describe it, and how it can be addressed?
- For typical annual income, what would be the simple payback period for purchasing a contra-rotating wind turbine (costing about $300,000) compared to a conventional wind turbine (costing $250,000)? Consider the contra-rotating wind turbine generating 30% extra power compared to the conventional one for the same rating (50 kW, for example).
- Determine the annual reduction in environmental emissions for the contra-rotating wind turbine compared to a conventional one. Make reasonable assumptions on the annual number of operating hours.
- Expand the environmental assessment carried out in the previous question by carrying out a comprehensive life cycle analysis of the contra-rotating wind turbine. Carry out any research necessary to find information needed to perform the analysis, and make reasonable assumptions.
- Develop a business case for the contra-rotating wind turbine, considering all pertinent factors. As part of the exercise, expand your answer to the previous questions to consider the issue of whether the ability of the contra-rotating wind turbine to increase energy production is sufficient on its own to change public and industry behaviour, or whether they need to be motivated by price, environment, and health impact.

Useful words: propose, design, create, formulate, generate, build, develop and plan.

Figure 9.32 Student's assignment on the last stage of creating.

9.10 *Knowledge acquisition*

Attempting to answer the following questions involves acquisition of knowledge from this book and other books, documents, and the Internet.

- What is a design? What makes a design a successful design?
- Differentiate between design theory and design methodology.
- What is the difference between design methodology and design process?
- What are the major stages in the engineering design process? Are these stages related to each other?
- What does a design process look like?
- What are the questions to ask before you design anything?
- What is meant by "need" in a design process?

- What is the importance of brainstorming on design?
- Give an example of how to apply problem-solving skills to a design challenge.
- What are some examples of good product design?
- What is meant by alternative design? How to choose among alternative designs?
- How can a proposed solution for a design problem be reasonably justified?
- What are the risks associated with PD project?
- List the advantages of computer-aided designs.
- What is the challenge in manufacturing products?
- Why is documentation of the engineering design process important?
- Describe the life cycle of a product.
- What are the stages of the PLC?
- What is the difference between PLC and technology life cycle?
- What is the main objective of PLC analysis?

9.11 Knowledge possession

Attempting to answer the following open-ended "not explicitly expressed" questions may require research and investigation beyond the scope of this book, mostly by engaging in conversation, class discussion, and Internet-based research.

- Consider a mechatronics system (e.g., a washing machine). Describe the types of engineering disciplines involved in designing and implementing the technology.
- Read and interpret an engineering problem and list its possible solutions. Use the design process to develop a solution to the problem.
- Create, test, and evaluate the mock-up or prototype of your class engineering design and have a design team check the specification.
- What are the differences between modular and integral design in terms of performance, cost, and required tools?
- Explain by example that products fail because they have faulty or poorly designed and built parts; are used in ways that exceed what was intended by the design; or were poorly designed.

9.12 Knowledge creation

Collaborate with peers on learning or you may work with others outside the class to narrow down the objectives of each activity. You may access class and online resources, analyze data and information to create new ideas and balanced solutions. High-level digital tools may be used to develop multimedia presentations, simulations or animations, videos and visual displays, digital portfolios (ePortfolios), reflective practice (online publishing and blogging), or well-researched and up-to-date reports.

9.12.1 Product design portfolio

The engineering design process is a series of steps that engineers use to help solve problems. There are many variations of this process and steps that are repeated as many times as necessary to solve the problem. Building anything takes several steps. Importantly is to decide what to build and what to do to build it. Materials to build are needed. To illustrate engineering solutions in the real world, the following video, which shows the steps in the process of building a new biomedical system, has been selected.

http://www.g9toengineering.com/MechatronicsStudio/Syringpump.htm

There is a story about each engineered solution. Students from the University of Ottawa in Canada paired up to build a syringe pump for the extrusion process of polymeric microspheres. Watch the video and try to realize the story of developing the above system. Think about what the developers considered when planning and applying their solution. Notice the problem and how they solved it. You may summarize the process by developing a design portfolio that describes the engineering development sequence.

9.12.2 Design portfolio of a smart self-driving vehicle

Automated vehicles (AVs) are a highly disruptive technology. The large-scale introduction of AVs will have major impacts on urban centers. Assume you are asked to design an AV and prepare a design portfolio (4–5 pages) that includes stages and corresponding stages by brainstorming the subject. In the process you may consider answering the following questions:

- How can a car drive itself? How would it look? What size would it be?
- How to master technologies that are going to shape the AVs?
- What would be the AV smart features?
- Would these vary, depending on the target users?
- What are the main selling points of smart and AVs?
- What is the single largest challenge that faces autonomous and smart cars?
- What are some of the major technological challenges that need to be overcome before AVs could be available to customers?

- When do you think we could realistically see a major shift toward this type of vehicle?
- Could you imagine future roads that can be built to be friendlier to automated cars?
- You may investigate the following innovative technical categories: safe driving, racing, knowledge of driving rules, road safety ideas, robotic car structure, and knowledge acquisition of vehicle structure using Solidworks software.
- What are the best active and complete AV open source hardware and software projects?
- Make a list of major AV projects.
- What are the policy impacts of AVs?
- What is the impact of AVs on the insurance industry?

9.12.3 Proposal for a system-based course with a designiette

Write instructional objectives for a course (with designiette) that incorporate both knowledge of content and mastery of the skills you wish the students to develop. At all levels of the engineering curriculum including the first year, include some higher level problem-solving skills (e.g., transdisciplinary analysis, design, critical thinking) and the soft skills (e.g., oral and written communication, teamwork, social and ethical consciousness). Make the objectives as detailed and specific as possible.

As a matter of involving students in planning this new, interdisciplinary course, teams of three to four students will be engaged in investigating and planning components of the course. For this task, write a proposal including outline, lesson plan, and evaluation protocol for a new system-based engineering course. Make class exercises, homework assignments, and tests consistent with the objectives. The course should include a designiette. The designiette may focus on certain stages of the product design process. Ideally, the designiette includes an innovation goal with the process steps of ideation, prototyping, and experimentation. The designiette may provide open-ended problems for linking ideation techniques and methods within one or more of the given topics.

9.12.4 Debate on design education

Traditionally issues considered in product design have related only to function, appearance, elegance, style, and financial concerns, but in recent years, designers and consumers have started to look beyond pure surface. This includes need, equity, ethics, social impact, and resource efficiency, and to develop more environmentally benign products and processes (Deniz 2002).

Objective	Introducing an open-ended debate in the classroom to help students understand argument on design education.
Time	15 min for debate and 15 min for review.
Format	For and against.
Learning Outcomes	Make an argument about a particular opinion; evaluate the arguments of peers; and understand the concept of counterarguments.
Capabilities Demonstrated	Developing skills on public speaking, research, teamwork, critical thinking, communication, and professional judgment.
Arrangement	Students are organized into two position groups of three speaking in a specific order. Three argue for an opinion and three argue against. One or two students might each work on the opening and closing statements while the group is investigating the subject; however, the entire group should revise the statements. Each group should read an opening and closing statement for the debate.
Ideas for the Topic	How should education teach design? Should education train designers to design things that people want to buy or should train designers to design things with the lowest possible environmental impact?
Assessment	Indicate what you consider the best three arguments in favor of the topic. How were they substantiated? Identify arguments that are based on poor facts, not ethical and/or sustainable or not well substantiated?

9.12.5 Reengineering competition

Reengineering is the action of improving the design of an engineering product or service, either by adding new features or improving its original purpose. In this competition, teams of four to five students are given a relevant design that is required to be reengineered under a new set of constraints. The teams will be given several hours to improve and redesign a solution to a problem to suit an alternate solution or application, fabricate the new design and repurpose of the process, and prepare technical presentation. They are required to take into account all relevant aspects of the engineering process such as components, materials, costs, and others. Each team's design is judged based on usefulness, originality, feasibility, and marketability. There should be no building component but competitors will create a presentation to the judges.

9.12.6 Hackathon design competition

Students from various disciplines will compete in teams of three to five to conceptualize, design, build, and communicate a solution to an engineering design challenge in a "hackathon format." Hackathons usually begin with a presentation about the competition and the specific subject. Then participants suggest ideas and form teams, based on individual interests and skills. Then the main work of the hackathon begins, which may last from several hours to several days. Hackathon is popular among the students of software, electrical, mechanical, computer, and mechatronics engineering. It is an opportunity to make an idea come to life within short period of time. For this task, the theme is "sustainable mechatronics." Teams should have access to a starter Arduino kit, basic building materials, workshop, and the Internet.

9.12.7 *Innovative design sustainability competition*

Tomorrow's innovations will need engineers who thoroughly understand how to apply their knowledge and skills to designing products and processes that did not exist before (Dym et al. 2005). In technology, innovation is often defined as "the use of new knowledge to offer a new product or service that customers want. It is invention+commercialization" (Afuah 1998).

This activity requires the competitors (teams of one to four students) to present an innovative and commercially viable solution to a real-world problem of their own choosing. Teams are expected to carry out market research and feasibility studies as well as draft a business proposal for their product. The design has to be innovative and fill a void in market's needs. Teams must choose a topic that is related to "sustainable engineering," which involves the various stages of product. It is expected that some of these innovative projects will lead to successful start-ups and the commercialization of new products and services.

9.12.8 *Write-up of a professional cover letter*

Consider a company that performs mechatronics engineering design. Look at the "career" section of the company website. Find two job descriptions for engineering positions and find the disciple background and skills of the engineer they intend to hire in these positions. Based on job description develop a cover letter for application to the above two jobs.

References

Adams, K. 2006. The sources of innovation and creativity. New Commission on the Skills of the American Workforce, National Center on Education and the Economy, Washington, DC.

Afuah, A. 1998. *Innovation Management*. New York: Oxford University Press.

Anderson, L. W., and D. R. Krathwohl. 2001. *A Taxonomy for Learning, Teaching and Assessing: A Revision of Bloom's Taxonomy*. New York: Longman Publishing.

Anderson, D., and J. Pine. 1997. *Agile Product Development for Mass Customization*. Chicago, IL: Erwin.

ANSYS. 2011. Parametric analysis: The key to rapid, robust design. White Paper. http://resource.ansys.com/staticassets/ansys/staticassets/resourcelibrary/whitepaper/wp-parametric-analysis.pdf.

Armstrong, P. 2016. Bloom's taxonomy. Center for Teaching. Vanderbilt University, Nashville, TN. https://cft.vanderbilt.edu/guides-subpages/blooms-taxonomy.

Asimow, M. 1962. *Introduction to Design*. Englewood Cliffs, NJ: Prentice-Hall.

Beck, D. F. 2013. *Technology Development Life Cycle Processes*. Albuquerque, NM: Sandia National Laboratories.

Bloom, B. S., M. D. Engelhart, E. J. Furst, W. H. Hill, and D. R. Krathwohl. 1956. *Taxonomy of Educational Objectives, Handbook I: The Cognitive Domain*. New York: David McKay Co.

Bloom, B. S., and D. R. Krathwohl. 1984. *Taxonomy of Educational Objectives. Book 1: Cognitive Domain*. New York: Addison-Wesley.

Boucher, M. 2008. Engineering evolved: Getting mechatronics performance right the first time. Aberdeen Group, Boston, MA.

Buchanan, R. 2001. Design research and the new learning. *Design Issues* 17(4): 3–23.

Cross, N. 2008. *Engineering Design Methods*. Fourth edition. Hoboken, NJ: Wiley.

Deng, J., X. Han, and X. Zeng. 2002. *Product Conceptual Design: Theories, Methods and Technologies*. Beijing, China: China Machine Press.

Deniz, D. 2002. Sustainability and environmental issues in industrial product design. Master Thesis in Industrial Design. Izmir Institute of Technology, Izmir, Turkey.

Derelöv, M. 2009. On evaluation of design concepts modelling approaches for enhancing the understanding of design solutions. Department of Management and Engineering, Linköpings universitet, Linköping, Sweden.

Ding, Y. 2009. Product maintainability design method and support tool based on feature model. *Journal of Software Engineering and Applications* 2: 165–172.

Dorta, T. 2008. Design flow and ideation. *International Journal of Architectural Computing Issue* 3(6): 299–316.

Dym, C. L., A. M. Agogino, O. Eris, D. D. Frey, and L. J. Leifer. 2005. Engineering design thinking, teaching and learning. *Journal of Engineering Education* 94(1): 103–120.

Eder, W. E., and S. Hosnedl. 2007. *Design Engineering: A Manual for Enhanced Creativity*. Boca Raton, FL: CRC Press.

Eppinger, S. D., C. F. Fine, and K. T. Ulrich. 1990. Interdisciplinary product design education. *IEEE Transactions on Engineering Education* 37(4): 301–305.

Forehand, M. 2010. Bloom's taxonomy. http://projects.coe.uga.edu/epltt/index.php?title=Bloom%27s_Taxonomy.

French, M. J. 1998. *Conceptual Design for Engineers*. Third edition. London: Springer.

Gabriel-Petit, P. 2010. Design is a process, not a methodology. http://www.uxmatters.com/mt/archives/2010/07/design-is-a-process-not-a-methodology.php.

Gass, S. I., and C. M. Harris. 1996. *Encyclopedia of Operations Research and Management Science*. Boston, MA: Kluwer Academic.

Habash, R. W. Y., Y. Yang, C. Blouin, and P. Guillemette. 2011. Experimental wind tunnel investigation of a contra-rotating small wind energy converter. *ISRN Mechanical Engineering* 1–10. https://www.hindawi.com/journals/isrn/2011/828739/.

Haik, Y., and T. Shahin. 2011. *Engineering Design Process.* Second edition. Stamford, CA: Cengage Learning.

Howard, T., S. Culley, and E. Dekoninck. 2007. Creativity in the engineering design process. *International Conference on Engineering Design, ICED'07,* Cite Des SCIENCES Et De L'industrie, Paris, France, August 28–31. http://www.ida.liu.se/~steho87/desres/buchanan.pdf.

Hubka, V., and W. E. Eder. 1995. *Design Science: Introduction to the Needs, Scope and Organization of Engineering Design Knowledge.* Second edition. New York: Springer.

ISO9000:2000. 2000. Quality management systems—Fundamentals and vocabulary. ISO, Geneva, Switzerland.

Johnson, J. K., S. H. Miller, and S. D. Horowitz. 2008. Systems-based practice: Improving the safety and quality of patient care by recognizing and improving the systems in which we work. In Henriksen, K., J. B. Battles, and M. A. Keyes et al. (Eds.), *Advances in Patient Safety: New Directions and Alternative Approaches* (Vol. 2: Culture and Redesign). Rockville, MD: Agency for Healthcare Research and Quality.

Jonassen, D. H. 2004. *Learning to Solve Problems: An Instructional Design Guide.* San Francisco, CA: John Wiley and Sons.

Kesselring, F. 1954. *Technische kompositionslehre.* Berlin, Germany: Springer.

Klein, J. T. 2004. Prospects for transdisciplinarity. *Futures* 36: 515–526.

Koskela, L., and P. Huovila. 1997. On foundations of concurrent engineering. In Anumba, C., and N. Evbuomwan (Eds.), *Concurrent Engineering in Construction CEC97,* July 3–4, pp. 22–32. London: The Institution of Structural Engineers.

Krishnan, V., and K. Ulrich. 2001. Product development decisions: A review of the literature. *Management Science* 47(1): 1–21.

Kroes, P. 2002. Design methodology and the nature of technical artefacts. *Design Studies* 23: 287–302.

Kuen Chang, S. K. 2016. *Understanding Industrial Design.* Newton, MA: O'Reilly Media, Inc.

Kukk, P., and S. Heikkinen. 2015. Assignments and grading in engineering graphics courses. *International Journal of Engineering Pedagogy* 5(3): 8–14. http://dx.doi.org/10.3991/ijep.v5i3.4661.

Kuo, T.-C., S. H. Huang, and H. C. Zhang. 2001. Design for manufacture and design for "X": Concepts, applications, and perspectives. *Computers and Industrial Engineering* 41(3): 241–260.

Littler, D., F. Leverick, and M. Bruce. 1995. Factors affecting the process of collaborative product development: A study of UK manufacturers of information and communications technology products. *Journal of Product Innovation Management* 12(1): 16–32.

Maxey, K. 2012. Product concept design—Past, present, and future. http://www.engineering.com/DesignerEdge/DesignerEdgeArticles/ArticleID/4986/Product-Concept-Design-Past-Present-and-Future.aspx.

McCahen, S., P. Anderson, M. Kortschot, P. E. Weiss, and K. A. Woodhouse. 2015. *Designing Engineers: An Introductory Text.* Hoboken, NJ: John Wiley and Sons.

McLaren, L. 2008. Approaches to the teaching of design. Higher Education Academy Engineering Subject Centre. Loughborough University Leicestershire, Loughborough, UK. https://www.heacademy.ac.uk/system/files/approaches-to-the-teaching-of-design.pdf.

Morris, R. 2009. *The Fundamentals of Product Design*. Singapore: AVA Publishing.

OEERE. 2000. From invention to innovation. Office of Energy Efficiency and Renewable Energy, US Department of Energy, Washington, DC.

Osborn, A. 1953. *Applied Imagination: Principles and Procedures of Creative Problem-solving*. New York: Charles Scribner's Sons.

Osorio, J., D. Romero, M. Betancur, and A. Molina. 2014. Design for sustainable mass-customization: Design guidelines for sustainable mass-customized products. The *20th International Conference on Engineering, Technology, and Innovation*, Bergamo, Italy, 23–25 June, pp. 1–9.

Pahl, G., W. Beitz, J. Feldhusen, and K.-H. Grote. 2011. Engineering design: A systematic approach. In Wallace, K., and L. T. M. Blessing (Eds.), *Translators and Editors*, third edition. Darmstadt, Germany: Springer.

Pugh, S. 1981. Concept selection: A method that works. In Hubka, V. (Ed.), *Review of Design Methodology*. Proceedings of the *International Conference on Engineering Design*, Rome. Zürich, Switzerland: Heurista, pp. 497–506.

Qamar, S. Z., A. Ramanathan, and N. Z. Al-Rawahi. 2016. Teaching product design in line with Bloom's taxonomy and ABET student outcomes. *IEEE Global Engineering Education Conference (EDUCON)*, Abu Dhabi, UAE, April 10–13.

Rittel, W. W. J., and M. M. Webber. 1973. Dilemmas in a general theory of planning. *Policy Sciences* 4: 155–169.

Salustri, F. A. 2005. An extremely abbreviated history of engineering design. http://deed.ryerson.ca/~fil/t/history0.html.

Sangarappillai, S., and H. Peter. 2001. Design for excellence. *Journal of Engineering Design* 12(4): 291–310.

Shaheen, R. 2010. Creativity and education. *Creative Education* 1(3): 166–169.

Suh, N. P. 1997. Design of systems. *Annals of the CIRP* 46(1): 75–80.

Suh, N. P. 1999. Introduction. In Kreith, F. (Ed.), *Engineering Design Mechanical Engineering Handbook*. Boca Raton, FL: CRC Press.

Suh, N. P. 2001. *Axiomatic Design: Advances and Applications*. New York: Oxford University Press.

Suh, N. P. 2005. *Complexity: Theory and Applications*. New York: Oxford University Press.

Telenko, C., B. Camburn, K. Holtta-Otto, K. Wood, and K. Otto. 2014. Designiette: New approach to multidisciplinary engineering design education. Proceedings of the *ASME International Design Engineering Technical Conferences and Computers and Information in Engineering Conference*, vol. 3, Buffalo, NY, August 17–20.

Ullah, L. 2012. *Project Based Learning, Common Core and Bloom's Revised Taxonomy: Putting It All Together*. Ullah Ventures, LLC. https://www.questar.org/services/rse-tasc-ii/presentations/instruction/Revised-Taxonomy-and-Project-Based-Learning.pdf.

Ullman, D. 2001. Toward the ideal mechanical engineering support system. *Research in Engineering Design* 13: 55–64.

Ullman, D. G. 2010. *The Mechanical Design Process*. Fourth edition. New York: McGraw-Hill.

Ulrich, K., and S. Eppinger. 2000. *Product Design and Development*. Second edition. Boston, MA: Irwin McGraw Hill.

VDI. 1987. Systematic approach to the design of technical systems and products (VDI Guideline 2221). The Association of German Engineers (VDI), Berlin, Germany.

Wood, K. L., R. E. Mohan, S. Kaijima, S. Dritsas, D. D. Frey, C. K. White, D. D. Jensen, R. H. Crawford, D. Moreno, and K.-L. Pey. 2012. A symphony of designiettes: Exploring the boundaries of design thinking in engineering education. The *ASEE Annual Conference and Exposition*, San Antonio, TX. https://peer.asee.org/20872.

chapter ten

Sustainability in engineering design

> Sustainable design is the set of perceptual and analytic abilities, ecological wisdom, and practical wherewithal essential to making things that fit in a world of microbes, plants, animals, and entropy. In other words, (sustainable design) is the careful meshing of human purposes with the larger patterns and flows of the natural world, and careful study of those patterns and flows to inform human purposes.
>
> **David Orr**
> *Oberlin College*

10.1 Objective

- Understand sustainable engineering product design and its key requirements.
- Understand the important role of engineering in sustainable development.
- Explore the transition path from typical design into sustainable design.
- Discuss the key requirements for sustainable engineering design (SED).
- Identify the role of technology in sustainable design.
- Name and briefly describe the steps for SED process.
- Discuss design through the 12 principles of green engineering.
- Explore measuring success tools including triple bottom line (TBL) and quality design.
- Discuss the various aspects of design for sustainability (DfS).
- Know about Hannover Principles which aim to provide a platform upon which designers can consider how to adapt their work toward sustainable ends.
- Show how green products are result of integrating sustainable manufacturing (SM) and clean technologies.

- Discuss the various assessment methods and tools for environmental and sustainability performance including life cycle thinking (LCT), carbon footprint (CF) and water footprint (WF), life cycle assessment (LCA), and life cycle sustainability analysis.
- Show the difference between eco-efficiency and eco-effectiveness assessment measures.
- Discuss the factors that support and contribute to SM.
- Understand the concept of closed-cycle manufacturing (CCM).
- Discuss the drivers and challenges for product remanufacturing.
- Discuss the challenges and approaches of building transdisciplinary sustainable education.
- Learn through a case activity how to incorporate life cycle sustainability and energy analysis in product remanufacturing process.
- Provide three end of chapter pedagogical knowledge strategies, namely knowledge acquisition, knowledge possession, and knowledge creation to help understand the above topics and generate new and alternative ideas and solutions.

10.2 Historical perspective

> I only feel angry when I see waste. When I see people throwing away things we could use.
>
> **Saint Teresa**

The sustainability concept is both very ancient and relatively modern. It has been an element design throughout history, although many consider the concept as contemporary movement. Before consumerist modern era, most civilizations lived in coherence with nature, their lifestyles, customs, and behavior aimed for stability and continuousness. However, the foreseeable change happened in the rising global economy; it led to an industrial consumerist monoculture that resulted in persistence of desperate poverty along with deep disparities as well as proliferation of risky technologies and degradation of essential ecosystems. Today, sustainability designers are not the first to consider the importance of utilizing reusing and recycling or renewable energy or green building where such practices were employed many centuries ago.

Throughout history, a wind catcher was introduced as an architectural device, which achieves thermal comfort inside buildings. It is believed that it is a traditional Persian architectural device, which was used for many centuries, but there is evidence that the idea of the wind catcher dates back to the early Pharaonic periods (El-Shorbagy 2010). Sustainable design was also prevalent during Roman Antiquity, where in addition to building aqueducts, the Roman used geothermal energy to heat homes and baths.

In 500 BC, Athens organized the first municipal dump program in the Western world. Local laws dictate that waste must be disposed of at least 1 mile from the city walls. And we think we have it rough today having to drag it out to the curb. During the seventeenth century, the recycled paper manufacturing process was introduced. The Rittenhouse Mill near Philadelphia made paper from fiber derived from recycled cotton and linen rags. In 1897, New York City created a material recovery facility where trash was sorted at "picking yards" and separated into various grades of paper, metals, and carpet. Burlap bags, twine, rubber, and even horse hair were sorted for recycling and reuse. In 1904, the first American aluminum can recycling plants opened in Chicago and Cleveland (Busch Systems 2014).

Both World War II and Great Depression brought back the desire to recycle goods both in Europe and the United States because of the obvious need. During the 1930s many people survived the Great Depression by selling scraps of metal, rags, and other objects. In 1964, the all-aluminum can was introduced and accordingly the aluminum industry began creating a massive system for recycling and redeeming used drink containers for making new cans.

The environmental movement during the 1960s and 1970s helped raise awareness about reduce, reuse, and recycling. In 1970, the first Earth Day (www.earthday.org) brought national attention to the problem of increasing waste and the importance of recycling. The idea for a national day to focus on the environment came to Gaylord Nelson, then a US Senator from Wisconsin, after witnessing the ravages of the 1969 massive oil spill in Santa Barbara, California. Inspired by the student anti-war movement, he realized that if he could fortify that energy with an emerging public awareness about air and water pollution, it would force environmental protection onto the national political agenda. Originally, the Earth Day was created and founded by entrepreneur John McConnell. He was rarely given the credit he deserves, for it was he who first used the term Earth Day and who was successful in establishing the first governmentally recognized observance on March 21, 1970 (Murdock 2012).

During the 1940s and 1950s, environmental and health concerns about the burning of fossil fuel became more pronounced as smog produced by the burning of these fuels in Europe and America was blamed for people illness and death. Activists rallied for the use of cleaner energy resources and practices by bringing the health hazards of fossil fuels to the audience.

Today, the various energy supply technologies that are used in a carbon-constrained future are extensively under review. This includes designing renewable energy by selecting appropriate technologies and advanced fossil-fuel systems with carbon capture and sequestration. The approach encourages explicit consideration of resilience in both engineered systems and the larger systems in which they are embedded. In addition, energy efficiency technologies are often cited as an essentially important and an often lower cost supplement to supply side developments.

10.3 Sustainable engineering design

> Design is critical today. It is the first signal of human intension.
>
> **William McDonough**
> *Stanford University*

10.3.1 The engineering factor

In the modern engineering culture, sustainable design has become dominant too and an application for engineers and users as society requirements and financial limitations mount. In all areas of engineering, engineers are advised to ensure that products and services have the maximum lifespan for their planned use and employ the least amount of natural recourses while still meeting client, economic, societal demands and code requirements.

Although engineering is not one of the three main components of sustainability as discussed in Chapter 2, it is indirectly linked to each. That is, engineering uses resources to drive much if not most of the world's economic activity, in virtually all economic sectors. Also, resources used in engineering, whether fuels, minerals, or water, are obtained from the environment, and wastes from engineering processes (production, transport, storage, utilization) are typically released to the environment. Finally, the services provided by engineering allow for good living standards, and often support social development (Rosen 2012). Given the intimate ties between engineering and the key components of SD, it is obvious that the accomplishment of sustainability in engineering is a significant aspect of achieving SD. In fact, Kreith (2012) writes on sustainability, "no subject is more important to the engineering profession or the wider world that we live in."

Most engineering activities utilize resources that are derived from nature. Such resources include water, materials (virgin and recycled), and energy. The degree to which resources are sustainable depends on many factors, including their scarcity and importance to ecosystems. An important requirement of sustainable engineering is the use of sustainable processes. This implies that the engineering processes utilized must exhibit sustainable characteristics in terms of the operations and steps they involve, and the energy and materials they utilize. High efficiency allows the greatest benefits, in terms of products or services, to be attained from resources, and thus aid efforts to achieve sustainability. Numerous environmental impacts associated with engineering processes are of concern and must be addressed in efforts to attain sustainability. Some important environmental impacts associated with engineering processes of concern include global climate change; ozone depletion (due to destruction of the atmospheric ozone layer and subsequent increases in ultraviolet reaching the earth's surface); acidification, and its impact on soil and water; abiotic resource depletion potential; eco-toxicity; and radiological impacts (Rosen 2012).

Various initiatives spurred during the last decade, simultaneously addressing environmental protection, economic development, and social equity, to constitute the foundations of what can be called a "sustainability revolution" (Edwards 2005). New approaches also appeared in the field of engineering, revolving around a redefinition and reorganization of the design process to better address sustainability issues.

Engineering forms an interface between main product development stages that include design, implementation and production. SED principles should be contemplated and applied early in the design stage (e.g., in conceptual design) to ensure that technology development and scale-up follow the sustainability benign route. It will be hard to turn back to redo and redesign things from later stages. In that sense, the SED principles should be taken into account in decision-making for both research and industrial projects as well as in policy-making and decisions regarding funding of technological research.

10.3.2 Sustainable design landscape

Design in the context of sustainability may be described as a new possibility, which is expected to allow the achievement of a preferred situation (Cowie 1993), while SED improves design by adding value, with specific attention to life cycle trade-offs between performance, economic realization, and the creation of healthy social and environmental advantages.

The phrase "sustainable design landscape" may be defined as the environment within which the design projects and in particular the sustainable design projects are implemented (Doepker 2010). Sustainable innovation and design is not essentially about new technologies only, but about reconsidering how to meet the need for development while at the same time reducing negative environmental, health, and social impacts. Sustainable design is often viewed as an essential tool for achieving sustainability. It is a typical term that involves multiple engineering disciplines including but not limited to electrical, mechanical, civil, structural, environmental, and architectural engineering. The aim of sustainable design is to produce products and services in a way that reduces use of nonrenewable resources, minimizes environmental impact, and relates people with the environment.

Sustainability has multiple dimensions and transcends multiple disciplines. Sustainable design in this scenario will require cross-disciplinary expertise covering energy and transportation systems, carbon emissions, social network effects, effects on family structures, but most importantly, the interaction between these and additional aspects (Becker et al. 2016). Design is increasingly being viewed as a significant facilitating factor for sustainability because the design function is a joint point for decisions about a large set of human and material resource flows. Additionally, design may have an enormous impact on the materialization or dematerialization of products. It is one among many other tasks granted to engineers including project and team management, operation and maintenance management, site supervision, quality control, research and development. Nonetheless, it remains a central part of engineering practice as the design of specific objects or projects is often entirely dedicated to engineers.

Today, sustainable design, as a process in which environmental attributes are treated as design objectives to realize SD of products, is regarded as the future of design. It reduces cost, energy consumption, material utilization, pollution at its source, and natural resource depletion. It also creates a healthier living environment. In general, it is hard to integrate sustainability into early stages of design where companies measure feasibility according to economic value, performance, and time metrics. Sustainability is commonly measured at a design cycle's end on finished products when design features cannot be easily modified for sustainability measures.

Moving from a traditional linear system of typical design to a closing loop system of sustainability is a growing idea in the world of sustainable design and manufacturing. Closing the loop means moving from traditional design, which looked at the linear model for design and production, make, use, and dispose to how the disposal stage could be fed back into the creation of a new product (Smith 2012). Figure 10.1 shows how typical design is shifted into sustainable design that reflects the social and environmental needs.

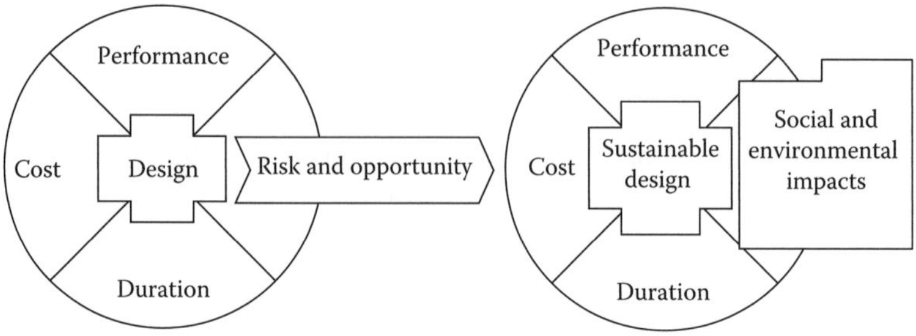

Figure 10.1 Shift from typical design into sustainable design.

Despite remarkable advances, the industry is not in a position to offer truly sustainable products and services yet. What leading industry is achieving now is an ability to continue evolving its design abilities so that, as the world around it changes, its ability to develop more sustainable products may advance along with it.

10.3.3 Key requirements of SED

Sustainable engineering is the process of utilizing energy and other resources at a rate that does not compromise the environment, or the ability of future generations to meet their life needs. Accordingly, SED is a product design where the resources used to make a product should not be depleted, and the usage and eventual disposal of the product should not damage the environment in which it functions.

There are several distinct technical components to the manner in which engineering can be practiced sustainably in society, each of which is a requirement for sustainable engineering. These include the availability of sustainable resources and processes; targeting efficiency and reduced environmental impact, and importantly the sustainable design thinking of all involved in the design process.

The design of products for sustainability is a complex issue that involves several different topics. It has been shown that significant advances toward sustainable product design can be gained by appropriate improvements in life cycle design (LCD) processes (Ping and Wang 2007). Existing approaches to supporting sustainable product design tend to focus on the later stages of product development, focusing on assessment of environmental impact costs after a design is selected, but not to

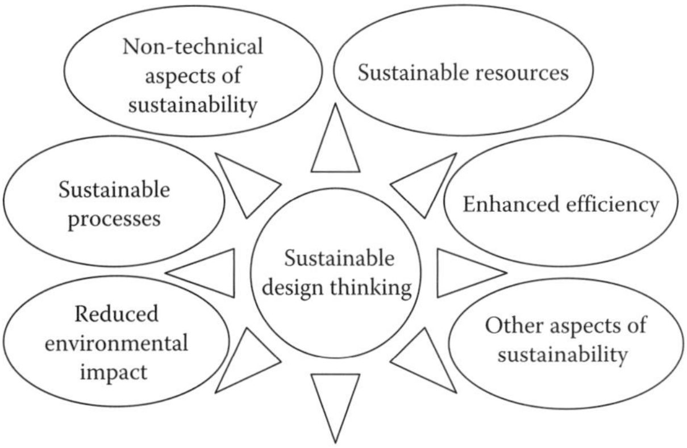

Figure 10.2 Key technical requirements of sustainable engineering.

include the early stages of design decision-making. However, support for more sustainable decisions during the conceptual design stages can lead to several advantages for enterprises (Eddy 2014).

At its core, sustainability is about how the flow of materials is managed to preserve and improve the quality of life for future generations (Ceridon 2011). Figure 10.2 shows the key technical requirements of sustainable engineering centered by sustainable design thinking.

10.4 Role of technology in sustainable design

> The 21st century is likely to become the solar-hydrogen-energy efficiency century.
>
> **Charles Secrett**
> *Friends of the Earth*

10.4.1 Innovation, technology, and design

Innovation is the core of sustainability, while sustainable technology is a means to assure that future actions are more sustainable and be an economic driver. Design on the other hand is an innovative decision-making process that intends to find a sustainability balance of trade-offs in the building of a product or service that best meets customer and other stakeholder preferences. Aspects of balance in design involve creativity, synthesis, and innovation; problem solving and technology selection; and planning for use, disuse, abuse, and reuse within all sectors.

To fully understand the role of technology in achieving sustainability, we must first acknowledge the role that technology had in providing us the means to stray off the pathway to sustainability in the first place. Instead of "leveraging technology for sustainability," we are caught in the cycle of using technology to mitigate the problems we caused with our prior increase in technological knowledge (Tlusty 2015).

The role of technology may be actually viewed as the interface that provides connection of an idea realized through design and engineering effort with practical and consumable outcomes, such as products or services. The latter would affect and shape societal lifestyle over time. Figure 10.3 presents a hierarchical view of these connections in the sustainability context.

10.4.2 Low-tech or high-tech?

A recurring question, and one which may arise in research in regard to technology, is the question "high-tech" or "low-tech" solution. While

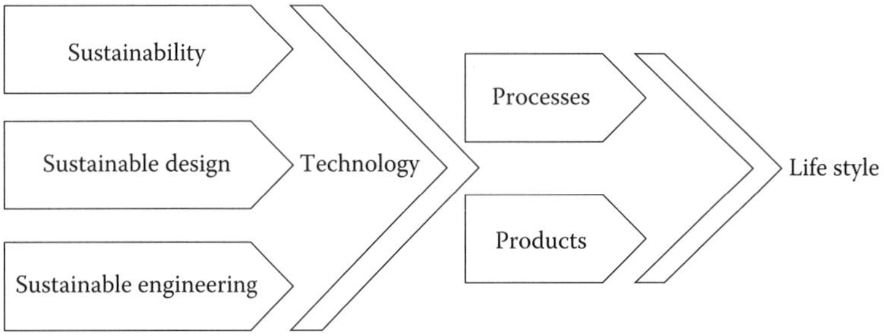

Figure 10.3 Hierarchy of sustainability guidelines.

high tech is often associated with development, it is also often accessible and realized by just few. However, low tech, which is correlated with the past, is often accessible and operable by the masses. Although it is good to remember that things called low tech are often relative to the things that are considered high tech within the same domain, in this regard the question is: which approach is more suited to help achieve a SD? Any answer to this question has to start by taking seriously the limits imposed by current situation, and by choices made in the decades already past. A certain tendency toward a preference for a low-tech approach may be discerned among many designers in practice and in research, and also among students in schools. This leaning toward low-tech would however seem more grounded on an emotional than rational level. Such development is somehow fascinating and, at the same time, somewhat confusing for a society which depends so much on technology in almost everything today.

Low-tech, accessible, local solutions offer an opportunity to address the poverty-ending objectives within the UN SDG for 2030. In order for development projects to be successful and sustainable, local communities should not be considered spectators in projects that are designed to help them. The main belief behind the low-tech, high-thinking movement is that it takes just as much creativity and skill to create inexpensive, simple solutions that may have a meaningful impact on a global scale. Understanding the systemic underlying causes, along with listening to and learning from the end-user, is a vital part of this design process (3P Contributer 2016).

In defense of high-tech tools for living a more sustainable life, which is in our pockets, on our desks, and in our cars, and regardless of the initial (and even ongoing) cost to the environment, some of it can and is helping us to make our lives more sustainable. One of the problems with using the word sustainability is that, because it is such a black-and-white issue for some people, it can be interpreted as meaning only that which is fully able to be sustained indefinitely with no external inputs, with no middle ground whatsoever. It is a case of the perfect being the enemy of the good, 100% sustainability is not really achievable, just due to the natural aging process of just about everything on the planet, and that what can and should aim for is living more sustainably (Markham 2013).

The big issue is to ensure that the technology used does actually reduce our environmental impact. It is critical to point out at once that the above argument of low-tech and high-tech options must not be interpreted to be a blanket rejection of appropriate hi-tech options. The key answer there, of course, is "appropriate." There is certainly a place for hi-tech innovations like renewable energy, computers, and the Internet as part of the good, sustainable, interconnected society.

10.4.3 Design-technology principles

The combination of sustainability and technology guide sustainable design process determines how things are supposed to be made and how they will function over their whole life cycle. Further, SED stage deals with technical implementation of ideas. Occasionally it is not an easy process, and certain aspects of design may be modified. When eventually the design and engineering routes practically converge technology may be created. Technology provides processes and products. Only then the benefits of new ideas and new engineering developments become available to society. Because of people's strong dependence on multiple technologies, those become the factors that may facilitate change in society and may even become tools of manipulation and initiation of global trends.

Designing systems to effectively meet the conditions and realities of sustainability will require a shift in current understanding of what constitutes good design and sound practice. Many of the practices that now are taken for granted, in future, may no longer be feasible. To address these needed changes in design and techniques, the design principles that might be taken into consideration have been shown in Table 10.1.

Table 10.1 Design-technology principles for sustainability

Low-carbon input materials	Design to minimize usage of any material that require large amounts of energy, or are derived from oil by-products in their manufacture.
Low external energy inputs	Design to use highly energy efficient technologies and systems.
Durability and robustness	Design to use materials and techniques that are durable in the face of more energetic conditions.
Local materials and products	Design to create a greater demand for locally produced materials and products.

10.5 Engineering approaches to sustainability design

> No subject is more important to the engineering profession or the wider world that we live in.
>
> **Frank Kreith (2012)**

10.5.1 Design through the 12 principles of green engineering

Although engineering is not directly one of the three components of sustainability, it is indirectly linked to each of them. Given the intimate ties between engineering and the key components of SD, it is evident that the attainment of sustainability in engineering is a critical aspect of achieving SD (Rosen 2012).

A design based on the 12 principles of green engineering moves beyond baseline engineering quality and safety specifications to consider sustainability factors and allow designers to consider them as fundamental factors at the earliest stages as they are designing a material, product, process, building, or a system. The 12 principles of green engineering have been developed by Anastas and Zimmerman (2003) to provide a framework for engineers to implement when designing new materials, products, processes, and systems. These principles have been developed with broad statements so they can be applied over a wide variety of engineering and science fields. Table 10.2 shows the principles as a toolbox that can be used systematically to optimize a system or its components. This approach builds on the technical excellence, scientific rigor, and systems thinking that have addressed the issue of science and technology for sustainability and SD in recent years (Anastas and Zimmerman 2003; McDonough and Braungart 2003).

Green engineering promotes innovative thinking toward sustainability which may not be achieved by applying the newest technology or process. The green engineer must redefine the project to evaluate the full life cycle of the inputs and outputs to achieve sustainability throughout the project. The 12 principles provide a framework to guide engineers into considering suitability through all stages of design. They encourage the redefinition of the task to consider the full life cycle, inputs, and outputs. A number of these principles are already implemented in the water treatment and wastewater treatment industries. The twelve principles aim to allow systematic incorporation of green engineering throughout the entire project to the benefit of the environment and society (Anastas and Zimmerman 2003).

Table 10.2 The 12 principles of green engineering

Principle 1	Designers need to strive to ensure that all material and energy inputs and outputs are as inherently nonhazardous as possible.
Principle 2	It is better to prevent waste than to treat or clean up waste after it is formed.
Principle 3	Separation and purification operations should be designed to minimize energy consumption and material use.
Principle 4	Products, processes, and systems should be designed to maximize mass, energy, space, and time efficiency.
Principle 5	Products, processes, and systems should be "output pulled" rather than "input pushed" through the use of energy and materials.
Principle 6	Embedded entropy and complexity must be viewed as an investment when making design choices on recycle, reuse, or beneficial disposition.
Principle 7	Targeted durability, not immortality, should be a design goal.
Principle 8	Design for unnecessary capacity or capability (e.g., "one size fits all") solutions should be considered a design flaw.
Principle 9	Material diversity in multicomponent products should be minimized to promote disassembly and value retention.
Principle 10	Design of products, processes, and systems must include integration and interconnectivity with available energy and materials flows.
Principle 11	Products, processes, and systems should be designed for performance in a commercial "afterlife."
Principle 12	Material and energy inputs should be renewable rather than depleting.

10.5.2 Conventional to SED

SED is often described in opposition to conventional design. The integration of relevant sustainability issues, the use of sustainability assessment tools, and the consideration of a range of sustainability criteria for decision-making should allow engineers to implement sustainable design and thus choose the most sustainable option among those considered.

The stages of a variety of proposed design processes are discussed in Chapter 9 to present an overview of the conventional design process (CDP) which is concerned with a relatively narrow set of economic and technical criteria, mainly related to costs, risk, environmental impact, and maintaining flexibility and build quality. In order to successfully implement the CDP, engineers can rely on a variety of tools such as functional analysis, creativity methods, quality function deployment matrixes, experiments, computer simulations, risk analysis, cost estimation, multicriteria analysis, etc. Various constraints must also be considered: budget, schedule, regulations, codes, patents, organizational strategies, public policies, etc. (Gagnon et al. 2008). The stages defined in the CDPs of Chapter 9 show many similarities upon which it is possible to build to propose the generic SED process.

Lu and Gu (2003) indicate that sustainable product development simultaneously considers functional, environmental, and economic requirements. These are respectively evaluated through LCA and life cycle costing (LCC). Assessments are performed on a process level and results are aggregated along four life cycle stages: extraction, production, operation, and retirement. Alternative solutions are finally compared according to their respective profiles.

Mulder (2006) phrased how SED process deeply contrasts with conventional one: "sustainability is not an add-on criterion: It is about all characteristics that a design should meet." Sustainability issues must thus be considered by engineers in all design phases of a project through well integrated complements to the conventional approach.

Boyko (2009) stated that the sustainable urban design decision-making process, developed in the VivaCity2020 (www.vivacity2020.co.uk) project, improves the CDP by integrating "sustainability tasks" and "sustainability reviews" along all design phases. Main sustainability tasks consist in (1) identifying and prioritizing of sustainability issues constituting a "sustainability agenda"; (2) developing sustainability advice on preliminary designs using relevant tools and highlighting trade-offs between sustainability issues; (3) assessing the performance of the design against the sustainability agenda; (4) developing a strategy for sustainability monitoring. Sustainability reviews are checkpoints between phases to ensure sustainability tasks are conducted in a satisfying manner.

Figure 10.4 shows the stages of sustainable engineering product development. At the center the steps of the product development process are outlined. On the path from step 1 to step 4, creative and effective designers and engineers develop a new product not by a flurry of innovation only, but relatively by implementing a series of systematic and continuous decisions to minimize potential downsides while improving the positive aspects of a developing concept and design. Design feasibility is commonly gauged using a series of weighted project metrics such as development cost, development schedule, product cost, and product features (performance).

Sustainable design is often described in opposition to conventional design. The integration of relevant sustainability issues, the use of sustainability assessment tools, and the consideration of a range of sustainability criteria for decision-making should allow an engineer to implement sustainable design and choose the most sustainable option among those considered. An example of this approach was developed

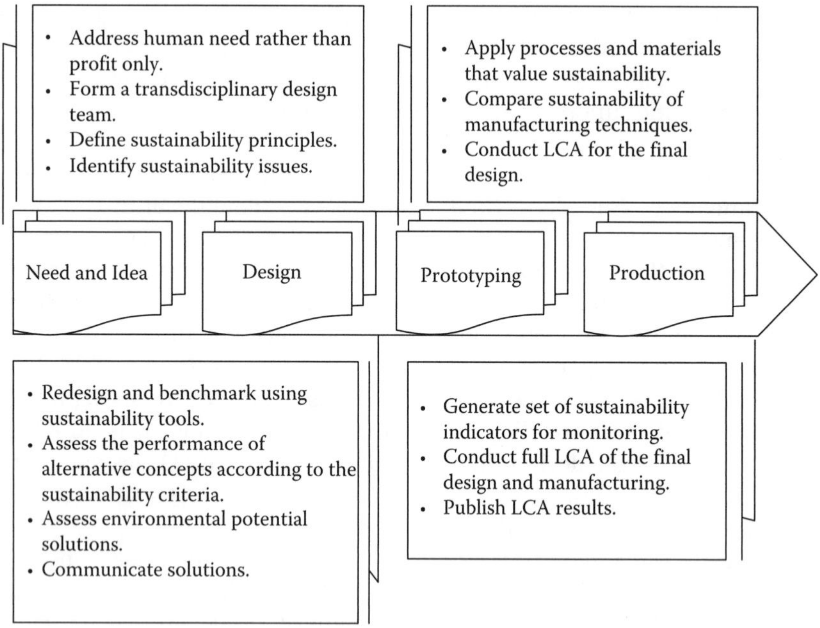

Figure 10.4 Stages of sustainable engineering product development.

by the USGBC LEED green building rating system. This rating system is a collaborative effort to first understand and address environmental issues in building design and then attempt to quantify the results of the achievement. The LEED rating system does have problems with quantification, but the system is an evolving process (Scheuer and Keoleian 2002).

10.5.3 Life cycle engineering

Life cycle engineering (LCE) is an engineering discipline that focuses on a systematic approach to design an entire life cycle of a product by incorporating the environmental aspect along with the economic, technical, and social aspects during the product development. This multidimensional view ensures that all important factors are assessed within the framework of decision-making with an interest in sustainability. The LCE methodology presents the results clearly, guaranteeing maximum transparency along with a solid basis for decision-makers (Fraunhofer IBP 2014). Several methods, such as LCA, LCC, and eco-design, are applied in order to analyze the data and make decision during the upstream and downstream of the product development. Moreover, LCE is an iterative process for continuous improvement of the sustainable performance of the product, process, system, and the facility involved during the PLC.

With an understanding of LCE, the product development process and sustainability metrics, a method for an improved product development process, is at hand. This process is the foundation of SED design with life cycle in mind. To do this, a return to the steps of the product development process outlined previously is needed with a set of consideration for each step.

Just as with the design process outlined earlier, sustainable products do not occur in a vacuum, but rather through the implementation of an iterative process of minimizing harmful ecological impacts while enhancing positive impacts as the design evolves. In this LCD activity, the designer recognizes and takes into account the various phases of a PLC, during the design of that product. The PLC activities are initially less intense than engineering and design activities. In earlier steps, the LCD activities are primarily research based with high-level screening to establish system boundaries and build the foundation for final analysis. As the process proceeds, LCD activities become more specific to inform design decisions, and ultimately, guide manufacturing decisions (Ceridon 2011).

10.6 *The triple bottom line*

> The conventional design criteria is tripod: Can we profit from it? The company asks. Will the customer find it attractive? And will it work? Champions of "sustainable development" like to use a "triple bottom line" approach based on the tripod of Ecology, Equity, and Economy.
>
> **William McDonough**

10.6.1 *Criteria for measuring success*

TBL and sustainability are two related constructs that are used interchangeably in the literature (Alhaddi 2015). In the late 1990s, John Elkington, the founder of a British consultancy called SustainAbility (Elkington 1997, 2004), coined the phrase TBL as a method for measuring sustainability. His argument was that companies should be preparing three different (and quite separate) bottom lines. One is the traditional measure of corporate profit, the "bottom line" of the profit and loss account. The second is the bottom line of a company's "people account," a measure in some shape or form of how socially responsible an organization has been throughout its operations. The third is the bottom line of the company's "planet" account, a measure of how environmentally responsible it has been (Onyali 2014) as shown in Figure 10.5. These elements are often called the "three Ps" of the TBL concept.

In Elkington's view the ecological, economic, and social factors had to be brought into a direct and balanced relationship to each other to achieve SD. The ecological measure of TBL refers to engaging in practices that do not compromise the environmental resources for future generations (Alhaddi 2015). The economic measure of TBL framework refers to the impact of the organization's business practices on the economic system (Elkington 1997). It includes the organization's financial performance, the flow of capital, and their economic involvement in society. The social measure of TBL refers to conducting beneficial and fair business practices to the labor, human capital, and to the community (Elkington 1997). Measuring performance against these measures is a complicated task. Shareholder value, market share, and customer satisfaction are relatively easy to quantify and measures developed by one organization are readily transferable to others, but social and environment performance are almost certainly unique to each organization, or at least each industry, and they are often very difficult to quantify (Hubbard 2009).

TBL establishes principles by which an organization should operate to concentrate on the total effect of their actions (both positive and negative) (Jackson et al. 2011). It is a concerted effort to incorporate economic, environmental, and social considerations into an organization's evaluation and decision-making processes (Wang and Lin 2007).

In the literature, there is no real consensus as to the exact dimensions used for the performance measures. Some other dimensions used are community improvement, environment, entrepreneurship and education (Sher and Sher 1994), stakeholder engagement and activism, and organizational integrity (Painter-Morland 2006). In all instances, performance is being measured based on the impact of companies on society as a whole, both now and into the future.

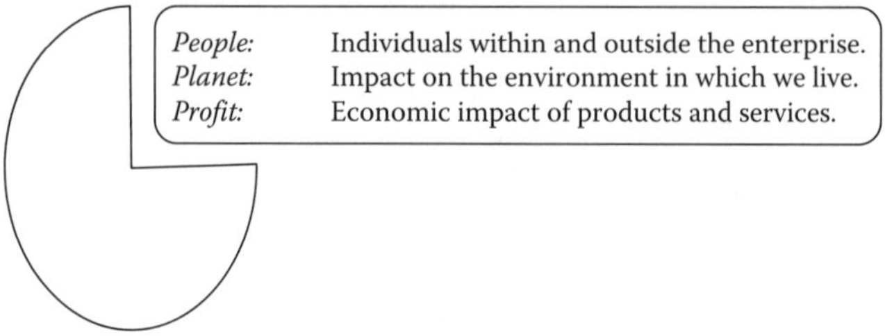

Figure 10.5 Triple P approach.

Inspired by Elkington's Triple P approach, Braungart and McDonough (2002) introduced the strategy of the TBL, better known as Economy, Ecology, and Equity (Triple E) as outlined in Figure 10.6.

Both these approaches bring ecological, economic, and social factors into a direct and balanced interrelationship. An important difference between the two approaches is in the concepts of people and equity. The Triple E strategy is not only about people, but about equality of people, animals, and plants.

10.6.2 Design quality

The design quality is not just to mean the "technical quality" of a product but also the less tangible "desirability" of a product, "pleasure of use" of a product, as well as the "attachment" of a user to a product. Designers can stimulate desirability, increase pleasure, and deepen attachment by designing products that not only function better, are more aesthetically pleasing than comparable products, but are also tailored to better suit the individual needs of the user. Govers and Mugge (2004) argue that if an object is highly desirable, its longevity is extended, and its negative impact on the environment is therefore reduced.

Products which are so well designed that they become lasting "objects of desire, pleasure and attachment" are more sustainable because they do not get disposed of in the way that lower quality designed products do. From this, one could argue that the E-type Jaguar, for example, is potentially more environmentally sustainable than a modern hybrid car because, if one looks at its complete life cycle, it performs superbly. This is because the quality of its design makes it such a great object of desire that it never gets scrapped as a conventional car possibly would. It is cherished by its owner, with great care being taken in its maintenance and, in all likelihood, could last for several generations (Diegel et al. 2010).

Design tools that allow engineers to choose inherently safer materials, improve energy and mass efficiency, and reduce emissions and exposures are certainly part of an emerging technology. However, these are

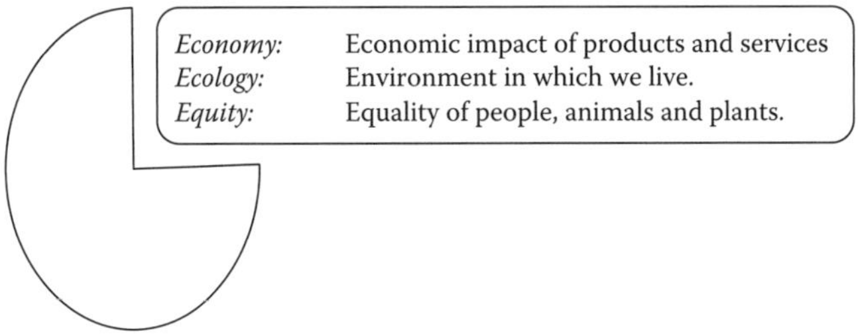

Figure 10.6 The Triple E approach.

not entirely new tools for engineers. Energy and mass efficiency are goals that have always been targeted in engineering design. What is new in fact is the need to thoroughly and concurrently integrate economic, environmental, and social objectives into engineering designs, at several levels. These tools and skills will be increasingly important, and challenging to apply, as engineers operate in diverse global environments, with varying social perspectives and environmental priorities.

TBL plus design quality divides sustainability into three major areas: environmental, economic, and social sustainability and one supporting factor, design quality, as shown in Figure 10.7. Sustainable design emphasizes the importance of using appropriate tools, especially for the analysis of potential solutions and the synthesis of the data gathered. Environmental analysis covers areas of ecological foot printing, energy or embodied energy (available energy), waste minimization, impact assessment, and LCA. In the social factor, impact assessment and LCA play an important role in product development (Gagnon et al. 2008). In the economic factor, it is clear that cost benefit analysis, economic impact analysis, inequality and equity analysis and LCC are key determinant in product development. LCC consists of an inventory and analysis of economic implication of environmental impact of a given product during its life cycle.

Sustainable projects need to be as technically sound as conventional projects. Sustainability tools must thus be used in conjunction with existing approaches associated with state-of-the-art engineering practice.

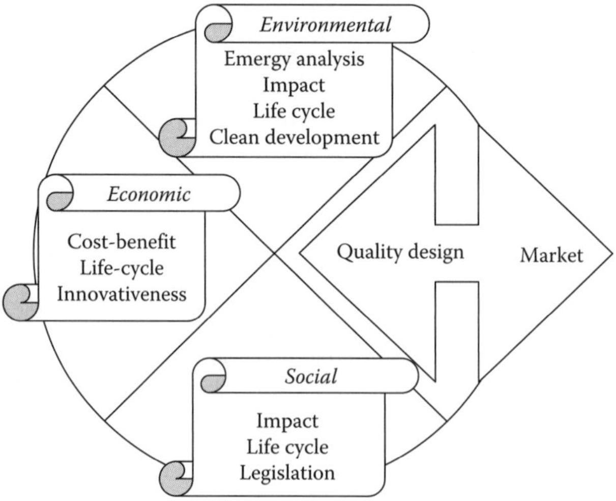

Figure 10.7 Triple bottom line plus design quality in sustainable product development.

10.7 *Design for sustainability*

People ignore design that ignores people.

Frank Chimero

10.7.1 *Hannover principles from 1992*

"The Hannover Principles" aim to provide a platform upon which designers can consider how to adapt their work toward sustainable ends. Designers include all those who change the environment with the inspiration of human creativity. Design implies the conception and realization of human needs and desires.

The Hannover Principles are a set of maxims that encourage the design professions to take sustainability into consideration. They are descriptive of a way of thinking not prescriptions or requirements. The guidelines shown in Figure 10.8 demonstrate the German City of Hannover's intention to apply these principles as elements of the overall design competitions associated with EXPO 2000 where the City of Hannover has been designated as the site of the world exposition in the year 2000. In order to insure that the design and construction related to the fair will represent sustainable development for the city, region, and world, the City of Hannover has commissioned "The Hannover Principles" to inform the international design competitions for EXPO 2000. They take the form of a framework, based on the enduring elements of Earth, Air, Fire, Water,

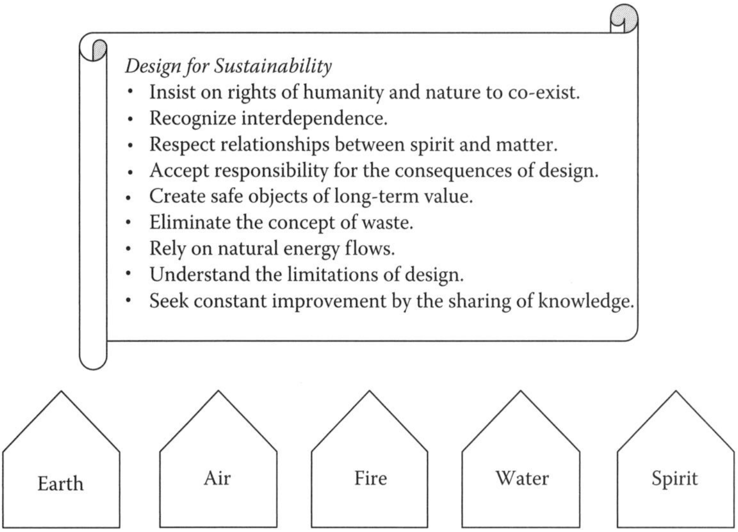

Figure 10.8 The Hannover Principles, 1992 based on the enduring of five elements.

and Spirit, in which design decisions may be reviewed and evaluated. The guidelines offer critical instruction on the responsibility of designers (McDonough and Partners1992). The Principles are to be considered by designers, planners, government officials, and all involved in setting priorities for the built environment.

10.7.2 Enabling DfS

DfS is an eco-design concept that has evolved to include both the social and economic elements of production. It has evolved from general cleaner production (CP) methods to focus on products and to include social, economic, and environmental elements of production. CP is defined as the continuous application of an integrated preventive environmental strategy to processes and products to reduce risks to humans and the environment (Clark et al. 2009).

Most of the environmental impacts of products, throughout their life cycle, are determined at the design phase. Thus concepts and methods such as design for recycling (DfR) or design for environment (DfE), and DfS in general have come to the fore. Their aim is to reduce material and energy consumption and waste throughout the product life cycle (PLC). The integration of sustainability in the design process has been reinforced by the use of sustainability assessment tools such as LCA which aims to integrate environmental considerations into product design. The assessment is used to guide the designer in evaluating different product alternatives with the same functional unit (FU). The evaluation of the product at design phase can be qualitative (questionnaires), semiqualitative (analysis grids), or quantitative (pure LCA with direct data collection or use of LCA data basis) (Medini et al. 2015).

The concept of DfS drives beyond how to make a "green" product and attempts to meet consumer needs through sustainability-oriented interventions in a systematic and complete way (Crul and Diehl 2006). It requires awareness of the full short- and long-term consequences of any transformation of the environment. These newly designed products and services offer increased functionality and ease of use, longer life spans, easy disassembly or recyclability, lower environmental impacts which can reduce cost and improve material sourcing and production which can positively impact communities. In other words, sustainability may offer added value through better quality and lower price.

DfS involves radically changing the principles which guide the process of design. It requires awareness of the full short- and long-term consequences of any transformation of the environment. It is the conception and realization of environmentally sensitive and responsible expression as a part of the evolving matrix of nature (McDonough and Partners 1992). DfS is an eco-design concept that has evolved to include both the social and economic elements of production. It integrates the three pillars of sustainability, people, profit, and planet (Triple P), but goes beyond simply, greening products to embrace how to meet consumer needs in a more holistic and sustainable way (Crul and Diehl 2006).

According to Crul and Diehl (2006), DfS considers a number of issues linked to people, planet, and profit. Modifications are suggested along all design phases and include (1) choosing team members from with various expertise; (2) defining goals and objectives through the analysis of drivers related to the three pillars of sustainability; (3) assessing the life cycle impact of a reference product; (4) selecting strategies to guide idea generation; (5) assessing solutions according to environmental, social, and economic criteria; (6) sustainability communication; and (7) sustainability performance monitoring after product launch.

The challenges for DfS are to generate knowledge supporting the innovation and design engineering of products and service systems with superior sustainability and to make optimal use of networking and entrepreneurship as success factors for implementation (Clark et al. 2009). The key DfS approaches are as follows: redesign, benchmarking (incremental design), new product design, product service system (PSS), and product innovation.

10.7.2.1 DfS redesign

Redesign is defined as the action of successive changes or improvements to a previously implemented design. The goal of redesign is to sustainably redesign an existing product for which the specific market and manufacturing conditions are already known, taking into account its primary function and the associated services provided. A product's improvement potential can be determined relatively easily as the product already exists, so market and manufacturing information is readily available.

It is important to realize that redesign for sustainability could again mean going back to the roots of a specific problem in design and then finding an improved solution, for example, practicing radical design strategy. In addition, redesign was recommended to be started by practicing radical design (Jarvenpaa and Stoddard 1998).

The redesign process uses a project team to harness expertise to incorporate sustainability aspects into products, but also company employees who can often provide valuable insight. When choosing an initial product for redesign, companies should focus on the interventions that have the potential for greatest impact while being simple and timely to implement, and in line with overall company goals. The finished, redesigned product should be compared against the initial product to consider and estimate the sustainability advantages of the new product versus the original; after the product is launched, the company must do follow-up to evaluate

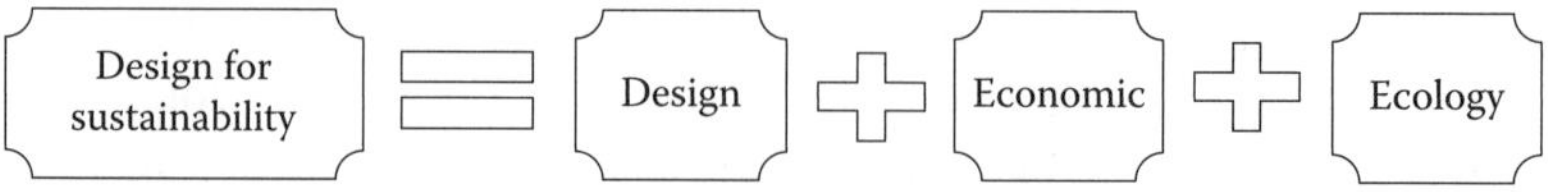

Figure 10.9 Major components of DfS.

overall sustainability, which will spawn new implementation ideas for future products. Figure 10.9 shows the major components of DfS (Clark et al. 2009; Crul and Diehl 2006).

Short-term incremental redesign of existing products, also called "inside-the-box" innovation, can typically lead to improvements up to a factor of 5. Incremental innovation and design improvements are known as the "bread and butter" of product innovation. To achieve long-term factors of 10, 20, or higher, or changes toward radical shifts in the whole of society, radical product innovation, or outside-the-box innovation, is necessary. This includes developing completely new products, improving products as well as the services connected to them, and developing entirely new functional systems of products and services. A high level of uncertainty is associated with radical innovation, especially at early stages (Crul and Diehl 2006).

10.7.2.2 Benchmarking

Benchmarking and other approaches based on replication are still the predominant way in which products are designed worldwide. DfS benchmarking is a structured approach to compare the environmental performance of a company's products against competitors' products and to generate improvement options. The goal of DfS benchmarking is to learn from the best practice of competitors by comparing one's product to those of competitors to determine how to make that product more sustainable. The methodology is a structured approach in which improvement options are generated by looking at the environmental, social, and economic aspects of a particular product. Benchmarking differs from DfS redesign in that it starts with comparing existing products in the market before moving into the design phase. Like redesign, DfS benchmarking also uses a project team to decide the goals for the process including, but not limited to, entering new markets, improving competitiveness, and making environmental improvements (Clark et al. 2009; Crul and Diehl 2006).

Since individual competitors often use different solutions to resolve the same design problems, like a different product architecture, components, or technology, DfS benchmarking offers a reflective approach and advises learning from others' products.

10.7.2.3 New product design

The DfS new product development (NPD) approach applies "out-of-the-box," or radical, innovation strategies, which can lead to more sustainable impacts while providing the breakthroughs necessary to ensure a company's continued competitiveness. NPD involves a higher level of technical, market, and organizational uncertainty than redesign but can be an inventive and iterative process where new ideas on how to meet needs are converted to products and services. Eco-friendly materials, sustainable development practices, and innovative information and communication technology are all concepts that can help inspire new product design (Clark et al. 2009).

The stages and processes involved with new product design can be viewed as three-fold: policy formulation, idea generation, and product development. Policy formulation addresses the company's goals and strategies; idea generation allows the company to brainstorm and develop ideas for new products, taking into account the ability to harness developing technologies, materials, and consumer needs; and finally, product development involves debating and testing concepts against the decisions in the idea finding phase. The key challenge with respect to new product design is market demand. Without a consumer need, even the most sustainable product will fail (Clark et al. 2009; Crul and Diehl 2006).

NPD comprises a higher level of technical, market, and organizational uncertainty than redesign but can be an inventive and iterative process where new ideas on how to meet needs are converted to products and services. Eco-friendly materials, sustainable development practices, and innovative information and communication technology are all concepts that can help inspire new product design (Crul and Diehl 2006).

10.7.2.4 Product service systems

PSSs illustrate the movement from eco-design to DfS because they use different ways of addressing at the design stage what a customer really needs and the way a product is designed, produced, used, and discarded. It can be an effective function-based strategy that concentrates on "satisfaction" as a product value instead of private ownership of physical products, a traditional standard of well-being that exists in many industrialized contexts.

PSSs already exist in today's society, especially in developing countries. Often these systems are a way of life, and are not perceived of in the sustainability context. In many developing countries where different members of a society cannot afford to own products, they divide the product to maximize the benefits among a wider group. These examples of

PSSs can be useful as illustrations, but the challenge is to find the benefits in cross-sharing of experience between developed and developing countries (Clark et al. 2009).

The concept of PSS proposes that companies move from merely selling products (or services) to designing and providing a system of products and services (and related infrastructure) which are jointly capable of fulfilling client needs or demands more efficiently and with higher value for both companies and customers than purely product-based solutions (Tukker and Tischner 2006).

10.7.2.5 Product innovation

Since DfS is based on a combination of product innovation and sustainability, understanding the underlying concepts of product innovation in terms of function and system can be helpful in implementing DfS projects. Innovation can be categorized into three levels: incremental, radical, and fundamental. Each category is progressively more significant and far-reaching.

A systematic approach for product innovation has been developed by Roozenberg and Eekles (1995). It consists of four basic steps: formulating goals and defining strategies for product development based on market perceptions; generating and selecting ideas for the new or improved product; developing these ideas into the blueprint of the new product; and transforming the plans into reality including production, distribution, sales, use, and end-of-life (EOL) assessment.

10.8 Life cycle-based sustainability assessment approaches

> There are no passengers on spaceship earth. We are all crew.
>
> **Marshall McLuhan**

Over the last decades numerous assessment methods and tools for environmental and sustainability performance have been developed. They are grouped in Figure 10.10 according to an adapted pyramid of needs (Maslow 1943). While the original pyramid of Maslow has the basic physiological needs at the bottom, followed by safety needs, love and belonging, esteem until self-actualization at the very top, the adapted version starts with the basic approach of LCT, followed by single-issue methods like carbon or water foot printing, LCA, resource or eco-efficiency assessment up to life cycle sustainability assessment (LCSA) at the top of the pyramid.

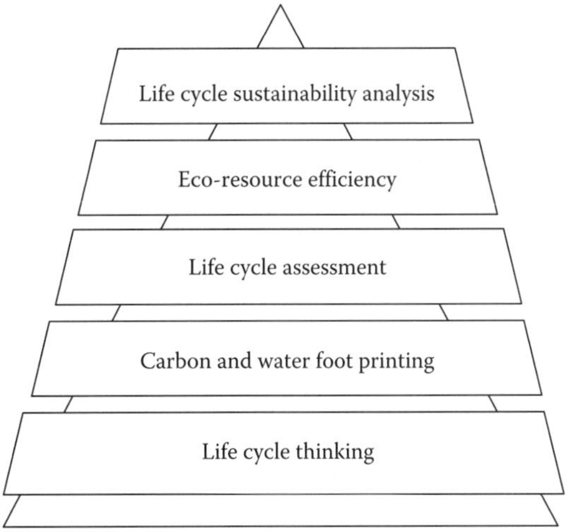

Figure 10.10 Adaptation of Maslow's pyramid of human needs for life cycle-based environmental and sustainability assessment approaches.

10.8.1 Life cycle thinking

LCT refers to going beyond the traditional focus on production and manufacturing processes and considers the environmental, social, and economic impacts of a product over its entire life cycle. The main objective of LCT is to reduce product's resource use and emissions to the environment as well as improve its socioeconomic performance throughout its life cycle. LCT is also a highly opportunity-driven thinking mode, because this approach can help identify important business opportunities by reducing resource consumption and improving the performance of products (Brazdauskas 2015).

Product LCT is essential in the path to sustainability by expanding the focus from the production site to the whole PLC. This facilitates the links between the economic and environmental dimensions within a company. LCT is about widening views and expands the traditional focus on manufacturing processes to incorporate various aspects associated with a product over its entire life cycle. The producer become responsible for the products from cradle to grave and has for instance to develop products, which have improved performance in all phases of the PLC (Jensen and Remmen 2006).

LCT is a qualitative concept that represents the basic concept of considering the whole PLC from "cradle to the grave." It aims to prevent individual parts of the life cycle from being addressed in a way that just results in the environmental burden being shifted to another part. LCT has for example been addressed as one of the five key principles of the Integrated Product Policy which is part of the sustainable development strategy of the European Union (EC 2003). The approach aims at incorporating LCT in products, working with the market by the use of incentives, involving stakeholders, using a variety of policies in order to cover most sectors and stakeholders, as well as aiming at the continuous improvement on the environmental impact of products.

LCT implies the understanding that materials are extracted from the earth, converted into process materials, combined with other materials to make parts, assembled into a finished product, shipped to customers who use the products and finally, the products are disposed of in some fashion. Energy and other natural, social, and economic resources are used, waste generated, and the related impacts, both positive and negative, are distributed across societies to varying degrees around the globe (UNEP 2012). LCT involves the following preferred action directions, which should be emphasized in order to develop student competencies for sustainable innovations (Brazdauskas 2015):

- Rethinking the product and its functions
- Reducing energy, material consumption, and socioeconomic impacts throughout a PLC
- Recycling or selecting materials that can be recycled
- Reusing or designing the product for disassembly so parts can be reused
- Replacing harmful substances with safer alternatives
- Repairing and making the product easy to repair

LCT is about getting reliable information about environmental, social, and economic impacts into people's hands at the time they are making decisions. The shift to LCT with an integration of life cycle approaches is simultaneously taking place in numerous sectors and on different levels (UNEP 2012). In this regard, LCT, due to its systemic approach, is considered to provide a significant support in integrating sustainability into innovation, design and evaluation of products and services.

10.8.2 Carbon and water footprint

The CF concept has become popular over the past few years since and is currently widely accepted and used by the public and media despite its lack of scientifically accepted and universally adopted guidelines: it describes greenhouse gas (GHG) emission measurement from the narrowest to the widest sense. Several calculation methods and approaches for CF accounting have been proposed and are being used. Since about 2008, WF has also become a popular term.

With the next level in the pyramid the approaches start to be quantitative. More recently, evaluation approaches for single environmental issues like CF (Berger and Finkbeiner 2010) and WF (Finkbeiner et al. 2006) have received considerable attention. The WF and CF concepts have similarities; however, their roots and intended purposes differ. The CF was formulated to quantify the contribution of various activities to climate change. The history of the WF lies in the exploration of water use along supply chains and in the search for a tool to understand the global dimension of water as a natural resource. Although each footprint has different roots and characteristics and addresses different research and policy questions, there is a tendency among practitioners in the fields of environmental policy to treat the WF in a similar way as the CF.

Concern about climate change started with the scientific recognition of the relationship between CO_2 emissions and global warming. Despite its popularity and use in commerce, there is no universally accepted definition of CF. Today it describes the narrowest to the widest interpretation of GHG emission measurement (East 2008; Finkbeiner 2009).

The WF concept is primarily rooted in the desire to illustrate the hidden links between human consumption and water use and between global trade and water resources management (Hoekstra 2008). Unlike the CF, which emerged in practice, the WF was born in science. The WF started to gain broad interest from about 2008, the year in which the Water Footprint Network (WFN) was established, a network of academic institutions, governments, nongovernmental organizations, companies, investors, and UN institutions. One of the aims of the WFN is to ensure the establishment of one common language and a coherent and scientifically sound framework for water footprint assessment that serves different interests.

CF and WF concepts complement each other, addressing different environmental issues: climate change and freshwater scarcity. Although there are similarities in the way both footprints are defined and calculated, they differ in important ways as well. In general, there is an acknowledgment that humanity's CF and WF have surpassed sustainable levels and that society must make efforts to reduce them, but it appears to be quite difficult to establish explicit and agreed upon maximum sustainable levels for these footprints. Knowing their ceilings is instrumental in formulating reduction strategies (Ercin and Hoekstra 2012).

10.8.3 Life cycle assessment

LCA is one of the most important techniques for the implementation of a process or product development in the framework of sustainability. It is an analytical tool designed to quantify the ecological impacts or sustainability performance of a system. It is a tool designed to account for all of the inputs and outputs of a system.

10.8.3.1 LCA defined

LCA is fundamentally a decision-making tool that includes extraction, processing, and quantifying materials and energy use, manufacturing, distribution, recycling, and final disposal. It is a complex environmental assessment tool that requires enormous amounts of data which are often hard to find or expensive to purchase. There are a few variants of LCA, namely cradle-to-grave, cradle-to-gate, cradle-to-cradle (C2C) or gate-to-gate (Zimoch 2012).

LCA is a methodology to evaluate the material flows and environmental impacts associated with the production of goods and provision of services over its full life cycle from extraction and processing of raw materials through manufacture, operation and, finally, disposal (ISO 1998). It characterizes the product according to input quantities, such as energy (in units of Joules), chemicals and raw materials, and output quantities, such as air, water, or land polluting elements (in weight- or volume-based units). The input and output quantities constitute the life cycle inventory (LCI) that define the impacts of the system (Ceridon 2011).

LCA is the process of evaluating the total effects that a product has on the environment over its entire existence, starting with its production and continuing through to its eventual disposal. Products can be made from any number of materials and components, each of which has a broad set of environmental impacts over their life cycle, starting with the extraction of raw materials from the ground to the end of the product's life. Some materials or designs may provide a reduction in one or more impact areas, but an increase in other impact areas. How do we decide which is better? LCA is a tool for answering exactly this question (Bonnema 2006).

LCA is a decision-making tool that is built around the principle of comprehensiveness and therefore aims to address all environmental interventions, not just one. It is a well-established environmental management tool for which international standards are available in their second generation (Finkbeiner et al. 2010).

10.8.3.2 LCA phases

The most widely used theory of the LCA analysis is an approach described in International Standards ISO 14040 that assumes four phases of the LCA: goal and scope definition, inventory analysis, impact assessment, and interpretation (Zimoch 2012). The first phase is goal and scope definition where goals, system boundaries, and intended uses are established. The second phase is called a LCI analysis. This is a data-based quantification of energy and raw material requirements, air emissions, waterborne and solid waste, and other environmental discharges through the life of a product or process. In this phase, the goal is to examine all the inputs and outputs in a PLC, beginning with what the product is made of, the source of the materials, the operations involved in making those materials, where they go, and all of the inputs and outputs related to those component materials during their lifetime. The third phase is life cycle impact assessment (LCIA) to evaluate the effects of the environmental information collected in the inventory. A full impact assessment addresses ecological and human health, as well as the range of social, cultural, and economic effects. The fourth phase is interpretation including identification of important issues arising from the LCI and LCIA phases, evaluation of completeness, sensitivity and consistency, and conclusions, limitations, and recommendations. Figure 10.11 shows the four phases of the LCA.

10.8.3.3 LCA analysis

LCA has been continuously developing over the past 30 years, with notable improvements at the modeling level both in the inventory and impact assessment. Nowadays, it is successfully used for continuous environmental improvements of products; internal strategic decision support; evaluating risks and opportunities along the supply chain; communication on strategic aspects with stakeholders at company and association level; communication with customers on products, for example via environmental product declarations and carbon labels, just to mention a few (Sala et al. 2012). Over the years, a shift has occurred from merely energy and environmental analysis to more comprehensive assessments, which include economic and social aspects (Benoit and Mazijn 2009). To carry out an LCA for a product, five stages of a PLC are analyzed (Skrainka 2012):

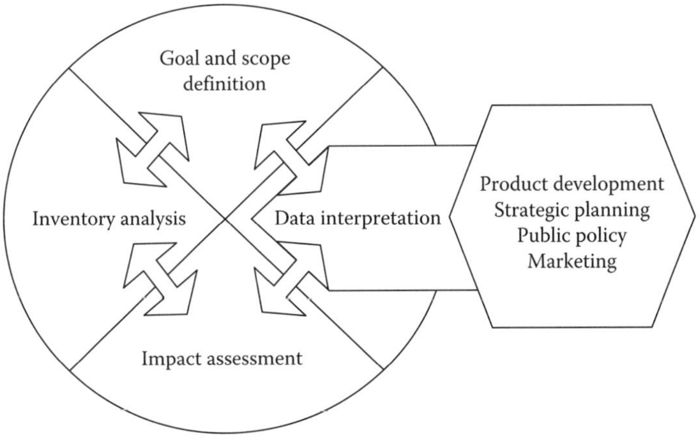

Figure 10.11 The four main phases of LCA.

- Premanufacturing: performed by suppliers who extract in most cases virgin materials employed for manufacture
- Manufacturing: processes related to the transformation of the virgin materials into usable products
- Product delivery: the transport of the product to the customer
- Use: the intensity and frequency of the use of the product
- Disposal: a product that is no longer needed could be reused or remanufactured to use it once again, recycled or incinerated if materials permit, or landfilled

Software tools are available to analyze CLA. For example, SimaPro is a LCA software tool used to analyze the environmental impacts of products following ISO 14040 guidelines. SimaPro allows the quantification of the burden carried at each stage of a PLC. SimaPro is a convenient tool because it makes a complex analysis such as an LCA, into a more straightforward analysis.

10.8.4 Eco-efficiency versus eco-effectiveness

Eco-efficiency is an important solution that leads to a more sustainable environment; it is only a partial solution. It starts with the assumption of a one-way, linear flow of materials through industrial systems: raw materials are extracted from the environment, transformed into products, and eventually disposed of. However, eco-efficient technique seeks total solution by minimizing the volume, velocity, and toxicity of the material flow system. The relation of eco-efficiency and eco-effectiveness is an important goal in sustainable design, as the relation between short-term strategy and long-term strategy is important in every product development project. Both eco-efficiency and eco-effectiveness are identified as important indicators in SD.

10.8.4.1 Eco-efficiency

Eco-efficiency is a key element for promoting fundamental changes in the way societies produce and consume resources, and thus for measuring progress in green growth. The concept of eco-efficiency can be traced back to 1970s as the concept of environmental efficiency (UN 2009). In the 1990s, Schaltegger and Synnestvedt (2002) introduced eco-efficiency as a "business link to SD." Later, it was popularized by the World Business Council for Sustainable Development (WBCSD) for the business sector in the course of the UN Conference on Environment and Development (UNCED) in 1992.

Eco-efficiency is based on the concept of creating more goods and services while using fewer resources and creating less waste and pollution. It is measured as the ratio between the added value of what has been

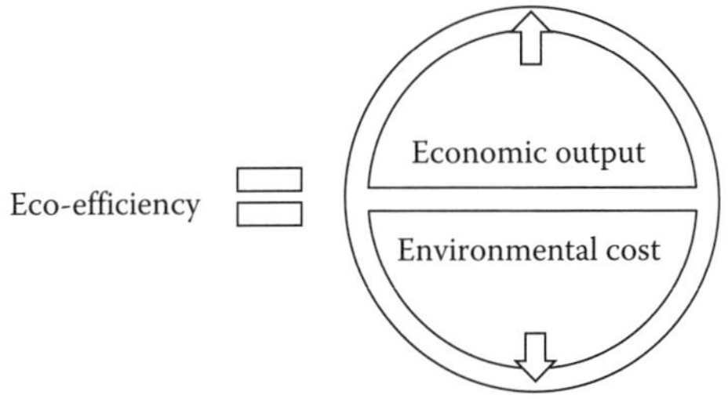

Figure 10.12 Conceptual definition of eco-efficiency.

produced (GDP per capita) and the added environment impacts of the product or service (pollution emissions (energy or water used; environmental burden) (Yadong 2013). Therefore, eco-efficiency can be described as the effective use of material and natural resources in processes that maximize efficiency and minimize environmental impacts. Organizations that consider eco-efficiency for assessment improve their effectiveness and productivity by reducing energy consumption, solid waste, wastewater, and GHG and air emissions. Figure 10.12 provides the conceptual definition of eco-efficiency. It may be derived by looking at the intensity of resource-use, intensity of environmental impacts, or both.

Conceptually, eco-efficiency corresponds to the CP approach. In both concepts, natural resources and energy use, toxic and hazardous chemical usage, waste materials and water generation, are to be reduced in a certain manner of controlling way.

Designers concerned with product performance and aesthetics must take into account the effect of design details on energy and material requirements for manufacturing, use, and secondary use. Companies should also pay closer attention to energy use and emissions. Major improvements in energy efficiency can often be achieved at little or no cost, even with net savings, through the use of targeted programs. Based on the approach of eco-efficiency, innovative strategies have been developed focusing on reduction and compensation of unsafe effects on the environment.

10.8.4.2 Eco-effectiveness

In contrast to eco-efficiency approach of minimization and dematerialization, the concept of eco-effectiveness means the transformation of products and their related material flows such that they create a caring relationship with ecological systems and future economic growth. The objective is not to minimize the cradle-to-grave flow of materials, but to generate cyclical, C2C metabolisms that assist materials to preserve their condition as resources for creative reuse (upcycling).

This C2C approach introduces the concept of eco-effectiveness to address the shortcomings of eco-efficiency. Eco-efficiency aims as far as possible to reduce and compensate the harmful effects on the environment. Eco-effectiveness aims for development without harmful effects on the environment. Out of concern about the lack of completeness of the concept of eco-efficiency, Braungart and McDonough (2002) came up with a response with the introduction of the concept of eco-effectiveness in 2002. Eco-efficiency delays environmental pollution and the exhaustion of natural resources. An eco-efficient approach would allow the use of fossil fuels to be minimized, but it will never be possible to eliminate their use completely. A total solution requires a new paradigm. Simply reducing the problem will never solve it completely, and will also limit freedom of trade and growth opportunities. Eco-effectiveness is based on a closed-cycle approach, in which materials are used in new products, processes and objects in a way that they are 100% reusable or can be recycled, and in which the energy for all activities must be renewable. Eco-effectiveness causes no adverse effects in relation to sustainability (Van de Westerlo 2011).

10.8.5 *Life cycle sustainability assessment (LCSA)*

Recently, social impacts are also being addressed in life cycles and supply chains, leading to the definition of LCSA. Andersson et al. (1998) examined the feasibility of incorporating the concept of sustainability principles into each phase of LCA. Four socio-ecological principles were identified: (1) substances from the lithosphere must not systematically accumulate in the ecosphere (i.e., the use of fossil fuels and mining must be radically decreased); (2) society-produced substances must not systematically accumulate in the ecosphere; (3) the physical conditions for production and diversity within the ecosphere must not systematically deteriorate; and (4) the use of resources must be efficient and must meet human needs.

Kloepffer (2008) put the LCSA framework into the conceptual following formula, which was improved into its current form:

$$\text{LCSA} = \text{LCA} + \text{LCC} + \text{SLCA}. \tag{10.1}$$

LCA represents the state of the art in science and application relating to the environmental dimension of sustainability. LCC is the total cost of a system or product, produced over a defined life time. The social dimension of sustainability captures the impact of an organization, product, or process on society. The social benefits can be estimated by analyzing the effects of the organization on stakeholders at local, national, and global

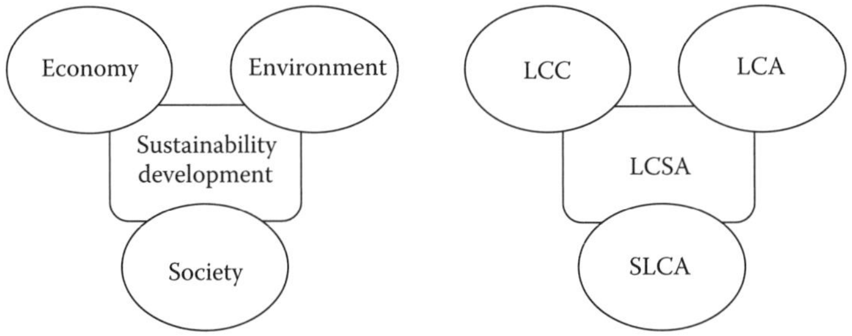

Figure 10.13 Domains of sustainability and life cycle sustainability assessment.

levels (GRI 2002). Based on the well-known depiction of sustainability, where the three dimensions of environment, economy, and society intersect, as depicted in Figure 10.13.

The majority of social indicators measure the degree to which societal values and goals in the particular areas of life or politics can be achieved. However, many social issues on which a performance measurement takes place are not easy to quantify. Therefore, a number of social indicators contain qualitative standards of systems and activities of the organization, including operating principles, procedures, and management practices. These indicators address needs specific to social issues such as forced labor, working hours, or existence of trade unions (Finkbeiner et al. 2010).

The LCSA framework is able to accommodate knowledge from different disciplines relevant to sustainability and to better link questions to models of analysis toward transdisciplinarity. LCSA is a conceptual framework, which needs to be made operational. Most of the present developments in LCA fit into the framework, and an excess of other methods and tools have already been identified as potentially useful (Jeswani et al. 2010). However, before any choice of methodologies and methods is made, it is necessary to investigate the extent to which the main characteristics and principles of sustainability are taken into account, what is still lacking and what is needed to develop a robust LCSA framework (Sala et al. 2012).

10.9 C2C design framework

> If we think about things having multiple lives, cradle to cradle, we could design things that can go back to either nature or back to industry forever.
>
> **William McDonough**
> *Stanford University*

10.9.1 C2C approach

A phrase invented by Walter R. Stahel in the 1970s and popularized by William McDonough and Michael Braungart in their 2002 book of the same name. In the book "Cradle to Cradle: Remaking the Way We Make Things," the authors present a "C2C" approach in which biological and technical cycles are closed without damaging effects on the environment. In August 2010 an exclusive, worldwide license was granted to the C2C Product Innovation Institute as a third-party not-for-profit organization to manage the certification program.

C2C is a biomimetic approach to the design of products and systems. It is a valuable design approach integrating multiple attributes: safe materials, continuous reclamation and reuse of materials, clean water, renewable energy, and social fairness. This approach pursues to establish production techniques that are not just efficient but are effectively waste free. C2C is a science- and values-based vision of sustainability successfully applied over the past decade that articulates a positive, long-term goal for engineers. It is a helpful design approach integrating numerous attributes, including safe materials, continuous reclamation and reuse of materials, clean water, renewable energy, and social fairness. C2C designs industrial systems to be commercially productive, socially beneficial, and ecologically smart. It is also a revolutionary approach to the redesign of human industry based on the conviction that thoughtful design, mirroring the safe, regenerative productivity of nature, can create an industry that is sustaining, not just sustainable.

10.9.2 Principles of C2C design

The C2C framework suggests a new way of designing human systems to eliminate conflicts between economic growth and environmental health resulting from poor design and market structure. In addition, the C2C design framework embraces the pursuit of maximum value (economic, ecological, and social) through the practice of intelligent design. It is the foundation of an emerging world in which all human industry is designed to celebrate interdependence with other living systems, transforming the making and consumption of things into a regenerative force.

C2C is a model of industrial systems in which material flows cyclically in appropriate, continuous biological, or technical nutrient cycles. All waste materials are productively reincorporated into new production and use phases, for example, "waste equals food." In this approach, waste materials are turned into "nutrients" for a following cycle. To achieve this, the C2C approach uses the following principles (McDounogh et al. 2003):

- Principle 1: Waste is food (everything is a nutrient for something else).
- Principle 2: Use the sun (use renewable energy only).

- Principle 3: Enjoy diversity (species, cultural, and innovative diversity).

Waste virtually does not exist in nature because each organism's processes contribute to the health of the whole ecosystem. The closed-cycle biological system of C2C has for millions of years led to a flourishing planet with a varied abundance of food. Every being on the planet has formed part of it, and it provided good conditions for growth. Resources are extracted from the earth's crust and concentrated, changed, and synthesized, leading ultimately to unlimited amounts of waste. This process means that valuable resources are lost (Cohen 2007).

Renewable energy sources have the potential to provide a thousand times more energy than what is used today (Van de Westerlo 2011). Human energy systems can be nearly as effective. C2C systems from buildings to manufacturing processes could directly collect solar energy or tap into passive solar processes, such as daylighting, where natural light can be piped into an indoor space. Wind power, thermal flows fueled by sunlight can also be captured. From a holistic perspective, natural systems thrive on diversity. Healthy ecosystems are complex communities of living things, each of which has developed a unique response to its surroundings that works in concert with those of other organisms to sustain the system.

Diversity means strength and monoculture weakness. It means a healthy and healing environment. But if diversity declines, the ecosystem becomes less stable. The more the diversity, the more productive the functions, both for the ecosystem and for the planet. Each individual in an ecosystem depends to a greater or lesser extent on the others (Van de Westerlo 2011). Paul Hawken (1992) explains biodiversity in his book "The Ecology of Commerce," in which he regards biodiversity as the source of all welfare. The author also predicts that in the 20 years following publication of his book, many species will disappear if man continues to use natural resources in the same way as at present.

10.9.3 C2C design reflection

The aim of the C2C approach of Braungart and McDonough (2002) is a delightfully diverse, safe, healthy, and just world, with clean air, water, soil, and power—economically, equitably, ecologically, and elegantly enjoyed (Van de Westerlo 2011). The C2C design principles provide a positive agenda for continuous innovation design and use of products and services. Specifically, the purpose of the product certification program is to improve the way we make, use, and reuse things recognizing two metabolisms, the biological metabolism and the technical metabolism, with a goal to leave a beneficial footprint for human society and the environment. C2C design offers a compelling alternative. It rejects the assumption that human industry inevitably destroys the natural world. Instead, C2C design embraces abundance, human ingenuity, and positive aspirations (MBDC 2002).

C2C is about choosing the right thing to do and then doing that thing the right way to achieve positive outcomes. In other words, to become "more good" not just "less bad." For example, while it makes sense to slow down the use of fossil fuels, this is not the goal. C2C is a continuous improvement process design tool that starts with the positive or beneficial end in mind and executes efficiently toward achieving this goal. The C2C goal is a move to renewable energy sources.

The short-term actions for product development start with complete identification of the materials and chemicals that make up the product and process in order to assess them for human and ecological impacts. In the medium term the goal is for designs that are positive or beneficial in terms of cost, performance, aesthetics, material health, and material (re) utilization potential with continuous use and reuse periods. Additionally, moving renewable energy forward in a cost-effective way, celebrating clean water as a human right, and honoring social systems are part of the holistic C2C approach. The long-term goals can be wholly positive and intended to support 10 billion people and other species. For example, McDonough and Braungart's long-term goal is as follows: a delightfully diverse, safe, healthy, and just world, with clean air, water, soil, and power—economically, equitably, ecologically, and elegantly enjoyed (MBDC 2012).

10.9.4 C2C product design criteria

Based on the C2C approach and the dual cycle principle, design criteria can be derived for the creation of new products, processes, and objects. The following 10 design criteria are regarded as crucial (Braungart and McDonough 2002):

- Design all products and processes so that after their initial use all materials can be fully reused in a biological or technological cycle.
- Make sure that no harmful or toxic substances are released or used in the production process and during the usage phase of the product or object.
- Make sure that the production process and the use of the product create added value for the stakeholders.
- Design on the basis of a Triple E (Economy, Ecology, and Equity) approach.
- Use renewable energy sources such as the sun, wind, and (ground) water.

- Respect the diversity of the location, species, innovation, and culture.
- Protect and maintain the quality of water reserves.
- Carry out the production process with social responsibility. That means no child or forced labor and no unhealthy workplaces.
- Follow a local approach in the production process.
- Make intentions transparent and translate them into measurable targets. Concrete targets can be reached by drawing up a roadmap with milestones.

10.10 *SM and sustainable production*

> In the 21st century, I think the heroes will be the people who will improve the quality of life, fight poverty and introduce more sustainability.
>
> **Bertrand Piccard**
> *Swiss Psychiatrist, and Balloonist*

10.10.1 *Definitions*

SM evolved from the concept of SD. It includes things such as making products using less energy and materials, producing less waste, and using fewer hazardous materials as well as products that have greener attributes such as recyclability or lower energy use. SM focuses on both how the product is made as well as the product's attributes. This includes the inputs, the manufacturing processes, and the product's design. Several definitions exist for SM. For instance, SM is defined by the US Department of Commerce as "the creation of manufactured products that use processes that minimize negative environmental impacts, conserve energy and natural resources, are safe for employees, communities, and consumers and are economically sound." However, the Lowell Center for Sustainable Production defines sustainable production (SP) as "the creation of goods and services using processes and systems that are nonpolluting, conserving of energy and natural resources, economically viable, safe and healthful for workers, communities, and consumers, socially and creatively rewarding for all working people" (Rosen and Kishawy 2012).

SP was introduced at the 1992 UN Conference on Environment and Development (UNCED) in Rio de Janeiro as a guide to help companies and governments transition toward SD (OECD 2009). Alting and Jørgensen (1993) define SP as products that are designed, produced, distributed, used, and disposed with minimal (or none) environmental and occupational health damages, and with minimal use of resources (materials and energy).

10.10.2 Sustainable trends in manufacturing

Usually, products and parts made by the industry are produced by taking pieces of raw material and cutting away sections to create the desired part or by infusing material into a mold; however, a relatively new process called additive manufacturing (AM) is beginning to take hold where material is aggregated together rather than formed in a mold or cut away. The initiation of AM technologies presents an opportunity that has the potential to greatly benefit designers, and contribute to the sustainability of products. The terms AM and 3D printing tend to be used interchangeably to describe the same approach to fabricating parts.

The Society of Manufacturing Engineers defines AM as the process of manufacturing a physical object through the layer-by-layer selective fusion, sintering, or polymerization of a material. The AM process begins by taking a 3D computer generated file and slicing it into thin slices (commonly ranging from 0.01 to 0.25 mm per slice depending on the technology used). The AM machine then builds the model one slice at a time, with each subsequent slice being built directly on the previous. As a result of the material deposition and processing operations, the digital electronic model is converted into a physical part or product (Diegel et al. 2010).

Unlike subtractive manufacturing, where material is removed from a larger block of material until the final product is achieved, most AM processes do not yield excessive waste material. As the part is made from material in a powder or liquid form, whatever powder or liquid does not get hardened by the process gets reused for the subsequent parts. AM typically also does not require the large amounts of time needed to remove unwanted material, consequently reducing time and costs, and producing very little waste (Wohlers 2009).

The adoption of AM and other advanced manufacturing technologies appears to herald a future in which value chains are shorter, smaller, more localized, more collaborative, and offer significant sustainability benefits (Gebler et al. 2014).

10.10.3 Green product and clean technologies

One major challenge in green product design faced by today's industry is how to deal with the trade-off between a product's traditional and environmental attributes. A company usually needs to deal with some difficult technical trade-offs when attempting to design an environment-friendly product. Notable examples include the trade-offs between vehicle fuel efficiency and weight/size or power as well as between the recycled material content and material consistency of a product (Hopkins 2010; Chen and Zhang 2013).

A green product can be any product that is designed to reduce its environmental impact. A key concept is that environmental concerns and impacts are taken into account from the beginning of the product design process. This is important because most of a product's environmental impact is determined in the design phase. The product may be made of recycled materials, designed so that it can be easily recycled, made without hazardous materials, or produced with less packaging (USDC 2011).

The link between manufacturing and its operations to the natural environment is gradually becoming recognized. Progress, profitability, productivity, and environmental stewardship are now seen as needing consideration by manufacturing organizations (Sarkis 2001). Improving environmental stewardship and sustainability, while maintaining profitability and productivity, are increasingly viewed as strategic goals of manufacturing companies.

Figure 10.14 shows the green products as a result of integrating SM and clean technologies. Clean technologies are technologies associated with environmental protection regulations, pollution control and prevention, waste management, remediation of contaminated property, design and operation of environmental infrastructure, and the provision and delivery of environmental resources. Examples of clean technologies include technologies for wastewater treatment, recycling, solid waste management, and renewable energy sources like solar panels and wind turbines. Many clean technologies can be used to green the manufacturing process and are therefore important to SM.

10.10.4 SM indicators

The importance of integrating sustainability with manufacturing and design is highlighted, along with the need to utilize appropriate tools, like DfE and LCA. Important contributors to SM, as illustrated in Figure 10.15, need to be considered. It is known that environmentally sound practices, approaches, and tools developed collaboratively by the manufacturing industry, academia, and others are beneficial and implementable. Also, manufacturing decision-makers that adopt a sustainability focus and establish a sustainability culture within companies are more likely to be successful in enhancing design and manufacturing sustainability. Along with competitiveness, profitability, and productivity, environmental stewardship and sustainability are likely to prove increasingly important for

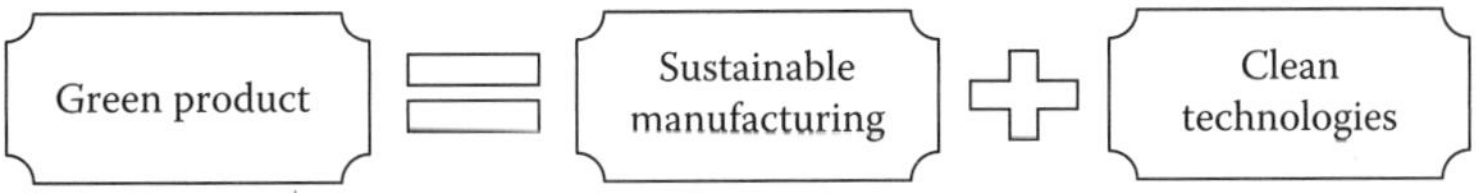

Figure 10.14 Green products as a result of integrating SM and clean technologies.

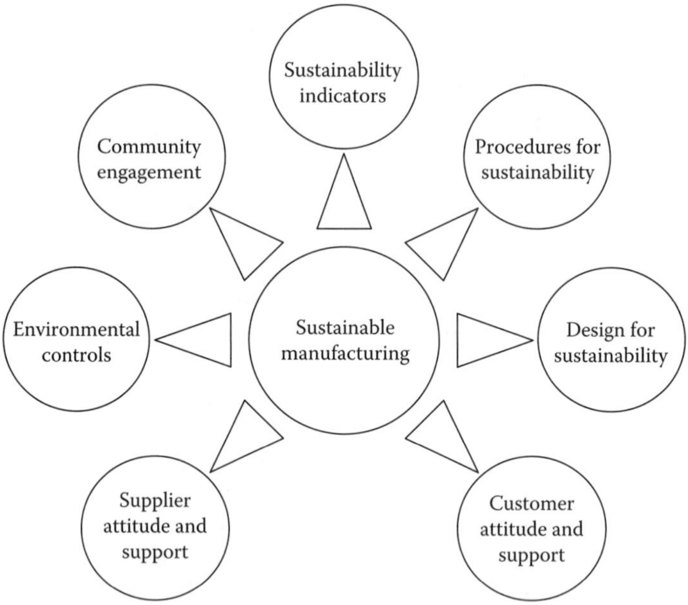

Figure 10.15 Key contributors to SM.

manufacturing in the future and in setting the main priorities for advancing manufacturing operations and technologies. Future prospects for SM are mixed, with improvements anticipated due to environmental pressures, while a focus on economics may dominate at the expense of sustainability due to the ongoing global financial crisis (Rosen and Kishawy 2012).

10.10.5 Closed-cycle manufacturing

CCM is a production process which has developed a system for the collection and recycle of scrap, wastewater, raw materials found in the water, and heat that is generated in some of the production phases. CCM thus means maintaining environmental standards throughout the industrial cycle so to limit the consumption of natural raw materials and reduce environmental impact. Applying CCM means that within the focus of the defined project objectives of a product, an appropriate solution will also have to be found for the reuse of the materials. A product is reused when it has not completed its life cycle and the user decides to stop its use, and the consumer sector is willing to accept it in its current use state, perhaps to its original purpose. For the design process, this means that the design should enable a transformation and/or dismantling of the product.

This industrial system is restorative or regenerative by intention and design. It is to keep resources in use for as long as possible, extracting maximum value from them during use, then recovering and regenerating them at the end of each service life. An advanced version of this would be an eco-industrial park where companies design their products and processes to use fewer virgin materials and use each other's byproducts, coproducts, or wastes as inputs (OECD 2010). When designing such a system, it is important to clearly define the inputs and outputs of the process. Inputs include energy and raw materials. Outputs include various types of products and wastes as shown in Figure 10.16 which depicts a typical CCM in the manufacturing process. According to Sassi (2008), the principle of CCM can be applied to the construction industry to establish list of materials and building components that can be recovered from buildings and considerably made from recycled materials naturally or industrially.

The term "reuse" generally means additional use of a component, part, or product after it has been removed from a clearly defined service cycle. It is applied to a product that has been used previously. The product will retain the problems it acquired during its previous life as it will not have been repaired (Gray and Charter 2006). Reused products usually have no warranty of any kind.

Repairing is simply the correction of specified faults in a product and returning it to useful service. When repaired products have warranties, they are less than those of newly manufactured equivalents. Repair makes a broken product operational again. An analysis of the root cause of the problem is generally not performed in the repair process which means the product may not perform like a new product (Gray and Charter 2006).

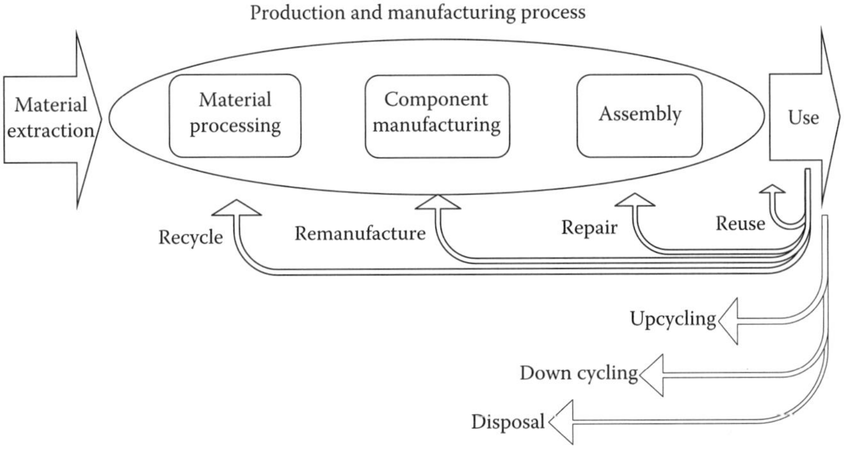

Figure 10.16 A typical closed-cycle principle in the manufacturing process.

Repairing a product minimizes the energy and material needed to keep it in use at the expense of not offering an updated or improved functionality (Barker and Andrew 2007).

Remanufacturing as an emerging industry with great potential presents a product EOL alternative to extend the number of PLC from one to multiple. Remanufacture may be understood as the refurbishment or upgrading of the product or of recoverable components. It is an industrial process whereby used products signified to as "cores" are refurbished to useful life. Cores are the essence of the remanufacturing process. Certain products have large parts with high material content which can be taken to the original standards by undergoing machining processes, and also have high-degradation subcomponents which can be substituted for new ones. The product generally matches its intended service for several years or decades.

Recycling returns a product to raw material form, which can be used as raw material for a future manufacturing process. The term "recycling" is generally applied to consumable goods such as newspapers, glass bottles, and aluminum cans but can also be applied to durable goods such as an engine (Gray and Charter 2006). Recycling involves the separation and collection of materials for processing and remanufacturing into new products, and the use of the products to complete the cycle (Barker and Andrew 2007). Care needs to be exercised to ensure that the material can genuinely be recycled through a system rather than be downcycled as a poor quality material for lesser uses. Usually, there are two major stages in recycling strategy: collection and processing. Both may consume resources and limit the process efficiency.

Analysis of multiple case studies showed that, aside from their environmental advantages, designing for adaptive reuse and deconstruction add short-term economic and possibly environmental costs to the project, but on a bigger scale of the life cycle of the project, the long-term benefits of utilizing those two concepts outweigh any extra initial costs (Chini and Saleh 2009).

The Canadian city of Surrey has chosen to build what the city claims will be the first closed-loop organic waste management system. The Surrey biofuels processing facility (www.surrey.ca/city-services/13015.aspx) processes 115,000 tons of residual kitchen and garden waste from the city of Surrey each year. The process will create a renewable natural gas which can then be used to power the city's natural gas waste collection trucks. The facility also produces a compost product that is suitable for landscaping and agricultural applications (Hower 2015).

10.10.6 *Design for remanufacturing*

Remanufacturing is returning a used product to at least its original performance with an assurance that is equivalent to or even better than that of the newly manufactured product. Remanufacturing typically applies to complex manufactured products that possess significant embedded material, energy, and labor resources; therefore, it fits well as a key strategy within the circular economy and represents an important component of a resource efficient manufacturing industry (ERN 2015). While it does not retain the value of the above approaches it does retain part of it; industry and public accept the value of the activity from household waste to production scrap to end of product life initiatives.

Remanufacturing describes the process of dismantling products, cleaning, repairing, or replacing parts, and then reassembling them to a good working condition. Remanufacturing process involves returning a used product to original specifications or even better through inspection, disassembly, cleaning, reprocessing, reassembly, and testing. It is a process of recapturing the value added to the material when a product was initially manufactured. The process can be performed on either entire products or the parts that constitute the product. A remanufactured product may be sold at a reduced price with a warranty that the remanufactured product is equal to or better than the original product. Figure 10.17 shows the remanufacturing process.

Remanufactured products include automotive and aircraft parts, compressors and electrical motors, office furniture, tires, toner cartridges, office equipment, machine tools, cameras, and still others (Lund 1996). Some products or components may be designated, by design, for single or multiple reuse, for single or multiple remanufacturing, for recycling, or for disposal. Key requirements for remanufacturing is that the retired products have significant residual value at the EOL, the remanufacturing firm can successfully capture the retired product, and product can be restored to like-new one.

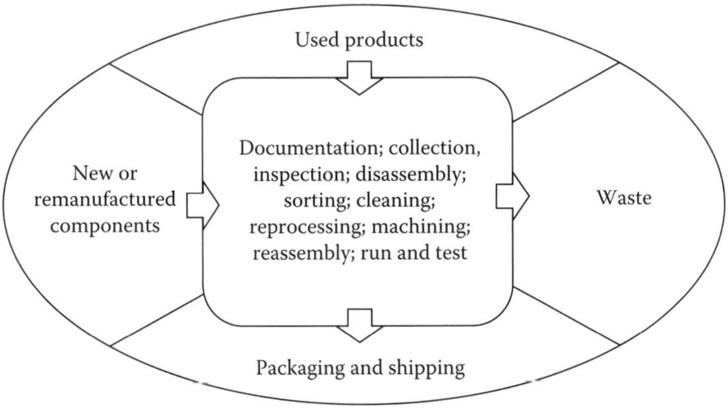

Figure 10.17 Remanufacturing and refurbishing process.

Remanufacturing also refers to total remanufacturing and partly remanufacturing. Total remanufacturing brings production equipment condition to as new-like and partly remanufacturing is less advantage than total remanufacturing. Under those two levels, two subcategories are identified: technical need and other need. From the technical need perspective, remanufacturing is categorized as mechanical remanufacturing, electrical remanufacturing, and control system upgrading (Yang 2014).

Remanufacturing, however, is the only method in which the performance of a used product is returned to at least the original equipment manufacturer's performance specification (Barker and King 2006). Remanufacture is affected by the physical characteristics specified during the design phase, whether the product has been designed for remanufacture or not. Product design for remanufacturing (DfRem) is enabled by business models which recognize the benefits of remanufacture. In general DfRem may be based on two levels (Gray and Charter 2006):

- Product strategy, including sales, marketing, service support, reverse logistics/core collection
- Detailed product design and engineering, including core collection and functional design

Although remanufacturing may involve many benefits including economic, environmental, technology, and social, it is important to note that remanufacturing is not always the best option. For example, from an environmental perspective it may be better in some circumstances to promote other strategies such as reuse, whereas fast-moving consumer goods may best designed for recycling.

Mukherjee and Mondal (2009) identify that the challenges of remanufacturing are related to the factors, such as managerial factors (product design, acquisition planning, logistics planning, inventory management, marketing, etc.), resource factors (technology and skills of workforce), and environment issues in the disposer market.

10.11 *Advancing sustainability through SED education*

Training teaches how. Education teaches why.

Nido Qubein
President of High Point University

10.11.1 *The challenge*

Sustainability is an important aspect concerning the impact of engineering as a profession. The perception and the capability to deal with and operate under this professional paradigm is something that must be fostered during the formation of new engineers. Commitment to this challenge requires that engineers acknowledge their professional obligation, extend their knowledge base, and contribute to all levels of policy decisions.

Educating future engineers may have the greatest positive impact on sustainability and green design. The next generation of engineering professionals must be prepared to solve complex and multidisciplinary problems in a sustainable and global context. Engineering education can provide students with the tools to approach these grand challenges of the twenty-first century while considering aspects that are keys for designing sustainable systems (Allen et al. 2006).

Today engineering education faces several challenges, including, but not limited to, addressing low diversity percentages, high attrition rates, and the need to better engage and prepare students for the role of a twenty-first century engineer (Davidson et al. 2010).

In sustainable engineering, students are taught to think of sustainability principles and DfE, which are essential in creating sustainable solutions to combat unsustainable operations across the world. LCA and industrial ecology are important part of the syllabus, which enhances students' understanding on sustainable operations in industrial plants. However, as quoted by many researchers, who are involved in sustainability education for engineering students, lack of student maturity in understanding the value of sustainability objectives together with course limitations often has negative impact on student attitude toward learning (Balan and Manickam 2013).

Design is the most powerful tool for solving complex and multidimensional problems. The need to embed sustainability in design education is linked to the idea that designers actually have more potential to slow environmental degradation than economists, politicians, businesses and even environmentalists (Fuad-Luke 2002). Hanks et al. (2008) suggest that designed things themselves shape us in complex ways, as much as we shape the world by means of our own designs, meaning that product design has the potential to modify behavior.

10.11.2 Approaches to teach sustainable design

Sustainability education encourages the use of integrative and creative approaches to learning; develop and apply critical thinking skills to complex local, regional and global issues; think creatively and apply creative problem solving; question established ways of doing things; and be self-directed at investigating and proposing creative solutions to sustainability problems. In turn, integration of sustainability within higher education implies many transitional shifts (Sterling 2004) including a shift from focus on accumulating knowledge and content orientation toward focus on self-regulative learning and "real issues" orientation.

Sustainability of products and services does not happen by itself; it has to be proactively taken into account in their design. In more technical terms: it has to be "engineered" using methods of integrated engineering design (Brissaud and Tichkiewitch 2001). To begin with, training elements on integrated design for instructors to shape an understanding of sustainability in the context of different types of organizations, standards, and, regulations are needed. Therefore, the notion of the product/system life cycle plays a central role. In integrated design the product/system life cycle comprises all phases that the product/system goes through from the idea to its EOL and revival. The principal stages that a product or system typically runs through in its life cycle are design, manufacturing, distribution, customization, and the EOL (Riela et al. 2015).

Many opportunities currently exist to infuse engineering curriculum with SED concepts, and the benefits are not only in terms of curricular augmentation alone, but they also provide a chance for faculty to develop new, innovative teaching materials. Engineering curricula can integrate specific courses and supplement or modify established engineering courses with sustainable practices, and the courses can be both quantitative and qualitative in nature.

Holmberg et al. (2008) draw from their experiences at three European universities to identify five areas in which a strong foundation will increase likelihood of success. The five areas for education for SD success are: legitimacy, commitment by university administration, responsibility is spread throughout the university, skilled teachers with SD experience, and an effective structure within the university (Stiver 2010).

A thoughtfully designed curriculum for engineering students would embed sustainability-related learning activities into a broad range of required courses using a coordinated approach. Sustainability-related learning activities should also be thoughtfully coordinated to build to effective outcomes through the curriculum. A diversity of teaching approaches and student learning activities provided a good method to reach synthesis-level cognitive skills (Bielefeldt 2013). For a sustainable future, it is needed to teach integrated design thinking. The complexity of modern products and services exceeds the expertise of any single profession. Components of such systems have a wide range of characteristics, which must be managed simultaneously.

Practically, there seem to be two ways to approach SED in the curriculum. First, is by introducing SED as one single course. It may complement an existing conventional design course. Second, the concepts of SED may be blended and embedded into existing engineering courses. In both approaches, the learning should focus on the life cycle concepts and assessment by presenting students with related notions such as LCA, energy audits, GHG, and CF.

Deciding what to include in an independent course may be challenging due to the many techniques and apparent differences of views on the subject. However, there are more similarities than differences between the various approaches. They all have the objective toward the environment and social welfare. A good way to assess what a course content should be is to see the courses given below:

- MECH 424: Sustainable Product Design, Queen's University
- ME 4171: Environmentally Conscious Design and Manufacturing, Georgia Tech

The content of this course should be flexible and the course may introduce both sustainable and traditional design and give students the opportunity to compare approaches. The use of case studies allows students to develop higher levels of cognitive understanding of sustainability, including applications, analysis, and design. Case study examples may include university campus itself, local hospital, visible projects, and technology-led success stories. In all case studies it is necessary that students should be presented with open-ended problems where students should define their own boundaries.

The use of specialized simulation tools can be a valuable way to help students analyze design aspects by assisting them in quantifying the various factors that relate to the sustainability. Most software tools focus on quantifying the environmental impacts of product or process. A possible set of core requirements for such course could include DFE, LCA, PLC, remanufacturing, environmental management standards, sustainable design, energy, risk analysis, CO_2, carbon trading, and other subjects according to the course objectives.

An attentively designed engineering curriculum would embed sustainability-related learning activities into a broad range of required courses using a coordinated approach (Bielefeldt 2013). Such process provides students with a meaningful way to connect more personally to their courses. A central challenge in incorporating sustainability in a greater range of engineering courses is the need to develop effective multidisciplinary and broad systems-based education material. Through the use of modules, engineering programs can integrate sustainability and experiential learning throughout a host of existing courses by threading individual sets of course skills together in an effort to reach higher levels of intellectual behavior via interdisciplinary concept connection (Warburton 2003). Modules can be designed to fit into one lecture or over a series of lectures. Modules typically include everything an instructor needs for implementation: a summary of learning objectives and module activities, lecture slides and notes, recommended readings, and an assignment for students. Using modules to teach sustainability concepts reinforces the broader applicability of sustainability to all engineering disciplines by connecting traditional engineering to impacts to, and solutions for, society, economy, and environment (Mckeown 2011).

Design projects are a potential learning approach to reach the synthesis level of Bloom's taxonomy for knowledge of sustainability as described in Table 9.3. Working on these design projects in a team setting can help students to appreciate differences of opinion and work toward a consensus on the value of sustainability considerations within the design process.

10.11.3 Building transdisciplinary education

The focus of sustainability is on engineering, more than on the sciences or on social science, because engineering is the foundation for the activities that drive the industrial state and the activities that implement scientific advance. It is obvious that engineering cannot do it alone but scientific as well as social and legal changes must happen as well. Engineering education today faces the challenge of coexisting of multidisciplinary and transdisciplinary research and teaching in a meaningful way in university structures. This raises a question of whether education relevant to SD requires its own protected environment to survive, or will it otherwise be marginalized by attempting to instill it throughout the traditional curriculum and disciplines. This is a major problem, especially if the majority of faculty are traditional to multidisciplinary and/or transdisciplinary work. It is no secret that faculty like to create "vest-pocket editions" of themselves. The insecurity of future employment also makes this a high risk venture for them (Ashford 2004).

The tremendous opportunity in calls for interdisciplinary collaborations and sustainability education is the possibility of a collective response producing still unknown green innovations. In contrast to the incremental innovations associated with ecological modernization (Mol and Spaargaren 2000), a focus on sustainability science- and project-based learning (PBL) requires the interdisciplinary synthesis of concepts and praxis, which is more likely to generate the radical or disruptive technologies that substantially improve efficiencies and equities shifting systems of production, provision and consumption and opening new possibilities (Metzger and Zare 1999).

Engineering is interdisciplinary by nature, however, sustainability is an inherently transdisciplinary concept covering the interplay between different disciplines and systems. Getting students to understand the extent of that transdisciplinarity requires that they appreciate the variety of perspectives involved, and building transdisciplinary

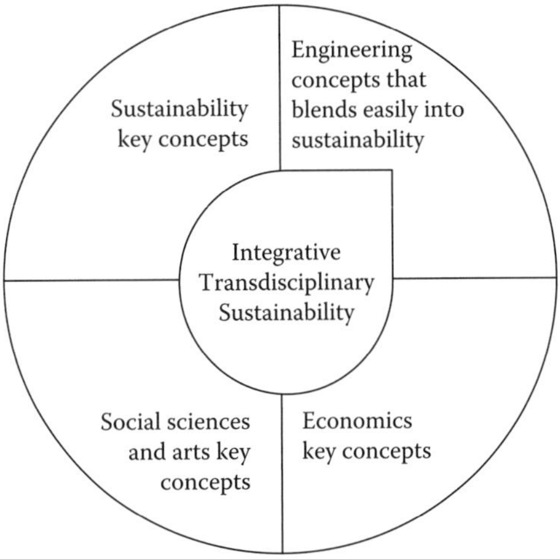

Figure 10.18 Transdisciplinary teaching module on engineering sustainability.

teaching into the learning process. Incorporating interdisciplinary teaching approaches such as inviting speakers, or team teaching a course is a good idea. Teaching a course to more than one discipline to bring together students from different viewpoints is another helpful idea. These approaches also help build a culture of collaboration that is needed as students tackle the transdisciplinary nature of sustainability challenges now and into the future. Figure 10.18 shows the integration of key sustainability concepts with major concepts from other disciplines for a transdisciplinary teaching module. In addition to engineering and sustainability, the model calls for broad knowledge of different fields in economics and social sciences.

10.12 Remanufacturing case: Wind turbine electric generator

Remanufacturing is not a widely understood concept.

Ron Giuntini
Remanufacturing Institute

There are many advantages in using wind energy especially by producing electric energy by clean means. Another benefit is that the life span of wind turbines goes between 20 and 30 years, in which time they can generate as much as 20 times the energy needed to manufacture them. Despite the advantages of using this technology, concerns arise from the large amount of materials used in this industry.

The wind turbine is an electromechanical system with a high remanufacturing rate. Remanufacturing can play an important role as a way to close the material cycles and thereby contribute to less material and energy use, which are important steps to realize SD.

However, in a life cycle perspective, not only the production or remanufacturing phase but also the use stage is needed to be taken into account. In the use stage of engines and generators, associated emissions and costs are of high concern (Schau et al. 2011). Up to now, few studies have looked at the whole life cycle of wind turbine parts that require energy in the use phase including remanufacturing of the used parts (Schau et al. 2012).

Because of the increased number of wind turbines and the materials used, this case provides an idea about the remanufacturing process as well as a comparison between the environmental impacts from remanufacturing the electric generator of multimegawatt wind turbines and wind turbines manufactured using new components. The study methodology is the following:

- Describe the life cycle and the materials and processes employed for the manufacture and remanufacture of components inside the nacelle and in particular the electrical generator.
- Focus on the energy and economic analysis of remanufacturing.
- Identify remanufacturing alternatives for the electrical generator at the end of the expected life time service of wind turbines.

10.12.1 Remanufacturing and energy needed

Remanufacturing is generally seen as the most environmentally friendly of "EOL" treatments for a retired product. If the remanufactured product can be considered a substitute for a new product, then a credit is usually claimed for the avoided resource use and emissions associated with the new product production. One of the primary requirements for remanufacturing is that the retired products have significant residual value at the EOL (Gutowski et al. 2011).

The energy consumption of manufacturing and remanufacturing systems results from the machines involved and is highly dynamic as a result of the fact that the energy consumption depends on the machine status and characteristics (Duflou et al. 2012). Remanufacturing can play an important role as a way to close the material cycles and thereby contribute to less material and energy use (Kim et al. 2009), which are the important steps to realize SD. It can offer a business model for sustainable prosperity, with reputed double profit margins alongside a significant reduction in carbon emissions (OHL 2004) and 15% of the energy required in manufacture (Gray and Charter 2006).

When determining if a product is or is not remanufactured, it is essential to consider the process utilized. Remanufacturing is a process of recapturing the value added to the material when a product was first manufactured (Gray and Charter 2006). A product begins a remanufacturing process when a company claims the core, which is the part that offers the structure where most of the individual parts are attached in a product. Parts disassembled from the core are cleaned, inspected, and tested to explore a reuse scenario for them. Some parts are disposed of while others may be repaired. At the end, a remanufactured product must meet quality standards similar to the original product (Skrainka 2012).

The projected use of materials for the manufacturing of wind turbines shows large quantities of resources employed. It is important to understand the overall environmental benefits of this technology, not only from the positive impact in the reduction of GHG but over the demand for raw materials. System components include blades, nacelle, hub, rotor shaft system, rotor brake, gearbox, electrical generator, yaw system, tower, and foundation system. The electrical generator is the topic of consideration in this case: Figure 10.19 shows the remanufacturing process and product under consideration.

10.12.2 Electrical generator

The electric generator is used to convert the mechanical movement transmitted by the gearbox into electric power. The remanufacturing option for a generator is carried out by rewinding the generator, replacing bearings and thermal protection, balancing dynamically the rotor, sandblasting the stator and end bells, cleaning all parts and applying coating (Skrainka 2012). Since the preferred materials for electric generators are copper for the coil and cast iron for the housing, they are considered valuable in case of recycling.

The life cycle of the generator is modeled as shown in Figure 10.19 and used for the LCA. Starting from the left part of the figure, the production phase consists of raw material extraction, material processing, and manufacturing. In the use phase, the alternator generates the necessary electricity for the wind turbine for about 30-year lifetime. Afterward, the generator is remanufactured in a factory and placed in a container for use.

In a life cycle perspective, not only the production or remanufacturing phase but also the use stage is needed to be taken into account. In the use stage of engines and generators, energy use, associated emissions, and costs are of high concern (Schau et al. 2011).

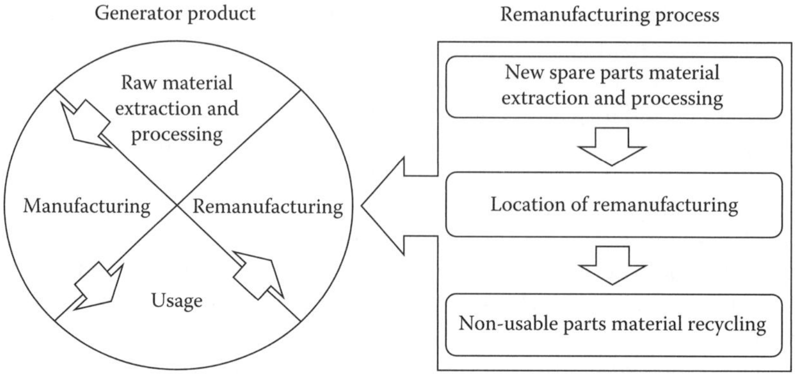

Figure 10.19 Remanufacturing process and product under consideration.

10.12.3 Energy analysis

In order to look at future concerns, LCA is employed as a tool of analysis of environmental impacts. LCA will allow a comparison of different remanufacturing scenarios for wind turbines and their environmental benefits. For example, the remanufacturing alternative has been proven effective in other industries such as the automotive industry where it is as an economical and environmentally friendly option. Therefore, using LCA to study the remanufacturing option for wind turbines could demonstrate additional environmental benefits in the current growing wind energy industry (Skrainka 2012).

The life cycle energy analysis of products is now a well-established field of study. Many software programs are available to help in this analysis, and international standards exist to guide the practitioner. The tool only requires a bill of materials (BOM) for the product and uses well-known estimates both for the embodied energy in materials (Smil 2008). An important result from studies shows that for most products the energy requirement for material production dominates the energy requirements for manufacturing.

The LCA for the generator is broken down into three primary phases: raw material processing, manufacturing, and use phase. The energy required to produce raw materials can be estimated starting with the material composition of the generator. As an example, the BOM are available for machines of sizes 1.1, 11, and 110 kW. At the same time BOMs for machines of size 22 and 200 kW are available from the Environmental Product Declarations by ABB (ABB 2016). Using these BOMs and specific energy to process the raw materials (Smil 2008), it is observed that both the weight and total energy to process the raw materials almost scaled linearly with size (kW rating of the motor). Each material requires its own energy to process. This energy encompasses extraction, processing, purification, and other steps to bring the raw materials to a usable condition.

The manufacturing energies can be obtained in the same way, for example by scaling the 11 kW motor manufacturing energy to 22 kW (linear extrapolation). In addition, the energy consumption incurred when choosing to rewind should also be estimated.

The energy to remanufacture the generator is assumed to be the sum of all processes. Though other processes like heating and refurbishing are also involved, it is needed to keep the calculation in favor of rewinding and remanufacturing. The upfront cost depends on the choice between rewinding and replacement with new. Rewinding cost is the fees given to the rewinding workshop to rewind the motor to the desired specifications. The operational cost is the cost to run the motor. This encompasses the total electricity cost (Sahni et al. 2011).

10.12.4 Economic analysis

The LCA of the three dimensions: environment, economy, and society should ideally use the same system boundary and the same reference unit (Kloepffer 2008) called the FU, a unit which all the results are related to and which quantify the performance (valuable main output) of the system (ISO 2006).

LCC is proposed for the assessment of the economic dimension of sustainability. The remanufactured alternator can be used again as an electrical generator in the wind turbine. In addition to the used generator, the remanufacturing process needs some new generator spare parts which are sourced out from other manufacturers. Similar to the new generator production, raw material extraction and material processing are needed for the new spare parts. The remanufacturing scenario will take place in the factory equipped with all necessary tools. The final stage, which is the EOL, is modeled as a part of the remanufacturing phase and also includes those fractions of the used alternators that cannot be used anymore (10%–100% cf. Table 10.3) (Schau et al. 2012).

Data for the LCA are mainly taken from PE (2009) and the characterization factor used was from Guinée et al. (2002), whereas the environmental LCC is estimated using literature (Schau et al. 2011). The data for the SLCA are from the social hotspot database (Norris et al. 2012) and other international database available online in addition to scientific literature.

Two different design alternatives are investigated by LCA and environmental LCC. The design alternative 1 is a conventional generator (weight; 6.069 kg) with belt fitting, fan and steel bearings, and cast iron housing. Design alternative 2 is an ultralightweight generator (3.952 kg), where also the belt fitting and bearings are replaced by lightweight parts (aluminum and plastic respectively). Table 10.3 shows the material, weight,

Table 10.3 Generator parts, materials, weights, and replacement probabilities

	Design 1: Conventional generator			Design 2: Lightweight generator		
Parts	**Material**	**Weight (kg)**	**Replacement probability (%)**	**Material**	**Weight (kg)**	**Replacement probability (%)**
Stator	Steel	0.773	20	Steel	0.773	20
Rotor coil	Copper	0.550	22	Copper	0.550	22
Rotor	Iron cast	1.094	19	Iron cast	1.094	19
Drive shaft	Steel	0.262	10	Steel	0.262	10
Belt fitting	Steel	0.519	10	Aluminum	0.180	75
Fan	Steel	0.138	10	Plastic/PP	0.016	100
Spacer	Aluminum	0.003	50	Aluminum	0.003	50
Bearing	Rolled steel	0.099	50	Plastic/PP	0.011	100
Slip rings N	Copper	0.033	100	Copper	0.033	100
Slip rings S	Copper	0.071	100	Copper	0.071	100
Housing	Iron cast	2.527	15	Aluminum	0.958	40

Source: Schau et al. 2011.

and replacement probability (the likelihood of a part being replaced within the generator by the remanufacturer) of the different parts of the generator for each design alternatives.

In terms of percentage savings, for both energy and economics the savings were less than 10% for all cases, and hence if the inherent error associated with LCA and LCC is 10%, then all comparisons are nuanced, and no conclusion can be drawn strongly. This calls for more detailed, case by case analysis. Overall it was shown that the common notion that remanufacturing leads to energy savings was challenged and it was shown that replacing with new in the case of electric generators is the energy-saving strategy.

10.12.5 Case research questions

- What are the main incentives customers would need to buy remanufactured products?
- What are the biggest barriers to remanufacturing and how could it be overcome?

- What change would be needed to the product design process to enable remanufacturing?
- What is the conclusion of the above case?

10.13 Acquisition knowledge

Attempting to answer the following questions involves acquisition of knowledge from this book and other books, documents, and the Internet.

- When and where is sustainability important?
- What is SED?
- What are the key requirements for SED?
- What does it mean for an engineering design to be sustainable and of high quality?
- Should engineers be committed to, engaged in, and leading sustainability efforts?
- What are the tools needed for the design of sustainable engineered systems?
- Which is better: a recycled material or a natural material?
- How is possible to determine if a material is green or not?
- What is the connection between TBL and sustainable engineering?
- What are the principles of the C2C approach?
- What is DfS and why do it?
- What does it mean to design sustained technology that becomes part of a user's life and not a disruption from it?
- What are the objectives and focal areas of DfS benchmark?
- How would nature solve green building challenges?
- Why is LCC important to a utility?
- Describe the difference between recycling and reuse.
- What does closed-cycle production mean?
- What are the principles of the C2C approach?
- Which sustainability criteria are relevant to the design phase of an engineering project?
- What is eco-efficiency?
- What is LCA? What is its purpose?
- Explain what is meant by life cycle considerations in the design of engineering applications.
- What is design in the context of remanufacture?

10.14 Knowledge possession

Attempting to answer the following open-ended "not explicitly expressed" questions may require research and investigation beyond the scope of this book, mostly by engaging in conversation, class discussion, and Internet-based research.

- How to incorporate sustainable principles and concepts into an engineering design?
- Which basic principles and reference framework are associated with efficient sustainable development?
- Do you think sustainable cars are feasible? Why or why not?
- What are the social benefits of sustainable design? What defines a socially sustainable product?
- Describe the concept of "CP" and investigate an engineering case of where this may be applied.
- Discuss the impacts of "waste materials" on the environment. Describe how various waste materials can be recycled, reused, or remanufactured into new products.
- How to combine LCA with other tools to obtain an added value for policy-making?
- To which extent is the combination of LCA components necessary for the generation of policy relevant and decision enabling information? How could a less complex LCA look like?
- How to incorporate recyclability principles into an engineering design?
- Compare the health risk in living close to a coal-fired power plant versus living close to a wind farm. In each situation, rate (high/medium/low) the relative magnitude of the two risk factors: consequence and likelihood.
- Compare the environmental impacts of installing and operating the following renewable energy technologies:
    - Hydroelectric
    - Solar photovoltaic (PV)
    - Wind turbine
    - Geothermal
    - Biofuel
- When is it worth remanufacturing? What makes a product more difficult to remanufacture than another?
- What are the drivers and challenges for both remanufacturing and new product acquisition? Is energy saving a factor to be considered when conducting remanufacturing project?

10.15 Knowledge creation

Collaborate with peers on learning or you may work with others outside the class to narrow down the objectives of each activity. You may access class and online resources, analyze data and information to create new ideas and balanced solutions. High-level digital tools may be used to develop multimedia presentations, simulations or animations, videos and visual displays, digital portfolios (e-portfolios), reflective practice (online publishing and blogging), or well-researched and up-to-date reports.

10.15.1 Designiettes on sustainable engineering

In these brief design projects students are introduced to the principles of sustainability, LCA, ethical design, and eco-design methods. The intent is to achieve an effective and appropriate engineering solution to address the challenges.

10.15.1.1 Designiette 1: Leaf mimicking solar cells

The PV system which harvests solar energy is a first step at mimicking the way leaf harvest energy (Vierra 2014). Investigate this topic and apply brainstorming to develop a creative design brief to build efficient solar cells that closely mimic nature. For example, these cells can be water-gel-based that couples plant chlorophyll with carbon materials, ultimately ending with more flexible and cost-effective solar cells. Other approaches can be also considered.

10.15.1.2 Designiette 2: Electronic waste

Consumer electronics have become an integral part of daily life and revolutionized the way we communicate, retrieve information, and view entertainment. As a result, electronic waste (e-waste), which is defined as any piece of electronic equipment which has reached the end of its useful life, has become the fastest growing component of the solid waste stream worldwide. This life fact requires effective solutions.

Question: *Is remanufacturing of consumer electronics consistent with the goals of sustainable design?*

In this empirical research proposal, investigate the electronics industry, examining the implications of domestic e-waste and considering the economic, environmental, and societal aspects of consumer electronics production, use, reuse, and remanufacturing. The research on economic and societal aspects is largely literature based. The research on environmental aspects is largely LCA based, drawing from direct observation of consumer electronics production and remanufacturing activities, as well as the literature. Apply sustainable design tools at early design concept stage, so that eco-design is integrated in the outcome. Design outcomes should be assessed for observance to the "Ten Golden Rules of Eco-Design" (Luttropp and Lagerstedt 2006).

10.15.1.3 Designiette 3: Recycling technologies

X, an E-waste processing company, is currently using a manual operation to dismantle and recycle computer hard disk drives (HDD), which recycles only the electronics board of the HDD. The remaining parts of the HDD are destroyed to liberate materials for resale. This is an inefficient technique as this process results in a lot of mixed materials, which greatly reduces the percentage and value of useful recycled materials. In order to increase value of the recycled materials, a proper nondestructive technique for dismantling HD components is needed. X plans to build a dismantling machine. The machine is meant to harvest parts and components such as permanent magnets and dc motors for reuse and resale at a high rate. Manual demanufacturing process is an inefficient way.

This situation poses an opportunity to collaborate with a university to develop the machine for dismantling and recycling of HDD. In this task, a team of students will play the role of the university researchers. The team should adopt the V-model for product development and create a concept of operations that describes user needs and the operating environment in order to define thorough system requirements. The team may reflect the generated brief design in a poster format.

10.15.1.4 Designiette 4: Water system

Apply brainstorming to develop an idea and/or solution for an innovative product or system for the human need that is inspired by imitating nature (design to model nature: biomimicry.org) to help better manage the water issue. The idea can be applicable or specific to region or climate. The design must either:

- Obtain water from the environment
- Produce usable water
- Promote efficient water use

Water accessibility projects are the focus of several organizations such as the alliance for water stewardship (www.allianceforwaterstewardship.org) and competitions such as the NAE's Grand Challenges (www.engineeringchallenges.org/cms/8996/9142.aspx). The project idea reflects the notion of how nature inspires innovative solutions to water management issues.

10.15.1.5 Designiette 5: Transportation challenge

A sustainable transport system must provide mobility and accessibility to all urban residents in a safe and environment friendly mode of transport. A sustainable system entails interaction of three components:

the economy, environment, and society. Applied to transportation, this reflects the necessity to determine interactive links of transportation with the economy, environment and, society over time. Considering the above requirements, apply brainstorming to develop an idea or solution to one or more of the following challenges:

- Reducing environmental impact of any form of transportation
- Bicycles, bus lanes, and efficient traffic flow
- Making public transportation, freight or individual transportation options safer and more responsive to user needs

10.15.1.6 Designiette 6: Trash can

Littering may have an impact on not just the environment's appearance but a lot of other things too including human health and safety. It may also have a big impact on plants and the wild life. For product designers, necessity is the source of invention. Trash cans are low-tech sustainable solutions: they stand at various places within cities and neighborhoods and collect trash and serve the noble purpose of keeping spaces clean. The objective of this project is to investigate this solution in order to make the trash can more attractive to use by community members. One way might be by utilizing high-tech solutions to make the trash can an appealing site of technological advancement that is going to do much more than collecting trash.

10.15.1.7 Designiette 7: Self-initiated and directed

Define your own problem and develop a concept to solve for it. Project outcomes must address humanitarian, environmental, medical, and/or sustainable needs. Students undertake extensive research to identify social needs and potential product solution, and then collaborate with industry partners to develop their designs. Project examples may include drinking water purification, solar and wind energy generators, a gray water toilet system, mobility aids for the visually impaired, music therapy, self-powered lighting for remote communities, a disaster relief cooking stove, a humanitarian aid air drop system, and models like Ecovative (www.ecovativedesign.com) for eliminating waste.

10.15.2 Design project on performance and life cycle cost analysis of a data center

A data center is a structure, or group of structures, dedicated to the centralized accommodation, interconnection, and operation of information technology and network telecommunications equipment that provides data storage, processing, and transport services. These data centers are packing together racks and racks of equipment and are usually located where power is affordable since they use a huge amount of energy. Inside data center can be classified into three parts (Wiboonrat 2014):

- Facility utilities including building infrastructure, power systems, cooling systems, telecommunication, water detection, fire protection.
- IT hardware including servers, storage, switching, routers, modems, etc.
- IT software including web applications, networking applications, database, storage application, virus scan, etc.

10.15.2.1 Thermal and energy performance

Consider a data center with 5000 server (200 W, each) producing 1 MW of heat to meet LEED gold requirements and setting high expectations for thermal and energy performance. Its most unique characteristic for this analysis, however, should be the inclusion of the heat energy management system, renewable energy supply, and dc electric power installation.

In your proposal consider assembling a multidisciplinary team consisting of electrical and mechanical engineers with expertise in electronic thermal analysis and network engineering ranging from the component to building system level to address this sustainable energy and infrastructure challenge. System-level tools should be designed to allow the analysis of the data center as macroscopic thermodynamic system and collateral subsystem models. The main objective is to develop models and algorithms to optimize workload and to minimize energy usage. In addition, investigate techniques that will permit a shut down or slowdown of a significant fraction of the system in times of lower than peak demand for data.

10.15.2.2 Life cycle cost analysis

Life cycle cost analysis (LCCA) is a data-driven tool that provides a detailed account of the total costs of a project over its expected life. By taking into account all costs of acquiring, owning, and disposing of a building or building system, LCCA can benefit all types of ownership. LCCA maximizes net savings by comparing project alternatives that fulfill the same performance requirements (Wiboonrat 2014).

The data center project transition is starting from planning, implementing, operating, and transforming as normal project management principle. The LCCA as a decision-making tool should be performed early in the preliminary project, design process while there is still a chance to refine the design to ensure a reduction in LCC. Establish the LCC process which can be first as simple as a table of expected annual costs and then it may be followed by a complex (computerized) model that allows for the creation of scenarios based on assumptions about future cost drivers.

10.15.2.3 What are the expected results and impact of this project?

The outcomes of this project are tools that can be used by data center designers and end users to perform parametric trade-off studies and ultimately provide optimized configurations. In addition, using this integrated approach may lead to the conceptualization of innovative energy demand, supply and space conditioning systems, and a financial decision-making tool.

10.15.3 Design portfolio on blending sustainability into control system principles

A control system is an interconnection of components forming a system configuration that will provide a desired system response. The basis for analysis of a system is the foundation provided by linear system, which assumes a cause effect relationship for the components of a system. A closed-loop control system (Figure 10.20) utilizes an additional measure of the actual output to compare the actual output with the desired output response. The measure of the output is called the feedback signal. A feedback is a control system that tends to maintain a relationship of one system variable to another by comparing functions of these variables and using the difference as a means of control.

10.15.3.1 Sustainable design guidelines

Currently, cities, communities, universities, and other organizations are developing sustainable design guidelines (SDG) to assist in advancing sustainable design. The guidelines are intended to be applied to new construction and major renovation projects. Designers, contractors, and

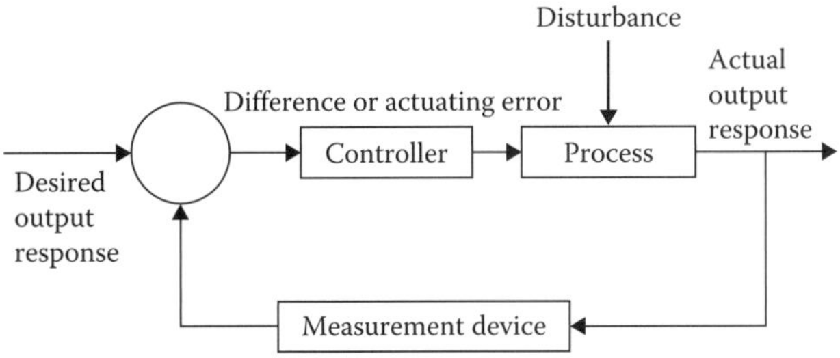

Figure 10.20 Closed-loop control system.

developers of projects shall be required to incorporate the guidelines into their projects. The goal is to meet as many of the guideline objectives as possible.

Teams of three to four students will team up to submit their project report of proposing a new energy efficient academic building at the campus to the coordinator of the Sustainable Design Advisory Committee (SDAC) of the University. The SDAC usually holds regularly scheduled meetings, but may call special meetings as needed depending on project schedule requirements. The SDAC will review all projects for compliance with the guidelines.

In the conceptual design phase, the team should use the SDG in addition to following related laws and regulations. First, when making the site plan and floor plans, they should consider spatial elements, such as specified spaces, areas relating to water use and hot water supply, and green areas. In addition to considering natural ventilation and day lighting, the team should plan position and area of windows. The team should also examine how to install a solar energy system and equipment for rainwater use.

At the beginning of the detailed design stage, the team should request the architects of the building to refer to the "guidelines" and to the team report. In the report the team should determine the site plan, floor plans, elevation, and fundamental specifications. After that, the architects will design the building's elements such as framework, exterior, windows and doors, interior, and lighting fixtures, so that as much as possible, the elements' variables meet their desired values.

10.15.3.2 Control system

In the basic control system for sustainability, "controlled objects" are human activities which need to be controlled. "Disturbances" are harmful influences on controlled objects, which are caused by environmental, social, or economic problems. Examples of the disturbances are damaging influences caused by environmental pollution, floods, or land (Fujihira 2016).

In this task, apply Figure 10.20 to redraw another block diagram that promotes the sustainable building design. In this new control system, incorporate "SDG and designers" as the "controller." Designers in this aspect include the team, architect, engineers, and the University. The "process" will involve "site and building drawing and specifications" as well as "controlled objects," which include environmental, economic, and social impacts. "Sustainability checklist" which the team should prepare in the form of a table will act as "measurement device." After the construction is finished, the new building can be inspected and evaluated against the "sustainability checklist." After the inspection and evaluation, the "designers" will make a "design for improvement" so that controlled variables meet their desired values as close as possible.

10.15.4 Design contest on using sustainability simulation tools

The use of specialized software tools can be a valuable way to help students analyze sustainability aspects by assisting them in quantifying the various factors that relate to the sustainability of a project. Most software tools focus on quantifying the environmental impacts of an engineered product or process. LCA is a method to assess the environmental impacts associated with a product or process from cradle-to-grave including raw material extraction and processing, manufacturing, transportation, use, upkeep, and disposal (SAIC 2005).

For this task, a class may form teams of three to four students to work on a design project using the West Point Bridge Designer software (bridgecontest.org/resources/download) (Bielefeldt 2013). This three-week project task requires students to design a bridge considering and balancing factors and metric related to sustainability. Students are encouraged to consider sustainability early in the design process. Incorporating sustainability approaches and methods in the design stage is important for achieving sustainability. Sustainability, or the ability to find an effective balance between the areas of technical details (the bridge must withstand a defined test-load with minimal deflection), economy (capital cost), environmental impacts (excavation and materials), and social issues (combination of aesthetics, culture, and safety), is the objective of this project.

A bridge's life cycle has an important role in verifying the sustainability of the bridge. Life cycles can be evaluated in terms of environmental or economic impacts. It is well known that a bridge construction project involves large number of products and processes. Accordingly, LCA which is a method to assess the environmental performance of the product or a process over its life cycle should also be investigated.

Finally and as a supplement, reduce the electrical consumption of the bridge and promote the use of electricity from renewable energy resources.

10.15.5 Piece of art on recycled and reused materials

Recycling is the reuse of waste material into the production process. Reuse of material refers to materials that can be reused after the deconstruction or demolition of products. The use of recycled and reused materials saves resources and primary raw material, reduces air and water pollution, and extends limited landfill life. Recycled and reused materials can also save financial resources through lower material costs and lower disposal costs or tipping fees (AASHTO 2012). The objective of this task is to create a digital piece of art that reflects the need to boost the demand for materials and products that include recycled content as well as material reuse, thus reducing impacts resulting from the extraction and processing of natural materials.

10.15.6 Poster on the 12 principles of green engineering

Read the 12 principles of green engineering given in Table 10.2. Create 12 logos and arrange them in a poster format that correspond to the 12 principles and arrange them in a table format.

10.15.7 Debate on design for sustainability

Objective	Introducing an open-ended debate in the classroom to help students understand argument on the concepts of design for sustainability
Time	15 min for debate and 15 min for review
Format	For and against
Learning outcomes	Make an argument about a particular opinion; evaluate the arguments of peers; and understand the concept of counterarguments.
Capabilities demonstrated	Developing skills on public speaking, research, teamwork, critical thinking, communication, and professional judgment
Arrangement	Students are organized into two position groups of three speaking in a specific order. Three argue for an opinion and three argue against. One or two students might each work on the opening and closing statements while the group is investigating the subject; however, the entire group should revise the statements. Each group should read an opening and closing statement for the debate.
Ideas for the topic	Identifying two examples of what is considered successful sustainable design projects, and two failure sustainable design projects.
Assessment	Indicate what you consider the best arguments in favor of the topic. How were they substantiated? Identify arguments that are based on poor facts, not ethical and/or sustainable or not well substantiated?

10.15.8 Video contest on designing out waste

Sustainability goals can be achieved through proper utilization of resources including energy, water, and materials. It is important that designers also focus initially on reducing waste, as this is where potentially larger impacts can be made. The efficient use of materials reduces the quantity of materials used in the first instance, lowers the material purchasing costs, minimizes waste, and eliminates the need for subsequent handling and disposal costs. Developing a strategy to reduce waste is one of the most effective ways to address waste in construction. Once effective waste reduction measures are in place, it is more appropriate to consider how to reuse, recycle, recover, or finally dispose of waste in a structured way (WRAP 2010).

This 3-min video creating contest aims to help students to explore their own values and viewpoints of the roles and responsibilities of a designer. The content of the video should focus on the following statement: designers should set examples by reusing waste materials within their design. Knowledge creation material should ascertain that values impact the choices designers make.

References

3P Contributer. 2016. Low-tech design can fuel the sustainable development goals. http://www.triplepundit.com/2016/05/low-tech-design-can-fuel-sustainable-development-goals/#.

AASHTO. 2012. Part 7.4, Bridge painting, coating, sealing and containment stewardship practices, Chapter 7. Bridge Maintenance Manual. http://environment.transportation.org.

ABB website. 2016. Internet Source: www.abb.com.

Alhaddi, H. 2015. Triple bottom line and sustainability: A literature review. *Business and Management Studies* 1(2): 6–10.

Allen, D., F. Murphy, C. F. B. Allenby, and C. Davidson. 2006. Sustainable engineering: A model for engineering education in the twenty-first century? *Clean Technologies and Environmental Policy* 8(2): 70–71.

Alting, L., and D. J. Jørgensen. 1993. The life cycle concept as a basis for sustainable industrial production. *CIRP Annals—Manufacturing Technology* 42(1): 163–167.

Anastas, P. T., and J. B. Zimmerman. 2003. Design through the 12 principles of green engineering. *Environmental Science and Technology* 37(5): 94A–101A.

Andersson, K., M. H. Eide, M. Lundqvist, and B. Mattsson. 1998. The feasibility of including sustainability in LCA for product development. *Journal of Clean Production* 6(3–4): 289–298.

Ashford, N. A. 2004. Major challenges to engineering education for sustainable development: What has to change to make it creative, effective, and acceptable to the established disciplines? *International Journal of Sustainability in Higher Education* 5(3): 239–250.

Balan, P., and G. Manickam. 2013. Promoting holistic education through design of meaningful and effective assignments in sustainable engineering. *IEEE International Conference on Teaching, Assessment and Learning for Engineering (TALE)*, Bali Dynasty Resort, Kuta, Indonesia, August 26–29, pp. 382–385.

Barker, S., and K. Andrew. 2007. Organising reuse: Managing the process of design for remanufacture (DFR). Department of Mechanical Engineering,

University of Bristol, Bristol, UK, Paper No. 007–0769. http://citeseerx.ist.psu.edu/viewdoc/download?doi=10.1.1.454.1651&rep=rep1&type=pdf.

Barker, S. G., and A. M. King. 2006. The development of a remanufacturing design platform model (RDPM): Applying design platform principles to extend remanufacturing practice into new industrial sectors. *13th CIRP International Conference on Life Cycle Engineering*, Leuven, Belgium, May 31st–June 2nd.

Becker, C., R. Chitchyan, L. Duboc, S. Easterbrook, B. Penzenstadler, N. Seyff, and C. C. Venters. 2016. Sustainability design and software: The Karlskrona manifesto, White Paper. http://www.cs.toronto.edu/~sme/papers/2015/Beckeretal-ICSE2015.pdf

Benoît, C., and B. Mazijn (Eds.). 2009. Guidelines for social life cycle assessment of products. Paris. 104pp. http://lcinitiative.unep.

Berger, M., and M. Finkbeiner. 2010. Water footprinting: How to address water use in life cycle assessment? *Sustainability* 2: 919–944.

Bielefeldt, A. R. 2013. Pedagogies to achieve sustainability learning outcomes in civil and environmental engineering students. *Sustainability* 5: 4479–4501.

Bonnema, M. 2006. *Sustainable Product Design: Just the Facts.* Haworth, Inc., Holland, MI. http://www.brookscorning.com/files/resources/Sustainable_Product_Design.pdf

Boyko, C. 2009. The urban design decision-making process: A new approach. In Cooper, R., G. Evans, and C. Boyko (Eds.), *Designing Sustainable Cities.* Chichester, UK: John Wiley & Sons, pp. 42–50.

Braungart, M., and W. McDonough. 2002. *Cradle to Cradle: Remaking the Way We Make Things.* New York: North Point Press.

Brazdauskas, M. 2015. Promoting student innovation-driven thinking and creative problem solving for sustainability and corporate social responsibility. *Journal of Creativity and Business* 1: 75–87.

Brissaud, D., and S. Tichkiewitch. 2001. Product models for life-cycle. *Annals of the CIRP* 50/1: 105–108.

Busch Systems. 2014. A brief timeline of the history of recycling. http://www.buschsystems.com/recycling-bin-news/2014/05/a-brief-timeline-of-the-history-of-recycling/.

Ceridon, K. 2011. Green design with life cycle mind. *The Journal of Engineering Entrepreneurship* 2(1): 39–53.

Chen, C., and J. Zhang. 2013. Green product design with engineering trade-offs under technology efficient frontiers: analytical results and empirical tests. *IEEE Transactions on Engineering Management* 60(2): 340–352.

Chini, A., and T. Saleh. 2009. Building green via design for deconstruction and adaptive reuse. School of Building Construction, University of Florida, Gainesville, FL.

Clark, G., J. Kosoris, L. N. Hong, and M. Crul. 2009. Design for sustainability: Current trends in sustainable product design and development. *Sustainability* 1: 409–424.

Cohen, D. 2007. Earth's natural wealth: An audit. *New Scientist* 2605: 35–41.

Cowie, R. L. 1993. A modelling framework for designing. Master's Thesis, Department of Mechanical and Aerospace Engineering, Ottawa-Carleton Institute for Mechanical and Aerospace Engineering, and the School of Industrial Design, Ottawa, ON.

Crul, M., and J. Diehl. 2006. Design for sustainability: A practical approach for developing Economics. UNEP and TU Delft, Paris, France.

Davidson, C. I., C. T. Hendrickson, H. S. Matthews, M. W. Bridges, D. T. Allen, C. F. Murphy, B. R. Allenby, J. C. Crittenden, and S. Austin. 2010. Preparing future engineers for challenges of the 21st century: Sustainable engineering. *Journal of Cleaner Production* 18(7): 698–701.

Diegel, O., S. Singamneni, S. Reay, and A. Withell. 2010. Tools for sustainable product design: Additive manufacturing. *Journal of Sustainable Development* 3(3): 68–75.

Doepker, P. E. 2010. Sustaining sustainable engineering design projects. *International Journal of Engineering Education* 26: 470–478.

Duflou, J. R., J. W. Sutherland, D. Dornfeld, C. Herrmann, J. Jeswiet, S. Kara, M. Hauschild, and K. Kellens. 2012. Towards energy and resource efficient manufacturing: A processes and systems approach. *CIRP Annals–Manufacturing Technology* 61(2): 587–609.

East, A. J. 2008. What is a carbon footprint? An overview of definitions and methodologies. Vegetable industry carbon footprint scoping study: Discussion Paper 1. Sydney, NSW: Horticulture Australia Ltd.

EC. 2003. *Communication on Integrated Product Policy*. Brussels, Belgium: European Commission.

Eddy, D. 2014. Sustainability-based product design in a decision support semantic framework. PhD Thesis, University of Massachusetts, Amherst, MA.

Edwards, A. R. 2005. *The Sustainability Revolution: Portrait of a Paradigm Shift*. Gabriola Island, BC: New Society Publishers.

Elkington, J. 1997. Cannibals with forks: the TBL of the 21st century business. Oxford, UK: Capstone Publishers.

Elkington, J. 2004. Enter the triple bottom line. Chapter 1: The triple bottom line: Does it all add up, pp. 1–16. http://www.johnelkington.com/archive/TBL-elkington-chapter.pdf.

El-Shorbagy, A. 2010. Design with nature: Windcatcher as a paradigm of natural ventilation device in buildings. *International Journal of Civil and Environmental Engineering IJCEE-IJENS* 10 (3): 21–26.

Ercin, A. E., and A. Y. Hoekstra. 2012. *Carbon and Water Footprint: Concepts, Methodologies and Policy Responses*. Paris, France: UNESCO.

ERN. 2015. Remanufacturing market study for horizon 2020. European Remanufacturing Network. https://www.remanufacturing.eu/wp-content/uploads/2016/01/study.pdf.

Finkbeiner, M. 2009. Carbon footprinting: Opportunities and threats. *The International Journal of Life Cycle Assessment* 14(2): 91–94.

Finkbeiner, M., A. Inaba, R. B. H. Tan., K. Christiansen, and H. J. Klüppel. 2006. The new international standards for life cycle assessment: ISO 14040 and ISO 14044. *The International Journal of Life Cycle Assessment* 11: 80–85.

Finkbeiner, M., E. M. Schau, A. Lehmann, and M. Traverso. 2010. Towards life cycle sustainability assessment. *Sustainability* 2: 3309–3322.

Fraunhofer IBP. 2014. Life cycle engineering. Fraunhofer Institute for Building Physics. Institute Stuttgart, Stuttgart, Germany. https://www.ibp.fraunhofer.de/content/dam/ibp/en/documents/Areas-of-Expertise/Life-cycle-engineering/IBP%20Abteilungsbrosch%C3%BCre_uk_neu.pdf.

Fuad-Luke, A. 2002. *The Eco-design Handbook*. London: Thames and Hudson.

Fujihira, K. 2016. System control for sustainability: Application to building design. In Thomas, C. (Ed.), *Complex Systems, Sustainability and Innovation*. InTech Open Access Book Publisher, Rijeka, Croatia. http://www.intechopen.com/books/complex-systemssustainability-and-innovation

Gagnon, B., R. Leduc, and L. Savard. 2008. From a conventional to a sustainable engineering design process: Different shades of sustainability. Working Paper. Groupe de Recherche en Économie et Développement International, University of Sherbrooke, Sherbrooke, QC.

Garner, A., and G. A. Keoleian. 1995. Industrial ecology: An introduction. National Pollution Prevention Center for Higher Education.

Gebler, M., J. M. Anton, S. Uiterkamp, and C. Visser. 2014. A global sustainability perspective on 3D printing technologies. *Energy Policy* 74(C): 158–167.

Govers, P. C. M., and R. Mugge. 2004. I love my jeep, because it's tough like me: The effect of product-personality congruence on product attachment. International Conference on Design and Emotion, Ankara, Turkey.

Gray, C., and M. Charter. 2006. Remanufacturing and product design designing for the 7th generation. The Centre for Sustainable Design University College for the Creative Arts, Farnham, UK. http://cfsd.org.uk/Remanufacturing%20and%20Product%20Design.pdf.

GRI. 2002. *Sustainability Reporting Guidelines*. Boston, MA: Global Reporting Initiative.

Guinée, J. B., M. Gorrée, R. Heijungs, G. Huppes, R. Kleijn, A. de Koning, L. van Oers, A. W. Sleeswijk, S. Suh, H. A. de Haes, U. de Bruijn, H. van Duin, and M. A. J. Huijbregts. 2002. *Handbook on Life Cycle Assessment-Operational Guide to the ISO Standards*. Dordrecht, the Netherlands: Kluwer Academic Publishers.

Gutowski, T. G., S. Sahil, A. Boustani, and S. C. Graves. 2011. Remanufacturing and energy savings. *Environmental Science and Technology* 45: 4540–4547.

Hanks, K., W. Odom, D. Roedl, and E. Blevis. 2008. Sustainable millennials: Attitudes towards sustainability and the material effects of interactive technologies. Indiana University at Bloomington, IN.

Hawken, P. 1992. *The Ecology of Commerce*. London: Scott London.

Hoekstra, A. Y. 2008. Water neutral: Reducing and offsetting the impacts of water footprints. Value of Water Research Report Series No. 28. Delft, the Netherlands, UNESCOIHE.

Holmberg, J., M. Svanström, D.-J. Peet, K. Mulder, D. Ferrer-Balas, and J. Segala`s. 2008. Embedding sustainability in higher education through interaction with lecturers: Case studies from three European technical universities. *European Journal of Engineering Education* 33: 271–282.

Hopkins, M. S. 2010. How sustainability fuels design innovation. *Sloan Management Review* 52(1): 75–81.

Hower, M. 2015. Canadian city moves toward first closed-loop organics waste management system. http://www.sustainablebrands.com/news_and_views/waste_not/mike_hower/canadian_city_moves_toward_first_closed-loop_organics_waste_mana.

Hubbard, G. 2009. Measuring organizational performance: Beyond the triple bottom line. *Business Strategy and the Environment* 18(3): 177–191.

ISO 14040. 1998. Environmental management—Life cycle assessment—Principles and framework. International Organization for Standardization, Geneva, Switzerland.

ISO 14040. 2006. Environmental management—Life cycle assessment—Principles and framework. Second edition. International Organization for Standardization, Geneva, Switzerland.

Jackson, A., K. Boswell, and D. Davis. 2011. Sustainability and triple bottom line reporting—What is it all about? *International Journal of Business, Humanities and Technology* 1(3): 55–59.

Jarvenpaa, S. L., and D. B. Stoddard. 1998. Business process redesign: Radical and evolutionary change. *Journal of Business Research* 41(1): 15–27.

Jensen, A. A., and A. Remmen (Eds.). 2006. *Background Report for a UNEP Guide to Life Cycle Management—A Bridge to Sustainable Products,* United Nations Environment Programme (UNEP). Paris, France: Life Cycle Initiative.

Jeswani, H. K., A. Azapagic, P. Schepelmann, and M. Ritthoff. 2010. Options for broadening and deepening the LCA approaches. *Journal of Cleaner Production* 18: 120–127.

Kim, H.-J., S. Skerlos, S. Severengiz, and G. Seliger. 2009. Characteristics of the automotive remanufacturing enterprise with an economic and environmental evaluation of alternator products. *International Journal of Sustainability Management* 1(4): 437–449.

Kloepffer, W. 2008. Life cycle sustainability assessment of products (with comments by Helias A. Udo de Haes, p. 95). *International Journal of Life Cycle Assessment* 13: 89–95.

Kreith, F. 2012. Bank for the buck. *Mechanical Engineering* 134: 26–31.

Lu, B., and P. Gu. 2003. Systematic life cycle design for sustainable product development. *Design Engineering Technical Conference and Computers and Information in Engineering Conference,* Chicago, IL, September 2–6. New York: ASME, pp. 85–94.

Lund, R. T. 1996. The remanufacturing industry: Hidden giant. White Paper, Boston University, Boston, MA.

Luttropp, C., and J. Lagerstedt. 2006. EcoDesign and the ten golden rules: Generic advice for merging environmental aspects into product development. *Journal of Cleaner Production* 14: 1396–1408.

Markham, D. 2013. In defense of high-tech tools for living a more sustainable life. http://www.treehugger.com/clean-technology/defense-high-tech-tools-living-more-sustainable-life.html.

Maslow, A. H. 1943. A theory of human motivation. *Psychology Review* 50: 370–396.

MBDC. 2002. Introduction to the cradle to cradle design framework. McDonough Braungart Design Chemistry (MBDC). http://www.chinauscenter.org/attachments/0000/0001/CradleDesign.pdf.

McDonough, W., and M. Braungart. 2003. Towards a sustaining architecture for the 21st century: The promise of cradle-to-cradle design. *Industry and Environment* 26(2): 13–16.

McDonough, W., and Partners. 1992. The Hannover principles. Design for sustainability. EXPO 2000. The World's Fair, Hannover, Germany.

McDounogh, W., M. Braungart, P. T. Anastas, and J. B. Zimmermann. 2003. Applying the principles of Green Engineering to cradle-to-cradle design. *Environmental Science and Technology* 37(23): 434A–441A.

Mckeown, R. 2011. Using rubrics to assess student knowledge related to sustainability a practitioner's view. *Journal of Education for Sustainable Development* 5(1): 61–74.

Medini, K., J. Le Duigou, C. Da Cunha, and A. Bernard. 2015. Investigating mass customization and sustainability compatibilities. *International Journal of Engineering, Science and Technology* 7(1): 11–20.

Metzger, N., and R. N. Zare. 1999. Science policy-interdisciplinary research: From belief to reality. *Science* 283(5402): 642–643.

Mol, A. P. J., and G. Spaargaren. 2000. Ecological modernisation theory in debate: A review. *Environmental Politics* 9(1): 17–49.

Mukherjee, K. and S. Mondal. 2009. Analysis of issues relating to remanufacturing technology-a case of an Indian company. *Technology Analysis and Strategic Management* 21(5): 639–652.

Mulder, K. 2006. *Sustainable Development for Engineers*. Sheffield, UK: Greenleaf Publishing.

Murdock, J. 2012. Earth day's forgotten founder. http://www.flourishonline.org/2012/03/earth-day-forgotten-founder/.

Norris, C. B., D. Aulisio, and G. A. Norris. 2012. Working with the social hotspots database-methodology and findings from 7 social scoping assessments. In Dornfeld, D. A., and B. S. Linke (Eds.), *Leveraging Technology for a Sustainable World*. Berlin, Germany: Springer, pp. 581–586.

OECD. 2009. Sustainable manufacturing and eco-innovation: Framework, practices and measurement. Synthesis Report. The Organization for Economic Co-operation and Development. OECD Publishing, Paris, France.

OECD. 2010. Eco-innovation in industry: Enabling green growth. OECD Innovation Strategy. OECD Publishing, Paris, France.

OHL. 2004. Remanufacturing in the UK: A significant contributor to sustainable development? Oakdene Hollins Ltd. Skipton, UK: The Resource Recovery Forum.

Onyali, C. I. 2014. Triple bottom line accounting and sustainable corporate performance. *Research Journal of Finance and Accounting* 5(8): 195–209.

Painter-Morland, M. 2006. Triple bottom-line reporting as social grammar: Integrating corporate social responsibility and corporate codes of conduct. *Business Ethics: A European Review* 15(4): 352–364.

PE. 2009. LBP: GaBi 4-Software-system and databases for life cycle engineering. Stuttgart, Echterdingen ver. 4.4.137.

Ping, G. C., and B. Wang. 2007. An activity-based modeling approach for assessing the key stakeholders' corporation in the eco-conscious design of electronic products. *Journal of Engineering Design* 18(1): 55–71.

Riela, A., A. Lelaha, G. Mandila, M. Rioa, S. Tichkiewitcha, F. Zhanga, and P. Zwolinskia. 2015. An innovative approach to teaching sustainable design and management. *Design Conference Innovative Product Creation*, pp. 29–34.

Roozenberg, N., and J. Eekles. 1995. *Product Design: Fundamentals and Methods*. Second edition. Chichester, UK: John Wiley and Sons.

Rosen, M. A. 2012. Engineering sustainability: A technical approach to sustainability. *Sustainability* 4: 2270–2292.

Rosen, M. A., and H. A. Kishawy. 2012. Sustainable manufacturing and design: Concepts, practices and needs. *Sustainability* 4: 154–174.

Sahni, S., A. Boustani, T. Gutowski, and S. Graves. 2011. Electric motor remanufacturing and energy savings. Environmentally Benign Laboratory, Laboratory for Manufacturing and Productivity, Sloan School of Management, Boston, MA.

SAIC. 2005. Life cycle assessment: Principles and practice; EPA/600/R-06/060. Scientific Applications International Corporation. US Environmental Protection Agency, Cincinnati, OH.

Sala, S., F. Farioli, and A. Zamagni. 2012. Life cycle sustainability assessment in the context of sustainability science progress (Part 2). *International Journal of Life Cycle Assessment* 18(9): 1686–1697.

Sarkis, J. 2001. Manufacturing's role in corporate environmental sustainability: Concerns for the new millennium. *The International Journal of Operations and Production Management* 21: 666–686.

Sassi, P. 2008. Defining closed loop material cycle construction. *Building Research and Information* 36(5): 509–519.

Schaltegger, S., and T. Synnestvedt. 2002. The link between "green" and economic success: Environmental management as the crucial trigger between environmental and economic performance. *Journal of Environmental Management* 65: 339–346.

Schau, E. M., M. Traverso, and M. Finkbeiner. 2012. Life cycle approach to sustainability assessment: A case study of remanufactured alternators. *Journal of Remanufacturing* 2: 5.

Schau, E. M., M. Traverso, A. Lehmann, and M. Finkbeiner. 2011. Life cycle costing in sustainability assessment—A case study of remanufactured alternators. *Sustainability* 3: 2268–2288.

Scheuer, C. W., and G. A. Keoleian. 2002. Evaluation of LEED using life cycle assessment methods. NIST GCR 02–836, Center for Sustainable Systems, University of Michigan, Ann Arbor, MI. National Institute of Standards and Technology, Technology Administration, US Department of Commerce.

Sher, J., and K. R. Sher. 1994. Beyond the conventional wisdom: Rural development as if Australia's rural people really mattered. *Journal of Research in Rural Education* 10(1): 2–43.

Skrainka, M. S. 2012. Analysis of the environmental impact on remanufacturing wind turbines. MSc Thesis, Rochester Institute of Technology, Rochester, NY.

Smil, V. 2008. *Energy in Nature and Society: General Energetics of Complex Systems.* Cambridge, MA: MIT Press.

Smith, T. 2012. Closing the loop: Designing for a sustainable future. http://www.climatechangenews.com/2012/02/28/closing-the-loop-designing-for-a-sustainable-future/.

Sterling, S. 2004. An analysis of the development of sustainability education internationally: Evolution, interpretation and transformative potential. In Blewitt, J., and C. Cullingford (Eds.), *The Sustainability Curriculum: The Challenge for Higher Education.* London: Earthscan.

Stiver, W. 2010. Sustainable design in a second year engineering design course. *International Journal of Engineering Education* 26(2): 378–383.

Tlusty, M. 2015. Why sustainability requires more than technological advances? http://www.triplepundit.com/2015/05/sustainability-requires-technological-advances/#.

Tukker, A., and U. Tischner (Eds.). 2006. *New Business for Old Europe. Product Services, Sustainability and Competitiveness.* Sheffield, UK: Greenleaf Publishers.

UN. 2009. Eco-efficiency indicators: Measuring resource-use efficiency and the impact of economic activities on the environment. Economic and Social Commission for Asia and the Pacific (ESCAP). Greening of Economic Growth Series. Bangkok, Thailand: United Nations Publications.

UNEP. 2012. Greening the economy through life cycle thinking. Ten years of the UNEP/SETAC life cycle initiative. United Nations Environment Program (UNEP), Paris, France.

USDC. 2011. Introduction to sustainable manufacturing. http://www.trade.gov/green/documents/introduction-to-sustainable-manufacturing.pptx.

Van de Westerlo, B. 2011. Sustainable development and cradle to cradle approach. A literature study of the opportunities to apply the Cradle to Cradle approach in the built environment. University of Twente, Enschede, the Netherlands.

Vierra, S. 2014. Biomimicry: Designing a model nature. Vierra Design and Education Services. https://www.wbdg.org/resources/biomimicry.php.

Wang, L., and L. Lin. 2007. A methodological framework for the triple bottom line accounting and management of industry enterprises. *International Journal of Production Research* 45(5): 1063–1088.

Warburton, K. 2003. Deep learning and education for sustainability. *International Journal of Sustainability in Higher Education* 4(1): 44–56.

Wohlers, T. 2009. Worldwide progress report on the rapid prototyping, tooling, and manufacturing state of industry. Wohlers Report. Wohlers Associates, Fort Collins, CO.

WRAP. 2010. Designing out waste: A design team guide for buildings. http://www.modular.org/marketing/documents/DesigningoutWaste.pdf.

Yadong, Y. 2013. Eco-efficiency trends in China, 1978–2010: Decoupling environmental pressure from economic growth. *Ecological Indicators* 24: 177–184.

Yang, Q. 2014. Remanufacturing versus new acquisition of production equipment: Definitions and decision-making checklist. Master Thesis, School of Innovation, Design and Engineering, Malardalen University, Västerås, Sweden.

Zimoch, E. 2012. Life-cycle assessment (LCA) of existing bridges and other structures. http://citeseerx.ist.psu.edu/viewdoc/download?

Index

D

E

H

I

Q

R

T